全国高等中医药院校中药学类专业双语规划教材
Bilingual Planned Textbooks for Chinese Materia Medica Majors in TCM Colleges and Universities

药用植物学

Pharmaceutical Botany

（供中药学类、药学类及相关专业使用）
(For Chinese Materia Medica, Pharmacy and other related majors)

主　编　王光志　高德民

主　审　严铸云（成都中医药大学）

副主编　严玉平　张新慧　陆　叶　林　莺　郭庆梅

编　者　（以姓氏笔画为序）

王光志（成都中医药大学）	方清影（安徽中医药大学）
任　艳（西南民族大学）	任广喜（北京中医药大学）
刘　钊（河北中医学院）	刘小芬（福建中医药大学）
刘湘丹（湖南中医药大学）	孙立彦（山东第一医科大学）
严玉平（河北中医学院）	李先宽（天津中医药大学）
李思蒙（南京中医药大学）	吴廷娟（河南中医药大学）
吴清华（成都中医药大学）	张　涛（长春中医药大学）
张新慧（宁夏医科大学）	陆　叶（苏州大学）
陈　莹（陕西中医药大学）	邵艳华（广东药科大学）
林　莺（滨州医学院）	赵　容（辽宁中医药大学）
俞　冰（浙江中医药大学）	秦　琴（成都医学院）
倪梁红（上海中医药大学）	高德民（山东中医药大学）
郭庆梅（山东中医药大学）	崔治家（甘肃中医药大学）
董丽华（江西中医药大学）	程丹丹［齐鲁工业大学（山东省科学院）］
童　毅（广州中医药大学）	樊锐锋（黑龙江中医药大学）

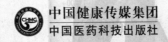

中国健康传媒集团
中国医药科技出版社

内 容 提 要

  本书是"全国高等中医药院校中药学类专业双语规划教材"之一。本书共分为3篇15章。第一篇为药用植物的形态，第二篇为药用植物的分类，第三篇为药用植物的组织结构。全书在语言内容的编排上充分重视双语的学习特征，希望读者在双语教材学习过程中既能提高语言能力，又能掌握专业内容，成为国际化中医药人才。本教材为书网融合教材，即纸质教材有机融合电子教材、教学配套资源（PPT、微课、视频等）、题库系统、数字化教学服务（在线教学、在线考试、在线作业），使教材内容更加立体、生动、形象，便教易学。

  本书供全国高等中医药院校中药学类、药学类及相关专业使用，亦可供相关领域的研究人员和科技工作者参考。

图书在版编目（CIP）数据

药用植物学：汉英对照 / 王光志，高德民主编 .—北京：中国医药科技出版社，2020.9

全国高等中医药院校中药学类专业双语规划教材

ISBN 978-7-5214-1884-2

Ⅰ.①药… Ⅱ.①王…②高… Ⅲ.①药用植物学 – 双语教学 – 中医学院 – 教材 – 汉、英 Ⅳ.① Q949.95

中国版本图书馆 CIP 数据核字（2020）第 102364 号

美术编辑 陈君杞

版式设计 辰轩文化

出版　**中国健康传媒集团** | 中国医药科技出版社

地址　北京市海淀区文慧园北路甲 22 号

邮编　100082

电话　发行：010-62227427　邮购：010-62236938

网址　www.cmstp.com

规格　889 × 1194 mm ¹⁄₁₆

印张　23

字数　649 千字

版次　2020 年 9 月第 1 版

印次　2023 年 12 月第 2 次印刷

印刷　三河市万龙印装有限公司

经销　全国各地新华书店

书号　ISBN 978-7-5214-1884-2

定价　71.00 元

获取新书信息、投稿、为图书纠错，请扫码联系我们。

出版说明

近些年随着世界范围的中医药热潮的涌动，来中国学习中医药学的留学生逐年增多，走出国门的中医药学人才也在增加。为了适应中医药国际交流与合作的需要，加快中医药国际化进程，提高来中国留学生和国际班学生的教学质量，满足双语教学的需要和中医药对外交流需求，培养优秀的国际化中医药人才，进一步推动中医药国际化进程，根据教育部、国家中医药管理局、国家药品监督管理局等部门的有关精神，在本套教材建设指导委员会主任委员成都中医药大学彭成教授等专家的指导和顶层设计下，中国医药科技出版社组织全国50余所高等中医药院校及附属医疗机构约420名专家、教师精心编撰了全国高等中医药院校中药学类专业双语规划教材，该套教材即将付梓出版。

本套教材共计23门，主要供全国高等中医药院校中药学类专业教学使用。本套教材定位清晰、特色鲜明，主要体现在以下方面。

一、立足双语教学实际，培养复合应用型人才

本套教材以高校双语教学课程建设要求为依据，以满足国内医药院校开展留学生教学和双语教学的需求为目标，突出中医药文化特色鲜明、中医药专业术语规范的特点，注重培养中医药技能、反映中医药传承和现代研究成果，旨在优化教育质量，培养优秀的国际化中医药人才，推进中医药对外交流。

本套教材建设围绕目前中医药院校本科教育教学改革方向对教材体系进行科学规划、合理设计，坚持以培养创新型和复合型人才为宗旨，以社会需求为导向，以培养适应中药开发、利用、管理、服务等各个领域需求的高素质应用型人才为目标的教材建设思路与原则。

二、遵循教材编写规律，整体优化，紧跟学科发展步伐

本套教材的编写遵循"三基、五性、三特定"的教材编写规律；以"必需、够用"为度；坚持与时俱进，注意吸收新技术和新方法，适当拓展知识面，为学生后续发展奠定必要的基础。实验教材密切结合主干教材内容，体现理实一体，注重培养学生实践技能训练的同时，按照教育部相关精神，增加设计性实验部分，以现实问题作为驱动力来培养学生自主获取和应用新知识的能力，从而培养学生独立思考能力、实验设计能力、实践操作能力和可持续发展能力，满足培养应用型和复合型人才的要求。强调全套教材内容的整体优化，并注重不同教材内容的联系与衔接，避免遗漏和不必要的交叉重复。

三、对接职业资格考试，"教考""理实"密切融合

本套教材的内容和结构设计紧密对接国家执业中药师职业资格考试大纲要求，实现教学与考试、理论与实践的密切融合，并且在教材编写过程中，吸收具有丰富实践经验的企业人员参与教材的编写，确保教材的内容密切结合应用，更加体现高等教育的实践性和开放性，为学生参加考试和实践工作打下坚实基础。

四、创新教材呈现形式，书网融合，使教与学更便捷更轻松

全套教材为书网融合教材，即纸质教材与数字教材、配套教学资源、题库系统、数字化教学服务有机融合。通过"一书一码"的强关联，为读者提供全免费增值服务。按教材封底的提示激活教材后，读者可通过PC、手机阅读电子教材和配套课程资源（PPT、微课、视频等），并可在线进行同步练习，实时收到答案反馈和解析。同时，读者也可以直接扫描书中二维码，阅读与教材内容关联的课程资源，从而丰富学习体验，使学习更便捷。教师可通过PC在线创建课程，与学生互动，开展在线课程内容定制、布置和批改作业、在线组织考试、讨论与答疑等教学活动，学生通过PC、手机均可实现在线作业、在线考试，提升学习效率，使教与学更轻松。此外，平台尚有数据分析、教学诊断等功能，可为教学研究与管理提供技术和数据支撑。需要特殊说明的是，有些专业基础课程，例如《药理学》等9种教材，起源于西方医学，因篇幅所限，在本次双语教材建设中纸质教材以英语为主，仅将专业词汇对照了中文翻译，同时在中国医药科技出版社数字平台"医药大学堂"上配套了中文电子教材供学生学习参考。

编写出版本套高质量教材，得到了全国知名专家的精心指导和各有关院校领导与编者的大力支持，在此一并表示衷心感谢。希望广大师生在教学中积极使用本套教材和提出宝贵意见，以便修订完善，共同打造精品教材，为促进我国高等中医药院校中药学类专业教育教学改革和人才培养做出积极贡献。

数字化教材编委会

药用植物学是中药学类、药学类专业的基础课程，是植物学与中药学、药学相互渗透又紧密联系的应用型学科。经过长期发展，形成了系统的理论体系、丰富的知识结构和实用的方法技术，在药材基源植物的鉴定、药用植物资源的评价利用和药用植物栽培育种等领域始终发挥着重要作用。

本书在继承历版教材经典内容的基础上，依据高校双语课程建设的特殊要求和教材编写基本原则，在教材基本内容的编写上坚持"必需、够用"的原则；在体例安排上，密切结合药用植物学理论、实验和实践教学环节，将全书分为三篇。第一篇药用植物的形态，重点介绍药用植物营养器官和繁殖器官的形态特征；第二篇药用植物的分类，分别介绍低等植物和高等植物的相关分类群特征和代表药用植物；第三篇药用植物的组织结构，重点介绍细胞、组织和根、茎、叶的组织结构特征。通过"学习目标"和"重点小结"引导学生把握课程主线、掌握核心知识点，设置的"目标检测"有助于学生课后复习和检测，达到课前预习、课间引导、课后复习的综合目标。

全书分15章。绪论由王光志编写；第一章由张新慧编写；第二章由赵容编写；第三章由方清影、李思蒙编写；第四章由刘小芬编写；第五章由高德民编写；第六章、第七章由陆叶编写；第八章、第九章由李先宽编写；第十章由崔治家编写；第十一章由任广喜编写；第十二章被子植物分类概述由严玉平编写，三白草科至马兜铃科由刘湘丹编写，蓼科至小檗科由倪梁红编写，防己科至十字花科由程丹丹编写，虎耳草科至大戟科由俞冰编写，无患子科至伞形科由刘钊编写，杜鹃花科至萝藦科由孙立彦编写，旋花科至紫草科由童毅编写，茜草科至菊科由林莺编写，禾本科至百合科由董丽华编写，石蒜科至兰科由吴廷娟编写；第十三章第一节由任艳编写；第二节由郭庆梅编写；第十四章第一节由樊锐锋编写，第二节由张涛编写；第十五章由陈莹、邵艳华、秦琴编写。数字教材由吴清华统稿。全书由王光志和高德民统一审改定稿。

本教材可供全国高等中医药院校的中药学类、药学类及相关专业的本科生和国内医药院校相关专业学习的留学生使用，亦可供相关领域的研究人员和科技工作者参考。各院校或专业使用教材时将二十大精神融入教材中，同时根据地区、专业、课时、教学内容和教学对象的差异，适当调整教材内容。

本书在编写过程中，得到各参编单位的鼎力支持；赵会礼先生在百忙之中对绪论的英文部分进行了修改完善；在图片的使用上得到严铸云教授的大力支持，在此一并致以诚挚谢意！编委会在双语教材的编写上属于第一次尝试，因受水平所限，书中难免存在缺陷和不妥之处，敬请读者和兄弟院校在使用过程中提出批评和建议，以便修订完善。

Pharmaceutical Botany is the basic course for students majoring in Chinese medicine and Pharmacology, and is an applied subject which is permeated and closely related with Botany, Chinese medicine and Pharmacology. After a long period of development, the subject has formed a systematic theoretical system, a rich knowledge structure and practical techniques. It plays an important role in the identification of original medicinal plants of Chinese *materia medica*, evaluation and utilization of medicinal plant resources and cultivation and breeding of medicinal plants.

On the basis of the classical contents of previous editions, this book adheres to the principle of "necessary and sufficient" in compiling the basic contents according to the special requirements of bilingual teaching and the basic principles of compilation. The book is divided into three parts in order to meet the needs of the theory, experiment and practice teaching of Pharmaceutical Botany. The first part is about the morphology of medicinal plants. The morphological characteristics of vegetative and reproductive organs were mainly introduced in this part. The second part is about the taxonomy of medicinal plants, in which the taxa of lower plants, the vascular plants and the representative medicinal plants were introduced. The third part is about plant anatomy, which mainly includes plant cells, plant tissues as well as the anatomical structure of roots, stems and leaves, and its application in the identification of Chinese *materia medica*. In addition, such modules as *"learning goals"* , *"summary"* are set up to guide students to master the main contents and key points of the subject. The " *question* " at the end of each chapter helps students review and test after class.

The book is divided into 15 chapters. The Introduction was compiled by Wang Guangzhi; Chapter 1-5 were compiled by Zhang Xinhui, Zhao Rong, Li Simeng, Fang Qingying, Liu Xiaofen and Gao Demin. Chapters 6 and 7, Chapters 8 and 9 were compiled by Lu Ye and Li Xiankuan, separately. Chapters 10 and 11 were edited by Cui Zhijia and Ren Guangxi. An Overview of the taxonomy of angiosperms of Chapter 12 was edited by Yan Yuping and other families from Trigonaceae to Apiaceae were edited by Liu Xiangdan, Ni Lianghong, Cheng Dandan, Yu Bing, Liu Zhao and families from Rhododendronaceae to Asteraceae were edited by Sun Liyan, Tong Yi and Lin Ying. Families from Poaceae to Orchidaceae were edited by Dong Lihua and Wu Tingjuan. The first section of Chapter 13 was compiled by Ren Yan and the second was edited by Guo Qingmei. The first section of Chapter 14 was compiled by Fan Ruifeng, and the second section was compiled by Zhang Tao. Chapter 15 was compiled by Chen Ying, Shao Yanhua and Qin Qin. The final draft was reviewed by Wang Guangzhi and Gao Demin.

This textbook is intended for undergraduates majoring in Chinese materica medica, pharmacy and other related majors as well as foreign students studying related majors in domestic medical colleges or universities. It can also be used as a reference for researchers and scientists in related fields. Those who use this textbook can arrange the contents properly according to the differences of their locations, majors, class hours, teaching contents and teaching objects.

In the process of compiling this book, we have received the full support of China Medical Science and Technology Press and the universities of the authors of this textbook. Mr. Zhao Huili revised and perfected the English content of the Introduction. Professor Yan Zhuyun also afforded strong support in the use of the pictures. We take this opportunity to express our heartfelt gratitude.

It is the first time for us to compile the bilingual textbook. Due to limited level and hasty time, the book is bound to have some defects and inappropriateness, so readers are kindly requested to submit criticisms and suggestions in the course of reading, so as to make amendments and improvements.

目录 | Contents

第一篇　药用植物的形态
Part I　Morphology of Medicinal Plants

3

第二篇　药用植物的分类

Part II　Taxonomy of Medicinal Plants

第三篇　药用植物的组织结构

Part III　Anatomy of Medicinal Plants

绪　论
Introduction

PPT

 学习目标｜Learning goals

1. **掌握**　药用植物和药用植物学的概念。
2. **熟悉**　药用植物学的发展简史和研究任务。
3. **了解**　药用植物学与先关学科的关系。

- Know the concepts of medicinal plants and Pharmaceutical botany.
- Be familiar with the brief history and research tasks of medicinal botany.
- Understand the relationship between medicinal botany and related subjects

我国是世界上生物多样性及其丰富的国家之一，有着极为丰富的药用植物资源。我国劳动人民在数千年同从疾病斗争过程中，积累了丰富的药用植物知识。《史记·补三皇本纪》记载："神农氏……始尝百草，始有医药"。"神农尝百草"尽管是传说，但反映了中药起源于药用植物的发展历程。先贤们经过探索实践，丰富和完善了利用药用植物防病、治病、保健、养生的认识，并逐步形成独具特色又博大精深的中医药理论体系。

As one of the mega-biodiversity countries in the world, China harbours plenty of medicinal plants. The Chinese have accumulated abundant knowledge and experience of medicinal plants during the thousand years of struggle against diseases. Chinese legends tell that "Shen Nong (literally: Divine Farmer) tasted a hundred herbs so began the use of medicaments." Ancient people in China improved their knowledge of medicinal plants through trial and error by preventing and treating illnesses, maintaining and promoting health. Based on that, a unique and profound traditional Chinese medicine (TCM) system was formed and developed gradually throughout the past 3000 years.

我国地域辽阔，气候多样，物种繁多，是世界上使用药用植物种类最多、范围最广、历史最悠久的国家之一。20 世纪 80 年代第 3 次全国中药资源普查数据显示，我国有药用植物 11146 种（包括亚种、变种或变型 1208 个），分属于 383 科 2313 属，约占中药资源总数的 87%。除中医学外，印度医学、希腊 - 阿拉伯医学、非洲和拉美的传统医学以及分布在世界各地的很多原住民也广泛使用药用植物。

China covers a vast territory with various climate and abundance of native species, is one of the countries in the world that has a tremendously long history on the use of medicinal plants. A wide range of species has been used for medicinal purpose. According to the third nationwide survey of TCM materia medica resources carried out in the 1980s, a total of 11146 species (including 1208 subspecies, varieties, forms), belonging to 2313 genera under 383 families, are medicinal plants, contributed 87% to total

TCM materia medica resources. Besides TCM, medicinal plants have been used by different traditional medicine systems, such as Ayurveda medicine, Unani medicine, traditional medicine in Africa and Latin America, or by various forms of indigenous medicine being practised traditionally.

一、药用植物及药用植物学的概念
1 Definition of medicinal plants and pharmaceutical botany

药用植物是指具有预防、治疗疾病，对人体有保健功能的植物种类的总称。其应用对象包括药用植物的整体、器官、组织或加工品等；从应用范围可分为中药植物、民族药植物、民间药植物、国外药用植物等。

Medicinal plants are plants that can be used to prevent disease, maintain health, or cure ailments. Whole plant or plant part (tissues, organs), or the preparations derived from them are used. According to the community where they are used, medicinal plants can be categorized into Chinese medicinal plants, ethnomedicinal plants, folk medicinal plants, foreign medicinal plants, etc.

药用植物学是利用植物学的理论和方法，研究药用植物分类鉴定、品质形成与调控、资源科学利用和调控的一门学科。它既是植物学的重要分支学科，也是植物学和药物学知识交叉融合的综合性应用型学科。药用植物学也是中药学和药学类专业的一门专业基础课，是中药鉴定学、生药学、中药资源学、中药栽培学、中药化学和天然药物化学等学科的基础。

Pharmaceutical botany is a discipline that studies the classification (taxonomy), identification, quality formation and regulation, resources utilization and regulation of medicinal plants. Pharmaceutical botany is not only a subset of botany but also a comprehensive applied science or multidiscipline that involves botany and pharmacy. Pharmaceutical botany is one of basic courses for TCM materia medica or pharmacy students. It serves as a foundation course for the study of other disciplines such as identification of TCM materia medica, pharmacognosy, resources of TCM materia medica, cultivation of TCM materia medica, chemistry of TCM materia medica, chemistry of natural products, etc.

二、药用植物学的发展简史
2 A brief history of pharmaceutical botany in China

我国对药用植物的认知和知识积累，有着悠久的历史，如3000多年前的《诗经》和《尔雅》中分别记载的200和300种植物中，其中约1/3为药用植物。

China has a long history of using and recording medicinal plants. In *Shi Jing* (*Classic of Poetry*) and the first ancient Chinese dictionary *Erya*, both were compiled about 3000 years ago, about 200 and 300 plants were recorded respectively, of which about one third are medicinal plants.

1. 古代药用植物知识的出现、积累和发展　千百年来，在向自然获取食物、寻求药物的实践中，人们逐渐认识并积累了丰富的药用植物知识。如神农"……尝百草之滋味，水泉之甘苦，令民之所辟就……一日而遇七十毒"，这样的传说从侧面反映了先民辨识和体验各种植物，并从中获得药用植物经验和知识的过程。最初，这种经验和知识仅以口传手授的方式得以积累并流传，但随着文字的出现，人们开始记录并总结以前口传手授的知识，久而久之，形成了关于药物知识的著作—本草。本草的出现，促进了药用植物学知识的传播和积累，丰富了药用植物学的原始体系。

2.1　Acquisition, accumulation, and growth of medicinal plants knowledge in ancient times　For thousands of years, humanity has learned more and more about medicinal plants during the process of obtaining food and medicine from nature. Chinese legends such as "Shen Nong … tasted a hundred herbs, water spring, let the people know what to avoid and what to accept…came across seventy poisonous herbs each day" reflect how ancestors identify and learn different plants and acquire experience. In the beginning, such experience and knowledge can only be transmitted through speech and can only be stored in memory. With the invention of writing, people began to collect, summarize and record their knowledge obtained earlier through word of mouth. Over a long time, books about TCM materia medica appeared. There is a Chinese word for such books maintaining TCM materia medica knowledge, Bencao (Ben means root or origin, Cao means grass or herbs). The emergence of Bencao facilitated the dissemination and accumulation of knowledge about medicinal plants, enriched the primitive system of pharmaceutical botany.

我国历代本草有400多部。《神农本草经》是迄今所知的我国第一部本草，共收载药物365种，其中植物药237种；该书总结了汉以前的药物知识，按照上、中、下品对药物分类，开创药物分类的先河。梁代陶弘景以《神农本草经》和《名医别录》为基础，编著了《神农本草经集注》，载药730种，将所载药物分为玉石、草、木、虫兽、果菜、米食及有名未用七类，首创了按药物自然属性对药物分类的方法。唐代出现了由政府组织编写的我国第一部官修本草——《新修本草》（唐本草），该书由苏敬等人编撰，载药844种，收录了不少外来药物，如豆蔻、丁香、诃子、胡椒、槟榔等，反映了中药理论的创新和兼容并包。该书被认为是我国和世界的第一部药典性质的本草。宋代唐慎微编著的《经史证类备急本草》，收载药物1746种，该书内容丰富，图文并茂，集宋以前本草之大成，是我国现存最早最完备的本草著作。明代"药圣"李时珍历经30多年的努力，完成了巨著《本草纲目》，全书52卷，200余万字，载药1982种，其中来源于植物的有1100多种，是集我国历代本草之大成的本草专著。《本草纲目》采用的药物自然属性分类方法，是当时最先进的分类方法。因其对植物描述、植物命名、植物分类做了超越前人的改进，对后来的植物分类贡献巨大，被誉为自然分类的先驱。清代吴其濬编写了《植物名实图考》和《植物名实图考长编》，共记载植物2252种，对每种植物的形态、产地、性味、用途等叙述颇详，尤其是书中的插图，是历代本草著作中最精美的。这两部书均是植物学专著，也是研究和鉴定药用植物的重要文献。

视频

There are more than 400 Bencao works throughout Chinese history. *Shen Nong Ben Cao Jing* (*Divine Farmer's Classic of Materia Medica*) is believed to be the first one. It is a compilation of herbal knowledge before Han dynasty, containing 365 drugs, of which 237 are from plants. This book marks the beginning of classifying drugs: its 365 entries are graded into three volumes, corresponding to properties referred to as "superior", "mediocre" and "inferior". Tao Hongjing of Liang dynasty completed *Shen Nong Ben Cao Jing Ji Zhu* (*Collected Commentaries to Classic of Materia Medica*) by combining the material of *Shen Nong Ben Cao Jing* and his manuscript *Ming Yi Bie Lu* (*Supplementary Records of Famous Physicians*). He launched the classification of drugs according to their elementary nature, composed 730 drugs into categories namely jades and stones, grasses, trees, insects and beasts, fruits and vegetables, grains, and a seventh category called "substances with a name but not used". At the request and organized by the Tang dynasty government, Su Jing and his team wrote the *Xin Xiu Ben Cao* (*Newly Compiled Materia Medica*). It provided details of 844 kinds of drugs, including many, such as cardamom, clove, myrobalan, pepper, betel nut, imported from other countries, reflecting the innovation and compatibility of Chinese herbology. Since it was sponsored by the government, it is considered to be the earliest officially published pharmacopoeia in the world. During the Song dynasty, Tang Shenwei wrote *Jing Shi Zheng Lei Bei Ji Ben Cao* (*Classified Materia Medica from Historical Classics for Emergencies*),

医药大学堂

recorded 1746 drugs. More than 240 Bencao works written in past ages were recompiled, cited, or excerpted in this book. It is the earliest and most complete Bencao work which still exists. Over 30 years in the making, Li Shizhen compiled his famous *Ben Cao Gang Mu* (*Compendium of Materia Medica*). This 52-volume masterpiece, contains about two million Chinese characters, 1982 drugs, including more than 1100 plants. Drugs in *Ben Cao Gang Mu* are grouped by their natural properties, which is regarded as the most advanced classification method at that time. For its surpassing improvement on plant description, nomenclature, classification, *Ben Cao Gang Mu* contributed greatly to the development of plant taxonomy, has been praised as pioneer work of natural classification. During the Qing dynasty, Wu Qijun compiled *Zhi Wu Ming Shi Tu Kao* (*Illustrated Plant Book*) and *Zhi Wu Ming Shi Tu Kao Chang Bian* (*Extensive Illustrated Plant Book*), covering a total of 2252 plants, with careful and detailed descriptions of morphological characters, habitat and distribution, taste and smell, usage, etc. It should be mentioned that the illustrations in these two books are delicate, much better than those in any previous Bencao works. These two books are entirely about plants, providing however valuable references for the research and identification of medicinal plants.

历代本草的知识积累，丰富了药用植物鉴定、采制和功用的知识，为我国药用植物学学科的形成和发展奠定了基础；由于当时科学认识和研究条件的限制，历代本草对药用植物的记载和研究忽视了植物的系统发育关系，没有形成体现植物自然特征的分类体系和专门的药用植物学知识体系。

For ages, knowledge about the recognition, collection, and usage of medicinal plants had been accumulated and recorded in Bencao works, provided valuable information for the origin and development of pharmaceutical botany in China. Admittedly, due to the restriction of knowledge and conditions in ancient China, the evolutionary relationship among the medicinal plants had never been revealed, natural classification systems had not been proposed, pharmaceutical botany as a scientific discipline had not been established.

2. 我国药用植物学的形成和发展　在我国，李善兰和韦廉臣共同编译的《植物学》（1858年）的问世，标志着我国现代植物分类学的出现。后来，从西方留学回国的一批植物学家，开创了我国本土植物学的研究和教育工作，开始用现代植物分类学的思想、理论和方法研究植物。现代分类学理论和方法的引入使人们对药用植物的认识产生了质的飞跃，药学工作者开始采用科学方法描述药用植物，采用系统发育方法对药用植物进行分类。药用植物学的研究进入了以形态、分类、组织结构和药用植物资源为研究内容，为中药品种整理和中药资源调查提供技术支撑的发展新阶段，并逐渐形成药用植物学学科。

2.2 Formation and development of pharmaceutical botany in China The 1858 publication compiled by Li Shanlan and Alexander Williamson, *Zhi Wu Xue* (*Botany*), signified the spread of modern western botany into China. Later, a band of Chinese botanists, who studied in the west and returned home, initiated the research and education in accordance with modern botanical thought, theory, and methods in China. Particularly the spread of the modern system of taxonomy from the West to China is a big leap for the Chinese to understand the medicinal plants. The pharmaceutical scientists began to describe medicinal plants using scientific terms, to classify the medicinal plants following the phylogenetic system of classification. Research of medicinal plants entered a new phase with most topics of morphology, classification, histology, and resources. All these provided technical support for the later species systematization and survey of TCM materia medica. Pharmaceutical botany has become an independent discipline in China.

1936年，韩士淑编译了我国第一部中文大学教材《药用植物学》，随后李承祐（1949年）、孙雄才（1963年）、丁景和（1985年）、杨春澍（1997年）等编写多部《药用植物学》教材，后来又相继出版了"十一五"、"十二五"、"十三五"《药用植物学》规划教材，推动了药用植物学学科发展和人才培养。常用中药材品种整理和质量研究以及中药资源调查工作的实施，明确了我国药用植物资源的基本现状，奠定了我国资源持续利用的基础。先后编写出版了《中国植物志》（80卷，126册）、《中药志》、《全国中草药汇编》、《中国中药资源志要》、《中华本草》、《中华人民共和国药典》等。以上专著和教材是我国中药和药用植物研究成功的结晶，推动了药用植物学的发展。

In 1936, Han Shishu compiled the first Chinese textbook of pharmaceutical botany for university students, followed by Li Chenghu (1949), Sun Xiongcai (1963), Ding Jinghe (1985), Yang Chunshu (1997), "11th 5-year planning textbook", "12th 5-year planning textbook", "13th 5-year planning textbook". Publishing of these different versions of textbooks played an important role in the development and education of pharmaceutical botany in the country. The carry out of a national project, species systematization and quality evaluation of commonly used TCM materia medica, clarified the status and laid the foundation for the sustainable use of the resources of medicinal plants in China. This project contributed the publications of *Zhong Guo Zhi Wu Zhi* (*Flora of China,* 80 Volumes, 126 books), *Zhong Yao Zhi* (*Chronicles of TCM Materia Medica*), *Quan Guo Zhong Cao Yao Hui Bian* (*National Compendium of Chinese Herbal Medicine*), *Zhong Guo Zhong Yao Zi Yuan Zhi Yao* (*Brief Chronicles of TCM Materia Medica Resources*), *Zhong Hua Ben Cao* (*Chinese Bencao*), *Zhong Hua Ren Min Gong He Guo Yao Dian* (*Pharmacopoeia of the People's Republic of China*). The above publications and textbooks represented the continuous improvement of TCM materia medica and pharmaceutical botany research, boosted, in turn, the development of pharmaceutical botany in China.

药用植物学通过和医学、药学、化学、农学、细胞生物学、分子生物学等学科密切联系相互渗透，分化出了中药鉴定学、中药化学、药用植物栽培学、中药资源学、分子生药学等学科，给药用植物学科增加了新的内容。

Pharmaceutical botany is deeply woven throughout medical science, pharmacy, chemistry, agriculture, cell biology, molecular biology; divided into various disciplines such as identification of TCM materia medica, chemistry of TCM materia medica, cultivation of medicinal plants, resources of TCM materia medica, molecular pharmacognosy, which have broadened the scope of pharmaceutical botany.

分子生物学理论及其技术的发展以及系统科学和信息技术的发展为药用植物学科的发展创造了有利条件，药用植物学在微观和宏观的研究上均取得突出成就，在研究深度和广度、研究模式方面，均达到了一个新的水平，进入一个新的发展阶段。如，一系列调控基因的发现，为了解药用植物发育过程及调控机制积累了大量研究基础；随着灵芝、丹参和人参基因组框架图的相继完成，研究结果为揭示药用植物活性成分生物合成途径的分子机制、道地药材形成机制以及中药新品种培育方面，开辟了新的天地，进一步促进了药用植物学学科的发展。

Developments of Pharmaceutical botany have been driven by theory and techniques of molecular biology, systems science, information technology. Great achievements have been obtained in pharmaceutical botany research at both macro and micro levels. In terms of depth, breadth, and pattern, pharmaceutical botany research reached a new level, entered a new stage. For instance, the discovery of a set of regulatory genes accumulated a large amount of information for the further research of the development process and regulatory mechanism of medicinal plants. Since the completion of the genomic framework of *Ganoderma lucidum, Salvia miltiorrhiza, Panax ginseng*, a new horizon has been opened up to reveal the molecular mechanism of the biosynthetic pathway of active ingredients in medicinal plants and the formation mechanism of *Daodi* TCM materia medica, to speed up the cultivation of new varieties of TCM materia medica.

三、药用植物学的研究任务
3 The task of pharmaceutical botany

1. **鉴定中药基源植物，确保临床用药准确** 中药和天然药物种类繁多、来源复杂，加之中药使用历史沿革的原因和不同地区的用药习惯差异，品种混乱、一药多源、同名异物、同物异名以及以假乱真的现象比较普遍，直接影响中药用药的准确、安全和有效。①同名异物现象：如贯众，《中国药典》规定为鳞毛蕨科植物粗茎鳞毛蕨 *Dryopteris crassirhizoma* Nakai 的根状茎，而据调查，同名为 "贯众" 的植物有 9 科 17 属 49 种及其变种，当做中药贯众使用的有 5 科 25 种。②一药多源问题：如中药大黄，《中国药典》（2015 版）规定其来源为蓼科植物掌叶大黄 *Rheum palmatum* L、唐古特大黄 *R. tanguticum* Maxim. ex Regel 和药用大黄 *R. officinale* Baill 的干燥根及根茎，而不同地区还将同属波叶组的藏边大黄 *R. australe* D. Donde 、河套大黄 *R. hotaoense* C. Y. Cheng et Kao 等的根及根状茎做大黄药用，来源不同，质量有别。③安全性问题：如，中药木通，除毛茛科植物绣球藤 *Clematis montana* Buch.-Ham. ex DC.、小木通 *C. armandii* Franch. 的茎（川木通）、木通科植物木通 *Akebia quinata*（Thunb.）Decne.、三叶木通 *A. trifoliate*（Thunb.）Koidz. 或白木通 *A. trifoliata*（Thunb.）Koidz.var.australis（Diels）Rehd. 的茎（木通）药用外，马兜铃科植物东北马兜铃 *Aristolochia manshuriensis* Kom. 的茎（关木通）一度曾被广泛使用，然而关木通具有肾毒性，在 2005 年之后，被禁止使用。④替代品或掺伪层出不穷，良莠不齐：如常用中药半夏，来源于天南星科植物半夏 *Pinellia ternate*（Tunb.）Breit. 的干燥块茎，但因资源不足，栽培产量低，市场紧缺时，许多地区以同科植物鞭檐犁头尖 *Typhonium flagelliforme*（Lodd.）的块茎（水半夏）代半夏用，水半夏植物形态、块茎性状和化学成分均与半夏有别，不可替代半夏使用。

3.1 Identification of original plant of TCM materia medica The greater part of TCM materia medica is from a wide range of plants, which means its source can be complicated. Due to its long history and different local usage all over China, species confusion of TCM materia medica is quite common. Mixed-use of plants of similar species, confusing nomenclature, i.e. different herbs sharing one name or one herb using different names, is part of reasons for this confusion. Accidental or deliberate substitution has always characterized Chinese *materia medica* trading. Accuracy, safety, and effectiveness of TCM materia medica are plagued by all these problems: (a), Problem of one name for different herbs. TCM materia medica Guan Zhong serves here as an example. In Chinese Pharmacopoeia, the official species of it is *Dryopteris crassirhizoma* (Family Lepidaceae). However, according to a thorough survey, 49 plants belonging to 17 genera under 9 families, are all assigned the same name: Guan Zhong. (b), Problem of mixed-use of plants of similar species. TCM materia medica Da Huang is defined in Chinese Pharmacopoeia as the root and rhizome of *Rheum palmatum*, *R.tanguticum*, *R. officinale*. However, other plants of genera *Rheum* such as *R. australe* and *R. hotaoense* are also used unproperly. The quality of these two substitutes is considered to be lower. (c), Safety problem. In addition to the stems of *Clematis montana*, *C. armandii*, *Akebia quinata*, *A. trifoliate* and *A. trifoliate var australis*, the stem of *Aristolochia manshuriensis* was once also widely used as TCM materia medica Mu Tong, but it was known later to have renal toxicity and was banned in 2005 by the Chinese national authority for drugs and medical products. (d), Problem of adulteration and substitution. Traditionally, TCM materia medica Ban

Xia is from the dried tubers of *Pinellia ternate*. Due to its low yield and insufficient supply, the tubers of *Typhonium flagelliforme* are often substituted to fill the demand. The morphological features of the plant, the shape of the tuber, and the chemical components of these two plants are quite different, should not be used alternatively.

混乱的中药名称和基源难以保证临床用药的准确、安全和有效。必须应用药用植物学分类鉴定的知识和方法，在对混乱品种的用药历史和基源植物考证的基础上，鉴定中药基源植种类，才能解决中药名实混乱问题，为中药生产、科研和临床服务。

Easily confused Chinese names or original plants invalidate the accuracy, safety, effectiveness of TCM materia medica. Therefore, pharmaceutical botany knowledge, in particular, plant taxonomy knowledge is essential, and give scientific names to individual original plant, combined with other methods such as literature review of usage history and plant description of the TCM materia medica, so as to provide reliable references for the industry, research, and clinic.

2. **调查研究药用植物资源，为资源的科学利用提供依据**　我国幅员辽阔，气候多样，地形地貌复杂，物种繁多，优越的自然条件和悠久的用药传统，孕育了极为丰富的药用植物资源。药用植物资源是临床用药的基础，开展药用植物资源调查，可为中药资源的合理利用和实施科学保护提供依据。我国分别于 1960 ~ 1962 年，1969 ~ 1973 年和 1983 ~ 1987 年进行了 3 次全国性的中药资源调查和地区性专项调查，基本摸清了我国中药资源的种类、数量、质量及分布和变化规律，并进行了科学的区划，为国家和地区制定中药资源科学利用、保护、管理的政策方针提供了依据。

3.2　Resources survey of medicinal plants　China covers a vast territory with various climate, complex geography, and abundant native species. Thanks to excellent natural conditions and long tradition of using the plant as medicine, China is rich in medicinal plant resources. Medicinal plants are the principal medicinal materials used in TCM. To establish a scientific basis for rational use and conservation of them, a nationwide survey is necessary. In China, three nationwide surveys together with a number of regional specific surveys were carried out during 1960–1962, 1969–1973, 1983–1987 respectively, regarding the species or varieties, amount, quality, distribution of medicinal plants and their trends. Resources regionalization was conducted based on the information collected, which provided evidence for the national and regional strategies related to the use, protection, and maintenance of medicinal plant resources.

在资源调查过程中，通过挖掘历代本草文献，收集整理民间用药经验和国内外药用植物研究新成果，发现了许多新的药用植物资源。如通过第一次中药资源普查，找到了能提取降血压成分利血平的萝芙木 *Rauvolfia verticillata*（Lour.）Bail.；根据《本草纲目》和《肘后备急方》对黄花蒿 *Artemisia annua* L. 的记载，从该植物中分离得到高效抗疟成分青蒿素（artemisinin）；从薯蓣科植物黄山药 *Dioscorea panthaica* Prain et Burkill、穿龙薯蓣 D. *nipponia* Makino 根状茎中分离的活性组分，研制出治疗冠心病的药物"地奥心血康胶囊"。此外如新疆紫草 *Arnebia euchroma*（Royle）Johnst.、新疆阿魏 *Ferula sinkiangensis* K.M.Shen、绿壳砂 *Amomum villosum* Lour. *var. xanthioides* T.L.Wu et Senjen、剑叶龙血树 *Dracaena cochinchinensis*（Lour.）S.C.Chen 等新资源的发现，为保证医药市场供应以及中医药防病治病提供了新的物质保障。

During these surveys, new resources were discovered by comprehensive approaches such as reviewing of Bencao literature, collecting the folk medicine experiences, learning from the latest research achievements. There are many known successful instances: *Rauvolfia verticillate* was discovered in China during the first survey. Reserpine isolated from this plant is used for the treatment of hypertension.

Artemisinin, a drug used to treat malaria, is isolated from *Artemisia annua,* enlightened by the record in *Ben Cao Gang Mu* and *Zhou Hou Bei Ji Fang* (*Prescriptions for Emergencies*). A Chinese herbal preparation with the name "*Di Ao Xin Xue Kang* Capsule" was developed to treat coronary heart disease by the active components extracted from the rhizomes of *Dioscorea panthaica* and *D. nipponia*. The discovery of some plants, for instance, *Arnebia euchroma*, *Ferula sinkiangensis*, *Amomum villosum.var. xanthioides*, *Dracaena cochinchinensis* provided new resources for the pharmaceutical industry and TCM clinic.

随着药用植物资源需求的不断增多，药用植物资源面临压力增大和物种数量减少的威胁，许多野生常用、名贵药用植物逐渐变为珍稀濒危物种，如红景天、川贝母、新疆雪莲等。据统计，中国处于濒危的 3000 种植物中，药用植物占 60% ~ 70%。国家公布的 398 种濒危植物中，有药用植物 168 种。基于此现状，药用植物资源亟待改善。需要利用多学科知识和技术，开展药用植物资源动态监测、优良种质保存与繁殖、药用植物栽培生产、资源野生抚育等，以确保药用植物资源的可持续利用。目前，正在进行的第四次全国中药资源普查，采用新技术，开展全国范围的中药资源的调查、评估和监测，这必将为中药资源的保护和科学利用提供重要支撑。

The escalating demand for medicinal plants is creating heavy pressure on some medicinal plant populations due to overharvesting. Many commonly used or high-value medicinal plants such as *Rhodiola rosea*, *Fritillaria cirrhosa* and *Saussurea involucrata* are becoming endangered. A highly conservative estimate states that among 3000 plants threatened with extinction in China, 60%–70% are medicinal plants; among 398 rare and endangered plant species published in *The Red Data Book of China's Plant*, 168 are medicinal plants. This status must be improved urgently. To ensure the sustainable use of medicinal plants, multidisciplinary knowledge and approaches are required, for their dynamic monitoring, germplasm conservation and propagation, domestic cultivation, wild tending. Currently, the fourth nationwide resources survey of TCM materia medica is ongoing. As a project to investigate, evaluate, and monitor the resources of TCM materia medica of nationwide, it will provide a reliable reference for their conservation and utilization scientifically.

3. 利用科学规律，寻找和开发新的药物资源 系统进化和植物化学分类揭示，亲缘关系越近的物种，除形态结构相似外，其生理生化特征和化学成分代谢类型也相近。用这一规律去寻找新的药用植物资源，有很多成功的例子。例如，通过对中国茄科 19 属 56 种莨菪类药用植物的系统分析研究，寻找到阿托品类生物碱含量很高的新资源植物矮莨菪 *Hyoscyamus pusillus* L.。紫金牛科植物紫金牛 *Ardisia japonica* (Horn.) Blume 的镇咳成分岩白菜素 (bergenin)，最初从虎耳草科植物岩白菜 *Bergenia purpurascens* (Hook.f.et Thomas) Engl. 中分得。通过重新筛选虎耳草科植物，发现落新妇属 (Astilbe) 多种植物中岩白菜素含量较高，使含岩白菜素的资源植物得以扩大。此外，由于同种植物不同器官的生物合成途径和遗传基础是相同的，可能产生相同或相似的化学成分。根据这一规律，可以扩大药用植物的非传统药用部位为新资源研究对象。例如人参、三七的叶和花中均含有与根中相同或相似的人参皂苷类成分；蒺藜 *Tribulus terrestris* 的全草含有保肝和心血管活性的皂苷类成分，且明显高于传统的药用部位果实。

3.3 New drug discovery from medicinal plants As revealed by phylogenetic and chemotaxonomic studies, closely evolutionary-related species are likely to have not only similar morphological characteristics but also physiological and biochemical characteristics, may contain similar chemical components, share similar metabolism. The last decades have seen many successful cases in drug discovery using this principle. With the help of phylogenetic analysis, atropine alkaloids with higher content were obtained from *Hyoscyamus pusillus* by screening 56 species plants belonging to 19 genera of the family Solanaceae distributed in China. Bergenin is the ingredient for cough in *Ardisia japonica*, isolated first but from *Bergenia purpurascens* of the family Saxifragaceae. By screening the plants of the family Saxifragaceae again, it was found that some plants belonging to genera *Astilbe* contain a higher amount of bergenin, the plant resources of which were therefore expanded. Furthermore, for one plant, biosynthetic pathways and the implicated genes of its different organs are identical, so it is reasonable to suppose that different organs of this plant can generate the same or similar chemical components. As an instance, leaves and flowers of *Panax ginseng* and *P. notoginseng* contain similar ginsenoside as in their roots. The entire plant of *Tribulus terrestris* contains saponins with hepatoprotective and cardiovascular activities, which are significantly higher than those in the fruit, even though the latter is the traditional part used in TCM.

4. 药用植物次生代谢产物的形成和调控 植物次生代谢产物是植物在次生代谢过程中产生的各种小分子有机化合物。次生代谢由初生代谢衍生而来。初生代谢是合成糖类、脂类、核酸和蛋白质等初生代谢产物的共有代谢途径。初生代谢经一系列酶促反应转化成为结构复杂的次生代谢产物，其产生和分布通常具有种属、器官、组织和生长发育的特异性。次生代谢产物广泛参与植物的生长、发育和防御等生理过程，在植物生命活动的过程中发挥重要作用。植物次生代谢产物种类丰富、来源多样，根据其基本结构可分为萜类、酚类和含氮化合物三大类。药用植物次生代谢产物是天然药物工业原料的重要来源。利用生物化学、分子生物学、功能基因组学、蛋白组学和代谢组学的方法和手段，通过对植物次生代谢产物合成途径的解析，在体外可以通过化学合成法或半合成法合成其有效成分；同时，通过对代谢途径中的关键酶和调控基因的功能分析，可以阐明药用植物的复杂代谢途径和代谢网络的分子机理，通过遗传工程技术对细胞代谢途径进行修饰、改造，生产特定需求的化合物，或进行药用植物的良种选育。

3.4 Research on the formation and regulation of secondary metabolites Secondary metabolites are a variety of low molecular weight natural compounds produced by secondary metabolic pathways in plants. It is generally believed that secondary metabolism is derived from primary metabolism. Primary metabolism is the common metabolic pathway for the biosynthesis of primary metabolites such as carbohydrates, lipids, nucleic acids, proteins. Secondary metabolites can be created from primary ones through a series of enzymatic reactions. Secondary metabolites are widely involved in the physiological processes of plants such as growth, development, and reproduction. Many of them evolved and function as defence compounds. Plant secondary metabolites are highly diversified, can be divided into three major classes by their chemical structure: terpenes, phenolics, and nitrogen-containing compounds. Secondary metabolites in medicinal plants are important resources of raw material for the "natural" pharmaceutical industry. By elucidating, analyzing, reconstructing biosynthetic pathways of plant secondary metabolites using biochemistry and molecular biology methods, as well as omic technologies (e.g. genomics, proteomics, and metabolomics), active ingredients can be synthesized or semi-synthesized from natural compounds. In addition, by function analysis of key enzymes and genes involved in the regulation, the complex metabolic pathway and the molecular mechanism of the metabolic network of medicinal plants may be clarified, modified or altered, resulting in successful engineering of certain compounds or breeding of improved varieties.

5. 药用植物分类和鉴定新方法研究 鉴定和分类是药用植物研究的核心工作。传统的鉴定方法主要根据植物的器官形态和组织结构特征而对植物进行分类鉴定；随着对染色体和化学成分研究的日益深入，出现了核型鉴定和化学分类鉴定。而数学和计算机的结合为药用植物的数值分类鉴定提供了可能。目前，基于 PCR 技术的分子标记技术（RAPD、RFLP、SCAR、ISSR、SNP

等）和 DNA 测序技术已广泛应用于药用植物的遗传多样性研究、分子谱系地理学、种质筛选物种鉴定中，解决了药用植物物种鉴定、种质评价和保护的困难，加深了植物分类学家对植物的系统进化及物种形成的认识。如，DNA 条形码技术已被引入《中国药典》用于中药鉴定。在分类系统的重构方面，以分子证据为主建立的 APG 系统已经成为最受人关注的被子植物的新的分类系统。

3.5 **New method development for identification and classification** To identify and classify the medicinal plants is the core subject of pharmaceutical botany. The traditional way to identify species is based mainly on the morphology of the organs and structure of the tissues. As the research on plant chromosomes and chemical components continue, karyotype and chemical characters have been used as tools to identify plant species. With the help of mathematics and computers, the possibility of using numerical taxonomy in classifying medicinal plants is explored. In recent years, PCR-based marker techniques such as RAPD (random amplified polymorphic DNA), RFLP (restriction fragment length polymorphism), SCAR (sequence characterized amplified region), ISSR (inter simple sequence repeats), SNP (single nucleotide polymorphisms) and DNA sequencing techniques are routinely being used in the genetic diversity, molecular phylogeographic and germplasm screening studies of medicinal plants. Due to these tools and techniques, the difficulties on species identification, germplasm evaluation and conservation of medicinal plants have been overcome and the taxonomists' understanding of evolution and systematics, species formation (speciation) has been increased rapidly. For instance, DNA barcoding technigue has been introduced in Chinese Phamacopoeia, for identifying certain TCM materia medica that is difficult to be identified by traditional methods. As a major advance in the reconstruction of the angiosperms phylogenetic relationships, the mostly molecular-based APG system (Angiosperm Phylogeny Group system) has been the much-concerned new taxonomy system.

在社会经济和科学技术高速发展的今天，药用植物学在中药及天然药物众多的研究领域中，仍然占据十分重要的地位，起着不可替代的作用。没有扎实的药用植物的基本理论知识，就无法进行中药的真伪优劣鉴定，无法开展药用植物资源调查、监测和管理，无法进行药用植物资源的科学保护与利用。

The modern age has witnessed the rapid improvement of society, economy, science, and technology. Pharmaceutical botany plays always an important and irreplaceable role in many research fields of TCM materia medica and natural drugs. Without solid elementary knowledge of pharmaceutical botany, it is difficult to make judgements to identify and evaluate the TCM materia medica, to perform surveys, to monitor and maintain the resources of medicinal plants, to promote their conservation and utilization scientifically.

四、药用植物学与相关学科的关系
4 Relationship between pharmaceutical botany and related disciplines

药用植物学是中药学类和药学类专业的专业基础课，在专业培养中，凡涉及中药或天然药物的品种、质量与资源的学科均与药用植物学密不可分。其中临床中药学是学习药用植物学的基础，而药用植物学是学习中药鉴定学和生药学、中药资源学、药用植物栽培学、中药化学和天然药物化学的基础，不掌握药用植物学的基本理论和研究方法，就不能很好学习这些课程。

Pharmaceutical botany is one of the professional basic courses for TCM pharmacy and pharmacy students. In their curriculum, pharmaceutical botany is inseparable from other disciplines concerning varieties, quality, resources of Chinese *materia medica* or natural drugs. Clinical TCM pharmacy is the basis to learn pharmaceutical botany and the latter is the basis to learn identification of TCM materia medica, resources of TCM materia medica, cultivation of medicinal plant, chemistry of TCM materia medica and chemistry of natural products. These courses cannot be learned well without understanding the concepts and methods of pharmaceutical botany.

五、药用植物学的学习方法
5 How to study pharmaceutical botany

药用植物学涉及药用植物的形态、组织结构和分类等内容，是一门实践性很强的学科。由于名词术语较多，理论系统性强，要求学习者在理解基本理论和知识的基础上，多加实践，才能更好地掌握其主要内容。

Pharmaceutical botany is a practical discipline that studies morphology, histology, classification (taxonomy), etc. of the medicinal plants. Because of its large amount of professional terms and coherent theoretical system, the students must practise again and again, to reinforce the theory and knowledge they learned, to grasp better the main content.

1. **以教材为蓝本，准确理解基本概念**　认真阅读教材，了解教材基本框架，熟悉教材的知识体系和专业名词术语的层次性。掌握教学大纲要求的基本知识和基本理论，凡涉及重要的名词、专业术语，必须做到概念准确清晰。

5.1　Understand the concepts accurately　Read the textbook carefully, to understand how the topics and chapters are organized, become familiar with the knowledge framework and arrangement of botanical terms, grasp the basic knowledge and theory described in the syllabus. Important botanical terms must be understood and memorized accurately.

2. **学习中贯穿系统进化的思想，纵横联系，系统比较**　药用植物学教材知识体系的编排，体现系统进化的思想。在学习的时候，要注重养成用系统进化的思维方式学习药用植物的形态、解剖和分类；学习过程中要注重掌握大的分类群的特征，注意纵横、系统比较，不能以点盖全。

5.2　Integrate evolutionary thought throughout the learning　Evolutionary thought is reflected in the arrangement of the knowledge system in this textbook. During the study, the students should develop the skills of learning morphology, anatomy, and taxonomy of medicinal plants with "evolutionary thinking", focus on the characteristics of principal taxonomic groups, their connection and comparison.

3. **理论联系实际是学好本学科的关键**　药用植物学实践性较强，对课程理论知识的理解和掌握必须以实践为基础。在听课的过程中，多观察多媒体课件；在课程实践环节，要利用好实验室和野外实习的实践机会，认真观察药用植物的器官形态和组织结构，在加深感性认识的基础上掌握理论知识，锻炼基本技能。通过理论联系实践，培养动手能力、分析和解决问题的能力，达到提高综合素质的目的。

5.3　Learn through practice　Pharmaceutical botany is a practical science. Practice is critical for the students to gain a deeper understanding of the knowledge. In theoretical lessons, multimedia presentation is helpful to enhance their understanding. The students are encouraged to take advantage of the practical lessons in the laboratory and field, observe the tissues, organs, and organ systems of the plants, to grasp the theoretical knowledge and train their basic skills through perceptual experiences. By integrating theory and practice, the students will develop their hands-on, analytical, and problem-solving skills, therefore comprehensive abilities will be improved.

重 点 小 结
Summary

药用植物是具有药用价值的植物的总称。药用植物学是药学、本草学和植物学交叉融合的学科；研究内容主要包括中药基源植物的鉴定、中药资源调查编目、药用植物活性成分的形成规律、药用植物分类鉴定新方法等。

Medicinal plants are the general term for plants with medicinal value. Pharmaceutical botany is a cross-disciplinary subject of pharmacy, herbal medicine and botany; the research content mainly includes the identification of origin plants of Chinese herbal materia medica, the survey of medicinal plant resources, the formation rules of active ingredients of medicinal plants, and the new methods of classification and identification of medicinal plants.

目 标 检 测
Questions

1. 基于药物发现的历史，怎样理解药用植物和植物的概念？

Based on the history of drug discovery, how to understand the concept of medicinal plants and plants?

2. 了解药用植物学发展简史有何现实意义？

What is the practical significance of understanding the brief history of medicinal botany?

3. 结合古代本草发展史和药用植物的现代研究情况，谈谈学习药用植物学的意义。

Discuss the significance of studying pharmaceutical botany in combination with the history of ancient herbal knowledge development and modern research on medicinal plants.

题库

第一篇 药用植物的形态
Part I Morphology of Medicinal Plants

第一章　根与茎
Chapter 1　Root and Stem

 学习目标 | Learning goals

　　1. **掌握**　根和根系的类型；根的变态特征及类型；茎的类型；茎的变态特征；茎与根的区别。

　　2. **熟悉**　根的功能。

　　3. **了解**　根际环境与药用植物质量和产量的关系。

- Know the types of root and root system, the abnormal characteristics of the root, the types of the stem, the abnormal character of the stem and the differences between the root and the stem.
- Be familiar with the functions of the root.
- Understand the relationship between rhizosphere and the quality and yield of medicinal plants.

第一节　根
1　Root

PPT

　　根通常是植物生长在土壤中的营养器官，具有向地性、向湿性和背光性。根有吸收、输导、固着、支持、贮藏和繁殖等功能。植物生活所需要的水分及无机盐，靠根从土壤中吸收和传输。有些植物的根还具有合成氨基酸、生物碱、生物激素等能力，如烟草的根能合成烟碱，橡胶草的根能合成橡胶等。有些植物的根可供食用，如番薯、胡萝卜等；有些植物的根是重要的中药材，如人参、党参、百部等。也有一些根连同上部的根状茎入药，如人参的"芦头"。

Roots are usually vegetative organs that grow in the soil. Roots mainly have functions of absorption, conduction, fixation, support, storage and reproduction. The water and inorganic salts needed for plant life are absorbed and transmitted from the soil by the roots. The roots of some plants also have the ability to synthesize amino acids, alkaloids, and biological hormones. For example, the roots of *Nicotiana tabacum* and *Taraxacum kok-saghyz* can synthesize nicotine and rubbers separately. The roots of some plants are edible, such as *Ipomoea batatas*, *Daucus carota* var. *sativa*, etc. The roots of some plants are important Chinese herbal medicines, such as *Panax ginseng*, *Codonopsis pilosula*, and *Stemona japonica*. There are also some roots with the upper rhizomes used for medicine, such as the rhizome of *Panax ginseng*.

医药大学堂
WWW.YIYAODXT.COM

15

一、根的形态
1.1 Root morphology

种子植物最初的根，由种子中的胚根发育形成。大多数双子叶植物和裸子植物的胚根继续发育，垂直于地面向下生长，形成主根，主根生长到一定长度后，生出许多分枝。因生长需要，在茎、叶、胚轴或老根上也会产生根。根无节和节间之分，一般不生芽、叶和花，细胞中不含叶绿体。

The initial root of a seed plant is formed by the development of the radicle in the seed. The radicles of most dicotyledons and gymnosperms continue to develop, grow perpendicular to the ground and form the main roots. After the main roots grow to a certain length, many branches emerge. Roots may also be produced on stems, leaves, hypocotyls or old roots due to growth needs. Roots have no nodes and internodes, generally do not produce buds, leaves and flowers, and cells without chloroplasts in cells.

二、根的类型
1.2 Root type

（一）根的类型
1.2.1 Root type

1. **主根、侧根和纤维根** 植物最初生长的，由种子的胚根直接发育来的根称为主根。多数双子叶植物和裸子植物如薄荷、菘蓝、银杏等都有一个主根。在主根的侧面生长出来的分枝，称为侧根，在侧根上形成的小分枝称纤维根（图1-1）。

(1) Primary roots, lateral roots and fibrous roots The first root of the plant originates in the embryo and is usually called the primary root, continues to grow, it develops branch roots, or lateral roots. These roots in turn may give rise to additional lateral roots, called the fibrous roots. This type is typically found in dicotyledonous and gymnospermous plants (Fig.1-1).

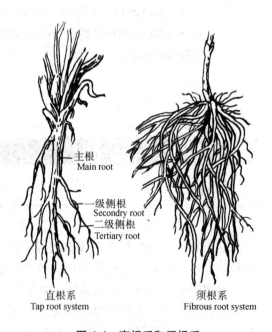

图 1-1 直根系和须根系
Fig. 1-1 Primary and fibrous root systems

2. **定根和不定根** 凡是直接或间接由胚根发育而成的主根、侧根及各级纤维根称为定根，它们有固定的生长部位，如桔梗、人参、棉花等的根。有些植物的根是从茎、叶或其他部位生长出来的，这些根的产生无固定位置，称不定根，如玉蜀黍、薏苡、麦、稻的大小、长短相似的须根，人参根状茎（芦头）上的不定根，都是不定根。

(2) Fixed roots and adventitious roots Primary roots, lateral roots, and fibrous roots developed directly or indirectly from the radicle are called fixed roots, and they have fixed growth sites, such as the roots of *Platycodon grandiflorum, Panax ginseng* and *Gossypium hirsutum*. The roots of some plants grow from stems, leaves, or other parts. These roots have no fixed location and are called **adventitious roots**, such as the fibrous roots of similar size and length from *Zea mays, Coix lacryma-jobi* L. var.

mayuen (Roman.), etc., and the rhizomes of *Panax ginseng*.

（二）根系的类型

1.2.2　Root　system

一株植物地下部分所有根的总体称为根系。根系常有一定的形态，按其形态的不同可分为直根系和须根系两类。

All the roots of a plant are called root system. Root system can be divided into two types.

1. 直根系　主根发达，主根和侧根界限非常明显的根系称直根系。

(1) Primary root system　The whole system is derived by growth and lateral branching of the seedling radicle, this type is typically found in dicotyledonous plants.

2. 须根系　主根不发达，或早期死亡，而从茎的基部节上生长出许多大小、长短相似的不定根，簇生呈须状，没有主次之分的根系称须根系。

(2) Fibrous root system　The primary root system is supplanted by an adventitious root system and is ubiquitous in the monocotyledons. An adventitious root develops from a root primordium arising in a stem or leaf. Some dicotyledonous plants possess both types.

三、根的变态
1.3　Metamorphosis of the root

根与植物的其他器官一样，在长期发展过程中，为了适应生活环境的变化，形态构造和生理机能产生了许多变态，常见的有以下几种（图 1-2、图 1-3）。

Like other organs of plants, root will change its morphological structure and physiological functions to adapt to changes in living environment in the long-term development process. The following are common (Fig.1-2, Fig.1-3).

1. 贮藏根　根的一部分或全部呈肥大肉质状，其内贮藏大量的营养物质，这种根称贮藏根。依来源及形态不同又可分为：

(1) Storage root　One or part of the root is fleshy and contains a large amount of nutrients. According to its origin and different shapes, storage root can be divided into the following categories.

肉质直根主要由主根发育而成，一株植物上只有一个肉质直根，其上部具有胚轴和节间很短的茎，其肥大部位可以是韧皮部，如胡萝卜，也可以是木质部，如萝卜。有的肉质直根肥大呈圆锥状，称圆锥根，如胡萝卜、白芷、桔梗；有的肥大呈圆柱形，称圆柱根，如菘蓝、丹参；有的

视频

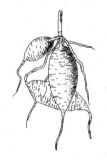

| 圆锥根
Conical root | 圆柱根
Cylindrical root | 圆球根
Spherical root | 纺锤状块根
Spindle tuber | 块状块根
Massive tuber |

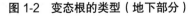

图 1-2　变态根的类型（地下部分）
Fig.1-2　Types of root metamorphosis

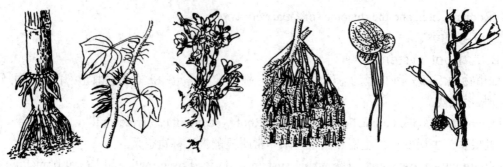

支持根（玉米）	攀援根（常春藤）	气生根（石斛）	呼吸根（红树）	水生根（青萍）	寄生根（菟丝子）
Prop root	Climbing root	Aerial root	Respiratory root	Aquaceous root	Parasitic root
(*Zea mays*)	(*Hedera nepalensis*)	(*Dendrobium sp.*)	(*Rhizophora apiculata*)	(*Lemna minor*)	(*Cuscuta sp.*)

图 1-3 根的变态（地上部分）

Fig.1-3 Types of root metamorphosis

肥大成圆球形，称圆球根，如芜青、珠子参。

Flashy tap root is mainly developed from the taproot. There is only one fleshy taproot on o plant. Its upper part has hypocotyls and stems with short internodes. Its hypocotyls can be phloem, such as carrots, or xylem, such as radish. Some fleshy taproots, such as carrots, *Angelica dahurica* and *Platycodon grandiflorum*, are thick and conical, which are called cone roots. Some are thick and cylindrical, named cylindrical roots, such as *Isatis indigotica* and *Salvia miltiorrhiza*. Some are thick and spherical and are called round bulbous roots, such as *Brassica rapa* and *Panax japonicus var. major*.

块根主要由不定根或侧根膨大而成，因此，在一株上可形成多个块根。药用块根有天门冬、郁金、何首乌、百部等。在不同的植物中，块状根的大小、色泽、质地都有许多不同，都可以作为识别植物的依据。

Tuberous roots are mainly formed by the expansion of adventitious roots or lateral roots. Therefore, multiple roots can be formed on one pant. There are many plants with medicinal tuberous roots, such as *Asparagus cochinchinensis*, *Curcuma wenyujin*, *Polygonum multiflorum* and *Stemona japonica*. There are many differences in the size, color and texture of tuberous roots, which can be used as the basis for indentify medicinal plants.

2. **支持根** 自茎上产生一些不定根深入土中，以增强支持茎干的力量，这种根称支持根，如玉蜀黍、高粱、薏苡、甘蔗等在接近地面的茎节上所生出的不定根。

(2) **Stilt roots** Adventitious roots arising from the lower nodes of the plant and penetrating the soil in order to give increased anchorage as in Zea mays, Pandanus and Rhizophora.

3. **气生根** 由茎上产生，不深入土里而暴露在空气中的不定根，称为气生根。它具有在潮湿空气中吸收和贮藏水分的能力，如石斛、吊兰、榕树等。

(3) **Aerial roots** Roots emerging from the plant wholly above the ground surface and growing in air. In epiphytes, the aerial roots termed epiphytic roots are found hanging from the orchids and are covered with a spongy velamen tissue.

4. **攀援根** 植物的茎细长柔弱，其上长出具攀附作用的不定根，能攀附石壁、墙垣、树干或其他物体上，使其茎向上生长，这种具攀附作用的不定根叫攀援根，如薜荔、络石、常春藤、凌霄等。

(4) **Prop roots** Elongated aerial roots arising from horizontal branches of a tree, striking the ground and providing increased anchorage and often replacing the main trunk as in several species of

Ficus. The large hanging prop roots of *Ficus* species are often used in bungee jumping sport.

5. 水生根　水生植物的根飘浮在水中呈须状，称水生根，如浮萍、睡莲、菱等。

(5) Respiratory roots　Negatively geotropic roots of some mangroves (e.g. *Avicennia*) which grow vertically up and carry specialized lenticels (*pneumathodes*) with pores for gaseous exchange. Such roots are also known as pneumatophores.

6. 寄生根　有一些植物的根，不是伸入土中，而是插入寄主植物体内，吸收寄主体内的水分和营养物质，以维持自身的生活，这种根称寄生根。具有寄生根的植物，叫寄生植物。寄生植物可分为两类：一类如菟丝子、列当等植物体本身不含叶绿素，不能自制养料而完全依靠吸收寄主体内的养分维持生活，这类植物称全寄生植物或非绿色寄生植物；另一类如桑寄生、槲寄生等植物，一方面由寄生根吸收寄主体内的养分，而同时自身含叶绿素可以制造一部分养料，这类植物称半寄生植物或绿色寄生植物。

寄生植物可以通过寄生根的吸收作用，把有毒寄主的毒性成分带入寄生植物体内，如马桑寄生等。

(6) Parasitic roots　Some plants, including dodders and broomrapes, have no chlorophyll and have become dependent on chlorophyll-bearing plants for their nutrition. They parasitize their host plants via peglike projections called haustoria, which develop along the stem in contact with the host. The haustoria penetrate the outer tissues and establish connections with the xylem and phloem. Some green plants, including *Indian warrior* and the *mistletoes*, also form haustoria. These haustoria, however, apparently aid primarily in obtaining water and dissolved minerals from the host plants, since the partially parasitic plants are capable of manufacturing at least so me of their own food through photosynthesis.

第二节　茎

2 Stem

PPT

茎是连接叶和根的轴状结构，主要起着输导和支持作用。茎将根部吸收的水分和无机盐以及叶制造的有机物质，输送到叶、花、果实中并支持其正常生长。茎还支持着叶、花和果实，并将它们有序地排放在一定的空间位置。茎还具有贮藏、繁殖和光合作用等功能。许多植物的茎贮藏有水分和营养物质，如仙人掌的茎贮存水分，甘蔗的茎贮存蔗糖，半夏的块茎贮存淀粉等。有些植物的茎上能产生不定根和不定芽，可作为扦插、压条、嫁接等繁殖材料。此外，绿色的幼茎和叶退化的变态茎能进行光合作用。

植物的茎有多种用途，包括食用、药用、工业原料、木材和竹材等。其中许多植物茎的全部或部分可以药用，如麻黄、桂枝等的茎枝，首乌藤、忍冬藤等的茎藤，杜仲、黄柏的茎皮，沉香、降香的心材，黄连、半夏的地下茎。

The principal functions associated with stems are support and conduction. The leaves—the principal photosynthetic organs of the plant—are supported by the stems, which place the leaves in favorable positions for exposure to light. Substances manufactured in the leaves are transported downward through the stems by way of the phloem to sites where they are needed, such as developing plant parts and storage tissues of both stems and roots. At the same time, water and minerals are transported upward in the xylem from the roots and into the leaves via the stem. In a number of plants, stems are modified in ways that allow specialized functions, such as climbing, the storage of food or water, reproduction, etc. The stems of plants have various uses, including edible, medicinal, industrial raw materials, wood and bamboo, etc. Stems of some plants are important Chinese herbal medicines, such as the shoots of *Ephedra sinica* Stapf, *Ephedra intermedia* Schrenk et C. A. Mey., *Ephedra equisetina* Bge., and *Cinnamomum cassia* Presl.

一、茎的形态
2.1 Morphology of the stem

茎的形状随植物种类而异，通常为圆柱形。但有些植物的茎形状特殊，是重要的鉴别依据，如薄荷、紫苏等唇形科植物的茎为方形，荆三棱、香附等莎草科植物的茎为三角形，仙人掌、竹节蓼等的茎为扁平形。茎常为实心，但芹菜、南瓜等植物的茎为空心；禾本科植物芦苇、麦、竹等的茎中空，且有明显的节，特称为秆。茎的高度差异大，从数厘米到百米以上。

茎顶端有顶芽，叶腋有腋芽。茎上着生叶和腋芽的部位称节，节与节之间称节间，节上叶柄和茎之间的夹角处称叶腋。一般植物茎的节部仅在叶着生处稍有膨大，而有些植物的节部膨大明显呈环状，如牛膝、石竹、玉米等；也有些植物的节部细缩，如藕。各种植物节间的长短也不一致，长的可达数十厘米，如竹、南瓜；短的还不到1mm，其叶在茎节簇生呈莲座状，如蒲公英。具有节与节间是茎在外形上与根的最主要区别。木本植物上着生叶和芽的茎称为枝条，枝条生长的强弱影响节间的长短，不同植物节间的长度不同。有些植物如苹果、梨和银杏等的茎具有两种枝条，一种节间较长，其上的叶螺旋状排列，称长枝；另一种节间较短，其上的叶多簇生，称短枝。一般短枝着生在长枝上，能生花结果，所以又称果枝。

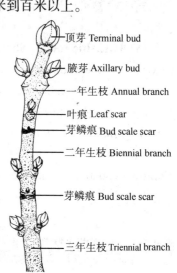

图 1-4　三年生枝条
Fig.1-4　Triennial branches

在木本植物的茎枝上，常见有叶痕、托叶痕、芽鳞痕等，分别是叶、托叶、芽鳞脱落后留下的痕迹；有些茎枝表面可见各种形状的浅褐色点状突起皮孔。这些特征常作为茎类、皮类药材鉴别的依据（图 1-4）。

The majority of aerial stems are more or less cylindrical in shape. However, it can take on one of a wide range of forms, some of which assist in the identification of a family, as in Labiateae, where the section is square, or may help to distinguish genera; for example many Carex species have stems with a triangular cross-section. Underground stems have a variety of shapes. The stems of succulent plants are typically swollen and in others the stem is flattened and mimics a leaf. The bases of leaves in some cases are extended some distance down the stem forming ridges. If particularly extended the stem becomes winged or pterocaul. A simple cylindrical shape may become elaborated by the formation of bark, or in the case of climbing plants develop a range of contortions and twistings. The old but living stems of some desert plants are similarly disrupted and split following the formation of longitudinal sections of cork within the wood. The trunks of some tropical trees become so deeply fluted that holes develop through from one side to the other, a condition known as fenestration.

There often (but not always) is a terminal bud present at the tip of each stem. A terminal bud usually resembles an axillary bud, although it is often a little larger. Unlike axillary buds, terminal buds do not become separate branches, but, instead, the meristems within them normally produce tissues that make the stem grow longer during the growing season. The bud scales of a terminal bud leave tiny scars around the stem when they fall off in the spring. In flowering plants (angiosperms), axillary buds may become branches, or they may contain tissues that will develop into the next season's flowers. Most buds are protected by one to several bud scales, which fall off when the bud tissue starts to grow. A stem consists of a series of nodes separated by internodes. Leaves are inserted on the stem at the nodes and commonly have buds in their axils. The area, or region, of a stem where a leaf or leaves are attached is called a node, and a stem region between nodes is called an internode. Internodes may be very short, in which case one node appears to merge into the next. The combined structure of stem and leaves is termed a shoot and thus each bud in a leaf axil represents an additional shoot. The sequences of shoot development give any plant its particular form. Stems can develop in a range of shapes and surfaces can become elaborated by bark development, emergences, adventitious roots, and adventitious buds. Generally, all stems have nodes, internodes, and axillary buds, these features distinguish them from roots and leaves, which do not have them.

Deciduous trees and shrubs (those that lose all their leaves annually) characteristically have dormant axillary buds with leaf scars left below them after the leaves fall. Tiny bundle scars, which mark the location of the water-conducting and food-conducting tissues, are usually visible within the leaf scars. There may be one too many bundle scars present, but more often than not, there are three. The shape and size of the leaf scars and the arrangement and numbers of the bundle scars are characteristic for each species. One can often identify a woody plant in its winter condition by means of scars and buds (Fig.1-4).

二、芽的类型
2.2 Types of the bud

芽是尚未发育的枝条、花或花序，包括茎尖的顶端分生组织及其外围附属物（图 1-5）。

Bud, a meristem (either apical or lateral) in early development or resting stages, with its protective coverings; immature shoot, usually protected by scales or prophyll (s), or immature flower, protected by bracts, bracteoles and/or perianth segments (Fig.1-5).

1. **定芽和不定芽** 根据芽的生长位置，芽可分为定芽和不定芽。定芽有固定的生长位置，又分为生于顶端的顶芽、生于叶腋的腋芽（侧芽）和生于顶芽或腋芽旁的副芽。不定芽的生长无固定位置，如生在茎的节间、根、叶及其他部位上的芽。

(1) Normal bud and adventitious bud Buds may be present in the axils of leaves, that is, between the leaf and the stem, close to where they join, called normal bud. Sometimes buds develop from other parts of the plant, these are called adventitious buds. There often (but not always) is a terminal bud

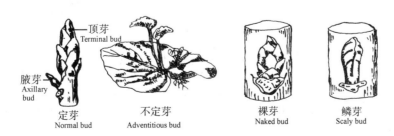

顶芽 Terminal bud
腋芽 Axillary bud
定芽 Normal bud
不定芽 Adventitious bud
裸芽 Naked bud
鳞芽 Scaly bud

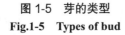

图 1-5 芽的类型
Fig.1-5 Types of bud

present at the tip of each twig. A terminal bud usually resembles an axillary bud, although it is often a little larger. Adventitious buds without an unusual place such as internodes, roots, leaves and other parts of the stem.

2. **叶芽、花芽和混合芽** 根据芽的性质分为发育成枝和叶的叶芽（枝芽）、发育成花或花序的花芽和同时发育成枝、叶和花的混合芽。

(2) Leaf bud, flower bud and mixed bud Leaf bud, also called branch bud, refers to the bud that develops into branches and leaves. A flower bud refers to a bud that develops into a flower or an inflorescence, and is a primitive body of a flower. Mixed buds refer to buds that can develop branches, leaves, and flowers or inflorescences at the same time, such as apples and pears.

3. **鳞芽和裸芽** 根据芽的外面有无鳞片包被分为鳞芽和裸芽。

(3) Scaly bud and naked bud The outer bread is called scaly bud by the scaly bud, whose scale is also called bud scale, such as the overwintering bud of most woody plants such as willow and camphor. The naked bud is the bud of most herbs, such as eggplant and mint, and a few woody plants.

4. **活动芽和休眠芽（潜伏芽）** 根据芽的活动能力分为活动芽和休眠芽（潜伏芽）。其中休眠芽的休眠期是相对的，在一定条件下可以萌发，如树木砍伐后，树桩上常见休眠芽萌发出的新枝条。

(4) Active bud and dormant bud The buds that form new branches, flowers, or inflorescences in the growing season of that year are called active buds. Usually the annual herb buds are active buds, while woody plants often have only the top buds and the side buds near the top buds are active buds. Buds that do not grow and develop during the growing season and maintain a state of rest are called dormant buds or latent buds. Dormant buds allow plants to adjust nutrients, control the growth of leaves and branches, and maintain sufficient reserve power to survive.

三、茎的类型
2.3 Type of stem

（一）按茎的质地分（图 1-6）
2.3.1 According to the texture of the stems（Fig. 1-6）

1. **木质茎** 茎的质地坚硬，木质部发达。具木质茎的植物称木本植物。木本植物可分为乔木、灌木、亚灌木或半灌木和木质藤本等类型。其中植株高大，主干明显，下部不分枝者称乔木，如银杏、杜仲、厚朴等；主干不明显，近基部分枝呈丛生状的为灌木，如枸杞、连翘、月季等；仅在基部木质化，上部草质的为亚灌木或半灌木，如草麻黄、草珊瑚等；茎细长且缠绕或攀

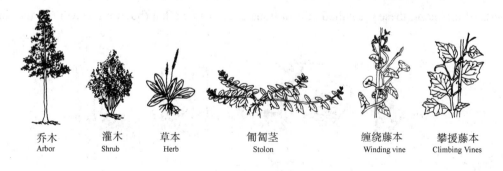

| 乔木
Arbor | 灌木
Shrub | 草本
Herb | 匍匐茎
Stolon | 缠绕藤本
Winding vine | 攀援藤本
Climbing Vines |

图 1-6 茎的类型
Fig.1-6 Types of the stem

附他物向上生长的为木质藤本，如五味子、密花豆等。

(1) Woody stem　Woody plants have a woody stem. In woody stems, the production of secondary xylem and phloem results in the formation of a cylinder of secondary vascular tissues and commonly, much more secondary xylem than secondary phloem is produced in the stem in any given year and thus characterized as hard. Plants are often classified according to their branching behavior as tree, shrub and subshrub. **Shrub**, self-supporting woody plant branching at or near the ground or with several stems from the base; **subshrub**, small shrub with partially herbaceous stems.

2. **草质茎**　茎的质地柔软，木质部不发达。具草质茎的植物称草本植物。草本植物根据其生命周期的长短可分为一年生草本、二年生草本和多年生草本等类型。多年生草本中地上部分每年枯萎，而地下部分仍保持生活能力的称宿根草本，如人参、黄精等；植物体保持常绿若干年不凋者称常绿草本，如万年青、麦冬等；茎缠绕或攀附他物向上生长的称草质藤本，如鸡矢藤、牵牛等。

(2) Herbaceous stem　Many herbaceous plants have a herbaceous stem. In herbaceous stems, primary phloem much more than primary xylem thus characterized as soft. Plants are often classified according to their seasonal growth cycles as annuals, biennials, or perennials. In the annuals— which include many weeds, wildflowers, garden flowers, and vegetables—the entire cycle from seed to vegetative plant to flowering plant to seed occurs within a single growing season, which may be only a few weeks in length. Most monocots are annuals, but many dicots (discussed next) are also annuals.

3. **肉质茎**　质地柔软多汁，肉质肥厚。如芦荟、仙人掌等。

(3) Succulent　A plant with thick, fleshy and swollen stems that is juicy and pulpy, adapted to dry environments, such as Aloe, Cactaceae or Stapelia.

（二）按茎的生长习性分

2.3.2　According to the growth habits of the stems

1. **直立茎**　茎直立生长于地面，不依附他物。如银杏、杜仲等。

(1) Erect stem　Stems growing erect and do not attach to anything else. Such as *Ginkgo biloba*, *Eucommia ulmoides*, etc.

2. **缠绕茎**　茎细长且不能直立生长，常缠绕他物作螺旋式上升。如五味子、何首乌、牵牛等。

(2) Twining stem　Stem coiling round the support due to special type of growth habit, as in Ipomoea and Convolvulus.

3. **攀援茎**　茎细长且不能直立生长，常依靠攀援结构依附他物上升。常见的攀援结构有茎卷须（如栝楼、葡萄等）、叶卷须（如豌豆等）、吸盘（如爬山虎等）、钩或刺（如钩藤、葎草等）、不定根（如络石、薜荔等）。

(3) Climbing stem　Climbing plants have stems modified in various ways that adapt them for their manner of growth. Some stems, called ramblers, simply rest on the tops of other plants, but many produce tendrils. In Boston ivy, the tendrils have adhesive disks. In English ivy, the stems climb with the aid of adventitious roots that arise along the sides of the stem and become embedded in the bark or other support material over which the plant is growing.

4. **匍匐茎**　茎细长，平卧地面，沿地面蔓延生长，节上生有不定根。如金钱草、草莓、积雪草等。

(4) Stolon　Creeping stem, a stolon is a stem growing along the substrate surface or through surface debris. It has long thin internodes and bears foliage, or occasionally scale leaves. Buds in the axils of the leaves will develop into inflorescences or additional stolons. Adventitious roots usually emerge at the nodes (nodal roots), sometimes only at nodes having a lateral stolon.

5. **平卧茎** 茎细长，平卧地面，节上无不定根，如地锦草等。

(5) Repent stem Vegetative shoot spreading along the surface of the ground and no rooting at the nodes.

四、茎的变态
2.4 Metamorphosis of stem

茎在长期适应生活环境变化的发展过程中，其形态结构和生理功能发生变化，出现了各种变态类型。根据茎的生长习性，分为地上茎的变态和地下茎的变态（图1-7、图1-8）。

The long-term adaptation process of the stem to changes in the living environment changes its morphological structure and physiological function, and various types of metamorphosis appear. According to the growth habit of stems, they are divided into metamorphosis of aerial stem and metamorphosis of subterraneous stem (Fig.1-7, Fig.1-8).

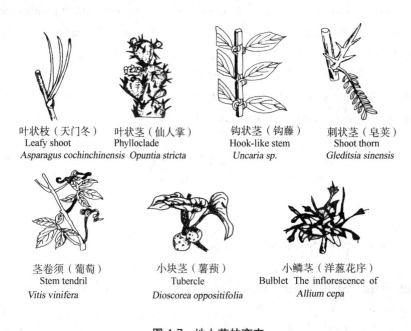

叶状枝（天门冬）　　叶状茎（仙人掌）　　钩状茎（钩藤）　　刺状茎（皂荚）
Leafy shoot　　Phylloclade　　Hook-like stem　　Shoot thorn
Asparagus cochinchinensis　*Opuntia stricta*　*Uncaria sp.*　*Gleditsia sinensis*

茎卷须（葡萄）　　小块茎（薯蓣）　　小鳞茎（洋葱花序）
Stem tendril　　Tubercle　　Bulblet The inflorescence of
Vitis vinifera　*Dioscorea oppositifolia*　*Allium cepa*

图 1-7 地上茎的变态
Fig.1-7 Metamorphosis of the above-ground stem

根状茎（玉竹）　　根状茎（生姜）
Rhizome of *Polygonatum odoratum*(left) and *Ziginber officinale*(right)

块茎（半夏）　　球茎（荸荠）　　鳞茎（洋葱）　　鳞茎（百合）
Tuber(*Pinellia ternata*)　Corm(*Eleocharis dulcis*)　Bulb(*Allium cepa*)　Bulb(*Lilium brownii*)

图 1-8 地下茎的变态
Fig.1-8 Metamorphosis of the underground stem

（一）地上茎的变态

2.4.1 Metamorphosis of aerial stem

1. 叶状茎或叶状枝 茎变为绿色扁平状或针叶状，代替叶的光合作用，叶退化为鳞片状、线状或刺状。如仙人掌、竹节蓼、天门冬等。

(1) Cladophyll (Phylloclade) The stems of some plants are flattened and appear leaflike which are green and photosynthetic and bear small scale leaves. Such flattened stems are called cladophylls (or cladodes or phylloclades). There is a node bearing very small, scalelike leaves with axillary buds in the center of each butcher's broom cladophyll. Such as greenbriers, certain orchids, prickly pear cacti.

2. 刺状茎（枝刺或棘刺） 枝变态成粗短坚硬分枝或不分枝的刺。如山楂、酸橙等的刺状茎不分枝；皂荚、枸橘等的刺状茎有分枝。而蔷薇、月季、花椒等的茎上均有较多的刺，是茎表皮或皮层细胞突起形成，易脱落，称皮刺。

(2) Shoot thorn The stems may be modified in the form of thorns, as in the honey locust, whose branched thorns may be more than 3 decimeters long, but all thornlike objects are not necessarily modified stems. For example, at the base of the petiole of most leaves of the black locust is a pair of spines. A stem spine is formed if the apical meristem of a shoot ceases to be meristematic and its cells become woody and fibrous. Spines are either lateral on longer usually indeterminate shoots, or terminal forming a determinate shoot. The prickles of raspberries and roses, both of which originate from the epidermis, are neither thorns nor spines.

3. 钩状茎 茎的侧枝变为钩状，粗短坚硬，不分枝。如钩藤。

(3) Tendril and hook Numerous climbing plants possess tendrils, or hooks acting in the manner of grappling irons. Prehensile stem tendrils can become secondarily thickened to form permanent woody clasping hooks. Alternatively a tendril will twine around the support and subsequently shorten in length by coiling up, the proximal and distal ends of the tendril often twisting in opposite directions. Stem tendrils may be branched; some have adhesive discs at their distal ends. Stem tendrils and hooks represent either modifications of axillary shoots or are terminations of a shoot, continued growth of that axis being sympodial.

4. 茎卷须 攀援植物的部分分枝或茎段变态为卷须状，柔软卷曲。如栝楼、丝瓜、葡萄等。

(4) Bulbil A bulbil is merely a small bulb, that is a short stem axis bearing fleshy scale leaves or leaf bases and readily producing adventitious roots. However, the term is also sometimes inaccurately applied to any small organs of vegetative multiplication such as axillary stem tubers found on the aerial stems of some climbers.

5. 小块茎和小鳞茎 有些植物的腋芽、叶柄上的不定芽可变态形成无鳞片包被的块状物，称小块茎。如山药的零余子、半夏的珠芽。有些植物由腋芽或花芽形成有小鳞茎，如卷丹腋芽形成的小鳞茎，洋葱、大蒜花序中花芽形成的小鳞茎。小块茎和小鳞茎均有繁殖作用。

(5) Bulblet and bubbet Also there are alternative terms, bulblet, bulbet, bulbel, which are variously given precise definitions or used indiscriminately as synonyms. Small bulbs are mostly found in one of two locations, either on an aerial stem, representing axillary buds and especially replacing flowers in an inflorescence, or developing in the axils of the leaves of a fully sized bulb.

6. 假鳞茎 一些植物茎基部肉质膨大呈卵球形至椭圆形，具贮存水分和养分功能，绿色的还能进行光合作用。常见于浮生兰类，如石仙桃、羊耳蒜等。

(6) Pseudobulb Short erect aerial storage or propagating stem of certain epiphytic orchids.

（二）地下茎的变态

2.4.2　Metamorphosis of subterraneous stem

1. **根状茎** （根茎）匍匐或直立生长在土壤中，节和节间明显，节上有退化的鳞片叶，具顶芽和腋芽。根状茎的形态及节间长短随植物种类而异。如人参、三七的根状茎短而直立，称芦头；姜、白术的根状茎呈团块状；白茅、芦苇的根状茎细长。黄精、玉竹等的根状茎上具有明显茎痕。

(1) Rhizome　A stem growing more or less horizontally below ground level is described as a rhizome. Rhizomes tend to be thick, fleshy or woody, and bear scale leaves or less often foliage leaves, or the scars when these leaves have been lost; they also bear adventitious roots most frequently at the nodes. Rhizome diameters vary from a few millimetres in some grasses up to half a metre or more as in the palm Nypci. A root bearing root buds can be distinguished from a rhizome by the lack of subtending leaves or leaf scars.

2. **块茎**　肉质肥大呈不规则块状，由茎基部腋芽伸入土壤形成，具顶芽和缩短的节间，节上具芽及鳞片叶或后期脱落，如天麻、半夏、马铃薯等。

(2) Tuber　A stem tuber is a swollen shoot usually underground and bearing scale leaves, each subtending one or more buds which give rise to vegetative shoots. The presence of leaves or leaf scars distinguishes a stem tuber from a root tuber. Typically a stem tuber forms by the swelling of the distal end of a slender underground rhizome and thus does not form one unit of a sympodial sequence as is commonly found in a rhizome.

3. **球茎**　肉质肥大呈球形或扁球形，节和节间明显，节上有较大的膜质鳞片，顶芽发达，腋芽常生于其上半部，基部生不定根，如慈菇、荸荠等。

(3) Corm　A corm is a short swollen stem of several internodes and nodes bearing either scale or foliage leaves. It develops at or below ground level in a vertical position. In favourable growing conditions the apical meristem of the corm or one of the buds close to the apex extends into an aerial flowering shoot usually bearing foliage leaves. The corm may be reduced in size at the expense of this shoot or may shrivel away altogether. One or more buds in the axils of leaves on the corm swell to form new corms during the growing season.

4. **鳞茎**　呈球形或扁球形，茎极度缩短为鳞茎盘，被肉质肥厚的鳞叶包围，顶端有顶芽，叶腋有腋芽，基部生不定根。洋葱鳞叶阔，内层被外层完全覆盖，称有被鳞茎；百合、贝母鳞叶狭，呈覆瓦状排列，外层无皮覆盖，称无被鳞茎。

(4) Bulb　Underground storage organ, are actually the bud (s) surrounded by fleshy scale leaves and/ or leaf bases, with a small stem at the lower end. Adventitious roots grow from the bottom of the stem, but the fleshy leaves comprise the bulk of the bulb tissue, which stores food. In onions, the fleshy leaves usually are surrounded by the scalelike leaf bases of long, green, above ground leaves. Other plants producing bulbs include lilies, hyacinths, and tulips. Tiger lilies produce small, aerial bulblets in the axils of their leaves.

重 点 小 结
Summaries

　　根具有吸收、输导、固着、贮藏及繁殖等功能。根分为定根和不定根。根系有直根系和须根系之分。根在功能及形态上常特化变态成贮藏根、支持根、气生根、攀援根、呼吸根、寄生根、水生根等。

The root has functions of anchorage, absorption, conduction, fixation, storage and clonal regeneration. Roots are divided into fixed roots and adventitious roots. The root system is divided into primary root system and fibrous root system. In terms of function and morphology, roots are often metamorphosed into fleshy roots, stilt roots, aerial roots, prop roots, respiratory roots and parasitic roots.

茎是种子植物重要的营养器官，有节和节间。茎具有输导、支持、贮藏及繁殖等功能。茎在功能及形态上常特化变态成叶状茎、刺状茎、茎卷须、钩状茎、小块茎和小鳞茎、假鳞茎、根状茎、块茎、球茎、鳞茎等。

Stems represent the main axes of plants, being distinguished into nodes and internodes, and bearing leaves and axillary buds at the nodes. The buds grow out into lateral shoots, inflorescences or flowers. Stems usually function supporting for leaves and flowers, transporting fluids between roots and shoots and storing nutrients. In terms of function and morphology, roots are often metamorphosed into cladophylls, shoot thorn, tendril and hook, bulbil, pseudobulb, rhizomes, tuber, corm, bulb.

目 标 检 测
Questions

题库

1. 如何从来源、形态和生长趋势方面，正确认识根的概念？

How to correctly understand the concept of root from the aspects of source and morphological growth trend?

2. 如何区别定根与不定根？

How to distinguish between normal roots and adventitious roots?

3. 什么叫根系？按形态不同根系分为哪几类？

What is the root system? What are the types of root system?

4. 何谓块根？试举出 5 种以块根入药的药用植物。

What is a tuberous root? Try to list 5 kinds of medicinal plants.

5. 如何从外形上区别根与根茎？

How to distinguish root and rhizome by morphological characteristics?

6. 如何从来源、形态生长趋势方面，正确认识茎的概念？

How to correctly understand the concept of root from the aspects of source and morphological growth trend?

7. 怎样区别缠绕茎与攀援茎？如何区别匍匐茎与平卧茎？

How to distinguish twining stem and climbing stem? How to distinguish stolon and repent stem?

8. 地下茎有哪几种变态类型？其共同特点是什么？

What are the types of metamorphosis of subterraneous stem? what are common characteristics of these subterraneous stem?

9. 如何区别块根与块茎？

How to distinguish tuberous and tubers?

10. 地上茎有哪几种变态类型？

What are the types of metamorphosis of aerial stem?

医药大学堂
WWW.YIYAODXT.COM

第二章　叶
Chapter 2　Leaf

学习目标 ┆ Learning goals

1. **掌握**　叶的组成和类型。
2. **熟悉**　叶的变态类型及叶序。
3. **了解**　叶的形态及叶的功能。

- Know the composition and type of the leaves.
- Be familiar with leaf metamorphosis and leaf arrangement.
- Understand the shape of leaves.

叶是植物制造有机养料的重要器官，一般为绿色扁平体，具有向光性。其主要生理功能是进行光合作用、气体交换和蒸腾作用。此外，叶还有繁殖、贮藏、吸收和合成等功能。除作为食物或饲料外，叶还具有观赏和药用价值。

Leaves are green photosynthetic organs of a plant arising from the nodes. Leaves are usually flattened with the function as photosynthesis, gas exchange and transpiration. In addition, leaves also have functions such as reproduction, storage, absorption and synthesis. Besides being used as food or fodder, leaves also have ornamental and medicinal value.

第一节　叶的组成与形态

1 Composition and morphology of the leaves

一、叶的组成
1.1　Leaf composition

叶由叶片、叶柄和托叶三部分组成（图 2-1）。这三部分俱全的叶称完全叶，如桃、柳、月季等；缺少其中的一或两部分，称不完全叶，其中最普遍的是缺少托叶，如女贞，还有些是同时缺少托叶和叶柄，如石竹、龙胆等。

A leaf is called complete leaf when it consists of three parts namely a blade, a petiole, and stipules (Fig. 2-1), such as leaves of peach, willow and *Rosa chinensis* Jacq. The lack of one or two parts is called

医药大学堂
WWW.YIYAODXT.COM

incomplete leaf. The most common one is the lack of stipules, such as *Ligustrum lucidum*, and some lack both stipules and petioles, such as *Dianthus sp*, *Gentiana sp*, etc.

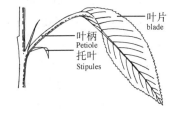

图 2-1 叶的组成
Fig.2-1 Composition of a leaf

1. **叶片** 是叶的主要部分，常为绿色薄的扁平体，有上表面（腹面）和下表面（背面）之分。叶片的全形称叶形，顶端称叶端或叶尖，基部称叶基，周边称叶缘，叶片内分布有叶脉。

(1) Blade The main part of the leaf is often green and flattened, with adaxial side (upper surface facing stem axis) and abaxial side (lower surface facing away from stem axis). The full shape, the tip, the base, the margin and the distribution of the veins often vary from plant to plant.

2. **叶柄** 叶片和茎枝相连接的部分称为叶柄，一般呈类柱形、半圆柱形或稍平，其形状随植物种类和生长环境的不同而异，例如凤眼莲、菱等水生植物的叶柄上具膨胀的气囊，以利于浮水。有的植物叶柄基部有膨大的关节，称叶枕，能调节叶片的位置和休眠运动，如含羞草。有的叶柄能围绕各种物体螺旋状扭曲，起攀缘作用，如旱金莲。

(2) Petiole The petiole is the part where the leaves and stems are connected. It is generally cylindrical, half-circle cylindrical or slightly flat. Its shape varies with plant species and growth environment. For example, *Eichhornia crassipes* and Water chestnuts and other aquatic plants have inflated air sac on the petioles to facilitate floating water.

Some plants have enlarged joints at the base of the petiole, called leaf cushion (pulvinus), which can adjust the position of the leaves and dormant movement, such as the leaves of *Mimosa pudica*. Some petioles can be twisted spirally around various objects, such as Nasturtium.

有的植物叶片退化，而叶柄变态成叶片状，如台湾相思树。有些植物的叶柄基部或叶柄全部扩大成鞘状，称叶鞘，如当归。有些禾本科植物叶鞘与叶片相接处还具有一些特殊结构，在其相接处的腹面的膜状突起物称叶舌，在叶舌两旁有一对从叶片基部边缘延伸出来的突起物称叶耳。此外，有些无柄叶的叶片基部包围在茎上，称抱茎叶，如苦荬菜。有的无柄叶基部或对生无柄叶基部彼此愈合，被茎所贯穿，称贯穿叶或穿茎叶，如元宝草。

Some plants have leaves that degenerate and the petioles become metamorphose like leaves, such as the *Taiwan acacia*. Some plants petioles or petiole expanded into all the sheath-like, named aid leaf sheath, such as *Angelica sinensis*. Some gramineous plants have a special structure at the junction of the leaf sheath and the leaf. The membrane-like protrusions on the ventral surface of the grass are called ligulate. On both sides of the ligulate, there are a pair of protrusions extending from the edge of the base of the leaf called auricle.

In addition, some sessile leaf bases surround the stem, called amplexicaul leaf, such as *Ixeris polycephala* Cass. Some sessile leaf bases or opposite sessile leaf bases heal each other and are penetrated by the stem, which is called the perforate leaf or perfoliate leaf, such as *Hypericum sampsonii* Hance.

3. **托叶** 托叶着生在叶柄与茎枝连接部位，常较小，起保护幼叶和芽的作用。托叶保持到叶片发育成熟后称托叶宿存，或托叶在叶片发育成熟前脱落，称早落；有些植物的托叶脱落后在茎上落下的疤痕，称托叶痕，如木兰属植物具有环状的托叶痕。托叶的形状多样，如豌豆的托叶呈叶片状，茜草、猪殃殃等与叶同形，大黄等蓼科植物扩大成鞘状并包围茎，称托叶鞘，金樱子与叶柄愈合成翅，菝葜属呈卷须状，三颗针呈刺状（图 2-2）。

(3) Stipules The stipules are born at the junction of the petiole and the stem and branch. They are usually smaller and protect the young leaves and buds. The stipules that keep remained until the leaves mature and are called persistent leaves, or the stipules fall off before the leaves mature and are called

视频

early fall; The scars that fall on the stems of some plants after the stipules fall off are called stipule scars. For example, *Mangolia* plants have ring-shaped stipule scars. The shape of the stipules is various. For example, stipules of *Pisum sativum* Linn is leaf-shaped, and that of *Rubia cordifolia* L. and bedstraw are the same shape as leaves. The stipules of the plants from Polygonaceae expand into sheaths and surround stems, they are called ocrea. The stipules of *Rosa laevigata*. heals wings with petiole, and that of *Smilax* is tendril-like, The stipules of *Berberis diaphana* are spiny (Fig. 2-2).

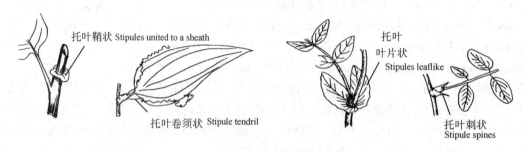

图 2-2 托叶的各种形态
Fig.2-2 Various forms of stipules

二、叶的形态
1.2 Morphology of leaf

叶的形态常指叶片的形状，叶片的形状包括叶形、叶端、叶基、叶缘、叶脉和脉序、叶片分裂状况、叶片质地和表面附属物等各部分形态特征。不同植物的叶形态差异较大，但同种植物叶形状与大小相对稳定，可作植物鉴别的特征。

Leaf morphology often refers to the shape of the blade, including the apex, base, margin, vein, texture, surface appendages, the split of the leaves and other morphological features. The leaf shape of different plants varies greatly, but the shape and size of the same plant are relatively stable, which can be used as a feature for plant identification.

1. 叶片的全形　常指整个叶片轮廓的几何形状，常按长宽比例及最宽处的位置来确定（图2-3）。叶的基本形状有：针形、条形（线形）、披针形、椭圆形、卵形、心形、肾形、圆形、剑形、盾形、带形、箭形、戟形等。此外，还有一些特殊的形态，如蓝桉呈镰刀形、杠板归呈三角形、菱呈菱形、车前呈匙形、银杏呈扇形、葱呈管形、秋海棠呈偏斜形等。植物叶常不是典型的几何形状，因此描述时常用"长"、"广"、"倒"等加以说明，如长圆形、倒卵形、广卵形等；或结合两种形状进行复合描述，如卵状椭圆形、椭圆状披针形等（图2-4）。

(1) The outline of lamina　The outline of lamina is often determined by the ratio of length to width and the widest position (Fig. 2-3).

The basic outlines of the Blades are as follows: acicular, linear, lanceolate, lliptial, ovate, cordate, reniform, orbicular, ensiform, peltate, banded, sailtate, and hastate etc. In addition, there are some special shapes, such as sickle-shaped in *Eucalyptus globulus*, triangular in *Polygonum perfoliatum*, diamond-like in *Trapa bispinosa*, spoon-shaped in *Plantago asiatica*, fan-shaped in *Ginkgo biloba*, tubular in *Allium fistulosum* and oblique in *Begonia grandis*. Plant leaves are often not typical geometric shapes, so terms as "long", "wide", and "inverted" are often used in descriptions, such as oblong, obovate, and wide-ovate; Or combine the two shapes for description, such as oval-oval, oval-lanceolate, etc (Fig. 2-4).

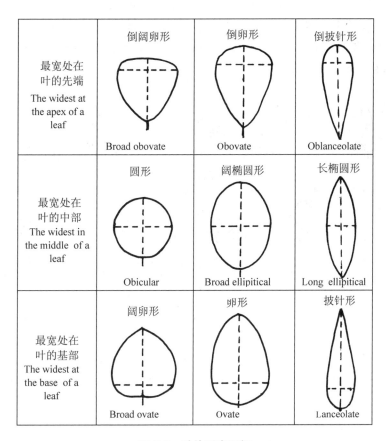

图 2-3　叶片形态图解
Fig.2-3　The figure of leaf morphology

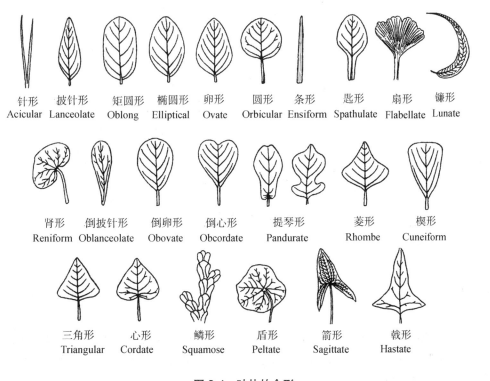

图 2-4　叶片的全形
Fig.2-4　The shape of leaf

2. 叶端 常见有芒尖、尾尖、渐尖、急尖、钝形、截形、微凹、微缺、倒心形等（图 2-5）。

(2) Leaf apex Commonly are aristate, caudate, acuminate, acute, obtuse, truncate, retuse, emarginate, obcordate etc. (Fig. 2-5).

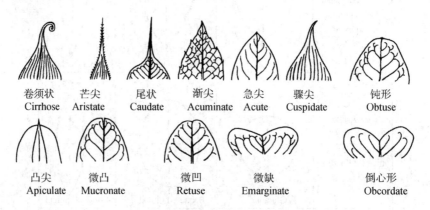

卷须状 Cirrhose　芒尖 Aristate　尾状 Caudate　渐尖 Acuminate　急尖 Acute　骤尖 Cuspidate　钝形 Obtuse

凸尖 Apiculate　微凸 Mucronate　微凹 Retuse　微缺 Emarginate　倒心形 Obcordate

图 2-5　叶端
Fig.2-5　Leaf apex

3. 叶基 常见有楔形、钝形、心形、耳形、渐狭、歪斜、抱茎、穿茎等（图 2-6）。

(3) Leaf base Commonly are cuneate, obruse, cordate, auriculate, attenuate, oblique, amplexicaul, perfoliate (Fig. 2-6).

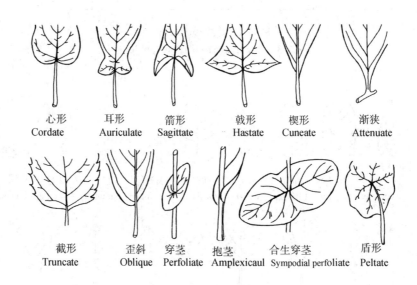

心形 Cordate　耳形 Auriculate　箭形 Sagittate　戟形 Hastate　楔形 Cuneate　渐狭 Attenuate

截形 Truncate　歪斜 Oblique　穿茎 Perfoliate　抱茎 Amplexicaul　合生穿茎 Sympodial perfoliate　盾形 Peltate

图 2-6　叶基
Fig.2-6　Leaf base

4. 叶缘 常见有全缘、波状、牙齿状、锯齿状、重锯齿、圆锯齿等（图 2-7）。

(4) Leaf margin Commonly are entire, undulata, dentate, serrate, double serrate, crenate (Fig. 2-7).

5. 叶脉和脉序 叶脉是贯穿叶内的维管束，起输导和支持作用。其中与叶柄相连最粗大的叶脉称主脉或中脉，主脉分枝形成侧脉，侧脉分枝形成细脉。叶片中叶脉的分布及排列式样称脉序，常见有三种类型（图 2-8）。

(5) Leaf vein and venation Leaf veins are vascular bundles that run through the leaf and play a

医药大学堂
www.yiyaodxt.com

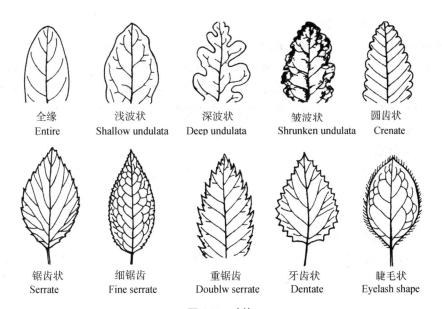

| 全缘 Entire | 浅波状 Shallow undulata | 深波状 Deep undulata | 皱波状 Shrunken undulata | 圆齿状 Crenate |
| 锯齿状 Serrate | 细锯齿 Fine serrate | 重锯齿 Doublw serrate | 牙齿状 Dentate | 睫毛状 Eyelash shape |

图 2-7 叶缘
Fig.2-7 Leaf margin

guiding and supporting role. The thickest veins connected to the petioles are called main veins or midribs. The main vein branches form lateral veins, and the lateral vein branches form veinlets. The distribution of vascular bundles that are visible on the leaf surface as veins constitutes venation, and there are three common types (Fig. 2-8).

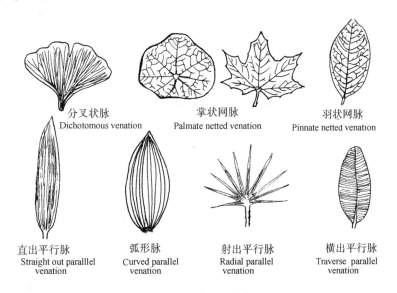

| 分叉状脉 Dichotomous venation | 掌状网脉 Palmate netted venation | 羽状网脉 Pinnate netted venation |
| 直出平行脉 Straight out parallel venation | 弧形脉 Curved parallel venation | 射出平行脉 Radial parallel venation | 横出平行脉 Traverse parallel venation |

图 2-8 叶脉和脉序
Fig.2-8 Vein and vein sequence

分叉脉序：一条叶脉分为大小相等的两个分枝，同一叶上有多级的二叉分支是较原始的脉序，如部分蕨类植物或裸子植物的银杏。

Dichotomous venation: The vein is divided into two branches of equal size, and a multi-stage binary branch on the same leaf is a more primitive vein sequence, such as some ferns or *Ginkgo biloba*.

平行脉序：各叶脉大致相互平行，主脉与侧脉及侧脉间相连细脉不成网状。单子叶植物多为平行脉序，按侧脉形状或主脉分支位置又分为直出平行脉（淡竹叶）、横出平行脉（芭蕉）、射出平行脉（棕榈）和弧形脉（玉竹）等。

Parallel venation: The veins are generally parallel to each other, and the main veins are connected to the side veins and the side veins are not meshed. Monocotyledons are mostly parallel veins. According to the shape of the side vein or the branch position of the main vein, it is divided into straight parallel veins as *Lophatherum gracile*, transverse parallel veins as *Musa basjoo*, shoot parallel veins as *Trachycarpus fortunei* and curved veins are emitted as *Polygonatum odoratum*.

网状脉序：叶脉有多级分支，主脉、侧脉和细脉区别明显，最小细脉连结成网状，是双子叶植物的特征。若主脉1条，分出的侧脉排列呈羽状，细脉与主、侧脉交织成网状称羽状网脉；主脉基部分出数条侧脉直达每一裂片，排列呈掌状，细脉与主、侧脉交织成网状称掌状网脉。此外，少数单子叶植物如薯蓣，天南星等也具网状脉序，但其叶脉末梢全部连接在一起，没有游离脉梢，有别于双子叶植物。

Reticulate venation: The veins have multi-level branches, and the main veins, lateral veins, and fine veins are clearly distinguished. The smallest fine veins are connected into a network, which is a characteristic of dicotyledons. If there is one main vein, the separated side veins are lined with feathers, and the fine veins are intertwined with the main and side veins to form a network called pinnate netted veins; If there are several side veins at the base of the main vein reaching each lobes, they are arranged in a palm shape, and the fine veins are intertwined with the main and side veins in a mesh shape called palm veins. In addition, a few monocotyledonous plants as in *Dioscorea oppositifolia* and *Arisaema heterophyllum* also have reticulate venation, but their veins are all connected together without free veins, which is different from dicotyledons.

6. 叶片的质地　常分为膜质、草质、纸质、革质和肉质等。膜质叶片薄而半透明如半夏，有的干薄而脆，不呈绿色称干膜质如麻黄的鳞片叶；草质叶片薄而柔软，如薄荷、商陆等；纸质叶片薄而柔韧，似纸张样，如糙苏；革质叶片厚而较坚韧，略似皮革，如枇杷、山茶等；肉质叶片肥厚多汁，如芦荟等。

(6) Leaf texture　Membaranceous leaves are thin and translucent as in *Pinellia ternata*, and some are thin and brittle, not green, which is called scarious, such as the scale leavess of *Ephedrae sp.* Herbaceous leaves are thin and soft as in *Mentha Canadensis*, *Phytolacca acinosa*. Chartaceous leaves are thin, flexible and paper-like as in *Phlomis umbrosa*. Coriaceous leaves are thick and tough, similar to leather as in *Eriobotrya japonica* and *Camellia japonica*. Succulent leaves are thick and juicy as in *Aloe vera*. etc.

7. 表面附属物　叶表面因附属物不同而表现出光滑、被粉、被毛、粗糙等特征。叶面光滑如女贞等；叶面粗糙如无花果、腊梅等；叶面被毛如茵陈、毛地黄等。

（7）Leaf surface　The appendages on the leaf surface show different characteristics such as smooth, powdered, coat, and rough. For example, smooth as in *Ligustrum lucidum* and rough in *Ficus carica* and *Chimonanthus praecox*. Sometimes the leaves are covered by hairs asin *Artemisia capillaris* and *Digitalis purpurea*.

第二节　叶的类型与叶序
2 Type of leaves and phyllotaxy

一、叶的分裂
2.1 Leaf incision

叶缘无缺刻或小齿称全缘叶，大部分植物的叶片边缘不整齐，具齿或深浅不一的缺刻，其凸出部分称裂片，凹入部分称缺刻。按叶裂片排列方式不同，分为羽状分裂，掌状分裂和三出分裂；按叶缺刻的深浅程度不同，又分为浅裂、深裂和全裂三种。缺刻的深度不足叶片宽度 1/4 称浅裂，如药用大黄、南瓜；缺刻的深度大于叶片宽度 1/4 称深裂，如唐古特大黄、荆芥；缺刻深度几达主脉基部或中脉，形成数个全裂片称全裂，如大麻、白头翁（图 2-9）。

A Simple leaf may be undivided or incised variously depending upon whether the incision progresses down to the midrib (pinnate) or towards the base (palmate). According to the different arrangement of leaf lobes, they are divided into pinnate division, palmate division and three out split; according to the depth of the incision, they are divided into three types: lobate, parted and divided. The depth of the incision is less than 1/4 of the leaf width termed lobate as in *Rheum sp.* and *Cucurbita moschata*. The depth of the incision is more than 1/4 of the leaf width, which is named parted as in *Rheum tanguticum* and *Nepeta cataria*. The depth of the incision is almost to the base or midrib, forming several full lobes termed divided as in *Cannabis sativa* and *Pulsatilla chinensis* (Fig. 2-9).

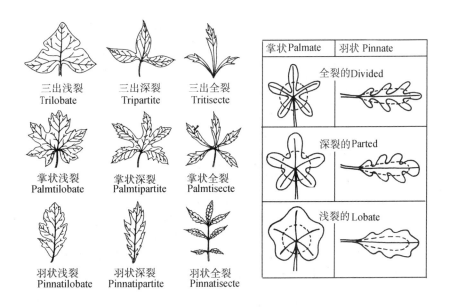

图 2-9　叶片的分裂
Fig.2-9　The split leaves

二、叶的类型
2.2　Type of leaves

1. **单叶**　一个叶柄上只着生一枚叶片，如厚朴、女贞等。

(1) Simple leaf　A leaf with a single blade (divided or not) is termed simple as in *Magnolia officinalis* and *Ligustrum lucidum*.

2. **复叶**　一个叶柄上着生2枚以上叶片，其上的叶称小叶，小叶的柄称小叶柄，如人参、野葛。复叶的叶柄称总叶柄，总叶柄以上着生小叶的轴状部分称叶轴。典型的复叶，小叶柄和叶轴之间具明显的关节。单叶全裂时与小叶柄不明显的复叶，区别在于单叶的裂片之间有或多或少的叶片相连。按小叶排列方式不同，复叶又分为以下四种类型（图2-10）。

(2) Compound leaf　A leaf with two or more distinct blades is termed the compound leaf. The blade on the compound leaf is termed leaflet, and the petiole of each blade is termed petiolule as in *Panax ginseng*, *Pueraria lobata*. The petiole of the compound leaf is termed common petiole, and the axial part of the petiole above the petiole is called rachis. When the single leaf is fully divided, the compound leaf is not obvious with the petiole, the difference is that there are more or less blades connected between the lobes of the single leaf. According to the arrangement of leaflets, compound leaves are divided into the following four types (Fig. 2-10).

图2-10　复叶的类型
Fig.2-10　Types of compound leaves

三出复叶：叶轴上的小叶为3枚。3枚小叶均生于叶轴顶端称掌状三出复叶，如半夏、酢浆草、大血藤等；若中央1枚小叶生于叶轴顶端，二枚小叶生于叶轴下端的两侧称羽状三出复叶，如大豆、野葛等。

Ternately compound leaf: There are three leaflets on the leaf axis. The three leaflets are all born at the top of the leaf axis and form palmately ternately compound leaf as in *Pinellia ternate*, *Oxalis corniculata* and *Sargentodoxa cuneata* (Oliv.) Rehd. et Wils.. If one central leaflet is born on the top of the leaf axis, two leaflets on both sides of the lower end of the leaf axis are termed pinnately ternately compound leaf, such as *Glycine max* (L.) Merr., *Pueraria lobata* (Willd.) Ohwi and so on.

掌状复叶：叶轴缩短，顶端聚生 3 枚以上小叶，呈掌状排列，如人参、三七等。

Palmately compound leaf: The leaf axis is shortened, and the top end gathers more than 3 leaflets in a palm-like arrangement, such as *Panax ginseng* C. A. Meyer and so on.

羽状复叶：叶轴较长，小叶在叶轴两侧排成羽状。顶生小叶 1 枚者称单数羽状复叶，顶生小叶 2 枚者称偶数羽状复叶。若叶轴不分枝，小叶直接着生其上称一回羽状复叶，如决明、皂角等；叶轴作 1 次分枝，其上着生小叶，称二回羽状复叶，如合欢、云实等；叶轴分枝 2 次为三回羽状复叶，如南天竹等；依此类推。

Pinnately compound leaf: The leaf axis is longer, and the leaflets are feathered on both sides of the leaf axis. One terminal leaflet is called singular pinnate compound leaf, two terminal leaflets are called even pinnate compound leaves. If the leaf axis is not branched, the leaflets are directly planted on it, which is called a pinnate compound leaf as in *Senna tora* and *Gleditsia sinensis*. Unipinnate compound leaf means the leaflets are borne directly along the rachis; The bipinnate means the pinnae (primary leaflets) are again divided into pinnules, so that the leaflets are borne on the primary branches of the rachis as in *Mimosa pudica* and *Caesalpinia decapetala*. The tripinnate means the dissection goes to the third order so that the leaflets are borne on secondary branches of the rachis as in *Nandina domestica*.

单身复叶：叶轴上具 2 枚叶片上下叠生在一起，形似单叶，而叶片之间具一明显的关节，如芸香科柑桔、柠檬等。

Unifoliate compound leaf: The leaf axis has two leaves stacked on top of each other, like a single leaf, and there is an obvious joint between the leaves, such as *Citrus reticulate* and *C. limon* of Rutaceae.

复叶与着生单叶的小枝有时难以区分，要区别它们，首先是分清叶轴与小枝。① 叶轴顶端无顶芽，小枝顶端有顶芽。② 小叶叶腋无侧芽，总叶柄腋内具芽，小枝上具腋芽。③ 小叶在叶轴上排在同一平面，小枝上的单叶常成不同的角度。④ 复叶脱落是整个脱落或小叶先落，然后叶轴连同总叶柄一起脱落，而小枝一般不脱落，仅其上的叶脱落。

The compound leaf and the branch with a single leaf are sometimes difficult to distinguish. As for the compound leaf, there is no apical bud at the top of the leaf axis and axillary leaflets have no lateral buds. In addition, the leaflets on the leaf axis of a compound leaf are arranged in the same plane. The whole leaf or the leaflets fall first, and then the leaf axis falls off along with the total petiole, while the twigs generally do not fall off, only the leaves above it fall off.

三、叶序
2.3　Phyllotaxy

叶序是叶在茎枝上的排列方式。主要有四种类型（图 2-11）。

The phyllotaxy is the arrangement of leaves on the stems and branches. There are four types as following (Fig. 2-11).

1. 互生　茎枝的每节仅着生一枚叶，交互而生，沿茎枝上呈螺旋状排列，如桑、榕等。

(1) Alternate　Bearing one leaf at each node.The successive leaves usually form a spiral pattern, in mathematical regularity so that all leaves are found to lie in a fixed number of vertical rows or orthostichies, such as *Morus alba* and *Ficus microcarpa*.

图 2-11　叶序的类型
Fig.2-11　Types of phyllotaxy

2. **对生**　茎枝的每节相对着生二枚叶，如相邻两节的叶排列呈十字形称交互对生，如薄荷、忍冬等；相邻两节的叶均排列在茎的两侧称二列状对生，如小叶女贞、水杉、醉鱼草等。

(2) Opposite　Bearing pairs of leaves at each node. The pairs of successive leaves may be parallel (superposed) as in *Ligustrum quihoui*, *Metasequoia glyptostroboides* and *Buddleja lindleyana* or at right angles (decussate) as in *Mentha canadensis* and *Lonicera japonica*.

3. **轮生**　茎枝的每节着生 3 枚或 3 枚以上叶，轮状排成，如夹竹桃、轮叶沙参等。

(3) Whorled　More than three leaves at each node as in *Nerium indicum* and *Adenophora tetraphylla*.

4. **簇生**　簇生是多枚叶着生在节间缩短的茎枝（短枝）上，密集成簇；如银杏、落叶松等。如多枚叶着生在根头部极度短缩且节间不明显的茎上，称基生叶；基生叶集生而成莲座状时，称莲座状叶丛，如蒲公英、车前等。

(4) Fascicled　Clusters of multiple leaves grow on shortened internodes (short branches) as in *Ginkgo biloba*, *Larix gmelinii*. If multiple leaves are placed on the stem with extremely short root head and inconspicuous internodes, it is called basal leaf. When the basal leaves gather into a rosette, it is called a rosette as in *Taraxacum mongolicum* and *Plantago asiatica*.

此外，有些植物具 2 种或 2 种以上的叶序，如栀子、桔梗等。无论哪种叶序，在枝条相邻节上的叶往往不相遮盖，通过叶柄伸展的角度和长度不同，形成镶嵌状的排列称叶镶嵌；叶镶嵌使各叶片能有效地接受阳光，有利于光合作用。

In addition, some plants have two or more kinds of leaf order as in *Gardenia jasminoides* and *Platycodon grandiflorus*. Regardless of the leaf order, the leaves on the adjacent nodes of the branches are often uncovered and form a mosaic-like arrangement through the petioles extending at different angles and lengths, which enable each leaf to effectively receive sunlight.

第三节　异形叶性和叶的变态

3 Leaf metamorphosis

一、异形叶性
3.1　Heterophylly

同种植物常具相对稳定的叶形，但一些植物在同一植株上具不同形状的叶，称异形叶性。异形叶性可发生在不同的生态环境，也可出现在植物的不同发育阶段。如慈姑沉水叶呈线形，浮水叶椭圆形，挺水叶箭形；一年生人参具 1 枚三出复叶，二年生有 1 枚掌状复叶，三年生有 2 枚掌状复叶，四年生有 3 枚掌状复叶，五年生有 4 枚掌状复叶，以后每年递增 1 叶，最多可达 6 枚掌状复叶。蓝桉幼枝上的叶对生，卵形，无柄；老枝叶互生，镰形，有柄。此外，薜荔、益母草、川贝母等也存在异形叶性（图 2-12）。

The shape of the leaves on the same plant is relatively stable, but sometimes is different, which is called heterotypic leaf. Heteromorphic leaves often occur in different ecological environments or at different growing stages of plant. As for *Sagittaria trifolia. var. sinensis*, leaves dipped into water are linear, leaves floating above the surface of the lake are oval, and emerging leaves are arrow-shaped. As for *Panax ginseng*, the annual plant has only one trifoliate compound leaf, while the biennial has one palmate, and the triennial, quadriennial and the five-year-old has two, three, and four, separately. Later, it increases by one leaf per year and can reach up to 6 palm-shaped compound leaves. The leaves on the young branches of *Eucalyptus globulus* are opposite, ovate, sessile, while alternate, sickle-shaped, pedunculate on the old branches. In addition, *Ficus pumila*, *Leonurus japonicus*, and *Fritillaria cirrhosa* also have heteromorphic leaf (Fig.2-12).

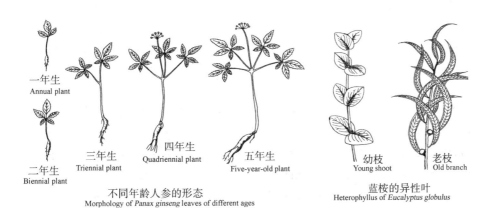

一年生
Annual plant

二年生
Biennial plant

三年生
Triennial plant

四年生
Quadriennial plant

五年生
Five-year-old plant

幼枝
Young shoot

老枝
Old branch

不同年龄人参的形态
Morphology of *Panax ginseng* leaves of different ages

蓝桉的异性叶
Heterophyllus of *Eucalyptus globulus*

图 2-12 异形叶性
Fig.2-12 The leaf of heterophylly

二、叶的变态
3.2 Leaf metamorphosis

叶与外界环境接触面最广，变态类型较多。

1. 苞片　多生于花下面的一种变态叶，有保护花的作用。苞片的形状、大小、质地，以及在花梗上着生的位置变化较大；苞片呈绿色、叶状，称苞叶或叶状苞片，如川贝母；轮状着生在花梗上端邻近花萼处称副花萼，如草莓、蜀葵等。着生于花序外围的 1 至多数苞片合称总苞片，花序中每朵小花下较小的苞片称小苞片；总苞的形状、大小、颜色、数目等变化较大，如鱼腥草的总苞片呈白色叶片状；天南星科植物呈佛焰状；苍耳呈囊状并具细刺。

(1) **Bract**　Bracts are metamorphosed leaves usually located below the flowers to protect them. The shape, size, texture, and position of the bracts on the pedicel vary greatly. Green, leaflike bracts can be seen in *Fritillaria cirrhosa*; whorl-shaped bracts near the calyx are termed epicalyx as in Strawberry and *Alcea rosea*. Involucre is 1 to many bracts born outside of the inflorescence collectively. The smaller bracts under each floret are called bractlets. The shape, size, color, and number of total bracts vary greatly,

for instance, the involucre of *Houttuynia cordata* is white leaf-like, while that of the Araceae plants are large in flame shaped; *Xanthium strumarium* is saclike and has spines.

2. **鳞叶**　特化或退化成鳞片状的叶。鳞叶肥厚多汁，贮藏大量营养物质称肉质鳞叶，如百合、洋葱等；鳞叶膜质、菲薄，常不呈绿色，如麻黄、姜、荸荠等。木本植物的冬芽（鳞芽）外常被有褐色鳞片叶，起保护芽的作用。

(2) Scale leaf　Fleshy scale leaves are thick and juicy, and store nutrients as in *Lilium brownii* and *Allium cepa*. Sometimes scales are membranous and thin, often not green as in *Ephedra sinica*, *Zingiber officinale* and *Eleocharis dulcis*. The winter buds (scale buds) of woody plants are often covered with brown scale leaves, which protect the buds.

3. **叶刺**　叶全部或一部分变成坚硬的刺状。托叶变成刺状，称托叶刺，如刺槐、酸枣；小檗属植物的托叶刺呈三叉针刺，俗称"三棵针"；有时叶尖、叶缘变成刺状，如红花、枸骨等。

(3) Leaf thorn　All or part of the leaves become spiny. For some plants as *Robinia pseudoacacia* and *Ziziphus jujuba var. spinosa*, their stipules are thorny and the stipules of *Berberis plants* are needle spines. Sometimes the leaf tip and leaf edge become spiny as in *Carthamus tinctorius* and *Ilex cornuta*.

4. **叶卷须**　叶全部或一部分变为卷须，适应攀援生长。如豌豆的前端几枚小叶变成卷须，菝葜属植物的托叶变成卷须。

(4) Leaf tendril　In order to adapt to climbing growth, all or part of the leaves become tendrils. For example, the pea's tendril is changed from a few leaflets in the front, and that of the *Smilax sp.* are from stipules.

5. **捕虫叶**　叶片变态成囊状、盘状或瓶状，上有腺毛能分泌黏液和消化液的叶，能诱捕昆虫并消化，从中获得营养，如捕蝇草、茅膏菜、猪笼草（图2-13）。

(5) Insectivorous leaf　The metamorphosis of the leaves is sac-like, disc-shaped or bottle-shaped, with glandular hairs capable of secreting mucus and digestive juices, to trap insects and digest them, and to obtain nutrients from them, such as *Dionaea muscipula*, *Drosera peltata* and *Nepenthes mirabilis* (Fig.2-13).

台湾相思树 Acacia confusa　　猪笼草 Nepenthes sp.　　捕蝇草 Dionaea muscipula

图 2-13　叶的变态
Fig.2-13　Leaf metamorphosis

6. **叶状柄**　叶片退化，而叶柄变成绿色扁平的叶状体。如幼苗期的台湾相思树的初生叶是羽状复叶，后生叶的小叶完全退化仅存叶状柄。

(6) Phyllode　The leaves often degenerate, and the petioles become green, flat and leaf-like. For example, at the seedling stage, the primary leaves of the *Acacia confusa* are pinnate compound leaves, while the leaflets that have grown later are completely degraded and only the petioles remain.

重点小结
Summaries

　　叶是光合作用、蒸腾作用的重要器官，由叶片、叶柄和托叶三部分组成；根据三部分是否完整分为完全叶和不完全叶。叶的形态常指叶片的形状，包括叶形、叶端、叶基、叶缘、叶脉和脉序、叶片分裂状况等。叶分为单叶和复叶，复叶分为三出复叶、掌状复叶、羽状复叶和单身复叶。叶在茎上的排列方式，有对生、互生、轮生和簇生。叶的变态类型有苞片、鳞叶、叶刺、叶卷须、捕虫叶和叶状柄等。

　　Leaves are important organs for photosynthesis and transpiration. A complete leaf includes blade, petiole and stipule; those lacking one or two parts are called incomplete leaves. Dicots exhibit a network of veins; whereas monocots usually have non-intersecting parallel veins. In ferns and Ginkgo, the venation is dichotomous with forked veins. A leaf with a single blade is termed simple, whereas one with two or more distinct blades (leaflets) is said to be compound. Pinnate compound leaves may be further differentiated into three-out, palmate, pinnate and unifoliate compound leaves. The arrangement of the leaves on the stem is opposite, alternate, whorled and clustered. The abnormal types of leaves include bracts, scale leaves, leaf thorns, leaf tendrils, insect trapping leaves, and stalks.

目 标 检 测
Questions

题库

　　1. 叶的生理功能主要有哪些？

What are the main physiological functions of leaves?

　　2. 什么是完全叶和不完全叶？

What are complete leaves and incomplete leaves?

　　3. 叶的类型有哪些？

What are the types of leaves?

　　4. 复叶有哪些类型？

How to describe compound leaves?

　　5. 如何区分复叶与小枝？

How to distinguish compound leaves from twigs?

　　6. 脉序有哪些类型？

What are the types of the veination?

　　7. 如何区别叶卷须和茎卷须？

How to distinguish between leaf tendril and stem tendril?

第三章　花
Chapter 3　Flower

学习目标 | Learning goals

1. **掌握** 花的组成和形态。
2. **熟悉** 花和花序的类型以及花程式。
3. **了解** 花图式、花的生殖。

- Know the composition and morphology of flowers .
- Be familiar with the type of flower, inflorescence and flower formula.
- Understand flower diagram, reproduction and tissue structure.

花是种子植物特有的繁殖器官，通过开花、传粉、受精而形成果实和种子，有繁殖后代、延续种族的作用。种子植物包括裸子植物和被子植物，裸子植物的花构造简单，被子植物的花高度进化，结构复杂，本章所说的花是指被子植物的花。

A flower is a highly modified shoot bearing specialized floral leaves. It is the unique reproductive organ of seed plants, which forms fruit and seed through the process of flowering, pollination and fertilization. It has the function of reproducing offspring and extending race. Seed plants include gymnosperms and angiosperms. The flower structure of gymnosperms is simple while the flower of angiosperms is highly evolved and rather complicate. The flowers mentioned in the chapter refer to the flower of angiosperms.

花的形态结构变化较小，较其他器官更稳定，因此在植物分类、药材的原植物鉴定及花类药材的鉴定方面有重要意义。

The morphology and structure of flowers are rarely changing, which are more stable than other organs. So it is of great significance for the study of the plant classification, identification of original plant of traditional Chinese medicine, and also medicinal flower materials.

常用的花类药材有：金银花、辛夷、槐米、菊花、旋覆花、番红花、蒲黄、款冬花等。

Commonly used flower herbs are Lonicerae Flos, Magnoliae Flos, Sophorae Flos, Chrysanthemi Flos, Inulae Flos, Croci Stigma, Typhae Pollen, Farfarae Flos and so on.

医药大学堂
WWW.YIYAODXT.COM

第一节　花的组成和形态
1 Flower composition and morphology

一、花的组成
1.1　Flower composition

典型的花由花梗、花托、花萼、花冠、雄蕊群和雌蕊群等部分组成。其中雄蕊群和雌蕊群具有生殖功能。花被常分为花萼和花冠，有保护花蕊和引诱昆虫传粉等作用。花梗和花托起支持作用（图3-1）。

A typical flower consist of pedicel, receptacle, calyx, corolla, androecium and gynoecium (Fig.3-1).

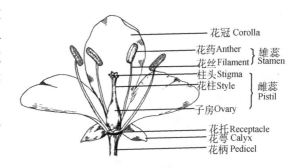

图 3-1　花的组成部分
Fig.3-1　Composition of a flower

二、花的形态
1.2　Flower morphology

（一）花梗
1.2.1　Pedicel
花梗又称花柄，通常绿色、圆柱形，是花与茎的连接部分。花梗的有无和形态因植物的种类而异。果实形成时，花梗便成为果柄。

It is usually green and cylindrical. Pedicels connect flower and stem. The morphology varies with plant species.

（二）花托
1.2.2　Receptacle
花托是花梗顶端膨大的部分。花托的形状随植物种类而异。一般植物的花托呈平坦或稍凸起的盘状，有的呈圆柱状，如木兰、厚朴；有的呈圆锥状，如草莓；有的呈倒圆锥状，如莲；有的凹陷呈杯状，如金樱子、蔷薇、桃。有的花托在雌蕊基部或在雄蕊与花冠之间形成肉质增厚，扁平垫状、杯状或裂瓣状结构，常可分泌蜜汁，称花盘，如柑橘、卫矛、枣等。有的花托在雌蕊基部向上延伸成一柱状体，称雌蕊柄，如黄连、落花生等，也有的花托在花冠以内的部分延伸成一柱状体，称雌雄蕊柄，如白花菜、西番莲等。

The enlarged portion of the bottom of pedicel. It is usually flat or slightly raised. Some are cylindrical, such as *Magnolia denudata*; some are conical, such as *Fragaria ananassa*; some are inverted cone, such as *lotus*.

（三）花被
1.2.3　Perianth
花被是花萼和花冠的总称。多数植物的花被分化为花萼与花冠。有一些植物的花被片无明显的分化，形态相似，称为花被，如厚朴、五味子、百合、黄精等。

Perianth is the general term for calyx and corolla. Most plants have calyx and corolla. Some plants have similar morphology of calyx and corolla as in *Magnolia denudata*.

1. 花萼 花萼是一朵花中所有萼片的总称，位于花的最外层。萼片一般呈绿色的叶片状，其形态和构造与叶片相似。

(1) Calyx Calyx is the general term for all sepals in a flower. They are located at the outermost layer of the flower. Sepals are generally green and leaf-like.

萼片彼此分离的称离生萼，如毛茛、菘蓝等植物的花萼；萼片中下部联合的称合生萼，如丹参、桔梗等，联合的部分称萼筒或萼管，分离的部分称萼齿或萼裂片。有些植物的萼筒一边向外凸起成伸长的管状，称距，如旱金莲、凤仙花等。一般植物的花萼在花枯萎时脱落。有些植物的花萼在开花前即脱落，称早落萼，如延胡索、白屈菜等。有些植物的花萼在花枯萎时不脱落并随果实一起增大，称宿存萼，如柿、酸浆等。萼片一般排成一轮，若在花梗顶端紧邻花萼下方另有一轮类似萼片状的苞片，称副萼，如棉花、蜀葵等。有的萼片大而颜色鲜艳呈花瓣状称瓣状萼，如乌头、铁线莲等。此外，菊科植物的花萼常变态成羽毛状，称冠毛，如蒲公英等；苋科植物的花萼常变成半透明的膜质状，如牛膝、青葙等。

视频

Separated sepals are termed chorisepalous calyx as in *Ranunculus japonicus*. Gemosepaloous calyx, the lower part of the sepals are united, as in *Platycodon grandiflorus*. The combined part is called calyx tube, the separated part is called calyx tooth. For some plants, the calyx tube bulges outward into an elongated tubular shape which called spur, such as *Impatiens balsamina*. Caducous calyx falls off before flowering, such as *Corydalis yanhusuo*. Persistent calyx increases with the fruit, such as *Lycopersicon esculentum*. If there is another round of sepal-like bracts below the calyx at the top of the pedicel, it is called epicalyx as in *Gossypium herbaceum*. Some sepals are large, brightly colored and petal-shaped as in *Aconitum carmichaelii*. Some calyx metamorphose into feather-like, called pappus as in Compositae.

2. 花冠 花冠是一朵花中所有花瓣的总称，位于花萼的内侧，常具各种鲜艳的颜色。

(2) Corolla Corolla is the general term for all petals in a flower, located on the inner side of the calyx, and has various bright colors in common.

花瓣彼此分离的称离瓣花冠，如甘草、仙鹤草等。花瓣彼此联合的称合瓣花冠，其下部联合的部分称花冠筒或花筒，上部分离的部分称花冠裂片，如丹参、桔梗等。有些植物的花瓣基部延长成管状或囊状，称距，如紫花地丁、延胡索等。有些花冠瓣片前端宽大，中部急剧缩窄并下延，下延的部分称爪，如油菜、石竹等。有些植物的花冠内侧或花冠与雄蕊之间生有瓣状附属物，称副花冠，如徐长卿、水仙等。

The petals which separated from each other are called choripetalous corolla. The petals which unite with each other are called synpetalous corollas. The combined part is called corolla tube. The separated part of the flowers is called corolla lobes. In some flowers, the base of petals is elongated into a tubular shape, which called spur, such as *Viola philippica*.

花冠有多种形态，可作为植物分类鉴定的重要依据，常见的花冠类型有以下几种（图 3-2）。

There are several types of corolla in common (Fig.3-2).

十字形：花瓣4枚，分离，常具爪，上部外展呈十字形排列，如菘蓝、油菜等十字花科植物的花冠。

Cruciform: 4 petals, separated, with clawed. The petals in a cross-shaped arrangement, such as the corolla of cruciferae plants.

蝶形：花瓣5枚，分离，上方1枚位于最外侧且最大，称旗瓣，侧方2枚较小称翼瓣，最下

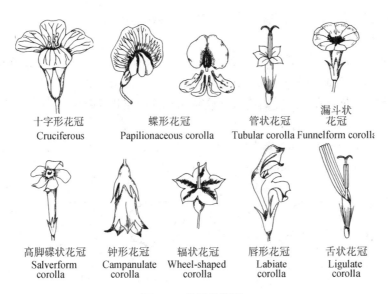

图 3-2　花冠的类型

Fig. 3-2　Types of the corolla

方 2 枚最小且位于最内侧，瓣片前端常联合并向上弯曲，称龙骨瓣，如甘草、槐花等豆植物的花冠。

Papilionaceous: 5 petals, separated. The top one is located at the outermost and largest, which is called flag petal; the side two are smaller, which is called wing petal; the bottom two are the smallest and located at the innermost side. The front end of the petal is often united and curved upward, which is called keel petal. Such as the corolla of Leguminosae plants.

唇形：花冠下部联合成筒状，上部为二唇形，上唇中部常凹陷，再分裂为 2 枚裂片，下唇常再分裂为 3 枚裂片，如益母草、丹参等唇形科植物的花冠。

Labiate: The lower part of the Corolla is combined into a tube shape, and the upper part is separate of two lips. The middle part of the upper lip is often sunken, and then divided into two lobes; the lower lip is usually divided into three lobes, such as plant of Lamiaceae.

管状：花冠合生，花冠筒细长管状，如菊科植物的管状花。

Tubular: Corolla connate, slender tubular shape, such as the tubular flowers of Compositae plants.

舌状：花冠基部联合呈一短筒，上部向一侧延伸成扁平舌状，如菊科植物向日葵的舌状花。

Liguliform: The base of the corolla united into a short tube, and the upper part extended into a flat tongue, such as the liguliform flower of Compositae plants.

漏斗状：花冠筒较长，自下向上逐渐扩大，上部外展呈漏斗状，如牵牛等旋花科植物和曼陀罗等部分茄科植物的花冠。

Funnelform: The corolla tube is longer and gradually enlarged from the bottom to the top, the upper part of the Corolla is funnel-shaped, such as Convolvulaceae plants.

高脚碟状：花冠下部细长呈管状，上部水平展开呈碟状，如水仙、长春花等植物的花冠。

Salverform: The lower part of the Corolla is long and thin and tubular, the upper part is horizontally unfolded, such as the corolla of *Narcissus tazetta*.

钟状：花冠筒阔而短，上部裂片扩大平缓外展似钟形，如沙参、桔梗等桔梗科植物的花冠。

Campanulate: The corolla tube is broad and short, the upper lobes expand smoothly like a bell, such as *Platycodon grandifloras*.

辐状或轮状：花冠筒甚短而广展，裂片由基部向四周扩展，形如车轮状，如龙葵、枸杞等部分茄科植物的花冠。

Wheel-shaped: The corolla is very short and widely spread. The lobes extend from the base to the surrounding like a wheel, such as *Lycium chinense*.

3. **花被卷叠式** 花被卷叠式是指花未开放时花被各片彼此的叠压方式，花蕾即将绽开时观察尤为明显。常见的花被卷迭式有（图3-3）。

(3) Aestivation of calyx and corolla parts Arrangement of sepals and petals in the flower bud. The following main types of aestivation are listed (Fig.3-3).

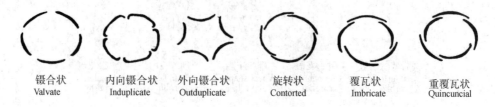

| 镊合状 | 内向镊合状 | 外向镊合状 | 旋转状 | 覆瓦状 | 重覆瓦状 |
| Valvate | Induplicate | Outduplicate | Contorted | Imbricate | Quincuncial |

图 3-3 花被的卷迭式
Fig.3-3 Aestivation of the perianth

镊合状：花被各片的边缘彼此互相接触排成一圈、但互不重叠，如桔梗、葡萄的花冠。若花被各片的边缘稍向内弯称内向镊合，如沙参的花冠；若花被各片的边缘稍向外弯称外向镊合，如蜀葵的花萼。

Valvate: Margins of sepals or calyx lobes not overlapping.

旋转状：花被各片彼此以一边重叠成回旋状，如夹竹桃、龙胆的花冠。

Contorted: The perianth overlap on one side to form a cyclotron such as *Nerium oleander*.

覆瓦状：花被边缘彼此覆盖，但其中有1片完全在外面，有1片完全在内面，如山茶的花萼、紫草的花冠。

Imbricate: The perianth overlap on one side to another, but one of them is completely on the outside, one is completely on the inside, such as *Camellia japonica*.

重覆瓦状：花被边缘彼此覆盖，覆瓦状排列的花被片中有2片完全在外面，有2片完全在内面，如桃、野蔷薇的花冠。

Quincuncial: The perianth covers each other, but two of them completely outside and two of them completely inside, as in *Amygdalus persica*.

（四）雄蕊群

1.2.4 Androecium

雄蕊群是1朵花中所有雄蕊的总称。少数植物的花一部分雄蕊不具花药或仅见其痕迹，成不育雄蕊或退化雄蕊，如鸭跖草的雄蕊；还有少数植物的雄蕊特化成花瓣状，如姜、美人蕉等的雄蕊。

1. **雄蕊的组成** 典型的雄蕊由花丝和花药两部分组成。

(1) The composition of the androecium

花丝为雄蕊下部细长的柄状部分，其基部着生于花托上，上部承托花药。花丝的粗细、长短随植物种类而异。

Filament: The lower part of stamen, stalk-like. It is base on the receptacle.

花药为花丝顶部膨大的囊状体，是雄蕊的主要部分。花药常分成左右两瓣，中间为药隔。雄蕊成熟时，花药自行裂开，花粉粒散出。

花药开裂的方式有多种，常见的有纵裂，即花粉囊沿纵轴开裂，如水稻、百合等；孔裂，即花粉囊顶端裂开1小孔，花粉粒由孔中散出，如杜鹃等；瓣裂，即花粉囊上形成1~4个向外展开的小瓣，成熟时瓣片向上掀起，散出花粉粒，如樟、淫羊藿等。此外还有横裂，即花粉囊沿中部横裂一缝，花粉粒从缝中散出（图3-4）。

花药在花丝上着生的方式也不一致（图3-5），常见的有：

丁字着药：花药背部中央一点着生在花丝顶端，与花丝略呈丁字形，如水稻、百合等。

个字着药：花药上部联合着生在花丝上，下部分离，花药与花丝呈个字形，如泡桐，玄参等。

广歧着药：花药两瓣完全分离平展近乎一直线，药隔着生在花丝顶端，如薄荷、益母草等。

全着药：花药自上而下全部贴生在花丝上，如紫玉兰等。

基着药：花药基部着生在花丝顶端，如樟，茄等。

背着药：花丝仅背部中央贴生于花丝上，如杜鹃、马鞭草。

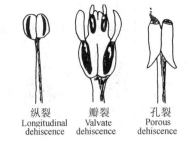

纵裂 Longitudinal dehiscence　瓣裂 Valvate dehiscence　孔裂 Porous dehiscence

图3-4　花药的开裂方式
Fig. 3-4　The ways of anther dehiscence

Anther: The expanded sac-like part at the top of the filaments. It is the main part of the stamen. There are many ways of anther dehiscence (Fig.3-4).

longitudinal dehiscence: Pollen sac dehiscence along the vertical axis.

Poricidal: The top of the pollen sac cracked a small hole, pollen grains scattered from the hole.

Valvular: 1–4 Portions of anther wall opening when mature.

Transverse: The pollen sac along the middle of a transverse crack. When the stamen matures, the pollen grains scatter.

The attachment of filament to the anther include (Fig.3-5).

Versatile: Filament attached nearly at the middle of connective so that anther can swing freely as, in *Lilium brownii*.

Adnate: Filament continues into connective which is almost as broad, as found in plant of Ranunculus.

Basifixed: The filament ends at the base of anther or at least base of connective. The resultant anther is erect, as in plant of Brassica.

Dorsifixed: Filament attached on the connective above the base. The resultant anther is somewhat inclined, as in plant of Sesbania

2. 雄蕊的类型　一朵花中雄蕊的数目、长短、离合、排列方式等随植物种类而异，形成不同的雄蕊类型。花中的雄蕊相互分离的，称离生雄蕊。有些植物雄蕊的花丝部分或全部联合在一起或长短不一。常见的有如下类型（图3-6）。

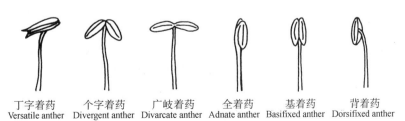

丁字着药 Versatile anther　个字着药 Divergent anther　广岐着药 Divaricate anther　全着药 Adnate anther　基着药 Basifixed anther　背着药 Dorsifixed anther

图3-5　花药着生方式
Fig.3-5　The attachment of filament to the anther

医药大学堂
www.yiyaodxt.com

(2) Androecium type The number and arrangement of stamens in a flower vary with plant species. Major description of androecium include (Fig.3-6).

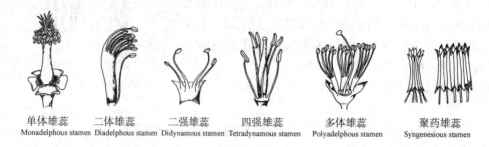

单体雄蕊　　　二体雄蕊　　　二强雄蕊　　　四强雄蕊　　　多体雄蕊　　　聚药雄蕊
Monadelphous stamen　Diadelphous stamen　Didynamous stamen　Tetradynamous stamen　Polyadelphous stamen　Syngenesious stamen

图 3-6　雄蕊的类型
Fig.3-6　Types of the stamen

单体雄蕊：花中所有雄蕊的花丝联合成 1 束，呈筒状，花药分离，如蜀葵、木槿等锦葵科植物和远志等远志科植物以及苦楝、香椿等楝科植物的雄蕊。

Monadelphous: The filaments of all stamens united in a single group, as in family Malvaceae.

二体雄蕊：花中雄蕊的花丝分别联合成 2 束，如延胡索、紫堇等罂粟科植物有 6 枚雄蕊，分为 2 束，每束 3 枚；甘草、野葛等许多豆科植物有 10 枚雄蕊，其中 9 枚联合，1 枚分离。

Diadelphous: The filaments of stamen united into two groups, as in family Fabaceae.

二强雄蕊：花中共有 4 枚雄蕊，其中 2 枚花丝较长，2 枚花丝较短，如益母草、薄荷等唇形科植物，马鞭草、牡荆等马鞭草科植物和玄参、地黄等玄参科植物的雄蕊。

Didynamous: 4 stamens, two shorter and two longer, as in family Lamiaceae.

四强雄蕊：花中共有 6 枚雄蕊，其中 4 枚花丝较长，2 枚较短，如菘蓝、独行菜等十字花科植物的雄蕊。

Tetradynamous: 6 stamens, two shorter and four longer, as in family Brassicaceae.

多体雄蕊：花中雄蕊多数，花丝联合成多束，如金丝桃、元宝草等藤黄科植物和橘、酸橙等部分芸香科植物的雄蕊。

Polyadelphous: Filaments united into more than two groups, such as *Hypericum perforatum*.

聚药雄蕊：花中雄蕊的花药联合成筒状，花丝分离，如蒲公英、白术等菊科植物的雄蕊。

Synandrous: The anthers of stamens in flowers united into a tube, filaments separated, such as *Taraxacum mongolicum*.

（五）雌蕊群

1.2.5　Gynoecium

All the pistils of a flower. It is in the central part of the flower.

雌蕊群是 1 朵花中所有雌蕊的总称，位于花的中心部分。

1. 雌蕊的组成　雌蕊由心皮构成。心皮是适应生殖的变态叶。裸子植物的心皮（又称大孢子叶或珠鳞）展开成叶片状，胚珠裸露在外，被子植物的心皮边缘愈合成雌蕊，胚珠包在雌蕊囊状的子房内，这是裸子植物与被子植物的主要区别。当心皮卷合形成雌蕊时，其边缘的愈合缝线称腹缝线，心皮中脉部分的缝线称背缝线，胚珠常着生在腹缝线上。

(1) The composition of the pistil　The pistil is composed of carpel. Carpels are abnormal leaves which are adapted to reproduction. Gymnosperms carpel (megasporophyll) spread into leaf shape, ovules exposed. Angiosperms carpel margin combined into the pistil, ovules wrapped in the pistil saclike ovary, this is the main difference between gymnosperms and angiosperms. When pistil formed, its edge of the healing suture is called ventral suture, the suture of carpel in middle part is called **dorsal suture**. The ovule is attached to the ventral suture.

　　雌蕊的外形似瓶状，由子房、花柱和柱头 3 部分组成。

A pistil is like a bottle, composed of ovary, style and stigma.

　　子房是雌蕊基部膨大的囊状部分，常呈椭圆形、卵形等形状，其底部着生在花托上。子房的外壁称子房壁，子房壁以内的腔室称子房室，其内着生胚珠，因此子房是雌蕊最重要的部分。

　　Ovary: The enlarged saclike part of the base of the pistil. The ovary contains ovules.

　　花柱是子房上端收缩变细并上延的颈状部位，也是花粉管进入子房的通道。花柱的粗细、长短、有无随植物种类而异，如玉米的花柱细长如丝，莲的花柱粗短如棒，而木通、罂粟则无花柱，其柱头直接着生于子房的顶端，唇形科和紫草科植物的花柱插生于纵向分裂的子房基部，称花柱基生。有些植物的花柱与雄蕊合生成一柱状体，称合蕊柱，如白及等兰科植物。

　　Style: The cervical part of the ovary. The passage of the pollen tube into the ovary.

　　柱头是花柱顶部稍膨大的部分，为承受花粉的部位。柱头常成圆盘状，羽毛状、星状，头状等多种形状。其上带有乳头状突起，并常能分泌黏液，有利于花粉的附着和萌发。

　　Stigma: The slightly enlarged part at the top of style. The site of pollen bearing.

　　2. 雌蕊的类型　根据组成雌蕊的心皮数及与心皮联合与否，形成不同的雌蕊类型。常见有如下类型（图 3-7）。

　　(2) Gynoecium type　Different pistil types are formed according to the number of carpels and whether they are united (Fig.3-7).

　　单雌蕊：是由 1 个心皮构成的雌蕊，如甘草、野葛等豆科植物和桃、杏等部分蔷薇科植物的雌蕊。

　　Single pistil: A pistil composed of one carpel.

　　复雌蕊：是由 1 朵花内的 2 个或 2 个以上心皮彼此联合构成的复合雌蕊，如菘蓝，丹参、向日葵等为二心皮复雌蕊；大戟、百合、南瓜等为三心皮复雌蕊；卫矛等为四心皮复雌蕊；贴梗海棠、桔梗、木槿等为五心皮复雌蕊；橘、蜀葵等的雌蕊则由 5 个以上的心皮联合而成。组成雌蕊的心皮数往往可由柱头和花柱的分裂数、子房上的主脉数以及子房室数等来判断。

　　离生雌蕊：是 1 朵花内有 2 至多数单雌蕊，彼此分离，聚集在花托上的雌蕊类型，如毛茛、乌头等毛茛科植物和厚朴、五味子等木兰科植物的雌蕊。

　　Syncarpous pistil: It is a compound pistil composed of two or more carpels in a flower.

　　Apocarpous pistil: It is a flower with two or more single pistil, separated from each other. They are gathered on the receptacle .

单心皮雌蕊	二心皮雌蕊	三心皮复雌蕊	三心皮单雌蕊	离生雌蕊
Single carpel pistil	Double carpel pistil	Tricarp compound pistil	Tricarp single pistil	Apocarpous pistil

图 3-7　雌蕊的类型
Fig.3-7　Types of the pistil

3. **子房的位置及花位**　由于花托的形状、结构不同，子房在花托上着生位置和愈合程度及其与花被、雄蕊之间关系也发生变化。常有以下类型（图 3-8）。

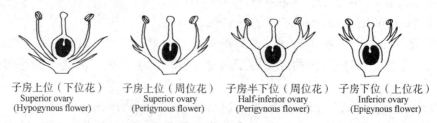

子房上位（下位花）　　子房上位（周位花）　　子房半下位（周位花）　　子房下位（上位花）
Superior ovary　　　　 Superior ovary　　　　 Half-inferior ovary　　　　 Inferior ovary
(Hypogynous flower)　　(Perigynous flower)　　(Perigynous flower)　　　(Epigynous flower)

图 3-8　子房的位置示意图
Fig.3-8　The position of the ovary on the receptacle

子房上位：花托扁平或隆起，子房仅底部与花托相连，称子房上位，花被、雄蕊均着生在子房下方的花托边缘，这种花称下位花，如油菜、金丝桃、百合等。若花托下陷为杯状，子房仅基部着生于杯状凹陷内壁的中央或侧壁上，亦为子房上位，花被、雄蕊则着生于杯状花托的上端边缘，称周位花，如桃、杏、金樱子等。

子房半下位：子房下半部着生于凹陷的花托中并与花托愈合，上半部外露，称子房半下位；花被、雄蕊均着生于花托四周的边缘，称周位花，如桔梗、党参、马齿苋等。

子房下位：花托凹陷，子房完全生于花托内并与花托愈合，称子房下位；花被、雄蕊均着生于子房上方的花托边缘，称上位花，如贴梗海棠、丝瓜等。

(3) Position of ovary and floral position　Due to the different shape and structure of the receptacle, the position of the ovary on the receptacle, the degree of healing and the relationship between the ovary and the perianth and stamen also changed. There are superior ovary, half-inferior ovary and inferior ovary in common (Fig.3-8).

4. **子房的室数**　子房室的数目由心皮的数目及其结合状态决定。单雌蕊子房只有 1 室，称单子房，如甘草、野葛等豆科植物的子房。合生心皮复雌蕊的子房称复子房，其中有的仅是心皮边缘联合，子房只有 1 室，称单室复子房，单室复子房侧壁上的腹缝线称侧膜，如栝楼、丝瓜等葫芦科植物的子房；有的心皮边缘向内卷入，在中心联合形成柱状结构，称中轴，形成的子房室数与心皮数相等，称复室复子房，复室复子房室的间壁称隔膜，如百合、黄精等百合科植物和桔梗、沙参等桔梗科植物的子房；有的子房室可能被次生的间壁完全或不完全地分隔，次生间壁称假隔膜，如菘蓝、芥菜等十字花科植物和益母草、丹参等唇形科植物的子房。

(4) Ovary number　The ovary number is determined by the number of carpel and its state. The ovary of monocarpellary (carpel one) called unilocular, such as the family Fabaceae. The ovary of the united carpel is complex. The number of carpels are represented as bicarpellary, tricarpellary, tetracarpellary, pentacarpellary, and multicarpellary. The number of chambers similarly are represented as unilocular, bilocular, trilocular, tetralocular, pentalocular and multilocular.

5. **胎座及其类型**　胚珠在子房内着生的部位称胎座。因雌蕊的心皮数目及心皮联合的方式不同，常形成不同的胎座类型。常见有如下几种（图 3-9）。

(5) Placenta and its type　Different placental types are formed due to the number of carpels and whether the carpels are combined. Common placental types are (Fig.3-9).

边缘胎座：单雌蕊，子房 1 室，多数胚珠沿腹缝线的边缘着生，如野葛、决明等豆科植物的胎座。

Marginal placenta: 1 pistil, 1 ovary with 1 row of ovules, as in family Fabaceae.

边缘胎座　　　　　　侧膜胎座　　　　　　　中轴胎座
Marginal placentation　Parietal placentation　　Axile placentation

中轴胎座　　　　特立中央胎座　　　　基生胎座　　顶生胎座
Axile placentation　Free-central placentation　Basal placentation　Apical placentation

图 3-9　胎座的类型
Fig.3-9　Types of the placentation

侧膜胎座：复雌蕊，单室复子房，多数胚珠着生在子房壁相邻两心皮联合的多条侧膜上，如罂粟、延胡索等罂粟科植物和栝楼、丝瓜等葫芦科植物的胎座。

Parietal placenta: 1 ovary with more than one discrete placental lines as in family Cucurbitaceae.

中轴胎座：复雌蕊，复室复子房，多数胚珠着生在各心皮边缘向内伸入于中央而愈合成的中轴上，其子房室数往往与心皮数目相等，如玄参、地黄等玄参科植物和桔梗、沙参等桔梗科植物以及百合、贝母等百合科植物的胎座。

Axile placenta: Compound pistil, compound ovary with compound chamber. Most ovule in each carpel edges inwards into the center axis. The number of ovary chambered often equals to the carpel number.

特立中央胎座：复雌蕊，单室复子房，来源于复室复子房，但子房室的隔膜和中轴上部消失，形成单子房室，多数胚珠着生在残留于子房中央的中轴周围，如石竹、太子参等石竹科植物和过路黄、点地梅等报春花科植物的胎座。

Free-central placenta: Compound pistil, compound ovary with single chamber. Ovules attached along the central column, as in family Caryophyllaceae.

基生胎座：子房 1 或多心皮，1 室，1 枚胚珠着生在子房室基部，如大黄、何首乌等蓼科植物和向日葵、白术等菊科植物的胎座。

Basal placenta: 1 ovary or several carpel, 1 chamber, with single ovule at the base of chamber, as in family Asteraceae (Compositae).

顶生胎座：子房 1 或多心皮，1 室，1 枚胚珠着生在子房室顶部，如桑、构树等桑科植物和草珊瑚等金粟兰科植物的胎座。

Epical placenta: 1 ovary, 1 chamber, with single ovule at the top, as in family Moraceae.

6. 胚珠的构造及其类型　胚珠是种子的前身，为着生在胎座上的卵形小体，受精后发育成种子，其数目、类型随植物种类而异。

(6) Ovule Structure and type　The ovule is the predecessor of the seed, which is the ovoid body on the placenta.

胚珠的构造：胚珠着生在子房内，常呈椭圆形或近圆形，其一端有一短柄称珠柄，与胎座相连，维管束从胎座通过珠柄进入胚珠。大多数被子植物的胚珠有 2 层包被，称珠被，外层称外珠被，内层称内珠被，裸子植物及少数被子植物仅有 1 层珠被，极少数植物没有珠被。在珠被的前端常不完全愈合而留下 1 小孔，称珠孔，是多数植物受精时花粉管达到珠心的通道。珠被内侧为一团薄壁细胞，称珠心，是胚珠的重要部分。珠心中央发育着胚囊。被子植物的成熟胚囊一般有 1 个卵细胞、2 个助细胞、3 个反足细胞和 2 个极核细胞等 8 个细胞（核）。珠被、珠心基部和珠柄汇合处称合点，是维管束到达胚囊的通道（图3-10）。

Ovule Structure: The ovules of most angiosperms have two layers of coating, called the integument, the outer layer is called the outer integument, the inner layer is called the inner integument. Gymnosperms and a few angiosperms have only one layer of integument, a very few plants do not have the integument. In the front of the integument, there is left a small hole, called a micropyle. It is channel of fertilization pollen tube to the nucellus for most plants. The inside of the parenchyma cells, called the nucellus, which is an important part of the ovule. The embryo sac develops in the center of the nucellus. The mature embryo sac of angiosperms generally has 8 cells (nucleus) such as one egg cell, 2 helper cells, 3 antipodal cells and 2 Polar nucleus cell (Fig.3-10).

胚珠的类型：胚珠生长时，由于珠柄、珠被、珠心等各部分的生长速度不同而形成不同的胚珠类型（图3-10）。

Ovule type: Different ovule types are formed due to the different growth rate of each part of the funicle, integument, nucellus and so on. Common examples are (Fig.3-10).

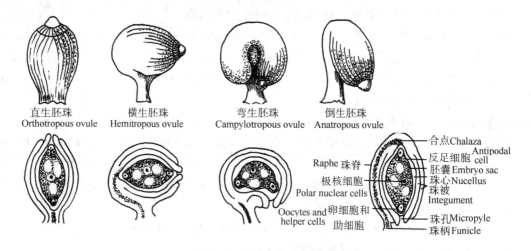

图 3-10　胚珠的构造及类型
Fig.3-10　The structure and type of ovule

直生胚珠：胚珠直立且各部分生长均匀，珠柄在下，珠孔在上，珠柄、珠孔、合点在一条直线上。如三白草科、胡椒科、蓼科植物的胚珠。

Atropous ovule: the ovule is erect and grows evenly in all parts.

横生胚珠：胚珠一侧生长较另一侧快，使胚珠横向弯曲，珠孔和合点之间的直线与珠柄垂直。如毛茛科、锦葵科、玄参科和茄科的部分植物的胚珠。

Hemitropous ovule: Body half inverted so that funiculus is attached near middle with micropyle terminal and at right angles.

弯生胚珠：胚珠的下半部生长速度均匀，上半部的一侧生长速度快于另一侧，并向另一侧弯曲，使珠孔弯向珠柄，胚珠呈肾形。如十字花科和豆科部分植物的胚珠。

Campylotropous ovule: The lower part of the ovule grows at a evenly rate, the upper part of the ovule grows at a faster rate than the other side, so bends to the slower side.

倒生胚珠：胚珠的一侧生长迅速，另一侧生长缓慢，使胚珠倒置，合点在上，珠孔下弯并靠近珠柄，珠柄较长并与珠被一侧愈合，愈合线形成一明显的纵脊称珠脊。大多数被子植物的胚珠属此种类型。

Anatropous ovule: Inverted ovule with micropyle facing and closer to funiculus.

第二节　花的类型
2 Types of flower

被子植物的花在长期演化中，各部发生不同程度的变化，使其呈现出了多姿多彩、形态多样的不同类型，常见的类型如下。

一、完全花和不完全花
2.1　Complete flower and incomplete flower

根据花的组成部分是否完整，可分为完全花和不完全花。

1. **完全花**　凡是花萼、花冠、雄蕊、雌蕊四部分俱全的称为完全花，如桃、桔梗等。

(1) Complete flower　with all the typically parts like sepals, petals, stamens, and pistils, as in *Amygdalus persica*, *Platycodon grandifloras*.

2. **不完全花**　若缺少其中一部分或几部分的花，称不完全花，如鱼腥草、桑等。

(2) Incomplete flower　lacking an expected part or series of parts of the floral whorls mentioned before (i.e. sepals, petals, stamens, or pistils), such as *Houttuynia cordata*, *Morus alba*.

二、无被花、单被花、重被花和重瓣花
2.2　Achlamydeous flower, Simple perianth flower, double perianth flower and double flower

根据花被有无及层次可分为无被花、单被花、重被花和重瓣花。

1. **无被花**　既没有花萼也没有花冠的花。这种花通常具有苞片，如鱼腥草、杨、柳、杜仲等（图 3-11）。

(1) Achlamydeous flower　lacking of perianth, usually with bracts, such as *Houttuynia cordata*, *Eucommia ulmoides* (Fig.3-11).

2. **单被花**　若只具花萼而无花冠，或花萼与花冠不分化的称单被花。这种花被常成 1 轮或多轮排列，且具鲜艳的颜色而呈花瓣状，如百合、玉兰、白头翁等。

(2) Simple perianth flower　with only one type of perianth member. The perianth is usually arranged in one or more whorls, which has bright colors and petal-shape, such as *Magnolia denudate*,

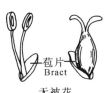

苞片　Bract

花萼　Calyx

花瓣　Petal
花萼　Calyx

无被花　Achlamydeous flower　　单被花　Simple perianth flower　　重被花　Double perianth flower

图 3-11　花的类型
Fig. 3-11　Types of the flower

Pulsatilla chinensis.

3. **重被花**　一朵花具有花萼和花冠的称为重被花，如桃、杏、萝卜等。

(3) Double perianth flower　with both calyx and corolla, such as *Amygdalus persica, Armeniaca vulgaris, Raphanus sativus.*

4. **重瓣花**　许多栽培型植物的花瓣常呈数轮排列且数目较多的花，如栽培樱花等。

(4) Double flower　some cultivated plants have a large number of petals than normal, i.e., cultivated cherry blossoms.

三、两性花、单性花和无性花
2.3　Bisexual flower, unisexual flower and asexual flower

根据花的性别可分为两性花、单性花和无性花。

1. **两性花**　一朵花中雄蕊与雌蕊都有的称两性花，如桃、桔梗、牡丹等。

(1) Bisexual flower　A flower with both male and female reproductive organs, such as peach, *Platycodon grandiflorum, Paeonia suffruticosa*, etc.

视频

2. **单性花**　一朵花中若仅具雄蕊或雌蕊的称单性花。其中只有雄蕊的称雄花，只具有雌蕊的称为雌花；若雄花和雌花在同一株植物上称单性同株或雌雄同株，如南瓜、蓖麻；若雄花和雌花分别生于不同植株上称单性异株或雌雄异株，如桑、银杏、杜仲等；若同一株植物既有单性花又有两性花，两者生于同一植株上称杂性同株，如厚朴；若两者分别生于不同植株上称杂性异株，如臭椿、葡萄。

(2) Unisexual flower　A flower with either male or female reproductive organs. Bearing stamens but not pistils, as a male flower which called staminate flower. A female flower only bearing pistils is called pistillate flower. If the staminate and pistillate flower borne on the same plant, they are called monoecism, such as pumpkin and castor. If male and female flowers are born on different plants, they are called dioecism, such as *Morus alba, Ginkgo biloba* and *Eucommia ulmoides.*

If one with unisexual and bisexual flowers on the same plant, it is called polygamo-monoecism, such as *Houpoea officinalis.* If unisexual and bisexual flowers are respectively born on different plants, they are called polygamo-dioecious, such as *Ailanthus altissima* and *Vitis vinifera* (grape).

3. **无性花**　一朵花中若雄蕊和雌蕊均退化或发育不全的称为无性花，如八仙花花序周围的花。

(3) Asexual flower　If both stamens and pistils in a flower degenerate or develop incompletely, it is called asexual flower, such as the flower around the inflorescence of hydrangea.

四、辐射对称花、两侧对称花和不对称花
2.4　Actinomorphic flower, zygomorphic flower and asymmetric flower

根据花的对称性，可分为辐射对称花、两侧对称花和不对称花。

1. **辐射对称花**　花被片的形状大小相似，通过花的中心可作两个以上对称面的花称辐射对称花或整齐花，如桃、桔梗、牡丹等。

(1) Actinomorphic flower　Radially symmetric. A line drawn through the middle of the structure along any plane will make a mirror image on either side, as in *Amygdalus persica, Platycodon grandifloras* and *Paeonia suffruticosa.*

2. **两侧对称花**　花被片各片形状大小不一，通过花的中心只能作一个对称面的称两侧对称花或不整齐花，如扁豆、益母草等。

(2) Zygomorphic flower　The perianth pieces have different shapes and sizes, and only one symmetry plane can be made through the center of the flower, as in *Lablab purpureus*, *Leonurus japonicus*.

3. **不对称花**　通过花的中心无对称面的花称为不对称花，如败酱、缬草、美人蕉等。

(3) Asymmetric flower　There is no symmetry plane through the center of the flower, such as *Patrinia scabiosifolia*, *Valeriana officinalis*, *Canna indica*.

五、风媒花、虫媒花、鸟媒花和水媒花
2.5　Anemophilous plants, entomophilous plants, ornithophilous plants and hydrophilous plants

根据花粉的传播媒介，可分为风媒花、虫媒花、鸟媒花和水媒花。

According to the transmission medium of pollen, it can be divided into wind pollinated, insect pollinated, bird pollinated and water pollinated.

第三节　花程式与花图式
3　Flower formula and diagram

为准确描述花各组成部分的数目、离合、排列方式等形态特征，可用花程式及花图式来记录。两种方法各有侧重与不足，可以选择其中一种或两种联用。

In order to describe the number, division, whorls and other morphological characteristics of the flower accurately, flower formula and flower diagram can be used to record. The two methods have their own emphases and shortcomings. One or both of them can be used.

一、花程式
3.1　Flower formula

花程式是用字母、数字和符号来表示花各部分的组成、排列、位置和彼此关系的公式。

It is a formula that uses letters, numbers and symbols to express the composition, whorls, the position of each part of flowers and the nature of gynoecium.

（一）以字母代表花的各部
3.1.1　Letters represent the parts of flowers

一般用花各部拉丁文或德文的第一个字母大写表示，花被（拉丁文 perianthium）用 P 表示；花萼（德文 kelch）用 K 表示；花冠（拉丁文 corolla）用 C 表示；雄蕊（拉丁文 androecium）用 A 表示，雌蕊（拉丁文 gynoecium）用 G 表示。

The first letter of the Latin or German words can be used to represent the parts of flowers. Accordingly, K = calyx, C = corolla, A = androecium, and G = gynoecium.

（二）以数字表示花各部的数目

3.1.2　Numbers express the amount of members in a whorl

在代表字母的右下方，用阿拉伯数字表示各部的数目。若超过 10 以上或数目不定用"∞"表示；若某部分缺少或退化以 0 表示；雌蕊右下角有三个数字，分别表示心皮数、子房室数、每室胚珠数，数字间用"："相连。

At the bottom right of the representative letters, Arabic numerals are used to indicate the number of parts. A very large number, more than ten, is shown by "∞", and "0" if absent. There are three numbers in the right corner of gynoecium, which represent the number of carpels, ovarys and ovules, respectively. The numbers are connected with symbol ":".

（三）以符号表示花的类型等其他情况

3.1.3　Symbols indicate the flower type

"*"表示辐射对称花；"↑"表示两侧对称花；"♀"表示雌花；"♂"表示雄花；"☿"则表示两性花；"（ ）"表示联合，"+"表示花部排列的轮数关系，"–"与"G"的方位代表子房的位置，即 $\underline{G}$、$\overline{G}$ 和 $\overline{\underline{G}}$ 分别表示子房上位、子房下位和子房半下位。

"*" indicates actinomorphic flower, "↑" indicates a zygomorphic flower, and "♀" means female flower, "♂" is for male flower, "☿" expressed as bisexual flower. "()" means members of the separate whorls are joined. The number of flower parts and whorls are linked up by "+". A bar above or below the gynoecium number indicates an inferior or superior ovary respectively. "$\overline{\underline{G}}$" represents half-inferior ovary.

（四）花程式书写举例

3.1.4　Examples

1. **百合花程式**：$☿*P_{(3+3)}A_{(3+3)}\underline{G}_{(3:3:∞)}$　表示百合花：两性；辐射对称；花单被，花被片 2 轮，每轮 3 片，分离；雄蕊两轮，每轮 3 枚；子房上位，3 心皮合生，子房 3 室，每室胚珠多数。

(1) **The floral formula of *Lilium brownie* is $☿*P_{(3+3)}A_{(3+3)}\underline{G}_{(3:3:∞)}$**　It means the flower of *Lilium brownie* have the following characteristics: bisexual, actinomorphic, single perianth with two whorls which have separated three pieces in each one, six stamens in two whorls with three stamens in each; ovary superior, three carpels connate, three ovary compartments, a numbe r of ovules in each room.

2. **玉兰花程式**：$☿*P_{(3+3+3)}A_∞\underline{G}_{(∞:1:2)}$　表示玉兰花：两性；辐射对称；花单被，花被片 3 轮，每轮 3 枚，分离；雄蕊多数，分离；雌蕊子房上位，心皮多数，离生单雌蕊，每子房 1 室，每室 2 枚胚珠。

(2) **The floral formula of *Magnolia denudate* is** $☿*P_{(3+3+3)}A_∞\underline{G}_{(∞:1:2)}$　It means the flower of *Magnolia denudate* have the following characteristics: bisexual, actinomorphic, single perianth with three whorls which have separated three pieces in each one. A number of stamens, separated, ovary superior, with numbers of capels. The ovary has single room and there is two ovules in each one.

3. **贴梗海棠花程式**：$☿*K_5C_5A_∞\overline{G}_{(5:5:∞)}$　表示贴梗海棠花：两性；辐射对称；萼片 5 枚，联合；花瓣 5 枚，分离；雄蕊多数，分离；雌蕊子房下位，5 心皮合生，子房 5 室，每室胚珠多数。

(3) **The floral formula of *chaenomeles speciosa*** 　$☿*K_5C_5A_∞\overline{G}_{(5:5:∞)}$　It means the flower of *Chaenomeles speciosa* have the following characteristics: bisexual, actinomorphic, Sepals 5, united; petals 5, separated; A number of stamens, separated; ovary inferior, 5 carpels connate, ovary 5-locular, ovules many per locule.

4. **桔梗花程式**：$☿*K_5C_5A_5\overline{G}_{(5:5:∞)}$　表示桔梗花：两性；辐射对称；萼片 5 枚，联合；花瓣

5 枚，联合；雄蕊 5 枚，分离；雌蕊子房半下位，5 心皮合生，复子房 5 室，每室胚珠多数。

(4) The floral formula of *Platycodon grandifloras* is $\male\female *K_5C_5A5\overline{G}_{(5:5:\infty)}$　It means the flower of *Platycodon grandifloras* have the following characteristics: bisexual; actinomorphic; five sepals, united; five petals, united, five Stamens, separated; half-inferior ovary; five carpels connate; five-locular; ovules majority in each locule.

二、花图式
3.2　Flower diagram

花图式是以花的横切面所绘出来的图解式。它可以表明花各部的形状、数目、排列方式和相互位置等情况（图 3-12）。

It's a graphic representation of flower, which can shows the number, whorl and position of each part in the flower (Fig.3-12).

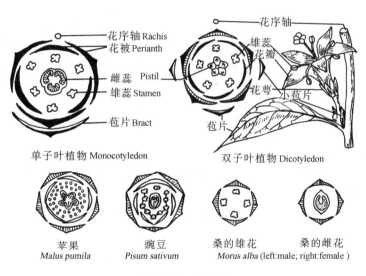

单子叶植物 Monocotyledon　　　双子叶植物 Dicotyledon

苹果　　　　　豌豆　　　　　桑的雄花　　　　桑的雌花
Malus pumila　*Pisum sativum*　*Morus alba* (left:male; right:female)

图 3-12　花图式
Fig.3-12　Flower diagram

花图式的绘制规则：先在上方绘一小圆圈表示花序轴的位置（如为单生花或顶生花可不绘出），在花序轴的下面自外向内按苞片、花萼、花冠、雄蕊、雌蕊的顺序依次绘出各部分，通常用部分涂黑带棱的新月形符号表示苞片，用由斜线组成或黑色带棱的新月形符号表示花萼；花萼内侧用黑色或空白的新月形符号表示花瓣；雄蕊和雌蕊分别用花药和子房的横切面轮廓表示绘于中央；一般雄蕊花药横断面蝴蝶图形、雌蕊子房横断面类圆图形。

The drawing rules of flower pattern: firstly, a small circle on the top to indicates the position of inflorescence axis (it is not necessary to draw the small circle if a solitary flower or a terminal flower). Then draw the bracts, calyx, corolla, stamen and pistil from the outside to inside at the bottom of inflorescence axis in sequence. Usually, the bracts are represented by crescent symbols partially painted with black edges. And the crescent symbols composed of diagonal lines or black edges are used to represents the calyx; the inner side of the calyx sets the petals with black or blank crescent symbols. In addition, the stamens and pistils are drawn in the center with the cross-sectional outline of the anther and ovary respectively. Generally, the cross-sectional butterfly pattern were used to indicate stamens anther, while the circle-like pattern were choosed to describe pistil and ovary.

花程式和花图式均能较简明地反映出花的形态、结构等特征，但均有不足之处，如花图式不能表明子房与花被的相关位置，花程式不能表明各轮花部的相互关系及花被卷叠情况等，所以两者结合使用才能较全面地反映花的特征。

Flower pattern and flower diagram can reflect the composition and whorls of flowers concisely, but they also have some shortcomings. They can be used completely to display the characteristics of flowers more comprehensively.

第四节 花序

4 Inflorescence

被子植物的花，有的是单独一朵生在茎枝顶端或叶腋部位，称单生花，如牡丹、木芙蓉等；有的是密集或稀疏地按一定方式有规律地生在花枝上形成花序。花序下部的梗称为花序梗（总花梗），总花梗向上延伸成为花序轴。花序轴可以不分枝或再分枝。花序上的花称小花，小花的梗称小花梗。小花梗及总花梗下面常有小型的变态叶分别为小苞片和总苞片。无叶的总花梗称花葶。

Some flowers solitary at the top of stems and branches or at the axils of leaves, are called solitary flowers, such as *Paeonia suffruticosa, Hibiscus mutabilis,* etc., while others are regularly grown on the flower branches to form inflorescence. The stalk of a solitary flower or of an inflorescence is called peduncle. The main axis of an inflorescence is rachis.

根据花在花序轴上排列的方式和开放的顺序，花序一般分为无限花序和有限花序两大类。

An individual flower within a dense cluster named floret, the stalk of a floret in an inflorescence is referred to pedicel. There are usually small abnormal leaves under the peduncle, which are named bracteoles and phyllary. A leafless peduncle arising from ground level in acaulescent plants called scape. Inflorescences are generally divided into indefinite inflorescence and definite inflorescence according to the maturing sequence on the axis.

一、无限花序
4.1 Indefinite inflorescence

花序轴在开花期内可继续伸长，产生新的花蕾，花的开放顺序是由花序轴从基部向顶部依次开放；或在缩短的花序轴上，由边缘向中心开放，这种花序称无限花序。根据花序轴有无分枝，有限花序又分为两类，花序轴不分枝的为单花序，花序轴有分枝的为复花序（图 3-13）。

The inflorescence axis can extending and producing new buds during the flowering period. The order of blooming is from bottom upwards along the inflorescence axis; The growth pattern in other plants is from the edge to the center along the shortened inflorescence axis. This kind of inflorescence is called indefinite inflorescence. According to the branches of inflorescence, indefinite inflorescence can be divided into two categories: simple inflorescence and compound inflorescence (Fig.3-13).

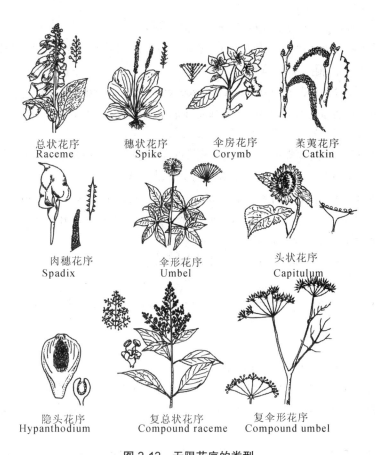

总状花序 Raceme	穗状花序 Spike	伞房花序 Corymb	柔荑花序 Catkin
肉穗花序 Spadix	伞形花序 Umbel	头状花序 Capitulum	
隐头花序 Hypanthodium	复总状花序 Compound raceme	复伞形花序 Compound umbel	

图 3-13 无限花序的类型

Fig.3-13 Type of indefinite inflorescence

（一）单花序

4.1.1 Simple inflorescence

1. **总状花序** 花序轴细长，其上着生许多花柄近等长的小花。如油菜、菘蓝等。

(1) Raceme An unbranched, elongated inflorescence with pedicellate flowers maturing from the bottom upwards (e.g. *Brassica chinensis, Isatis tinctoria*).

2. **穗状花序** 花序轴细长，着生多数花梗极短或无花梗小花。如车前、牛膝、知母等。

(2) Spike An unbranched, elongated inflorescence with sessile or subsessile flowers or spikelets (e.g. *Plantago asiatica, Achyranthes bidentate, Anemarrhena asphodeloides*).

3. **柔荑花序** 形似穗状花序，花序轴柔软下垂，其上着生许多无梗的单性或两性小花，花后整个花序脱落，如杨、柳、核桃等。

(3) Catkin Spike-like inflorescence, an inflorescence consisting of a dense spike, unisexual flowers (e.g. *Populus simonii, Salix babylonica*).

4. **肉穗花序** 与穗状花序相似，花序轴肉质粗大呈棒状，其上密生许多无梗的单性小花，花序外常具有 1 个大型苞片称佛焰苞。如天南星，半夏等天南星科植物。

(4) Spadix Spike-like inflorescence, inflorescence usually erect, bisexual, with a thickened, a large bract which called spathe (e.g. Araceae).

5. **伞房花序** 略似总状花序，但小花梗不等长，靠花序轴下部的花梗长，向上逐渐缩短，所有小花上部在一个平面上，如山楂、绣线菊等。

(5) Corymb Similar to raceme, inflorescence flat-topped or rounded, pedicels not arising from a common point (e.g. *Crataegus pinnatifida*, *Spiraea salicifolia*).

6. 伞形花序　花序轴缩短，在总花梗顶端着生许多放射状排列、花柄、花梗近等长的小花，形如张开的伞，如人参、刺五加、葱等。

(6) Umbel　A flat-topped or convex inflorescence with the pedicels arising more or less from a common point, like the struts of an umbrella (e.g. *Panax ginseng*, *Eleutherococcus senticosus*, *Allium fistulosum*).

7. 头状花序　花序轴极度短缩成头状或盘状的花序托，其上密生许多无梗的小花，外围有 1 至多层苞片密集成的总苞，如向日葵、红花、菊花、蒲公英等。

(7) Capitulum　A head-shaped cluster, flowers enclosed on a flat or convex receptacle. Inflorescence axis extremely short and shrunk into a receptacle, many sessile florets on it, more than one bracts on the periphery (e.g. Compositae).

8. 隐头花序　花序轴肉质膨大而向下向内凹陷呈囊状，其内壁着生多数无梗单性小花，如无花果、薜荔等。

(8) Hypanthodium　An inflorescence with flowers borne on the walls of a concave capitulum (e.g. *Ficus carica*, *Ficus pumila*).

（二）复花序

4.1.2　Compound inflorescence

凡是由两种相同类型的单花序所组成的花序就叫复花序，有如下类型。

1. 复总状花序　又称圆锥花序。花序轴产生多数分枝，每一分枝各为一总状花序，整体呈圆锥状，如女贞等。

(1) Compound raceme　panicle, many branches, racemose inflorescence with flowers maturing from the bottom upwards (e.g. *Ligustrum lucidum*).

2. 复穗状花序　花序轴产生分枝，每一分枝为一穗状花序。如小麦、香附等。

(2) Compound spike　many branches, each branch has a spike.

3. 复伞形花序　花序轴顶端产生许多等长分枝，每一分枝上着生一个小伞形花序，如柴胡、当归、小茴香等伞形科植物。

(3) Compound umbel　inflorescence branches arising from a common point, each branch has a umbel (e.g. Umbelliferae).

4. 复头状花序　由许多小头状花序组成的头状花序，如蓝刺头。

(4) Compound capitulum　composed by numbers of capitulums, as in *Echinops davuricus*.

二、有限花序（聚伞花序类）
4.2　Definite inflorescence

有限花序与无限花序开花顺序相反，花自花序轴顶端向下，或自中心向外依次开放。由于顶端花先开放，从而限制了花序轴的继续生长。根据有限花序花序轴的数目，有限花序分为以下几种类型（图 3-14）。

The flowering sequence of definite inflorescence is completely different with that of the indefinite inflorescence. In definite inflorescence, its terminal flower blooms first, halting further elongation of the main axis. They can be divided into the following types according to the number of axis (Fig.3-14).

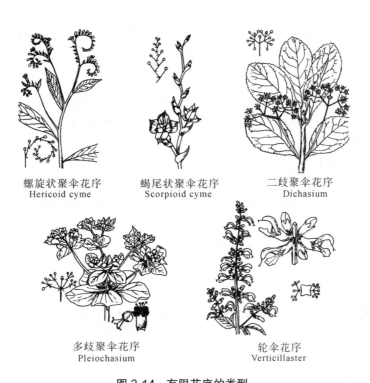

| 螺旋状聚伞花序 Hericoid cyme | 蝎尾状聚伞花序 Scorpioid cyme | 二歧聚伞花序 Dichasium |

多歧聚伞花序 Pleiochasium

轮伞花序 Verticillaster

图 3-14 有限花序的类型
Fig.3-14 Type of definite inflorescence

1. **单歧聚伞花序** 花序轴顶端生一花，然后在顶花下面依次产生一个侧枝，侧枝上又顶端生花，如此连续分枝则为单歧聚伞花序。若花序轴下分枝均向同一侧生出而呈螺旋状弯转，称螺旋状聚伞花序，如紫草、附地菜等。若分枝成左右交替生出，则称蝎尾状聚伞花序，如射干、唐菖蒲等。单歧聚伞花序多见于单子叶植物，双子叶植物中少见。单歧聚伞花序初形成时为螺旋形或曲折形，以后由于各侧轴向上生长，花序逐渐变成直立。直立的单歧聚伞花序要注意与总状花序区分。

(1) Monochasium A type of cymose inflorescence with only a single main axis. Among them, a determinate cymose inflorescence with a zigzag rachis called **scorpioid cyme,** as in *Lithospermum erythrorhizon*. On the contrary, an one-sided cymose inflorescence coiled like a spiral or helix named helicoid cyme, which are usually borne on monocotyledonous plant, as in *Belamcanda chinensis*.

2. **二歧聚伞花序** 花序轴顶花先开，后在其下两侧同时产生两个等长的侧轴，每分枝以同样方式继续开花分枝，如石竹、大叶黄杨、卫矛等。二歧聚伞花序通常在生有对生叶的植物上。

(2) Dichasium A cymose inflorescence in which each axis produces two opposite or subopposite lateral axes, as in *Dianthus chinensis, Buxus megistophylla*. A dichasium usually borne on plants with opposite leaves.

3. **多歧聚伞花序** 花序轴顶花先开，其下两侧同时发出数个侧轴，侧轴多比主轴长，各侧轴又形成小的聚伞花序，称多歧聚伞花序。

(3) Pleiochasium A cymose inflorescence with more than two branches from the main axis. A pleiochasium usually borne on plants with opposite leaves.

4. **轮伞花序** 聚伞花序生于对生叶的叶腋中并呈轮状排列称轮伞花序，如薄荷、益母草等唇形科植物。

(4) Verticillaster A pair of axillary cymes arising from opposite leaves or bracts and forming a false whorl, such as *Labiatae*.

医药大学堂
WWW.YIYAODXT.COM

5. 杯状聚伞花序　杯状聚伞花序外观似一朵花，花序轴下面生有杯状总苞，总苞内含一朵无被的雌花和几朵无被雄花，雌花先开，之后雄花由内向外开放。如京大戟、甘遂、泽漆等大戟属植物。

(5) Cyathium　A cymose inflorescence with cyathiform, cup-shaped involucres under the inflorescence axis and metandrous, while the female flowers maturing before the male flowers, as in *Euphorbia*.

此外，有的植物的花序既有无限花序又有有限花序等不同类型特征，称混合花序。如丁香，其主花序轴为总状花序，侧花序轴则为聚伞类花序式，这种混合花序也叫做聚伞圆锥花序。

In addition, some plants possess the mixed characteristics both indefinite inflorescence and definite inflorescence, as in *Syzygium aromaticum*. The main inflorescence axis is raceme, while the lateral inflorescence axis is cymose, also named as thyrse.

第五节　花的生殖
5 Flower reproduction

花由花芽发育而成，主要功能是进行生殖，通过开花、传粉、受精等过程来完成。

Flower develop from flower bud, the function is reproduction. It can succeed through flowering, pollination, fertilization and other processes.

一、开花
5.1 Bloom

开花是种子植物发育成熟的标志，当雄蕊的花粉粒和雌蕊的胚囊成熟时，花被由包被状态逐渐展开，露出雄蕊和雌蕊，呈现开花。

Flowering is a sign of the maturity of seed plants. When the pollen grains and the embryo sac of the pistil are mature, the perianth gradually expands. The stamen and pistil are exposed.

二、传粉
5.2 Pollination

花开放后花药裂开，花粉粒通过风、水、虫、鸟等不同媒介的传播，到达雌蕊的柱头上，这一过程称为传粉。植物传粉有自花传粉和异花传粉两种方式。

After the bloom, the anther splits and the pollen grains travel to the stigma of the pistil through different media. There are two ways of plant pollination:

1. 自花传粉　是雄蕊的花粉自动落到同一朵花的柱头上的传粉现象，如小麦、棉花、番茄等。若花在开放之前就完成了传粉和受精过程，称闭花传粉，如豌豆、落花生等。自花传粉植物的特征是：两性花，雄蕊与雌蕊同时成熟，柱头可接受自身的花粉。

(1) Self pollination　The pollen of a stamen falls automatically on the stigma of the same flower, as in *Gossypium herbaceum*. It is cleistogamy if the flower has completed the pollination and fertilization process before opening.

2. 异花传粉　是雄蕊的花粉借助风或昆虫等媒介传送到另一朵花的柱头上的现象。异花传粉比自花传粉进化，是被子植物有性生殖中一种极为普遍的传粉方式。

(2) Cross pollination　The pollen from the stamen to the stigma of another flower by a medium such as wind or insects. It is common pollination methods in the sexual reproduction of angiosperms.

植物花粉可以借助风、昆虫等多种媒介完成传粉。

The media of the pollen grains travel to the stigma are pollinators, wind, etc.

虫媒花：以昆虫为传粉媒介的花称虫媒花，大多数植物采用此方式。虫媒花通常具备以下特点：两性花，花较大，花被颜色鲜艳，雄蕊和雌蕊不同时成熟，有蜜腺，散发特殊气味，花粉粒大，量小，表面粗糙或附有黏性物质。

Entomophilous flower: Entomophilous flower pollinate by pollinators. Most plant through this way. Entomophilous flower has the following characteristics: bisexual flowers, big, colorful, stamen and pistils are not mature at the same time, with nectary, special smell, pollen grain is big, less, rough surface or a sticky substance.

风媒花：花粉借助风力随机传播到雌蕊柱头上的植物称风媒花，如小麦、杨等。风媒花的结构特点为：穗状花序或荑葇花序，单性花，雌雄异株，无花被或花被不显著，花粉量大，花粉粒小，表面光滑或具延展的翅等结构，柱头较长，多呈羽毛等形状，面积大并有黏液质。此外，还有鸟媒、水媒等。

Anemophilous flowers: Pollen is carried by the wind to the stigma at random. Anemophilous flower structure characteristic for: spike or catkin inflorescence, unisexual flowers, dioecious, apetalous flowers or not significant, much of pollen, small, smooth surface or extension of the wing structure, stigma is longer, in the shape of feathers, the area is large and phlegmatic.

三、受精
5.3　Fertilization

被子植物的受精全过程包括受精前花粉在柱头上萌发，花粉管生长并到达胚珠，进入胚囊，释放2枚精子，其中1枚精子与卵结合的过程称受精，另1枚精子与中央细胞（或2个极核）结合，亦称受精，所以又称为双受精。

The fertilization of angiosperm including several processes. They are pollen germination on stigma, pollen tube growth and get to the ovule, into the embryo sac, release two sperm. The process of one sperm combine to egg is called fertilization. The other sperm combine to the central cells (or two polar nucleus), which is also called fertilization, known as double fertilization.

成熟花粉粒经传粉后落到柱头上，因柱头上有黏液而附于柱头上。花粉粒在柱头上萌发，自萌发孔长出若干个花粉管，其中只有1个花粉管能继续生长，经由花柱伸入子房。大多数植物的花粉管到达胚珠时通过珠孔进入胚珠，称珠孔受精；少数植物如核桃的花粉管由合点进入胚珠，称合点受精；还有少数植物的花粉管从胚珠中部进入胚囊，称中部受精。花粉管进入胚珠后穿过珠心组织进入胚囊，先端破裂，释放精子进入胚囊，此时营养细胞大多已分解消失。精子与卵受精后的二倍体受精卵（合子）发育成胚；精子与中央细胞（或2个极核）结合，形成三倍体的初生胚乳核，以后发育成胚乳。双受精是被子植物特有的现象。在双受精过程中，合子的产生恢复了植物体原有的染色体数目，保持了物种的相对稳定性，分别来自父本和母本遗传物质的重组，为后代提供了变异的基础；合子在同源的三倍体胚乳中孕育，不仅保证了二者亲和一致，还提供了合子发育中坚强的物质和信息保障，极大增强了后代的生活力和适应性。

The mature pollen grains are fall onto and attach to the stigma because of the mucus on the stigma. Pollen grains germinate on the stigma, and a number of pollen tubes grow from the germ pore, only one of them can continue to grow, through the style into the ovary. Most plants are porogamy, means the pollen tube reaches the ovule through micropyle. If the pollen tube enters ovule by chalaza, it is called chalazogamy. There are a few plants which pollen tube into the embryo sac from the middle of the ovule called mesogamy. After entering the ovule, the pollen tube passes through the nucellus to enter the embryo sac. The tip ruptures, and the sperm is released to enter the embryo sac. Spermatozoa combine with central cells (or two polar nucleus) to form triploid primary endosperm nucleus, which develop into endosperm. Double fertilization is a unique phenomenon in angiosperms.

重 点 小 结
Summaries

花是种子植物特有的繁殖器官，通过开花、传粉、受精过程形成果实和种子。花由花梗、花托、花被（花萼和花冠）、雄蕊群、雌蕊群组成。花冠有不同的结构和形态，主要分为离生花冠与合生花冠两大类。雄蕊由花丝和花药构成，常见的雄蕊类型有单体雄蕊、二体雄蕊、多体雄蕊、聚药雄蕊、二强雄蕊、四强雄蕊。雌蕊由柱头、花柱和子房组成，常见的雌蕊类型有单雌蕊、复雌蕊和离生心皮雌蕊。根据花的不同分类方法，可分为完全花、不完全花、无被花、单被花、重被花、重瓣花、两性花、单性花、无性花、辐射对称花、两侧对称花、不对称花。为准确描述花各组成部分的数目、离合、排列方式等形态特征，可用花程式及花图式来记录。花序可分为无限花序和有限花序两大类，其中，无限花序包括总状花序、穗状花序、柔荑花序、肉穗花序、伞房花序、伞形花序、头状花序、隐头花序、复总状花序、复穗状花序、复伞形花序、复头状花序；有限花序包括单歧聚伞花序、二歧聚伞花序、多歧聚伞花序、轮伞花序、杯状聚伞花序。

Flower is the unique reproductive organ of seed plants, which forms fruit and seed through the process of flowering, pollination and fertilization. A flower consist of pedicel, receptacle, perianth include calyx and corolla, androecium and gynoecium. Choripetalous corolla and synpetalous corolla are two types of corolla in common. The composition of the androecium is filament and anther. The androecium types are: Monadelphous, Diadelphous, Didynamous, Tetradynamous, polyadelphous. The gynoecium composition is consisted of ovary, style, stigma. The common types are: single pistil, syncarpous pistil and apocarpous pistil. Flowers can be divided into the followings: complete flower, incomplete flower, achlamydeous flower, simple perianth flower, double perianth flower, double flower, bisexual flower, unisexual flower, asexual flower, actinomorphic flower, zygomorphic flower, asymmetric flower, anemophilous plants, entomophilous plants, ornithophilous plants and hydrophilous plants. Flower formula and diagram can be used to describe the number, division, whorls and other morphological characteristics of the flower. Indefinite inflorescence can be divided into the followings types: raceme, spike, catkin, spadix, corymb, umbel, capitulum, hypanthodium, compound raceme, compound spike, compound umbel, compound capitulum. Definite inflorescence has the following types: monochasium, dichasium, pleiochasium, verticillaster, cyathium.

题库

目 标 检 测
Questions

1. 花的组成包括哪些部分？

What are the components of a flower?

2. 雄蕊有哪些类型？各有什么特点？

What are the types of stamens? What are their characteristics?

3. 被子植物的花在长期演化中，各部发生不同程度的变化，常见的分类方法与类型有哪些？请详细描述并举例。

What's the classification method of flower type? Please describe in detail and give examples.

4. 无限花序的定义及分类，试举例。

Please describe the definition and classification of infinite inflorescence, then give examples.

5. 螺旋状聚伞花序与蝎尾状聚伞花序是如何定义的，请举例。

Please describe the definition of scorpioid cyme and helicoid cyme. Please give examples.

6. 绘制桑花程式与花图式。

Draw formula and diagram of *Morus alba* flowers.

医药大学堂
WWW.YIYAODXT.COM

PPT

第四章 果实和种子
Chapter 4 Fruit and Seed

学习目标 | Learning goals

1. **掌握** 果实的结构与类型；种子结构与类型。
2. **熟悉** 常见果实种子类药材。
3. **了解** 果实种子的传播方式。

- Know the shape and type of fruit and seed.
- Be familiar with identification characteristics of common medicinal fruit.
- Understand how fruit and seeds spread.

果实是被子植物的特有结构，是有性生殖的产物。果皮包裹着种子，起到保护和参与传播种子的作用。果实与种子的形态特征和类型，可作为植物分类与果实种子类药材鉴定的依据。

A fruit is a matured and ripened ovary, wherein the ovary wall gets converted into the fruit wall pericarp (differentiated into outer epicarp, middle mesocarp and inner endocarp), and the ovules into seeds. Fruit is the product of sexual reproduction of angiosperms and is a characteristic structure.

第一节 果实
1 Fruit

一、果实的形成和结构
1.1 Fruit formation and structure

（一）果实的形成
1.1.1 Fruit formation
被子植物的花经传粉和受精后，花梗发育成果梗，花萼、花冠、雄蕊及雌蕊的柱头、花柱常枯萎或脱落，子房或子房外花的其他部分，如花被、花托及花序轴等，共同发育形成果实。

有些被子植物不经传粉或受精作用也能发育成果实，但果实不含种子，称为单性结实或无籽结实。自然发生的单性结实称为自发单性结实，如香蕉。通过人为诱导形成的单性结实称诱导单性结实，可用同类植物或亲缘相近的植物的花粉浸出液喷洒到柱头上而形成无籽果实，如用马铃

医药大学堂
WWW.YIYAODXT.COM

薯的花粉刺激番茄的柱头发育形成无籽番茄。植物受精后胚珠发育受阻亦可成为无籽果实，或由四倍体与二倍体杂交产生的不孕性三倍体植株生成无籽果实，如无籽西瓜。

After pollination and fertilization, those parts of the flower as sepals and petals, stamen, stigma and style often wither or fall off. The ovary will develop into fruit alone or together with perianth, receptacle, and inflorescence axis. Some angiosperms can develop into fruit without pollination or fertilization, and the fruit containing no seeds is called parthenocarpy. Naturally occurring parthenocarpy is called spontaneous parthenocarpy, as in bananas. The parthenocarpy could be induced by artificial, thus the seedless fruit can be formed by spraying the pollen extract of the same or related species on the stigma, such as potato pollen to stimulate the stigma development of tomatoes to form a seedless tomato. A plant whose ovule development is blocked after fertilization may also become a seedless fruit, or a infertile triploid plant produced by a tetraploid-diploid hybrid, as in seedless watermelon.

（二）果实的结构

1.1.2　The structure of the fruit

果实由果皮和种子组成。果皮通常可分为外果皮、中果皮、内果皮3部分，分层明显或不明显。外果皮是果皮的最外层，外面常有角质层、蜡被、毛茸、气孔、刺、瘤突、翅等附属物。中果皮是果实的中层，多由薄壁细胞组成，具有多数细小维管束，有的含石细胞、纤维、油细胞、油室及油管等。内果皮是果皮的最内层，具1至多层细胞，如核果的内果皮（即果核）由多层石细胞组成；伞形科植物的内果皮由5~8个长短不等的扁平细胞镶嵌状排列。

A fruit consists of fruit wall pericarp and seeds. The pericarp is usually divided into outer epicarp (Exocarp), middle mesocarp and inner endocarp. The character of the outer epicarp is of great evidence in the identification of medicinal fruit.

Exocarp is often covered with cuticle, wax, hair, stomata, spines, nodules, wings and other appendages. Mesocarp is the middle layer of fruit, mostly composed of parenchyma cells and small vascular bundles, some containing stone cells, fibers, oil cells, oil cavity and oil tubes. Endocarp is the innermost layer of the pericarp, with 1 to many layers of cells. The endocarp of Umbelliferae plants is arranged by 5~8 flat cells of varying lengths.

二、果实的类型
1.2　Type of fruit

（一）根据果实的来源分类

1.2.1　According to the origin of the fruit

根据是否有子房以外部位参与果实形成，可分为真果和假果。真果指仅由子房发育而成的果实，如商陆。由子房和其他部分如花被、花托及花序轴等共同参与形成的果实称假果，假果的结构比较复杂，如苹果、梨主要的食用部分是由花托和花被筒合生部分发育而来；南瓜、冬瓜较坚硬的皮部是花托和花萼发育成的部分及外果皮，主要食用部分是中果皮和内果皮；无花果、凤梨肉质化部分是由花序轴和花托等发育而成。

If the fruit is developed only from the ovary, then it is called true fruit, as *Phytolacca acinosa*. The fruit formed by ovary and other parts as perianth, receptacle and inflorescence axis is called false fruit. The structure of false fruit is relatively complex. For example, the main edible part of apple and pear is developed from the union part of receptacle and perianth tube. The hard pericarp of pumpkin and white gourd is the part formed by receptacle, calyx and the exocarp, and the edible parts are from mesocarp and endocarp. The fleshy parts of figs and pineapples are developed by the inflorescence axis and receptacle.

（二）根据心皮与花部的关系分类

1.2.2　According to the relationship between carpel and flower

单雌蕊或复雌蕊的子房发育形成的果实称单果，即1朵花只形成1个果实。离生雌蕊的每枚雌蕊形成1个小单果，聚生于同一花托上称聚合果，如乌头、莲等。整个花序发育而成的果实称聚花果或复果，其中每朵花发育成1个小单果，聚生在花序轴上，成熟后从花序轴基部整体脱落。如凤梨是由多数不孕的花着生在肥大肉质的花序轴上所形成的果实，其肉质多汁的花序轴成为果实的可食部分；桑葚由雌花序发育而成，每朵花的子房各发育成1个小瘦果，包藏于肥厚多汁的肉质花被内；无花果由隐头花序发育而成，称为隐头果，花序轴肉质化并内陷成囊状，囊的内壁上着生许多小瘦果，肉质化的花序轴是可食部分（图4-1）。

A single fruit develops from a flower having a single carpel or several united carpels so that the flower has a single ovary. Such a fruit may be dehiscent opening by a suture exposing seeds or remain indehiscent. Aggregate fruits develop from multi-carpellary apocarpous ovary. Each ovary forms a fruitlet, and the collection of fruitlets is known as etaerio. Common examples are etaerio of achenes of *Aconitum carmichaelii* in Ranunculaceae, etaerio of follicles in Calotropis, etaerio of drupes in raspberry (Rubus) and etaerio of berries in Polyalthia. A multiple fruit, or collective fruit involves ovaries of more than one flower, commonly the whole inflorescence. Common examples are: ①Sorosis: Composite fruit develops from the whole inflorescence and floral parts become edible, as seen in Morus (having fleshy perianth but dry seeds) and Artocarpus (with fleshy rachis, perianth and edible seeds). ②Syconium (syconus): Fruit developing from hypanthodium inflorescence of figs. There is a collection of achenes borne on the inside of fleshy hollow receptacle（Fig.4-1）.

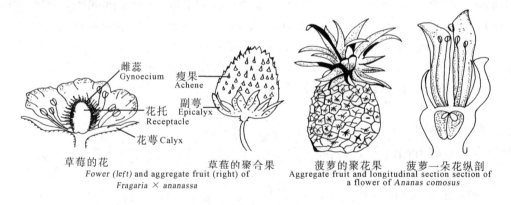

图 4-1　聚合果和聚花果

Fig.4-1　Aggregate fruit and collective fruit

（三）根据果实成熟时果皮的性质分类

1.2.3　According to the nature of the pericarp at maturity

依据果皮质地不同，分为肉质果和干果。

1. **肉质果**　成熟时果皮肉质多浆，不开裂。常见有以下5个类型（图4-2）。

(1) Fleshy fruit　The pericarp of the fruit is pulpy and does not crack. There are five common types (Fig. 4-2).

浆果：单雌蕊或复雌蕊子房发育形成，外果皮薄，而中果皮和内果皮肥厚、肉质多浆，内含1至多粒种子，如葡萄等。

Berry: Fruit with uniformly fleshy pericarp with numerous seeds inside, as seen in Solanum, tomato

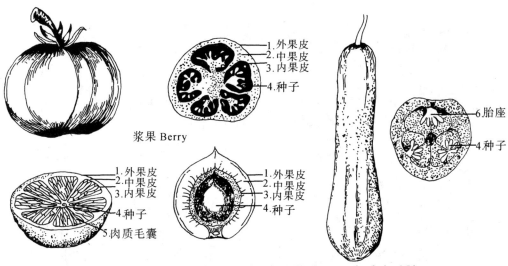

浆果 Berry

1.Exocarp；2.Mesocarp；3.endocarp；4.Seeds；5.Fleshy saccular hair；6.Placenta

图 4-2　肉质果
Fig.4-2　Fleshy fruit

and grapes.

　　柑果：复雌蕊上位子房发育形成，外果皮较厚且革质，内含多数油室；中果皮与外果皮结合而界限不明显，常疏松呈白色海绵状，具多数分支的维管束（橘络）；内果皮膜质，分隔成多室，内壁上生有许多肉质多汁、可食的囊状毛。柑果是芸香科柑橘属和金橘属特有的果实，如佛手。

　　Hesperidium: Fruit developing from superior ovary with axile placentation, epicarp and mesocarp forming common rind and endocarp produced inside into juice vesicles, as seen in citrus fruits. The pericarp is usually thick and leathery, containing oil cavities.

　　核果：单雌蕊上位子房发育形成，外果皮薄，中果皮肉质肥厚，内果皮木质坚硬，内含 1 粒种子，如桃。核果有时也泛指具有坚硬果核的果实，如人参。

　　Drupe: Fruit with usually skinny epicarp, fibrous or juicy mesocarp and hard stony endocarp, enclosing single seed, as seen in panax ginseng.

　　瓠果：假果，3 心皮复雌蕊具侧膜胎座的下位子房与花托一起发育而成，花托与外果皮形成外果皮，中、内果皮与肉质胎座，成为果实的可食部分。瓠果是葫芦科特有的果实类型，如栝楼等。

　　Pepo: Fruit formed from inferior ovary and lateral membrane placenta of cucurbits with epicarp forming tough rind. Pepo is a special fruit in Cucurbitaceae, as *Trichosanthes kirilowii*.

　　梨果：假果，2~5 个心皮复雌蕊下位子房与花筒一起发育而成，分隔成 2~5 室，每室常含 2 粒种子。花筒与外、中果皮一起发育成肉质可食部分，彼此界限不明显，内果皮革质或木质。梨果是蔷薇科梨亚科特有的果实类型，如山楂。

　　Pome: Fruit developing from inferior ovary, an example of accessory (false) fruit, wherein fleshy part is formed by thalamus and cartilaginous pericarp isinside, Pome is a king of special fruit in Maloideae, as seen in *Crataegus pinnatifida*.

　　2. 干果　果实成熟时果皮干燥。根据果熟时果皮开裂与否，可分为裂果和不裂果（图 4-3）。
　　(2) Dry fruit　Such fruits have dry pericarp at maturity, split open or not at maturity (Fig.4-3).

　　裂果：果实成熟后果皮自行开裂，散出种子。依据雌蕊心皮数与开裂方式不同，可分为以下类型。

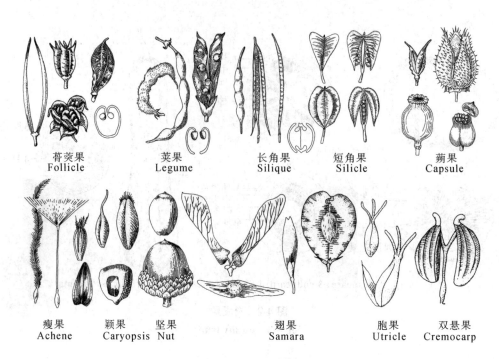

菁葖果　荚果　长角果　短角果　蒴果
Follicle　Legume　Silique　Silicle　Capsule

瘦果　颖果　坚果　翅果　胞果　双悬果
Achene　Caryopsis　Nut　Samara　Utricle　Cremocarp

图 4-3　干果的类型
Fig.4-3　Type of the dry fruit

Dehiscent fruit: Such fruits split open at maturity.

菁葖果：单雌蕊或离生雌蕊发育形成的果实，成熟时沿腹缝线或背缝线一侧纵向开裂，如牡丹沿腹缝线开裂，厚朴沿背缝线开裂。离生雌蕊形成的菁葖果可 1 至多个，夹竹桃科常为双生菁葖果。

Follicle: Fruit developing from superior monocarpellary ovary and dehiscing along one suture, as *Paeonia suffruticosa* dehiscing along the ventral suture, *Magnolia officinalis* dehiscing along the dorsal suture. Follicles formed by separate carpel may be 1 to more than one, as seen in Apocynaceae, which is often double follicles.

荚果：单雌蕊发育形成的果实，果熟时沿腹缝线和背缝线同时开裂，果实裂成 2 片。荚果是豆科植物特有的果实类型，如白扁豆。荚果形态多样，少数荚果成熟时不开裂，如落花生；有的荚果成熟时在种子间节节断裂，每节含 1 种子，如含羞草；有的荚果呈螺旋状，并具刺毛，如紫苜蓿；还有的荚果肉质，在种子间缢缩呈念珠状，如槐。

Legume or pod: Fruit developing like follicle from monocarpellary superior ovary but dehiscing along two sutures, as in legumes like *Lablab purpureus*. Sometimes a legume does not dehisce, as *Arachis hypogaea*; Some pods, as *Mimosa pudica*, have one seed per node that breaks at maturity between the seeds; Some pods are spiral-shaped and spiny, as *Medicago sativa*; There are also pods fleshy, constricted between seeds in the form of rosary beads, as *Sophora japonica*.

角果：二心皮复雌蕊具侧膜胎座的子房发育而成，子房 1 室，但在形成过程中由二心皮边缘合生处生出假隔膜，将子房分隔成 2 室，种子着生于假隔膜两侧。果实成熟时果皮沿两侧腹缝线开裂，成 2 片脱落，假隔膜仍留在果柄上。角果是十字花科特有的果实类型。果实长宽比大于 3 的角果称为长角果，如萝卜；长宽比小于 3 的称为短角果，如菘蓝。

Siliqua and silicuela: Fruit developing from bicarpellary syncarpous superior ovary, which is initially one chambered but subsequently becomes two chambered due to the formation of a false septum, visible on the outside in the form of a rim known as replum. The fruit dehisces along both sutures from the base upwards, valves separating from septum and seeds remaining attached to the rim (replum), characteristic of the family Brassicaceae. Silique is narrower and longer, at least three times longer than broad, as in *Raphanus sativus*. Silicula similar to siliqua but shorter and broader, less than three times longer than broad as seen in *Isatis tinctorial*.

蒴果：复雌蕊发育而成的果实，子房1至多室。成熟果实具多种开裂方式，一是纵裂：开裂时沿心皮纵轴开裂，其中沿腹缝线开裂的称室间开裂，如马兜铃；沿背缝线开裂的称室背开裂，如百合；沿背、腹缝线同时开裂，但子房间隔仍与中轴相连的称室轴开裂，如牵牛。二是孔裂：果实成熟时顶端呈小孔状开裂，种子由小孔散出，如罂粟。三是盖裂：成熟果实中部环状横裂，上部果皮帽状脱落，如马齿苋。四是齿裂：果实顶端齿状开裂，如王不留行。

Capsule: Fruit developing from syncarpous ovary and dehiscing in a variety of ways. Circumscissile (pyxis) dehiscence: transverse so that top comes off as a lid or operculum, as in *Portulaca oleracea*. Poricidal: Dehiscence through terminal pores as *Papaver somniferum* in poppy (Papaver). Denticidal: Capsule opening at top exposing a number of teeth as in *Vaccaria hispanica*. Septicidal: Capsule splitting along septa and valves remaining attached to septa as in *Linum*. Loculicidal: Capsule splitting along locules and valves remaining attached to septa, as in family Malvaceae. Septifragal: Capsule splitting so that valves fall off leaving seeds attached to central axis as in Datura.

不裂果（闭果）果实熟后果皮不开裂，或分离成几部分，但种子仍包被于果皮中。常分以下几种类型。

Indehiscent fruit: Such fruits do not split open at maturity.

瘦果：单室、单种子的小干果，成熟时果皮与种子极易分离，如何首乌等；菊科植物的瘦果由下位子房与萼筒共同发育形成，称连萼瘦果或菊果，如蒲公英。

Achene and Cypsela: Achene is single seeded dry fruit developing from a single carpel with superior ovary. Fruit wall is free from seed coat. Achenes are often aggregated, as *polygonum multiflora* in family Ranunculaceae. **Cypsela** is also single seeded dry fruit, similar to (and often named achene) but developing from bicarpellary syncarpous inferior ovary, as *Taraxacum mongolicum* in family Asteraceae.

颖果：果皮与种皮愈合不易分离，内含1粒种子，是禾本科特有的果实类型，如小麦、薏苡等。农业生产中常把颖果称为"种子"。

Caryopsis: Fruit similar to above two but fruit wall fused with seed coat as seen in Poaceae family.

坚果：果皮坚硬，成熟时果皮和种皮分离，内含1粒种子。有的坚果较小，称小坚果，如唇形科的果实。壳斗科的坚果外常有总苞发育成的壳斗附着于基部。

Nut: One-seeded, generally large fruit developing from multicarpellary ovary and with hard woody or bony pericarp, as seen in Quercus and Litchi. Some nuts are small as in Lamiacea.

翅果：单室、单种子，果皮一端或周边向外延伸成翅状，如杜仲。

Samara: Fruit with one-seeded, winged in one end or surrounding whole, as *Eucommia ulmoides*.

胞果：又称囊果，复雌蕊上位子房发育形成，果皮薄且膨胀疏松地包围种子，与种皮极易分离，如青葙。

Utricle: Similar to nut but with papery often inflated pericarp as *Celosia argéntea*.

双悬果：二心皮复雌蕊发育而成，果实成熟后分离成2个分果，悬挂在心皮柄上端，心皮柄的基部与果柄相连，分果内含1粒种子，为伞形科特有的果实类型，如当归。

Cremocarp: Fruit developing from bicarpellary syncarpous inferior ovary and splitting into two one seeded segments known as mericarps, as *Angelica sinensis* in umbellifers.

第二节 种子

2 Seeds

种子由植物的胚珠受精后发育而成，其主要功能是繁殖。种子是种子植物特有的器官，被子植物种子包被于果皮之内，裸子植物胚珠裸露，种子外无果皮包被。

Seed is formed by the development of plant ovule after fertilization, and its main function is reproduction. Seeds are the special organs of seed plants. Angiosperms seeds are encased in pericarp, gymnosperm ovules are naked, and seeds are encased without pericarp.

一、种子的形态与结构
2.1　Seed morphology and structure

（一）种子的形态
2.1.1　Seed morphology

种子的形状、大小、色泽、表面纹理等随植物种类不同而异，是种子类药材重要的鉴别依据。种子常呈圆形、椭圆形、肾形、卵形、圆锥形、多角形等。其大小悬殊，较大的如椰子；较小的如菟丝子；兰科植物的种子极小，呈粉末状，如白及。种子的颜色亦多样，绿豆为绿色，赤小豆为红紫色，白扁豆为白色，藜属植物的种子多为黑色；相思子一端为红色，另一端为黑色；蓖麻种子的表面由一种或几种颜色交织组成各种花纹和斑点。种子表面的特征也不相同，有的光滑、具光泽，如红蓼；有的粗糙，如长春花；有的具皱褶，如乌头；有的密生瘤刺状突起，如太子参；有的具翅，如木蝴蝶；有的顶端具毛茸，称种缨，如白前。

The shape, size, color and surface texture of seeds vary with different plant species, which is an important identification basis for medicinal seed materials. Seed is quite big as in seen in coconut. Small as *Cuscuta chinensis*. Extremely small in powder form as seen in *Bletilla striata* of Orchidaceae family. The color of seeds is also diverse in green, red, or purple. The seeds of *chenopodium* are black. The seed *Acacia* is red to one end while black to another. The surface of castor seeds consists of one or more colors interwoven into a variety of patterns and spots. The character of the seed surface is also different, some smooth, with luster, such as *Polygonum orientale*. Some rough, such as *Catharanthus roseus*. Some have folds, such as *Aconitum carmichaelii*. Some dense tumor spines, such as *Pseudostellaria heterophylla*. Some have wings, such as *Oroxylum indicum*. Some of the top with pilose, called species of tassel, such as *Cynanchum glaucescens*.

（二）种子的结构
2.1.2　Seed structure

种子一般由种皮、胚、胚乳 3 部分组成。有的种子不具胚乳，有的种子具外胚乳。

The seed generally consists of seed coat, embryo and endosperm. Some seeds have no endosperm, others have ecosperm.

1. **种皮**　种皮由胚珠的珠被发育而来，包被于种子表面起保护作用。种皮通常 1 层，如大豆；有的种子具 2 层种皮，即外种皮和内种皮，外种皮常坚韧，内种皮较薄，如蓖麻。在质地方面，有干性种皮，如豆类；有肉质种皮，如石榴籽外的肉质可食部分。有的种子在种皮外具假种皮，由珠柄或胎座部位的组织延伸而成，有肉质假种皮，如龙眼；有膜质假种皮，如砂仁。

种皮上常可看到以下结构。

(1) Seed coat　Developed from the integument of the ovule, which is wrapped on the surface of the seed to protect it. Seed coat usually 1, as in soybean. Sometimes with 2 layers, namely outer seed coat and inner seed coat, outer seed coat often tough, inner seed coat is thinner, as seen in *Ricinus communis*. Some seed coat is dry as in beans, but some fleshy as in *Punica granatum*. Some seeds have an aril outside the testa, extending from tissue at the funicle or placenta, with a fleshy aril, as in *Litchi chinensis*; or a membranous aril, as in *Amomum villosum*.The following structures are often seen on the seed coat.

种脐是种子成熟后从种柄或胎座上脱落后留下的疤痕，常呈圆形或椭圆形。

Hilum: A scar left after a seed is shed from the stalk or placenta as it maturing, often round or elliptic.

种孔来源于胚珠的珠孔，种子萌发时吸收水分和胚根伸出的部位。

Micropyle: The micropore from the ovule that absorbs water as the seed germinates and where the radicle sticks out.

合点来源于胚珠的合点，是种皮上维管束汇合之处。

Chalaza: The chalaza of an ovule where the vascular bundles meet in the seed coat.

种脊来源于珠脊，种脐到合点之间的隆起线，内含维管束，倒生胚珠发育的种子种脊较长，弯生或横生胚珠形成的种子种脊短，直生胚珠发育的种子无种脊。

Raphe: Originated from the bead ridge, the ridge between the seed umbilicus and the confluent point, containing vascular bundles, the seed ridge with inverted ovule development is longer, the seed ridge formed by curved or horizontal ovules is short, and the seed with straight ovule development has no seed ridge.

种阜是种孔处由珠被扩展形成的海绵状突起物，在种子萌发时可以帮助吸收水分，如巴豆。

Caruncle: A spongy bulge formed by spreading beads at the micropyle that helps to absorb water during seed germination, as in a crocus.

2. **胚**　由卵细胞受精后发育而成，是种子中尚未发育的幼小植物体，由4部分组成。

(2) Embryo　Developed from fertilized egg, also an undeveloped young plant body in a seed and consists of four parts.

胚根是幼小未发育的根，正对着种孔，将来发育成植物的主根。

Radicle: A young undeveloped root facing the micropyle that will develop into the taproot of a plant.

胚轴又称胚茎，连接胚根与胚芽的部分，以后发育成为连接根和茎的部分。

Hypocotyl: The part of the embryonic stem that connects the radicle to the germ and later develops into the part that connects the root to the stem.

胚芽是胚的顶端未发育的地上枝，以后发育成植物的主茎和叶。

Plumule: The undeveloped aerial shoot at the apex of the embryo that later develops into the main stem and leave of the plant.

子叶：一般单子叶植物具1枚子叶，双子叶植物具2枚子叶，裸子植物具多枚子叶。

Cotyledon: Organ that absorbs and stores nutrients in the embryo, and photosynthesis after germination. Generally monocotyledons with 1 cotyledon and dicotyledons with 2, gymnosperms with multiple cotyledons.

3. 胚乳 被子植物的胚乳由极核细胞受精发育而成，常位于胚的周围，呈白色，含丰富的淀粉、蛋白质、脂肪等，是种子内的营养组织，供胚发育时所需的养料。少数植物珠心或珠被的营养组织，在种子发育过程中未被完全吸收而形成营养组织，包围在胚乳和胚的外部，称外胚乳，如肉豆蔻，其外胚乳内层细胞向内伸入，与类白色的胚乳交错形成错入组织。

(3) Endosperm Formed by the development of polar nucleus cells after fertilization. It is usually located around the embryo. It is rich in starch, protein and fat.

The nucellus or integument which is not fully absorbed during seed development will develop into perisperm surrounding the endosperm and the outside of the embryo. When interlacing with endosperm, it will form interlacing tissue as in *Myristica fragrans*.

二、种子的类型
2.2　Seed type

被子植物的种子常依据胚乳的有无，分为两类（图 4-4、图 4-5）。

The seeds of angiosperms are often divided into two groups, depending on the presence or absence of the endosperm (Fig.4-4, Fig.4-5).

1. 有胚乳种子 种子中有发达的胚乳，则胚相对较小，子叶薄，如掌叶大黄。

(1) Albuminous seed If there is a developed endosperm in the seed, the embryo is relatively small

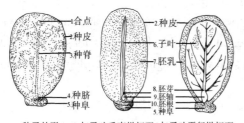

a.种子外形　b.与子叶垂直纵切面 c.与子叶平行纵切面
Seed shape and longitudinal section perpendicular /parallel to cotyledons
1. Chalaza;2. Seed coat;3.Raphe; 4. Hilum;5. Caruncle; 6. Cotyledon; 7. Endosperm;
8.Plumule; 9. Embryonal axis;10.Radicle

图 4-4　有胚乳种子
Fig. 4-4　Albuminous seed

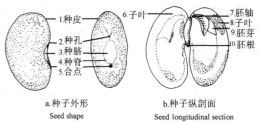

a.种子外形　　　　b.种子纵剖面
Seed shape　　　　Seed longitudinal section
1. Seed coat; 2. Micropyle; 3. Hilum; 4. Raphe; 5. Chalaza; 6. Cotyledon;
7. Hypocotyl; 8.Cotyledon; 9. Plumule; 10. Radicle

图 4-5　无胚乳种子
Fig. 4-5　Exalbuminous seed

and the cotyledons are thin, as seen in *Rheum palmatum*.

2. 无胚乳种子 种子中胚乳不存在或仅残留一薄层，常具发达的子叶，如泽泻。

(2) Exalbuminous seed The endosperm of the seed does not exist or only a thin layer remains, often with developed cotyledons, as seen in *Alisma plantago-aquatica*.

重 点 小 结
Summaries

果实是被子植物特有的繁殖器官。果实由果皮和种子构成，果皮分为外果皮、中果皮和内果皮。按果实的来源不同分为真果和假果；根据心皮与花部关系分为单果、聚合果和聚花果；根据成熟时果皮质地分为肉质果和干果，肉质果包括浆果、柑果、核果、瓠果、梨果等；干果分裂果、闭果两种，其中裂果包括菁葖果、荚果、角果、蒴果，闭果包括瘦果、颖果、坚果、翅果、胞果、双悬果。种子由种皮、胚、胚乳三部分组成；种子的胚包括胚芽、胚轴、胚根和子叶；根据胚乳的是否存在，分为有胚乳种子和无胚乳种子。

Fruit is the special reproductive organ of angiosperms. The fruit consists of pericarp and seed, and the pericarp is divided into exocarp, mesocarp and endocarp. According to different sources, the fruit can be divided into true fruit and false fruit. According to the relationship between carpels and flowers, the fruit can be divided into single fruit, aggregate fruit and collective fruit. According to the texture of the pericarp, the fruit can be divided into fleshy fruit and dry fruit, the former includes berry, hesperidium, drupe, pepo and popme, the latter includes dehiscent fruit (follicle, legume, silique, silicle) and indehiscent fruit (achene,caryopsis, nut, samara, utricle, and cremocarp).

The seed is composed of seed coat, embryo and endosperm. There are two types of seeds: endosperm seeds and non-endosperm seeds.

目 标 检 测
Questions

题库

1. 植物花的哪些部位参与果实的形成？真果与假果有什么区别？

What parts of a plant flower are involved in fruit formation? What's the difference between true fruit and false fruit?

2. 果实的类型有哪些？请举例说明。

What are the types of fruit? Please give an example.

3. 请统计以果实或种子入药的中药，说明入药的种类、入药的部位。

Please make statistics of the traditional Chinese medicine which uses fruit or seed as medicine, and explain the kinds and parts of the medicine.

4. 草莓、桑葚、菠萝、无花果、荔枝、石榴的可食用部分分别是什么？

What are the edible parts of strawberries, mulberries, pineapples, figs, litchi, and pomegranates?

第二篇　药用植物的分类

Part II　Taxonomy of Medicinal Plants

第五章 药用植物分类概述
Chapter 5 Introduction to Medicinal Plant Taxonomy

PPT

学习目标 | Learning goals

1. **掌握** 药用植物分类的单位和植物命名方法。
2. **熟悉** 药用植物分类的目的和方法，检索表的使用。
3. **了解** 植物命名法规和植物分类系统。

- Know the classification and nomenclature of medicinal plants.
- Be familiar with the purpose and method of classification of medicinal plants and the use of the key.
- Understand plant nomenclature and plant classification system.

第一节 植物分类学的目的和任务
1 Task of plant taxonomy

自然界的生物种类繁多，分布广泛。据全球生物多样性评估，目前有175万种生物，其中已记载的绿色植物约33.5万种。人类在长期的实践活动中对其中众多的植物加以应用，必然要对它们进行识别、命名和分门别类，由此便产生了植物分类学。

At present, according to the global biodiversity assessment, it is estimated that there are about 1.75 million species of organisms, including about 335000 recorded green plants. It is necessary to identify, name and classify many plants in the long-term practice of human beings, which leads to plant taxonomy.

植物分类学是一门研究整个植物界不同类群的起源、亲缘关系，以及进化发展规律的学科。掌握了植物分类学的理论和方法，就可以对自然界极其繁杂的植物进行鉴定、分群归类、命名并按系统排列，便于认识、研究和利用。植物分类学的主要任务是：

Plant taxonomy is a discipline that studies the origin, relationship and evolution of different taxa of the entire plant kingdom. Only by understanding and comprehension the theories and techniques of plant taxonomy, can we identify, classify, name and arrange the plants in a systematic way, which is convenient for recognition, research, and utilization.The main tasks of modern plant taxonomy are as follows.

一、植物分类群的描述和命名
1.1 Description and nomenclature of plant taxa

植物分类学的首要任务是运用植物形态学、解剖学、遗传学、分子生物学等知识，对植物个体间的异同进行比较研究，将类似的一群个体归为"种"层级的分类群，并对各分类群进行描述，按照《国际植物命名法规》确定植物的拉丁学名。

The primary task of plant taxonomy is to use plant morphology, anatomy, genetics, molecular biology and other knowledge to compare and study the similarities and differences among plant individuals, classify a group of similar individuals into "species" level taxa, describe the characters of each taxa, and determine the Scientific names of plants in accordance with The International Code Of Botanical Nomenclature.

二、建立自然分类系统
1.2 Establish a natural classification system

借助植物生态学、植物地理学、古植物学、生物化学、分子生物学等学科的研究资料，探索植物"种"的起源和类群分化路径，进而根据各类群之间的亲缘关系，确立其等级和排列顺序，从而建立能反映历史发展真实情况的植物自然分类系统。

Based on the research data of plant ecology, plant geography, palaeobotany, biochemistry, molecular biology and other disciplines, explore the origin of "species" and the path of group differentiation, and then establish the rank and order of "species" according to the relationship between various groups, so as to establish a natural plant classification system that can reflect the real situation of historical development.

三、编写和修订植物志
1.3 Compilation and revision of flora

运用植物分类学知识，根据不同的需要对不同区域或不同范围的植物进行采集、鉴定、描述，并按照分类系统编排，编纂不同的植物志或植物名录，大到《中国植物志》《中国药用植物志》等，小到一个植物园、一个校园的植物名录，可用于研究、教学或科普教育等不同用途。开展各种植物志修订性研究也是植物分类学家持之以恒的不懈工作，如药用植物分类学的深入研究，不仅可以更加准确、便利地鉴别药用植物，澄清中药材基原，保证临床用药安全有效，同时还能依据植物类群间的亲缘关系、物种地带性分布规律、中药材道地性形成规律等，深入发掘和开发中药资源，促进中药资源的可持续性利用。

Using the knowledge of plant taxonomy, collecting, identifying and describing plants in different regions or ranges according to different needs, and arranging them according to the classification system, compiling different flora or plant lists, ranging from Flora of China and Flora of Medicinal Plants of China, to a botanical garden or a campus plant list, which can be used for research, teaching or popular science education Etc. It is also the persistent and unremitting work of plant taxonomists to carry out the revision research of all kinds of flora. For example, the in-depth study of medicinal plant taxonomy can not only identify medicinal plants more accurately and conveniently, clarify the basic sources of traditional Chinese medicine, ensure the safety and effectiveness of clinical medication, but also according to the relationship between plant groups, the law of species zonal distribution and the local

视频

医药大学堂
WWW.YIYAODXT.COM

production of traditional Chinese medicine materials explore and develop the resources of traditional Chinese medicine and improve the sustainable utilization of traditional Chinese medicine.

药用植物分类学是利用植物分类学的原理和方法，对有药用价值的植物进行鉴定、命名、分类，促进药用植物合理开发利用的一门应用学科。通过学习药用植物分类学的知识和方法，可以鉴定中药基源、评价种质资源、发掘中草药资源，为中药的研究、生产以及临床安全有效使用提供重要支撑。

Medicinal Plant taxonomy is an applied discipline that uses the principles and methods of plant taxonomy to identify, name and classify medicinal plants, and to promote the rational development and utilization of medicinal plants. By learning the knowledge and methods of plant taxonomy, we can identify the source of Chinese medicine, evaluate the germplasm resources, and discover new resources of Chinese herbal medicines as well as provide important support for the research, production and effective use of Chinese materia medica.

第二节 植物分类的单位
2 Taxonomic hierarchy

人们根据植物类群范围大小和阶层高低设定了一些分类的名称，即植物分类等级单位（阶元）。林奈首先提出界、门、纲、目、属、种的分类法，被人们采用至今。按照国际植物命名法规记载，有关绿色植物和真菌命名有 12 个主要等级，按照其高低和从属亲缘关系，分别为：门、纲、目、科、族、属、组、系、种、亚种、变种、变型。其中常用的七阶分类单位排列见表 5-1。

According to the International Code of Botanical Nomenclature (ICBN), there are 12 major taxonomic categories of plant kingdom as follows: Divisio, Classis, Ordo, Familia, Tribus, Genus, Sectio, Series, Species, subspecies, Varietas, Forma. The commonly 7 taxonomic units are listed in Table 5-1.

表 5-1 植物分类的单位

中 文	英 文	拉丁文
界	Kingdom	Regnum
门	Division	Divisio（phylum）
纲	Class	Classis
目	Order	Ordo
科	Family	Familia
属	Genus	Genus
种	Species	Species

Table 5-1 Main taxonomic units of the plant kingdom

English	Latin
Kingdom	Regnum
Division	Divisio (phylum)

continued

English	Latin
Class	Classis
Order	Ordo
Family	Familia
Genus	Genus
Species	Species

在各级单位之间，有时因范围过大，不能完全概括其特征与系统关系时，常增设亚级单位，如亚门、亚纲、亚目、亚科、亚属及亚种。

Among the taxonomic units, their circumscription is sometimes large enough to set up some sub-units, such as subdivision, subclass, suborder, subfamily, subgenus and subspecies, etc.

物种，简称种，是植物分类的基本单位，也是植物体分类等级中最为客观的一个单位。物种指有一定自然分布区，具有实际或潜在交配并繁衍后代能力的个体群（居群）。同种的个体具有许多共同特征，表现为性质稳定的繁殖群体，非同种个体不能交配进行有性生殖，即使产生后代，其个体也不育，这种现象称生殖隔离。

The species has different meaning for different botanists. It's hard to define a species perfectly. According to ICBN, which has attempted to clarify the meaning of the word species, "species is a convenient classificatory unit defined by trained biologists using all information available". According to the concept morphological species (taxonomic species concept), the species is regarded as an assemblage of individuals with morphological features in common, and separable from other such assemblages by correlated morphological discontinuity in a number of features. The biological species concept defines species as a community of cross-fertilizing individuals linked together by bonds of mating and reproductively isolated from other species by barriers to mating.

一个物种往往由多个居群组成，一个居群又由许多个体组成，每个居群的分布常常不连续。不同居群生长环境存在差异，因而会产生一些或大或小的变异。这些种内各居群之间若出现比较大的变异，分类上会设立种以下分类等级，即亚种、变种和变型。

A species is often composed of multiple populations. A population is composed of many individuals, and the distribution of each population is often discontinuous. There are some differences in the growth environment of different populations, so there will be some large or small variations. If there is a large variation among the populations within these species, the following taxonomic categroies will be set up, namely subspecies, varieties and forma.

亚种（subspecies，缩写为 subsp. 或 ssp.）：指一个种内的居群具有地理分布、生态或季节的隔离，形态上亦有变异的居群。

Subspecies (subsp. or ssp.): It refers to populations within a species that is different in geographical, ecological, seasonal or morphological.

变种（varietas，缩写为 var.）：是指种内不同居群有共同的分布区，但形态上多少有变异，且变异可以稳定遗传，互称变种。

Varietas (var.): It refers to the common distribution area of different populations within the species, but there are some morphological variations, and the variation can be inherited stably.

变型（forma，缩写为 f.）：是一个种内不同居群的个体间有细小变异，遗传稳定性不如变种居群，也没有一定分布区。有时将栽培植物中的品种也视为变型。

Forma (f.): it is a small variation among individuals in an intraspecific non-cohabiting group, and its genetic stability is not as good as that of the mutant population, and there is no certain distribution area. Sometimes varieties in cultivated plants are also considered as forma.

品种（cultivar，缩写为 cv.）：指栽培植物的种内变异的居群。往往和人类干预、培育有关，常表现在形态上或经济价值上的差异，如色、香、味、形状、大小、植株高矮和产量等。如果品种没有了经济价值，也就失去了品种的实际意义，终将被淘汰。药材中一般所称的品种，是一种约定俗成的叫法，指分类学上的"种"，或指中药的个药。

Cultivar (Cv.): It refers to the intraspecific variation population of cultivated plants. It is often related to human intervention and cultivation, often showing the differences in shape or economic value, such as color, fragrance, taste, shape, size, plant height and yield, etc. If a variety has no economic value, it will lose its practical significance and will eventually be eliminated. The commonly called variety in TCM is a conventional name, which refers to the "species" in taxonomy or sometimes refers to a kind of specific Chinese materia medica.

在教学和实践应用中，科、属、种三个分类等级常常会被特别强调。任何生物学研究均需基于同一类群的生物体作为实验取材的对象，得出的结论才有说服力，否则取得的成果将受质疑。

In teaching and practical application, three classification levels of family, genus and species are often emphasized. Any biological research needs to be based on the same group of organisms as the object of experiment, the conclusion is convincing, otherwise the results will be questioned.

第三节 植物的命名

3 Plant nomenclature

植物物种学名必须遵从《国际植物命名法规》，用拉丁文来命名。种的命名采用瑞典植物学家林奈倡导的"双名法"，即前者是属名，第二个是种加词，最后是命名人。

The scientific name of a plant must be written in Latin in accordance with the ICBN. Casper Bauhin introduced the concept of Binomial under which the name of a species consists of two parts, the first the name of the genus to which it belongs and the second the specific epithet. Linnaeus firmly established this system of naming in his Species plantarum.

一、科的命名
3.1 Family

按 ICBN 规定，植物科的拉丁名词尾加 -aceae，如蓼科为 Polygonaceae。但有些科的植物分布广泛，研究历史久远，长期以来学术界运用的拉丁语习用名也被保留使用，同时按国际植物命名法规制定的规范化科名也一并使用。8 个有保留科名的科见表 5-2。

According to the ICBN, the name of a family is ended with the termination -aceae in the genera of classical Latin or Greek origin (Family Polygonaceae from genus *Polygonum*). The following eight families of angiosperms, however, whose original names are not in accordance with the rules but the use of these names has been sanctioned because of old traditional usage. The alternate names of these families which are in accordance with the ICBN rules need to be encouraged. Each family is listed in Table 5-2。

表 5-2　8 个科的保留科名和规范化科名

中文科名	保留拉丁科名	规范拉丁科名
十字花科	Cruciferae	Brassicaceae
豆　科	Leguminosae	Fabaceae
藤 黄 科	Guttiferae	Hypercaceae
伞 形 科	Umbelliferae	Apiaceae
唇 形 科	Labiatae	Lamiaceae
菊　科	Compositae	Asteraceae
棕 榈 科	Palmae	Arecaceae
禾 本 科	Gramineae	Poaceae

Table 5-2　Traditional names and alternate names of 8 families

Traditional Latin name	Alternate Latin name
Cruciferae	Brassicaceae
Leguminosae	Fabaceae
Guttiferae	Hypercaceae
Umbelliferae	Apiaceae
Labiatae	Lamiaceae
Compositae	Asteraceae
Palmae	Arecaceae
Gramineae	Poaceae

科一级分类单位在必要时也可分亚科。亚科的拉丁名词尾加 -oideae，如蔷薇科分为四个亚科，绣线菊亚科为 Spiraeoideae；蔷薇亚科为 Rosoideae；李亚科 Prunoideae；苹果亚科为 Maloideae 等。

When necessary, the family can also be divided into subfamilies. For example, Rosaceae is divided into four subfamilies: Spiraeoideae, Rosaoideae, Prunoideae and Malooideae.

二、属的命名
3.2　Genus

属名是植物各级分类群中重要的组成之一，不仅是植物学名的主体，也是构成科拉丁名的基础。植物属名来源广泛，如形态特征、生活习性、用途、地方俗名、神话传说等，无论其来源如何，均做拉丁语名词对待。

The name may be based on any source, but the common sources for generic names are as following: to commemorate of a person, based on a place or on an important character, sometimes even based on myths and legends, aboriginal names taken directly from a language other than Latin without alteration. regardless of their sources, the generic name is a uninomial singular word treated as a noun.

三、物种的命名
3.3 Species

植物的种加词用于区别同属不同种，多用形容词，也有用同格名词或属格名词，种加词的所有字母小写。当形容词作为种加词时，其性、数、格要与属名一致。如掌叶大黄 *Rheum palmatum* L，种加词 *palmatum*，表示该植物的叶掌状分裂，与属名同为中性、单数、主格。当种加词为名词时，有同格名词和属格名词两种类型。同格名词，其数、格与属名一致，性可不一致，如薄荷 *Mentha haplocalyx* Brig.，种加词 *haplocalyx* 为阳性单数主格，而 Mentha 为阴性单数主格。属格名词，大多引用人名姓氏，有单数普通名词和复数属格。

The name of a species is a binomial: consisting of two words, a generic name followed by a specific epithet. Plant species epithet distinguishes different species in the same genus, the most is adjective, and some are nouns. The letter of species epithet all must be a lower case initial letter. When an adjective is used as species epithet, its gender, number and case must be consistent with that of the genus, for instance, *Rheum palmatum* L, its species epithet 'palmatum', indicating that the plant is also palmately. Species epithet and the genus name is negative, neutral, singular and nominative. When noun is used as species epithet, there are two types: nominative noun or genitive noun. If it is a nominative noun, the number and case are the same as the genus name, but not the gender. For example, *Mentha haplocalyx* Brig., the species adjunctive *haplocalyx* is masculine singular nominative, while *Mentha* is feminine singular nominative. If it is a genitive noun, it is mostly a surname with a singular and plural genitive, such as *Alpinia officinarum* Hance, and *officinarum* is the plural genitive of *officinlis*.

四、种下分类群的命名
3.4 Infraspecies taxon

植物种下的类群有亚种、变种和变型，在命名时，在原种名后，加表示种下单位的缩写词，再加亚种、变种或变型名即可。如鹿蹄草 *Pyrola rotundifolia* L.*subsp. chinensis* H.Andces。

Plant infraspecies taxon have subspecies (subsp./ssp.), varietas (var.) and forma (f.). For example: *Pyrola rotundifolia* L.*subsp. chinensis* H. Andces is a subspecies of *Pyrola rotundifolia* L.,its scientific name is made of scientific name plus subsp., subspecies epithet (chinensis) and subspecies author H.Andces).

五、栽培植物的命名
3.5 Nomenclature of cultivated plants

栽培植物的命名，必须遵从《国际栽培植物命名法规》。栽培植物名称在种加词后加上栽培品种加词，首字母大写，外加单引号，后不加命名人。如菊花 *Dendranthema morifolium*（Ramat.）Tzvel. 经栽培后，培育出不同的品种，分别命名为亳菊 *Dendranthema morifolium* 'Boju'，滁菊 *D. morifolium* 'Chuju'，贡菊 *D. morifolium* 'Gongju'，小黄菊 *D. morifolium* 'Xiaohuangju' 等。

Nomenclature of cultivated plants must comply with 'International Code for Nomenclature for cultivated plants'. The name of a cultivated plant is capitalized after the species epithet, with a single quotation mark, and nominators of the plant are not added. For instance, medical chrysanthemum cultivated in a long time, produced different varieties and named them, *Dendranthema morifolium* 'Boju', *D. morifolium* 'Chuju', *D.morifolium* 'Gongju', and *D. morifolium* 'Xiaohuangju' etc.

六、命名人
3.6 Nominators

命名人的名字必须用拉丁字母拼写，每个单词的首字母必须大写。当作者的名字较长时，可以缩写，加上缩略号"."，如桑 *Morus alba* Linn.，"Linn." 是 Linnaeus 名字的缩写，也可缩写为 L.。共同发表的学名，把不同的作者用 et 连接，如紫草 *Lithospermom erythrorhizon* Sieb et Zucc.。替代发表的学名，在两作者之间用 ex 连接，如延胡索 *Corydais yanhusuo* W.T.Wang ex Z.Y.Su et C. Y. Wu。植物学名属于新组合的，将被组合的作者的名字用圆括号括起来，如紫金牛 *Ardisia japonica*（Thunb.）Blume。

The names of the authors are commonly spelled in Latin and the first letter of each word must be capitalized. If the author's name is too long, it can be abbreviated, e.g. *Morus alba* L. "L." for Carolus Linnaeus. When two or more authors publish a new species or propose a new name, their names are linked by "et" as in *Lithospermom erythrorhizon* Sieb et Zucc. The names of two or more authors are linked by ex when the first author had proposed a name but was validly published only by the other author, the first author failing to satisfy all or some of the requirements of the Code. e.g.*Corydais yanhusuo* W. T. Wang ex Z. Y. Wu et C. Y. Wu. Epithet of plant a botanical name is followed by a bracket indicating that the scientific name has been recombined. For example, *Ardisia japonica* (Thunb.) Blume. former author is in the bracket who had ever establishd *Bladhia japonica* Thunb. as scientific name, then Karl Ludwig von Blume put it into Ardisia and recombine new name.

第四节 植物界的分门
4 Division of the plant kingdom

按照恩格勒系统，植物界分为 16 门（图 5-1）。通常按照植物是否产生种子分为种子植物（显花植物）和孢子植物（隐花植物）两类。前者包括裸子植物和被子植物，后者包括藻类、菌类、地衣、苔藓萼蕨类植物。根据植物体是否有根、茎、叶的分化，分为原植体植物和茎叶体植物，前者包括藻类、菌类和地衣，后者包括其余各类；按合子是否发育成胚，分为无胚植物和有胚植物，藻类、菌类和地衣属于无胚植物，苔藓、蕨类、裸子植物和种子植物属于有培植物；按照植物体有无维管组织的分化，分为维管植物和无维管植物。苔藓植物、蕨类植物和裸子植物在繁殖过程中产生了颈卵器这一结构，称为颈卵器植物。茎叶体植物和有胚植物又称为高等植物；原植体植物和无胚植物又称为低等植物。

According to Engler's taxonomical system, the plant kingdom is divided into 16 phylum (Fig.5-1). According to whether they produce seeds or not, plants are usually divided into seed plants (flowering plants) and spore plants (cryptogamous plants). The former includes gymnosperms and angiosperms, and the latter includes the other 14 phylum plants. According to the differentiation of roots, stems and leaves, they are divided into protophytes and cormophytes. The former includes algae, fungi and lichens, while the latter includes other plants. According to whether the zygote develops into embryo, they are divided into anemophytes and embryophytes. The former includes algae, fungi and lichens, while the latter includes other kinds of plants. According to whether the plant has vascular tissue or not, they are divided into vascular plants and non vascular plants. Bryophytes, ferns and gymnosperms produce the structure of archegonia in the process of reproduction, which is called archegonia. Cormophytes and embryophyte are also called higher plants; protophytes and non-embryophyte are also called lower plants.

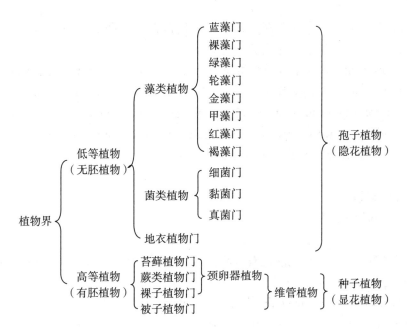

图 5-1　植物界的分门

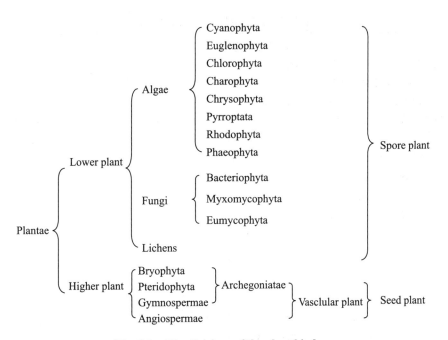

Fig. 5-1　The divisions of the plant kindom

第五节　植物分类检索表

5　Key to plant classification

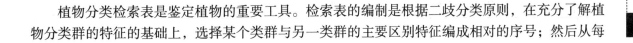

　　植物分类检索表是鉴定植物的重要工具。检索表的编制是根据二歧分类原则，在充分了解植物分类群的特征的基础上，选择某个类群与另一类群的主要区别特征编成相对的序号；然后从每

个类群中找出相对的特征再区分为两类，依次下去直到所需要的分类单位为止。常见的植物分类检索表有定距式、平行式和连续平行式三种类型。

A key is an important tool in plant identification. Based on the dichotomous principle and basis of fully understanding the morphology characters of plant groups, the main distinguishing features of one group from another are sorted and coded as relative numbers, then the relative traits from each group are selected and separated into two groups, and then the desired taxonomical unit appears. There are three types of keys: fixed distance, parallel and continuous parallel.

一、定距式检索表
5.1　Fixed distance key

检索表中每对相对特征，在距左边有等距离处编写同样序号，下一序号后缩一格排列，如此编排，直到编制终点为止（表 5-3）。

Fixed distance key requires each pair of relative traits to write the same sequence number at an equal distance from the left, and the next sequence number will be reduced by a lattice. It is arranged in turn until the end of the plant taxa (Table 5-3).

表 5-3　植物类群检索表（定距式）

1. 植物体构造简单，无根、茎、叶的分化，无胚
　　2. 植物体不为藻类和菌类的共生体
　　　　3. 植物体含叶绿素或其他光合色素，自养生活 ………………………………… 藻类植物
　　　　3. 植物体不含叶绿素或其他光合色素，异养生活 ………………………………… 菌类植物
　　2. 植物体为藻类和菌类的共生体 ………………………………………………… 地衣植物
1. 植物体构造复杂，有根、茎、叶的分化，有胚
　　4. 植物体内无维管组织，生活史中配子体世代占优势 ……………………… 苔藓植物
　　4. 植物体内有维管组织，生活史中孢子体世代占优势
　　　　5. 植物无花，以孢子繁殖 …………………………………………………… 蕨类植物
　　　　5. 植物有花，以种子繁殖
　　　　　　6. 胚珠裸露，不形成果实 ……………………………………………… 裸子植物
　　　　　　6. 胚珠被心皮包被，形成果实 ………………………………………… 被子植物

Table 5-3　A key to some divisions of plant kingdom (Fixed distance key)

1. Plants without differentiation of roots, stems and leaves, no embryo
　　2. Plant body is not a symbiont composed of algae and fungi
　　　　3. Plants contain chlorophyll or other photosynthetic pigments, autotrophic life ……………　Algae
　　　　3. Plant body does not contain chlorophyll or other photosynthetic pigment, heterotrophic life ……Fungus
　　2. Plant is a symbiont composed of algae and fungi……………………………………………　Lichens
1. Plants with differentiation of roots, stems and leaves, and embryo present
　　4. Plant without vascular tissue, and gametophyte have an advantage in life cycle…………　Bryophyta
　　4. Plant with vascular tissue, and sporophyte have an advantage in life cycle
　　　　5. Plant without flowers, and propagate with spores　………………………………Pteridophyta
　　　　5. Plant with flowers, and propagate with seeds
　　　　　　6. Ovule is naked, and not form fruit ………………………………………　Gymnosperm
　　　　　　6. Ovule is closed with carpel, and form fruit………………………………………　Angiosperm

二、平行式检索表
5.2 Parallel key

将每一对相区别的特征编以同样的序号并列在相邻两行，数字号码均写在左侧第一格中，不同的序号排列不退格，每行后面标明应查的序号或分类群（表5-4）。

Each pair of distinguishing features is arranged in the adjacent two lines with the same sequence number, the numbers are written in the first lattice on the left. The different sequence numbers are arranged without backspace, and each line is followed by the sequence numbers or taxonomic group (Table 5-4).

表 5-4 高等植物分门检索表（平行式）

1. 植物体有茎、叶，而无真根 ·······················	苔藓植物门
1. 植物体有茎、叶和真根 ·····························	2
2. 植物以孢子繁殖 ·································	蕨类植物门
2. 植物以种子繁殖 ·································	3
3. 胚珠裸露，不为心皮包被 ····················	裸子植物门
3. 胚珠被心皮构成的子房包被 ·················	被子植物门

Table 5-4 A key to divisions of higher plants (The Parallel key)

1. Plant with stems and leaves, without real roots ····················	Bryophyta
1. Plant with stems, leaves and real roots ····························	2
2. Plants propagate with spore ····································	Pteridophyta
2. Plants propagate with seed·····································	3
3. Ovule is naked, not closed with carpels····························	Gymnosperm
3. Ovule is closed with ovary composed of carpels ····················	Angiosperm

三、连续平行式检索表
5.3 Continued parallel key

编制时将每对相对的特征分列在检索表的不同位置，具有相同特征的分类群邻近排列，依照排列顺序为特征编号。每对特征用两个不同的序号表示，其中后一序号加括号，以表示是相对比的性状（表5-5）。

The relative features of each pair are listed in different positions in the continued parallel key, and the adjacent arrangement of the taxon with the same features was numbered in accordance with the order of arrangement. Each pair of features is represented by two different sequence numbers, and the latter of which is bracketed to indicate traits that are relative to each other (Table 5-5).

表 5-5　高等植物分门检索表（连续平行式）

1.（2）植物体有茎、叶，而无真根 ………………………………………………	苔藓植物门
2.（1）植物体有茎、叶和真根	
3.（4）植物以孢子繁殖………………………………………………………………	蕨类植物门
4.（3）植物以种子繁殖	
5.（6）胚珠裸露，无果实……………………………………………………………	裸子植物门
6.（5）胚珠被心皮构成的子房包被，有果实……………………………………	被子植物门

Table 5-5　A key to divisions of higher plants (The continued parallel key)

1. (2) Plant with stems and leaves, but witout real roots ………………………	Bryophyta
2 .(1) Plant with stems, leaves and real roots	
3 .(4) Plant propagate with spore ………………………………………………	Pteridophyta
4 .(3) Plant propagate with seed	
5 .(6) Ovule is naked, and not form fruit ……………………………………	Gymnosperm
6 .(5) Ovule is closed with ovary, and form fruit …………………………	Angiosperm

　　要正确检索一种药用植物，首先要选择合适的植物分类检索表。其次要掌握检索对象的详细特征，包括营养器官和繁殖器官，特别是花的构造，与检索表进行逐步核对确认种的名称。

　　To correctly identify a medicinal plant, the first step is to select the appropriate key. Secondly, the detailed characters of the retrieval plant need to be mastered, including the vegetative and reproductive organs, especially the flowers, and one by one check the name of the species against the characters described on the key.

重 点 小 结
Summary

　　植物分类的单位主要有界、门、纲、目、科、属、种等；其中界是最大的分类单位，种是基本的分类单位。物种的概念，有分类学物种和生物学种的区别。植物的命名采用"双名法"，并遵从 ICBN 和 ICNCP 和规定。植物分类检索表常见的有定距式、平行式和连续平行式3种。

　　The unit of plant classification mainly includes division, class, order, family, genus and species etc.; the kindom is the largest classification unit, and species is the basic unit. The concept of species has the difference between taxonomic species and biological species. The naming of plants adopts the "binomial method" and complies with the regulations of ICBN and ICNCP. There are three types of plant classification key: fixed distance key, parallel key and continuous parallel key.

　　植物界分门有不同分法，本教材采用二界分类系统，将植物分为低等植物和高等植物，或孢子植物和种子植物进行讲解。

　　There are different ways to divide the plant kingdom. This textbook uses a two-world system to classify plants into lower plants and higher plants, or spore plants and seed plants.

目 标 检 测
Questions

1. 药用植物分类学的主要任务是什么？

What are the main tasks for medicinal plant taxonomy?

2. 什么是双名法？检索表的类型有哪些？

What's the binomial nomenclature? What are the types of the key?

3. 植物界有哪些类群？是如何进化的？其规律有哪些？

What groups are there in the plant kingdom? How did they evolve? What are the classification rules?

4. 调查校园植物的种类及分布，编写校园植物名录。

Investigate the species and distribution of campus plants, and compile a key of campus plants.

第六章 藻类植物
Chapter 6　Algae

学习目标 ┊ Learning goals

1. **掌握** 藻类植物的特征及分类。
2. **熟悉** 常见的药用藻类。
3. **了解** 藻类的经济用途。

- Master the characteristics and classification of algae.
- Be familiar with common medicinal algae.
- Understand the economic uses of algae.

PPT

第一节　藻类植物概述

1　Brief introduction to algae

藻类是一类含光合色素的自养型原植体植物，属于低等植物。藻类植物现约有 3 万种，广布世界各地，大多数生活于淡水或海水中，少数生活于潮湿的土壤、树皮和石头上。

Algae are a group of autotrophic organisms with photosynthetic pigment, belonging to lower plants. About 30,000 species are widely distributed all over the world. Most live in freshwater or seawater, and a few live in moist soil, bark, and stones.

一、形态结构
1.1　Morphology of algae

藻类植物是一类构造简单，没有真正根、茎、叶分化的原植体植物。其植物体大小差异很大，但通常较小，最小者只有几微米，也有较大的，如生长在太平洋中的巨藻，长可达 100m 以上。藻类植物形态上有单细胞、多细胞、丝状体、管状体、叶状体或枝状体等。单细胞藻类如小球藻、衣藻、原球藻等；多细胞藻类丝状的如水绵、刚毛藻等；叶状的如海带、昆布等；树枝状的如马尾藻、海蒿子、石花菜等。

Algae have a simple structure, without real root, stem, or leaf differentiation. Algae plants have the enormous range in size, the smallest algae being only a few microns and the largest algae are found

医药大学堂
WWW.YIYAODXT.COM

only in the Pacific Ocean, some species of which may reach 100m in length. Some algae consist of a single cell, such as single-cell chlorella, Chlamydomonas, Protococcus, etc.; whereas other algae are multicellular or occur as colonies or filaments, such as multicellular filaments, spirogyra, etc.; such as *Laminaria japonica*, *Ecklonia kurome*, etc. with multi-celled leaves; such as *Sargassum enerve*, *S. confusum*, and *Gelidium amansii* that are dendritic.

二、细胞结构
1.2 Cell structure

藻类植物的细胞有原核细胞（蓝藻和绿藻）和真核细胞（其余藻类）两类。具有和高等植物类似的叶绿素、胡萝卜素、叶黄素，还含有藻蓝素、藻红素、藻黄素、藻褐素等其他色素。各种色素构成了藻类特有的色素体（载色体），不同藻类色素体的形状、大小均不同。因色素成分及比例的不同，不同种类的藻体呈现不同颜色。各种藻类通过光合作用制造的养分以及所贮藏的营养物质也是不相同的，如蓝藻贮存蓝藻淀粉、蛋白质粒；绿藻贮存淀粉、脂肪；红藻贮存的是红藻淀粉、红藻糖；褐藻贮存的是褐藻淀粉、甘露醇等。

There are two types of algae cells: prokaryotic cells (blue and green algae) and eukaryotic cells (the rest of the algae). Algae have the same chlorophyll, carotene, and lutein as higher plants, in addition, algae also contain other pigments, such as phycocyanin, phycoerythrin, phycoxanthin, and fucoxanthin. Various pigments form the algae-specific pigment bodies (chromosomes), the shape and sizes of different algae pigment bodies are all different. Due to different pigment components and ratios, different types of algae show different colors. The nutrients produced by photosynthesis of various algae and the stored nutrients are also different. For example, blue algae store cyanophycean starch and protein granules; green algae store starch and fat; red algae store floridean starch and floridose; brown algae store laminarin and mannitol, et al.

视频

三、繁殖和生活史
1.3 Reproduction and life history

藻类植物的繁殖方式一般分为营养繁殖、无性生殖和有性生殖三种。营养繁殖是藻类的主要繁殖方式，有细胞分裂、藻体断裂或营养繁殖小枝、珠芽等。无性生殖产生孢子，产生孢子的囊状结构细胞叫孢子囊。孢子不需要结合，一个孢子可长成一个新个体。孢子主要有游动孢子、不动孢子（又叫静孢子）、拟亲孢子、四分孢子、单孢子、果孢子和厚壁孢子等。有性生殖产生配子，产生配子的囊状结构细胞叫配子囊。在一般情况下，配子必须两两结合成为合子，由合子萌发长成新个体，或由合子经减数分裂产生孢子再长成新个体。根据相结合的两个配子的大小、形状、行为又分为同配、异配和卵配。同配指相结合的两个配子的大小、形状、行为完全一样；异配指相结合的两个配子的形状一样，但大小和行为有些不同，大的不太活泼，叫雌配子，小的比较活泼，叫雄配子；卵配指相结合的两个配子的大小、形状、行为都不相同，大的圆球形，不能游动，特称为卵，小的具鞭毛，活泼，特称为精子。卵和精子的结合叫受精，受精卵即形成合子。合子不在性器官内发育为多细胞的胚，而是直接形成新个体，故藻类植物是无胚植物。

The reproduction of algae is generally divided into three types: vegetative, asexual and sexual reproduction. There are four methods of vegetative reproduction: binary fission, fragmentation, hormogonia and propagules. Asexual reproduction produces spores; a type of sac-like structural cell that produces spores is called sporangia. Spores do not need to bind, and a spore can grow into a new individual. There are three types of spores: planospores (zoospores); aplanospores: hypenospores, statospores, autospores, endospores, exospores; Resting spoers: akinetes. Sexual reproduction produces gametes; a type of sac-like cell that produces gametes is called gamete sacs. Under normal circumstances, gametes must be combined in pairs to form zygotes, which can sprout into new individuals, or spores produced by meiosis through meiosis to grow into new individuals. According to the size, shape, and behavior of the two gametes combined, they are further divided into isogamy, anisogamy and oogamy, these three sexual reproduction types represent three stages of evolution in plants. Male gametes are called sperm cells and female gametes are called egg cells. Gametes are found in different sizes. Some gametes are flagellated and hence they are motile. They are produced in gametangium. Male gametes are produced by spermatogenesis while female gametes are produced by oogenesis. During sexual reproduction, fusion of haploid male and female gametes results in diploid zygotes, which give rise to a new organism. The main difference is that anisogamy is the fusion of gametes in dissimilar size, while isogamy is the fusion of gametes in similar size, and oogamy is the fusion of large, immotile female gametes with small, motile male gametes. Zygotes do not develop into multicellular embryos in the sex organs, but instead form new individuals directly. Therefore, algae plants are embryoless plants.

四、藻类的药用价值
1.4 Medicinal value of algae

藻类植物分布广泛，经济价值较高。许多海洋藻类资源丰富，生长繁殖快，含有丰富的蛋白质、脂肪、碳水化合物、各种氨基酸、多种维生素、抗生素、高级不饱和脂肪酸以及其他活性物质。近年来，从藻类中寻找新的药物或先导化合物，成为研究的热点。陆续发现了具有抗肿瘤、抗病毒、抗真菌、降血压、降胆固醇、防止冠心病和慢性气管炎、抗放射性等广泛生物活性的化合物。海洋藻类将是人类寻找新的药物资源、发展保健食品等的重要资源。我国已知的药用藻类115种。

Algae are widely distributed and have high economic value. Many marine algae are not only rich in resources and growing fast, but also rich in proteins, fats, carbohydrates, various amino acids, multivitamins, antibiotics, higher unsaturated fatty acids and other active substances. In recent years, new drugs or lead compounds from algae have become a focus of research. Compounds with antitumor, antibacterial, antiviral, and antifungal, lowering blood pressure, lowering cholesterol, preventing coronary heart disease and chronic bronchitis, and anti-radiation activities have been discovered successively. Marine algae will be an important resource for humans to find new medicine resources and develop health food. There are 115 species of medicinal algae known in China.

PPT

第二节　藻类植物的分类

2　Classification and representive medicinal plants

根据藻类细胞内所含色素的种类、贮藏物，以及植物体的形态构造、细胞壁的成分、鞭毛数目、着生位置及类型、繁殖方式和生活类型等方面的差异，分为八个门：蓝藻门、裸藻门、绿藻门、轮藻门、金藻门、甲藻门、红藻门、褐藻门。药用植物较多的蓝藻门、绿藻门、红藻门和褐藻门的特征见表6-1. 其主要药用植物介绍如下。

According to the differences in the types of pigments contained in algae cells, their storaged nutrients, and the morphology and structure of plant bodies, cell wall components, flagellum numbers, locations and types, reproduction and life types, they are divided into eight divisions: Cyanophyta, Euglenophyta, Chlorophyta, Charophyta, Golden algae, Dinoflagellate, Rhodophyta, and Brown Algae. Those four divisions as Cyanophyta, Chlorophyta, Rhodophyta and Brown Algae, which have more medicinal plants, are introduced as follows (Table 6-1).

表 6-1　藻类植物 4 个常见门的主要特征

门	细胞核	色素种类	贮藏物质	细胞壁成分	繁殖方式	鞭毛	生境	种类
蓝藻门	无细胞核；无染色体；有裸露的环状 DNA	叶绿素 a；β-胡萝卜素；叶黄素；藻胆素	蓝藻淀粉	肽聚糖；果胶酸；黏多糖	营养繁殖；无性生殖	无	分布广泛	1500~2000
绿藻门	真核	叶绿素 a，b；α，β-胡萝卜素；叶黄素	淀粉	纤维素；果胶	营养繁殖；无性生殖；有性生殖（同配，异配，卵配）	2~4根等长鞭毛，顶生	分布广泛，多在淡水中	约8600
红藻门	真核	叶绿素 a，d；α，β-胡萝卜素；藻红素；叶黄素	红藻淀粉；红藻糖	纤维素；果胶	营养繁殖；无性生殖；有性生殖（卵配）	无	多分布于海水中	约4000
褐藻门	真核	叶绿素 a，c；β-胡萝卜素；墨角藻黄素	褐藻淀粉；甘露醇	纤维素；藻胶酸；褐藻糖胶	营养繁殖；无性生殖；有性生殖（同配，异配，卵配）	2根不等长鞭毛	多分布于海水中	约1500

Table 6-1　Main characteristics of 4 common phylums of algae

Division	Nucleus	Types of pigments	Storaged nutrients	Components of the cell wall	Reproduction	Flagellum	Habitat	Species
Cyanophyta	No nuclear envelope or nucleolus, contains naked circular DNA. No chromosomes	chlorophyll a, β-carotene, lutein and phycobilin	cyanophycean starch	Peptidoglycan, pectic acid and mucopolysaccharide	Vegetative/asexual reproduction	/	widely distributed	1500~2000
Chlrophyta	Eukaryotic cells	Chlorophyll a and b, α-carotene, β-carotene, and some lutein	Starch	Cellulose, pectin	Vegetative/asexual/sexual reproduction (anisogamy, Isogamy and oogamy)	2 or 4 terminally-equipped	widely distributed, mostly in freshwater	About 8600
Rhodophyta	Eukaryotic cells	chlorophyll a and d, α- and β-carotene, lutein, phycoerythrin	Red floridean starch, floridose	Cellulose, pectin	Vegetative/asexual/sexual reproduction (oogamy)	/	Mostly distributed in seawater	About 4000
Phaeophyta	Eukaryotic cells	chlorophyll a and c, β-carotene, fucoxanthin	laminarin and mannitol	Cellulose, algin, algin fucoiin	Vegetative/asexual/sexual reproduction (anisogamy, Isogamy and oogamy)	2 unequal flagella	Mostly distributed in seawater	About 1500

一、蓝藻门
2.1　Cyanophyta

葛仙米 *Nostoc commune* Vauch. 念珠藻科。植物体念珠状。分布于全国各地，生于湿地或地下水位较高的草地上，民间习称"地木耳"。食用和药用，能清热、收敛、明目。钝顶螺旋藻 *Spirulina platensia*（Nordst.）Geitl. 颤藻科。藻体卷曲状，富含蛋白质、维生素等多种营养物质，可用于治疗营养不良症及增强免疫力，是生产保健品的重要原料。目前许多国家和地区已进行人工养殖。

Nostoc commune Vauch**.** Nostocaceae, the plant body shaped like prayer beads. Distributed all over the country in wetlands or grasslands with high groundwater levels. It can be used as food and medicine. It can clear heat, astringent and improve eyesight.

Spirulina platensia (Nordst.) Geitl. Oscillatorae. The algal body is curled and rich in protein, vitamins and et al. It can be used to treat malnutrition and enhance immunity. It is an important raw material for the production of health products. At present, many countries and regions have carried out artificial breeding.

其他常见的药用植物还有：海雹菜 *Brachytrichia quoyi*（C.Ag.）Born.et Flah.、苔垢菜 *Calothrix crustacea*（Chanv.）Thur. 均具有解毒、利水之功效。发菜 *Nostoc flagilliforme* Born. et Flah. 产于我国西北地区，具有解毒清热、理肺化痰、调理肠胃的作用。为我国一级重点保护野生植物。

Other common medicinal plants include: *Brachytrichia quoyi* (C. Ag.) Born. Et Flah, and *Calothrix crustacea* (Chanv.) Thur., all of which have detoxifying and water-relieving effects. *Notoc flagilliforme* Born. Et Flah is distributed in northwestern China and has the functions of detoxifying, clearing heat, regulating lungs and phlegm, and regulating gastrointestinal tract. It is a first-level key protection of wild plants in China.

二、绿藻门
2.2　Chlrophyta

蛋白核小球藻 *Chlorella pyrenoidosa* Chick（图 6-1）小球藻科，为淡水单细胞植物。藻体富含蛋白质、维生素 C、维生素 B 和抗生素（小球藻素），药用能治疗水肿、贫血、肝炎等，也可作营养品。同时也是研究光合作用的实验材料。石莼 *Ulva lactuca* L. 石莼科。供食用，叫"海白菜"。药用能软坚散结、清热解毒、祛痰、利水。水绵 *Spirogyra nitida*（Dillow）Link. 双星藻科。

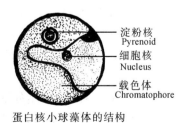
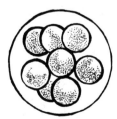

淀粉核　Pyrenoid
细胞核　Nucleus
载色体　Chromatophore

蛋白核小球藻体的结构
The structure of *Chlorella pyrenoidosa*

拟亲孢子的形成　　拟亲孢子的释放
The formation(*left*) and release(*right*) of quasiphilic spores

图 6-1　蛋白核小球藻
Fig. 6-1 *Chlorella pyrenoidosa* Chick.

能治疮疡、烫伤。

Chlorella pyrenoidosa Chick. Chlorellaceae. It is a freshwater single-celled plant. The algal body is rich in protein, vitamin C, vitamin B and antibiotics (chlorella). It can treat edema, anemia, and hepatitis, etc., and be used as nutrients. It is also an experimental material for studying photosynthesis (Fig.6-1).

Ulva lactuca L. Ulvaidae. It is edible and is called "sea cabbage". It can soften and remove hard mass, clear heat and remove toxicity, expectorant, and induce diuretic. *Spirogyra nitida* (Dillow) Link. Zygnemataceae., used to cure sores and scald.

三、红藻门
2.3 Rhodophyta

石花菜 *Gelidium amansii* Lamouroux（图 6-2）石花菜科。藻体扁平直立，丛生，4~5 羽状分枝，小枝对生或互生。分布于渤海、黄海、台湾北部。藻体具有清热解毒和缓泻作用，亦可食用。还可供提取琼胶（琼脂）用于医药、食品和作细菌培养基。

Gelidium amansii Lamx. Gelidiaceae. The algae body is flat and erect, clustered, 4 to 5 pinnate branches, opposite or alternate branchlets. Distributed in the Bo-hai Sea, the Yellow Sea, and the northern coast of Taiwan. Algae body has heat-clearing and removing toxicity and laxative effects, and is also edible. It can also be used for extracting agar (agar) for medicine, food and bacterial culture (Fig.6-2).

甘紫菜 *Porphyra tenera* Kjellm.（图 6-2）红毛菜科。藻体为薄叶片状、卵形、竹叶形或不规则圆形。分布于辽东半岛至福建沿海，现有大量栽培。藻体能清热利尿、软坚散结、消痰。也供食用。

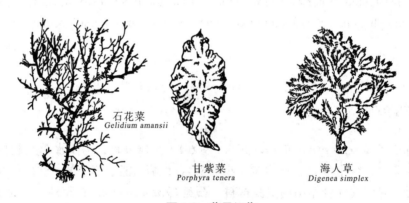

石花菜
Gelidium amansii

甘紫菜
Porphyra tenera

海人草
Digenea simplex

图 6-2　药用红藻
Fig. 6-2　Medicinal rhodophyta plants

Porphyra tenera Kjellm. Bangiaceae.Thin leaf-shaped, oval, bamboo leaf-shaped or irregularly round. Distributed from the Liaodong Peninsula to the coast of Fujian, there is a large amount of cultivation. Algae can clear heat and diuretic, soften and resolve hard mass, and reduce phlegm, can also serve as food.

其他常见药用植物还有：琼枝 *Eucheuma gelatinae*（Esp.）J. Ag. 藻体能清热解毒、缓泻、降脂，也可食用及作微生物培养基。鹧鸪菜（美舌藻）*Caloglossa leprieurii*（Mont.）J. Ag. 藻体能驱蛔、化痰、消食。海人藻 *Digenea simplex*（Wulf.）C. Ag. 藻体能驱蛔虫、鞭虫、绦虫。

Other common medicinal plants are: ***Eucheuma gelatinae* (Esp.) J. Ag.** can clear away heat and detoxify, relieve diarrhea, reduce fat, and can also be used as edible and microbial culture medium.

The algae body of **_Caloglossa leprieurii_** (Mont.) J. Ag. can drive roundworms, reduce phlegm, and indigestion. The algae body of **_Digenea simplex_ (Wulf.) C. Ag.**can kill roundworms, whipworms and tapeworms.

四、褐藻门
2.4　Phaeophyta

海带 *Laminaria japonica* Aresch. 海带科。孢子体较大，褐色，长 2~4m，由固着器、带柄和带片三部分。固着器呈分枝的根状，固着于岩石或其他物体上；柄不分枝，圆柱形或略侧扁，内部组织分化为表皮、皮层和髓三层；带片生长于柄的顶端，不分裂，没有中脉，幼时常凸凹不平，内部构造和柄相似，也分为三层。分布于辽宁、河北、山东沿海。目前海带人工养殖已推广到长江以南的浙江、福建、广东等省沿海。植物体（昆布）能软坚散结、消痰、利水消肿；也可食用或作提取碘和褐藻淀粉的原料。翅藻科的昆布 *Ecklonia kurome* Okam. 其功效与海带相同，作昆布入药。

Laminaria japonica Aresch. Laminariaceae. There are two types of the frond: sporophyte and gametophyte. The sporophyte is large, brown, 2–4m long, and consists of three parts: holdfast, stipe, and blade. The holdfast are branched roots, which holds the seaweed to rocks and other objects; the stalks are unbranched, cylindrical or slightly flattened, and the internal tissues are differentiated into epidermis, cortex, and medulla; the blade grows at the apex of the stipe, does not divide, has no midribs, and is often uneven when young. The internal structure is similar to the stipe, and it is also divided into three layers. Distributed along the coasts of Liaoning, Hebei and Shandong. At present, kelp culture has been extended to the coasts of Zhejiang, Fujian, and Guangdong provinces in south of the Yangtze River. Kelp (Kunbu) can soften and resolve hard mass, eliminate phlegm, and induce diuresis to alleviate edema. It is edible, can also be used as a raw material for extracting iodine and laminarin. *Ecklonia kurome* Okam. has the same medicinal effect as kelp.

海蒿子 *Sargassum pallidum*（Turn.）C. Ag.（图 6-3）马尾藻科。藻体高 30~60cm，褐色。固着器盘状，主干多单生，圆柱形，两侧有羽状分枝。藻"叶"形态变化较大，初生"叶"披针形至倒披针形；次生"叶"线形至披针形，有时羽状分裂。小枝末端常有圆球形的气囊。分布于我国黄海、渤海沿岸，生于潮线下 1~4m 的岩石上。藻体（海藻）能软坚散结、消痰、利水，药材称"大叶海藻"。羊栖菜 *S. fusiforme*（Harv.）Setch. 分布于辽宁至海南，长江以南沿海较多；藻体也作海藻入药，药材称"小叶海藻"。

Sargassum pallidum (Turn.) C. Ag. Sargassaceae. The algal body is 30~60cm tall and brown. The holdfast is disc-shaped; the trunk is solitary mostly, cylindrical, with feathery branches on both sides. The morphology of the "leaf" of the algae varies greatly. The primary "leaf" is lanceolate to oblanceolate, 5–7 cm long and 2–12cm wide, with sparsely serrated margin; the secondary "leaf" is linear to lanceolate, sometimes pinnate lobed. The side branches grow from the axils of the secondary "leaf". Distributed on the coasts of the Yellow Sea and Bohai Sea in China, and is born on rocks 1~4m below the tide line. It can soften and resolve hard mass, eliminate phlegm, and induce diuresis. The crude drug is called "Daye Haizao". **_S. fusiforme_ (Harv.) Setch.** are distributed from Liaoning to Hainan. Algae body is also used as "Xiaoye Haizao" (Fig.6-3).

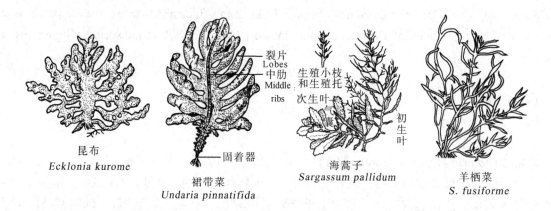

裂片
Lobes
中肋
Middle
ribs

生殖小枝
和生殖托
次生叶

初生叶

昆布
Ecklonia kurome

固着器

裙带菜
Undaria pinnatifida

海蒿子
Sargassum pallidum

羊栖菜
S. fusiforme

图 6-3　常见药用褐藻
Fig. 6-3　Medicinal phaeophyta plants

重 点 小 结
Summary

　　藻类是一群含光合色素的自养型原植体植物，属于低等植物。藻类形体较小、构造简单，常为单细胞、多细胞群体、丝状体、叶状体或枝状体，没有真正的根、茎、叶的分化，仅有少数具有组织分化和类似根、茎、叶的构造。《中国药典》（2020 版第一部）收载的药用藻类有 4 种，常用中药有海藻和昆布。

　　Algae are a group of autotrophic organisms with photosynthetic pigment, belonging to lower plants. Algae are small and have simple structures. They are usually single cells, multicellular populations, filaments, thallus, or dendrites. There is no differentiation of true root, stem or leaf. Only a few have tissue differentiation and similar structures of roots, stems and leaves. There are four kinds of medicinal algae included in the Chinese Pharmacopoeia (2020 edition). Commonly used medicinal algae include seaweed and kelp.

目 标 检 测
Questions

1. 药用藻类的四个门主要区别点有哪些？
What are the main differences between the four divisions of medicinal algae?
2. 药用藻类植物有什么开发利用前景？
What are the development prospects of medicinal algae?

题库

第七章 菌类植物
Chapter 7　Fungi

学习目标 | Learning goals

1. **掌握** 真菌门的特征及代表药用真菌。
2. **熟悉** 子囊菌亚门和担子菌亚门的区别及营养方式。
3. **了解** 细菌、黏菌和真菌的区别。

- Know the characteristics and representative medicinal fungi.
- Be familiar with the differences between Ascomycota and Basidiomycota and their nutritional methods.
- Understand the differences between bacteria, slime molds, and fungi.

　　菌类与藻类一样，没有组织和根、茎、叶的分化。但菌类不含光合作用色素，是通过异养方式生活的一类低等植物。其异养生活方式多样，有腐生、寄生、共生等，多数种类营腐生生活。菌类具有形体微小、生长旺盛、繁殖快、代谢类型多、适应能力强、易变异、种类多和分布广的特点，在土壤、水体、空气、人和动植物体里都有它们的踪迹。现已知的黏菌类植物约 10 万种。在两界分类系统中，菌类属于植物界，分为细菌门、真菌门和黏菌门。这三门植物的形态、结构、繁殖和生活史差别很大，彼此并无亲缘关系。在四、五界系统中，它们分属于原核生物界、真菌界和原生生物界。

　　Fungi, like Algae, are lower plants without differentiation of tissue, root, stem or leaf. However, fungi do not contain photosynthetic pigments, and cannot perform photosynthesis, so the nutrition of fungi is heterotrophy. Heterotrophic lifestyles of fungi include saprophytism, parasitism, symbiosis, et al., most species live on saprophytism. Those who draw nutrients from living animals and plants are called parasitism; those who draw nutrients from dead animals and plants or other inanimate organisms are saprophytism; those who obtain nutrients from living organisms while providing the living body with favorable living conditions are symbiosis. Symbiosis is a close relationship between two organisms of different kinds which benefits both organisms. Due to the diversity of lifestyles, fungi are widely distributed. In addition, the fungi have the characteristics of small size, vigorous growth, and fast reproduction, multiple metabolic types, strong adaptability, easy mutation, et al. They can be found in the soil, water, air, plants, even in animals and human. There are about 100,000 species of existing fungi. According to the Two Kindom System, fungi are grouped into Plant Kindom and can be divided into Bacteriophyta, Eumycophyta and Myxomycophyta. Those three kinds of fungi differ from morphology, structure, reproduction and life history, and are not related to each other. In the Four or Five Kindom

PPT

System, they belong to the prokaryote, fungal and protozoa kingdom.

第一节　细菌、黏菌和放线菌
1 Bacteria, myxomycophyta and actinomyces

一、细菌
1.1　Bacteria

细菌是微小的单细胞有机体，无细胞核，属于原核生物，营养方式有异养和自养两种，异养包括寄生和腐生；自养包括化能自养（如硝化细菌、硫细菌等）和光合自养（如着色杆菌、外硫红螺菌、紫色非硫细菌等）。细菌按基本形态分为球菌、杆菌和螺旋菌。杆菌是自然界最常见的细菌。细菌按照细胞壁组成的差异可分为革兰阳性菌和革兰阴性菌。广布于空气、土壤、水和其他生物的体内、体外。细菌已在微生物学中详细讲述，本书不再叙述。

Bacteria are tiny single-celled organisms without nuclear structure. They belong to prokaryotes and have two types of nutrition: heterotrophic and autotrophic. Heterotrophic includes parasitism and saprophytism; autotrophic includes chemoautotrophic (such as nitrifying bacteria, sulfur bacteria, etc.) and photosynthetic autotrophic (such as Bacillus chromobacterium, Rhodospirillum exophytica, purple non-sulfur bacteria, etc.). Bacteria are divided into Coccus, Bacillus, and Spirilla according to their basic forms. Bacillus is the most common bacteria in nature. Bacteria can be divided into Gram-positive and Gram-negative bacteria according to the difference in cell wall composition. They are widely distributed in the air, soil, water and other organisms. Bacteria have been described in detail in Microbiology and will not be described in this book.

二、黏菌
1.2　Myxomycophyta

黏菌是介于真菌和原生动物之间的真核生物，在营养期为裸露的无细胞壁而具多核的原生质团，不含叶绿素，细胞质流动使黏菌呈现类似变形虫的运动状态。但在繁殖期产生具纤维素细胞壁的孢子，孢子再发育成新的个体。黏菌营养期的结构、行为和吞噬方式摄食的特征与原生动物相似，其繁殖方式又与真菌类似。大多数黏菌为腐生菌，无直接的经济意义。

Myxomycophyta are eukaryotic organisms between fungi and protozoa. They are bare cell wall-free and multinucleated protoplasts during the growth phase or vegetative period. They do not contain chlorophyll, and the cytoplasmic flow makes the slime mold behave like amoeba. However, spores produced during reproduction have cellulose cell walls, and the spores develop into new individuals. The structure, action and ingestion of myxomycetes are similar to that of protozoa, and their reproduction is similar to that of fungi. Most myxomycetes are saprophytism and have no direct economic value.

医药大学堂
WWW.YIYAODXT.COM

三、放线菌
1.3 Actinomyces

放线菌是细菌与真菌之间的过渡类型，也是单细胞的丝状菌类，大多数有发达的分枝菌丝，其结构和细胞壁化学组成与细菌类似（图7-1）。在显微镜下，放线菌呈分枝丝状。放线菌在自然界分布很广，空气、土壤、水源中都有放线菌存在；在土壤中，尤其是富含有机质的土壤里存在较多。

Actinomyces are single-celled filamentous fungi and a transitional form between bacteria and fungi. Most of them have developed branched hyphae. Their structure and chemical composition of cell wall are similar to those of bacteria. Under the microscope, actinomyces are branched and filamentous. Actinomyces are widely distributed in nature and can live in air, soil and water. They are abundant in soil, especially in soil rich in organic matter (Fig. 7-1).

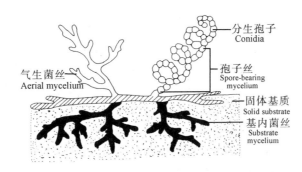

图 7-1 链霉菌的一般形态和构造
Fig. 7-1　The general morphology and structure of Streptomyces

视频

放线菌绝大多数为腐生，少数寄生，易引起动、植物的病害。一些放线菌是抗生素的重要产生菌，它们能产生多种抗生素。迄今为止，已发现的4000多种抗生素中，有约70%是放线菌产生的。如红霉素由红霉素链霉菌 *Streptomyces erythreus* Waksman et Henrici 产生，链霉素由灰色链霉菌 *S. griseus*（Krainsky）Waksman et Henrici 产生，氯霉素由委内瑞拉链霉菌 *S.venezuelae* Ehtlich et Al. 产生，卡那霉素由卡那霉素链霉菌 *S. kanamyceticus* Okami et Umezawa 产生，金霉素、四环素由金色链霉菌 *S.aureofaciens* Duggar 产生，达托霉素由玫瑰孢链霉菌 *S. roseosporus* 产生，万古霉素有东方拟无枝酸菌 *Amycolatopsis orientalis* 产生；抗真菌药制霉菌素由诺尔斯链霉菌 *S. noursei* 产生，两性霉素 B 由结节链霉菌 *S. nodosus* 产生。近年来，还从放线菌的次生代谢产物中筛选出一些抗癌药、酶抑制剂和免疫抑制剂等，如抗癌药新生霉素由浑球链霉菌 *S. sphaeroides* 产生，博来霉素由轮枝链霉菌 *S. reticillus* 产生，丝裂霉素 C 由头状链霉菌 *S. caespitocus* 产生；免疫抑制剂雷帕霉素由吸水链霉菌 *S. hygroscopicus* 产生等。

Most actinomyces are saprophytism, and a few are parasitism, which often cause animal and plant diseases. Actinomyces are important producers of antibiotics. They can produce a wide variety of antibiotics. So far, about 70% of the 4,000 antibiotics found are produced by actinomyces. For example, Erythromycin is produced by *Streptomyces erythreus* Waksman et Henrici, Streptomycin is produced by *S. griseus* (Krainsky) Waksman et Henrici, and Chloromycetin is produced by *S. venezuelae* Ehtlich et Al. Kanamycin is produced by *S. kanamyceticus* Okami et Umezawa, Chlortetracycline and Tetracycline are produced by *S. aureofaciens* Duggar, and Daptomycin is produced by *S. roseosporus*, Vancomycin is produced by *Amycolatopsis orientalis*; The antifungal drug Nystatin is produced by *S. noursei*, and Amphotericin B is

PPT

produced by *S. nodosus*. In recent years, some anticancer drugs, enzyme inhibitors, and immunosuppressive agents have also been selected from the secondary metabolites of actinomyces. For example, the anticancer drug Novobiocin is produced by *S. sphaeroides*, Bleomycin is produced by *S. reticillus*, Mitomycin C produced by *S. caespitocus*; immunosuppressive agent Rapamycin is produced by *S. hygroscopicus*, etc.

第二节　真菌门

2 Eumycophyta

一、主要特征
2.1　General features

真菌属真核异养生物，常由分枝繁茂的多细胞细丝（菌丝）组成，称菌丝体。大多具有坚硬的由几丁质组成的细胞壁，但真菌细胞壁成分极其复杂，可随年龄和环境条件而变化。有些真菌的细胞壁因含化学成分种类不同，使细胞壁呈黑色、褐色或其他颜色，因此菌体也呈现各种颜色。真菌营养体除大型菌外，分化很小，高等大型真菌有定形的子实体。

The fungus is a kind of heterotrophic eukaryote with branched, multicellular filamentous body, most of which have solid cell wall composed of chitin. However, the composition of the cell wall is extremely complex and can change with age and environmental conditions. The cell walls of some fungi are black, brown, or other colors due to the different types of substances they contain, so the cells also show various colors. Except for macrofungi, other fungal vegetative bodies have very little differentiation, and higher macrofungi have shaped sporophores.

1. 营养体　无定型的菌丝体即为真菌的营养体，其贮存的养分主要是肝糖，还有少量蛋白质、脂肪以及微量的维生素。除少数单细胞真菌（如酵母）外，绝大多数真菌的营养体由菌丝构成，菌丝按结构不同分为无隔菌丝和有隔菌丝。无隔菌丝是一个多核长管形细胞，有分枝或无，低等真菌的菌丝一般为无隔菌丝；有隔菌丝中由隔膜把菌丝隔成许多细胞，每个细胞内含 1 至多个核，是高等真菌的菌丝类型（图 7-2）。

(1) Vegetative body　The storage nutrients of fungus are mainly glycogen, as well as a small amount of protein, fat and trace vitamins. Except for a few single-celled fungi (such as yeast), the vegetative body of most fungi is composed of slender tubular hyphae. According to structure, hyphae include nonseptate hyphae and septate hyphae. Nonseptate hyphae is a long tubular cell with branches or without, most of which are multinucleated cell. The hyphae of lower fungi are generally nonseptate, and only produce a fully-closed septum when injured or produces reproductive structures. There is a septum in the septate hyphae to separate the hypha into many cells; each cell containing one or more nucleus, the hyphae of higher fungi is mostly septate (Fig. 7-2).

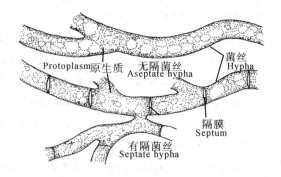

图 7-2　菌丝
Fig. 7-2　Different kinds of hypha

医药大学堂
WWW.YIYAODXT.COM

真菌在繁殖或环境条件不良时，菌丝常相互密结，形成两种组织：拟薄壁组织和疏丝组织，再构成菌丝组织体，常变态为四种形态。

根状菌索是菌丝体密结呈绳索状，外形似高等植物的根，具有促进菌体蔓延和抵御不良环境的功能，一般生于树皮下或地下。在引起木材腐朽的担子菌最普遍。

菌核由菌丝密结成颜色深、质地坚硬的核状休眠体。

子座是容纳子实体的褥座，是从营养阶段到繁殖阶段的过渡形式。

子实体是高等真菌在繁殖时期形成的能产生孢子的菌丝组织体。

When the fungus reproduces or the environmental conditions are poor, the hypha are often densely formed with each other, and forming two kinds of tissues: pseudoparenchyma and prosenchyma, and then forming a mycelium, which usually changes into three morphology: ① Rhizomorph, mycelium is a long bundle of tightly-shaped ropes, which looks like roots of higher plant. It has the function of promoting the spread of fungus body and resisting the bad environment. It is usually born under the bark or underground. Rhizomorph is most common among the Basidiomycotina that cause wood decay. ② Sclerotium, the hypha is densely formed into a dormant body with dark color and hard texture. ③ Stroma, a mattress that contains sporophore, a transitional form from the vegetative stage to the reproduction stage. ④ Sporophore, it is also a kind of mycelium, which can produce spores.

2. 营养方式　多数真菌营腐生或寄生生活，部分真菌属于共生真菌，少数为捕食真菌。真菌主要利用菌丝吸收养分，腐生菌可由菌丝直接从基质中吸收养分，也可产生假根用于吸收养分；寄生的真菌通过直接与寄生细胞的原生质接触而吸收养分。胞间寄生的真菌则利用从菌丝体上特化产生的吸器伸入寄主细胞内吸取养料。吸取养料的过程是首先借助于多种水解酶（均是胞外酶），把大分子物质分解为可溶性的小分子物质，然后借助于较高的渗透压吸收。寄生真菌的渗透压一般比寄主高2~5倍，腐生菌的渗透压更高。

(2) Nutrition　Most of fungi live by saprophytism or parasitism, some fungi are symbiosis, and a few are predacious. Fungus mainly uses hyphae to absorb nutrients. Saprophytic fungi can absorb nutrients directly from the matrix by hyphae, and can also produce rhizoid to absorb nutrients; fungi that are parasitic in host cells absorb nutrients by directly contacting the protoplasm of the parasitic cells. The intercellular parasitic fungus uses a specialized haustorium from the mycelium to reach into the host cells to absorb nutrients. The process of absorbing nutrients is to first break down macromolecules into soluble small molecules by various hydrolytic enzymes (extracellular enzymes), and then absorb them through higher osmotic pressure. The osmotic pressure of parasitic fungus is generally two to five times higher than that of the host, and the osmotic pressure of saprophytic fungus is higher.

3. 细胞结构　真菌细胞由细胞壁、细胞膜、细胞质、细胞核组成，细胞质中分散着线粒体、核糖体、内质网、高尔基体、液泡、溶酶体等多种细胞器。细胞壁的主要成分为己糖或氨基己糖构成的多糖，如几丁质、纤维素、葡聚糖、甘露聚糖、半乳聚糖等。此外，还有一些特殊构造，如鞭毛、膜边体、微体、壳质体、伏鲁宁体等。

(3) Cell structure　Fungal cells are composed of cell walls, cell membranes, cytoplasm, and nucleus. Mitochondria, ribosomes, endoplasmic reticulum, Golgi body, vacuoles, and lysosomes are dispersed in the cytoplasm. The main component of the cell wall is hexose or aminohexose, such as chitin, cellulose, dextran, mannan, galactan, etc. In addition, there are some special structures, such as flagella, membrane border, microbody, chitin, fulunin, etc.

4. 真菌的繁殖　真菌繁殖的方式多种多样，并涉及很多不同类型的孢子。主要的繁殖方式有营养繁殖、无性生殖和有性生殖。

(4) Reproduction of fungus The reproduction of fungi is various and produces many different types of spores. The main reproduction methods are vegetative reproduction, asexual reproduction and sexual reproduction.

营养繁殖：少数单细胞真菌如裂殖酵母菌属（*Schizosaccaromyces*）主要通过细胞分裂产生子细胞进行繁殖。大部分真菌可以通过产生芽生孢子、厚壁孢子或节孢子等进行营养繁殖。芽生孢子是从一个细胞的一定部位突起形成出芽，芽生孢子脱离母体后即长成一个新个体；厚壁孢子是由菌丝中个别细胞膨大形成的休眠孢子，其原生质浓缩，细胞壁加厚，渡过不良环境后，再萌发为菌丝体；节孢子是由营养菌丝断裂形成的。

Vegetative reproduction: A few single-celled fungi like *Schizosaccharomyces*, mainly produce daughter cells through cell division, and most fungi undergo vegetative reproduction by producing blastospore, chlamydospore or arthrospores. Blastospore sprout from a certain part of a cell, and the blastospore grow into a new individual when they leave the mother body. Chlamydospore is dormant spores formed by the expansion of individual cells in the hypha, concentrated protoplasts and thickened cell walls of chlamydospore. After poor environment, they germinate into mycelium; arthrospores are formed by breaking vegetative mycelium.

无性生殖：真菌在无性生殖过程中主要形成多种不同类型的孢子，包括游动孢子、孢囊孢子和分生孢子等（图 7-3）。游动孢子是水生真菌产生的借水传播的孢子，无壁，具鞭毛，能游动，在游动孢子囊中形成；孢囊孢子是在孢子囊内形成的不动孢子，借气流传播；分生孢子是由分生孢子囊梗的顶端或侧面产生的一种不动孢子，借气流或动物传播。

Asexual reproduction: The fungal asexual reproduction is also extremely developed, and many different types of spores are formed during the asexual reproduction process, including zoospores, sporangiospore, and conidium (Fig.7-3). Zoospores are water-borne spores produced by aquatic fungi. They are wall-less with flagella, can swim, and form in zoosporangium. Sporangiospores are immobile spores formed in sporangium, which spread by air flow; conidia are an aplanospore produced by the top or sides of conidiophores, and it dispersed by air currents or animals.

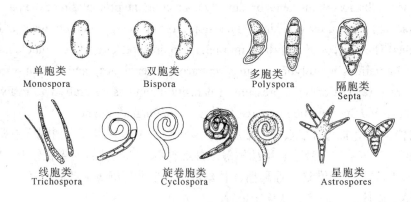

单胞类 Monospora   双胞类 Bispora   多胞类 Polyspora   隔胞类 Septa

线胞类 Trichospora   旋卷胞类 Cyclospora   星胞类 Astrospores

图 7-3　各种类型的分生孢子
Fig. 7-3　Various types of conidia

有性生殖：有些真菌产生单细胞的配子，以同配或异配的方式；一些真菌通过两性配子囊的结合形成"合子"，这种类型的合子习惯上称之为接合孢子或卵孢子。子囊菌有性配合后，形成子囊，在子囊内产生子囊孢子。担子菌有性生殖后，在担子上形成担孢子。担孢子和子囊孢子是有性结合后产生的孢子和无性生殖的孢子完全不同。

Sexual reproduction: The sexual reproduction of fungi is also extremely diversified. Some fungi can produce single-cell gametes and reproduce through isogamy or heterogamy. Other fungi are combined to "Zygote" by bisexual gametagium, this zygote is customarily called zygospore or oospore. After sexual production of ascomyces, ascus is formed and ascospores are produced in the ascus. In the basidiomycete, the basidiospores are borne on the tips of the sterigmata, which are outgrowths of the basidium. Basidiospores and ascospores are spores produced after sexual combination, and are completely different from those of asexual reproduction.

5. 真菌的分布 广布于世界各地，从大气、土壤、水体到动植物及其残体均有其踪迹。目前已知约 12 万种，其中药用真菌约 500 种，对人类健康有害的真菌有 300 多种。

(5) Distribution of fungus They are distributed all over the world widely, from the atmosphere, soil, water bodies to animals, plants and their debris. About 120,000 species are currently known, of which about 500 are medicinal fungi, and more than 300 are harmful to human health.

二、真菌的主要分类群
2.2 Main taxa of fungi

真菌分类常依据形态学、细胞学、生理学、生态学等特征，尤其是有性繁殖阶段的形态特征。真菌分类系统较多，本教材采用安斯沃兹等（Ainsworth 1973）系统，将真菌分为五个亚门，即鞭毛菌亚门、接合菌亚门、子囊菌亚门、担子菌亚门、半知菌亚门（表 7-1）。药用真菌以子囊菌亚门和担子菌亚门为多见，少数为半知菌亚门。

Fungal classification is often based on morphological, cytological, physiological, ecological and other features, especially the morphological characteristics of the sexual reproduction stage. There are many fungal classification systems. This textbook uses Ainsworth system (Ainsworth 1973) to divide the fungi into five subdivisions, namely the Mastigomycotina, Zygomycotina, Ascomycotina, Basidiomycotina, Deuteromycotina (Table 7-1). The most common medicinal fungi are from Ascomycotina and Basidiomycotina, and a few are Deuteromycotina.

表 7-1 真菌亚门检索表

1. 有能动孢子；有性阶段的孢子典型为卵孢子 ················· 鞭毛菌亚门
1. 无能动孢子
  2. 有性阶段
    3. 有性阶段孢子为接合孢子 ················· 接合菌亚门
    3. 无接合孢子
      4. 有性阶段孢子为子囊孢子 ················· 子囊菌亚门
      4. 有性阶段孢子为担孢子 ················· 担子菌亚门
  2. 缺有性阶段 ················· 半知菌亚门

Table 7-1 The Key of fungal subphylum

1. Having motile spores; spores of the sexual stage are typically oospores ·········· Mastigomycotina
1. Without motile spores
  2. Having sexual stage
    3. The spores are zygotes at the sexual stage ·········· Zygomycotina
    3. Without zygotes
      4. The spores in sexual stage are ascospores ·········· Ascomycotina
      4. The spores in the sexual stage are basidiospores ·········· Basidiomycotina
  2. No sexual stage ·········· Eduteromycotina

（一）子囊菌亚门

2.2.1 Ascomycotina

子囊菌亚门是真菌中种类最多的一个亚门，全世界有2720属，28000多种，除少数低等子囊菌为单细胞（如酵母菌）外，绝大多数为有隔菌丝组成的菌丝体。子囊菌的无性生殖发达，裂殖、芽殖或形成各种孢子，如分生孢子、厚垣孢子等。有性生殖产生子囊，内生子囊孢子，这是子囊菌亚门最主要的特征。不同真菌子囊孢子的数目不等，通常8个。

Ascomycotina is the most abundant subdivision of fungi. There are 2720 genera and more than 28,000 species in the world. Except for a few lower ascomycetes that are single cells (such as yeast), most have developed hypha. The hypha has septum and is tightly bound together to form a certain structure. Asexual reproduction of ascomycetes is particularly developed, with fission, budding, or formation of various spores, such as conidium, akinete, etc. Sexual reproduction produces ascus. Ascospore is produced in ascus, which is the most important characteristic of the Ascomycotina. Ascospores are produced after karyogamy and meiosis. The number of ascospores of different fungi varies, and is usually eight.

大多数子囊菌都产生子实体，也称子囊果，子囊包于子实体内。子囊果的外壁由营养菌丝交织形成，子囊之间的菌丝称为侧丝。子囊果的形态是子囊菌分类的重要依据，常见的有三种类型（图7-4）：①子囊盘，子囊果盘状、杯状或碗状，子囊和侧丝平行排列在一起形成子实层；②闭囊壳，子囊果完全闭合成球形，无开口，待其破裂后子囊及子囊孢子才能散出；③子囊壳，子囊果瓶状或囊状，先端开口。

Most ascomyces produce sporophore, and ascus is enclosed in sporophore (Fig. 7-4). The sporophore of ascomyces is also called ascocarp. The outer wall of the ascocarp is interweaved vegetative hyphae, and the hypha between the ascus is called paraphysis. The morphology of ascocarp is an important basis for the classification of ascomyces. There are three common types. ① Apothecium: ascocarp is disc-shaped, cup-shaped or bowl-shaped. There are many ascuses and paraphysis in apothecium, which are arranged in parallel to form hymenium. The hymenium is completely exposed to the outside, such as peziza. ② Cleistothecium: The ascocarp is completely closed into a sphere, without openings. After they are ruptured, the ascus and ascospore can be scattered, such as the ascocarp of erysiphaceae. ③ Perithecium: ascocarp is bottle-shaped or sac-like, with opening in apex. This ascocarp is mostly buried in stroma, such as ergot and cordyceps.

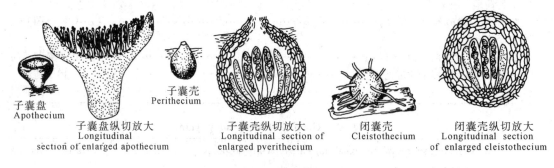

子囊盘 Apothecium　　子囊壳 Perithecium

子囊盘纵切放大
Longitudinal section of enlarged apothecium

子囊壳纵切放大
Longitudinal section of enlarged perithecium

闭囊壳 Cleistothecium

闭囊壳纵切放大
Longitudinal section of enlarged cleistothecium

图7-4　子囊果类型

Fig.7-4　Types of ascocarp

【药用真菌 / Representative medicinal fungi】

酿酒酵母菌 *Saccharomyces cerevisiae* Hansen 酵母菌科。单细胞，球形或卵形。无性繁殖方式为芽殖；有性繁殖阶段可形成子囊和子囊孢子。除传统用于酿酒、制作面包和馒头外，其菌体可制成酵母菌片治疗消化不良，或用以提取核酸、谷胱甘肽、细胞色素C、烟酰胺腺嘌呤二核苷酸等。

Saccharomyces cerevisiae Hansen Saccharomycetaceae. Single-celled, spherical or ovate, asexual reproduction is budding. The sexual reproduction can form ascus and ascospores. In addition to traditionally used for brewing, making bread, its fungus body can be made into yeast tablets for the treatment of indigestion, or used to extract nucleic acids, glutathione, cytochrome C, nicotinamide adenine dinucleotide, etc.

麦角菌 *Claviceps purpurea*（Fr.）Tul. 麦角菌科。寄生在禾本科植物的子房内，菌核形成时伸出子房外，呈紫黑色，坚硬，角状，故称麦角。在繁殖期，孢子借助气流、雨水或昆虫传播到麦穗上，萌发形成芽管，侵入植物子房，长出菌丝，菌丝充满子房而发出极多的分生孢子，再传播到其他麦穗上。当菌丝体继续生长，最后不再产生分生孢子时，会形成紧密坚硬紫黑色的菌核即麦角。菌核（麦角）含麦角新碱、麦角胺、麦角毒碱等多种活性成分，麦角胺和麦角毒碱可治偏头痛。麦角制剂已用作子宫收缩及内脏器官出血的止血剂。

Claviceps purpurea (Fr.) Tul. Clavicipitaceae. It is parasitized in the ovary of Graminea plants. When the sclerotium formed, it is exposed outside from the ovary. It is purple-black, hard, and shaped like an animal horn, so it is called ergot. After the spores are liberated, they spread to the ears of wheat by means of air flow, rain or insects, germinate into germ tubes, invade the ovary, grow hyphae, and the hyphaes fill the ovary and emit a large number of conidium, which will be transmitted to other wheat ears. The mycelium continues to grow, and not produces conidium finally, and forming a tight, hard purple-black sclerotium, which is ergot.The ergot contains ergonovine, ergotamine, ergotoxine and other active alkaloids. Ergotoxine and ergotoxine can cure migraine. Ergot has been used as a hemostatic for uterine contractions and internal organ bleeding.

冬虫夏草 *Cordyceps sinensis*（Berk.）Sacc.（图 7-5）麦角菌科。寄主为鳞翅目蝙蝠蛾科昆虫的幼虫。夏秋季节，冬虫夏草的子囊孢子成熟后由子囊散出，侵入寄主幼虫体内，萌发形成菌丝，并以酵母状出芽方式进行繁殖，充满虫体直至幼虫死亡。冬季来临，菌丝体变态形成坚硬的菌核。翌年夏初，从虫体头部长出棒状子座，伸出土层。子座上部膨大，表层下有一层子囊壳，壳内生有许多长形子囊，子囊各产生 8 个线形多细胞子囊孢子。子囊孢子散发后，断裂成许多节段，重新侵染其他蝙蝠蛾寄主幼虫。冬虫夏草多分布于海拔 3500m 以上的高山草甸区。现已能人工培养或通过薄层发酵工艺，大量繁殖其菌丝体。冬虫夏草能补肺益肾，止血化痰。

Cordyceps sinensis (Berk.) Sacc. (Fig.7-5) Clavicipitaceae. The host is insect larva of Hepialidaceae of Lepidopteran. In the summer and autumn, the ascospores of *C. sinensis* are emitted from the ascus after maturity and invade the healthy larvae of Hepialidaceae in the soil. Ascospores germinate to form hypha, and reproduce by yeast-like budding, filling the larvae body until it dies. As winter comes, mycelium turns into hard sclerotium. In the early summer of the following year, a rod-shaped stroma emerges from the head of the larvae

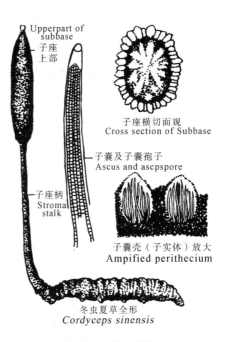

子座上部 Upperpart of subbase

子座横切面观 Cross section of Subbase

子囊及子囊孢子 Ascus and ascpspore

子座柄 Stroma stalk

子囊壳（子实体）放大 Ampified perithecium

冬虫夏草全形 *Cordyceps sinensis*

图 7-5　冬虫夏草
Fig. 7-5　*Cordyceps sinensis* (Berk.) Sacc.

109

and extends out of the soil layer. There is a layer of perithecium under the surface. There are many long ascuses in the perithecium, and there are 8 linear multicellular ascospores in the ascus. After the ascospores are emitted, they broke into many segments and reinfect other healthy larvae of Hepialidaceae. *C. sinensis* is mostly distributed in alpine meadows above 3500 m. Now it can be cultivated artificially or through thin-layer fermentation technology to produce its mycelium in large numbers.It can invigorate the lungs and nourish the kidney, stop bleeding and reduce phlegm.

（二）担子菌亚门

2.2.2　Basidiomycota

担子菌是真菌中最高等的一个亚门，已知有1100属，2万余种。它是食用和药用菌的重要来源，也是筛选抗肿瘤药物的重要资源。绝大多数担子菌具有发达的菌丝体，菌丝均具横隔膜。菌丝体有初生菌丝体、次生菌丝体和三生菌丝体之分。由担孢子萌发形成具有单核单倍体的菌丝，叫初生菌丝，生活时间短；初生菌丝接合进行质配，核不配合，而保持双核状态，生活时期长，这是担子菌的特点之一，主要行营养功能；三生菌丝次生菌丝特化的特殊菌丝，也是双核的，它常集结成特殊形状的子实体，也称担子果。

Basidiomycotina is the highest subdivision of fungi, with 1,100 genera and more than 20,000 species known. It is an important source of edible and medicinal fungi, and an important resource for screening antitumor drugs. Most basidiomycetes have developed mycelium, and the hypha has septum. The mycelium of most basidiomycetes can be divided into three types: primary mycelium, secondary mycelium, and tertiary mycelium. The hyphae with mononuclear haploid germinated by basidiospores is called primary hyphae, with short life time. Primary mycelium is joined by plasmogamy without caryogamy, while maintaining the state of dual nucleus, called secondary mycelium, which life time is quite long. This is one of the characteristics of basidiomycetes, which mainly perform nutritional functions. The tertiary mycelium is a tissue-specialized hypha, and it is also a dual-nucleus. It often forms many various sporophore, also known as basidiocarp.

担子菌最大特点是有性阶段形成担子和外生担孢子。担孢子、典型的双核菌丝、锁状连合是担子菌的三个明显特征。双核菌丝细胞分裂前，在两核之间的细胞壁一侧形成一个喙状突起，突起向下弯曲，其中1核移入突起中之后，两核同时分裂形成4个子核，新分裂产生的2个子核移动到细胞的一端，细胞产生横隔，将上下两个细胞隔开，同时突起基部产生一个隔膜，突起向下与原细胞壁融合沟通，并将留在突起中的1个子核移入下面的细胞。此时，一个双核细胞产生两个双核细胞，此过程即为锁状连合（图7-6）。

绝大多数担子菌以锁状连合方式发育成繁茂的次生菌丝体，子实体成熟后，双核菌丝顶端膨大成担子，担子顶端伸出4个小梗，减数分裂产生的4个单倍体的核分别移入小梗内，发育成4

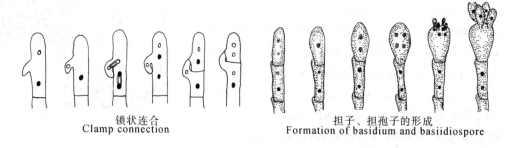

锁状连合
Clamp connection

担子、担孢子的形成
Formation of basidium and basiidiospore

图7-6　锁状联合、担子、担孢子的形成
Fig. 7-6　Clamp connection, formation of basidium and basidiospore

个担孢子。产生担孢子的复杂结构的菌丝体叫担子果，就是担子菌的子实体。其形态、大小，质地、颜色各不相同，如伞状、耳状、菊花状、笋状、球状等，其特征是担子菌分类的重要依据之一。

The most important feature of basidiomycetes is the formation of basidium and external basidiospore (Fig.7-6). Basidiospore, typical dinuclear hypha, and clamp connection are three obvious characteristics of basidiomycetes.e protrusion, the two nuclei divide at the same time to form four daughter nucleus. Two daughter nucleus produced by the new division move to one end of the cell, and the cell forms a septum, separating the upper and lower cells. At the same time, the base of the protrusion creates a septum, and the protrusion is fused with the original cell wall, and the daughter nucleus remaining in the protrusion is moved into the cell below. If the growth of the hyphae of compatible mating types happens to bring them close together, cells of each mycelium may unite, initiating a new mycelium in which each cell has two nuclei. Such a mycelium is said to be dikaryotic. Dikaryotic mycelia sometimes have little walled-off bypass loops called clamp connections between cells on the surface of the hyphae. The clamp connections develop as a result of a unique type of mitosis that ensures each cell will have one nucleus of each original mating type within it. Most basidiomycetes develop into luxuriant secondary mycelium in a clamp connection. After the sporophore matures, the top of the dinuclear hypha is expanded into a basidium, with four sterigmas protruding from the tip of the basidium. The haploid nuclei were transferred into sterigma and developed into four basidiospores. The mycelium that produces the complex structure of basidiomycetes is called basidiospores, which is the sporophore of basidiomycetes. Its shape, size, texture, and color are different, such as umbrella-shaped, ear-shaped, chrysanthemum-shaped, bamboo-shaped, spherical, etc., and its characteristics are one of the important bases for the classification of basidiomycetes.

担子菌亚门分为 3 个纲，即冬孢菌纲 Teliomycetes，纲内锈菌目和黑粉菌目的真菌一般为害植物；层菌纲 Hymenomycetes，如银耳、黑木耳、灵芝等；腹菌纲 Gasteromycetes，如马勃、地星、鬼笔等。许多食用和药用的大型担子菌多属于层菌纲，最常见的是伞菌类。伞菌的担子果主要包括伞盖、菌褶、菌柄等结构。菌盖是伞菌担子果最明显的部分，形态多样，常见的有伞状、半球形、斗笠形等。菌盖下面辐射状排列的片状结构称菌褶，其上布满子实层，子实层内有担子、担孢子、侧丝等。菌盖下面的柄称菌柄，部分伞菌子实体幼小时，菌盖边缘与菌柄之间连有一层菌膜，称内菌幕或菌环；另一类群在幼小子实体外面包被有一层膜，称外菌幕，当子实体扩大，菌柄伸长时，外菌幕破裂，在菌柄基部留下一个杯状的菌托（图 7-7）。这些结构特征是鉴别伞菌的重要依据。

Basidiomycotina is divided into 3 classes, namely Teliomycetes, which includes the order Uredinales and Ustilaginales. The fungi of Uredinales and Ustilaginales are harmful to plants of the family Granineae and Cyperaceae. The fungus of Hymenomycetes has such as Tremella, Black Fungus, Ganoderma, etc.; Gasteromycetes, such as puffball, earth star, stinkhorn, etc. Many edible and medicinal large basidiomycetes belong to the hymenomycetes, the most common is agaric. Basidiocarp of agaric mainly includes structures such a pileus, gills, and stipe. The pileus is the most obvious part of the basidiocarp of agaric, and it has various forms. The common ones are umbrellalike, hemispherical, and bamboo hat-shaped. The slice structure arranged in a radial pattern under the pileus is called gill. The gills are covered with hymenium, and there are basidium, basidiospores, paraphysis, etc in the hymenium. The stalk below the pileus is called stipe. There is a layer of fungi membrane between the edge of the pileus and the stipe, called annulus, when sporophore of some agaric is young. The membrane in the outer side of other young sporophore was called universal veil. When the sporophore expands and the stipe is elongated, the universal veil breaks, leaving a cup-shaped volva at the base of the stipe. These structural characteristics are important basis for identifying agaric (Fig.7-7).

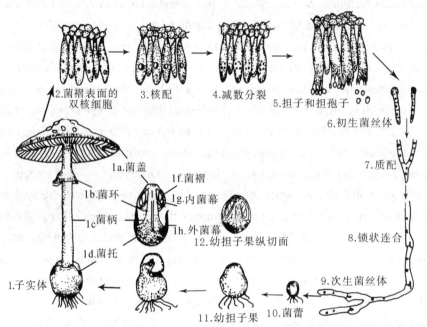

1. Sporophore（1a.Pileus，1b. Collarium，1c. Stipe，1d.Teleblem，1f. Lamella，1g.Inner veil，1h.Universal veil）；2.Binucleate cells on the surface of fimbria；3.Caryogamy；4. Meiosis；5.Basidiospores and basidiospores；6.Primary mycelium；7.Plasmogamy；8.Clamp connection；9. Secondary mycelium；10.Button；11.Juvenile basidioma；12.Longitudinal section of juvenile basidioma

图 7-7　伞菌的形态和生活史

Fig.7-7　Agaric morphology and life history

【**药用真菌 / Representative medicinal fungi**】

　　灵芝（赤芝）*Gandoerma lucidum*（Leyss. ex Fr.）Karst.（图 7-8）多孔菌科。子实体木栓质，菌盖半圆形或肾形，初生黄色，后渐变成红褐色，有漆样光泽，具环状棱纹及辐射状皱纹，子实层生于菌盖下的菌管内。菌柄近圆柱形，侧生，长度通常长于菌盖的长径，与菌盖同色，具光泽。担孢子宽卵圆形，壁有两层，内壁褐色，布有无数小疣，外壁光滑、透明无色。分布于全国多省区，多生于阔叶树木基部，现已人工栽培。子实体（灵芝）能补气安神，止咳平喘。紫芝 *G. Sinense*　Zhao Xu et Zhang 的干燥子实体与赤芝同作灵芝入药，分布于浙江、江西、福建、湖南、广东、广西等。

***Gandoerma lucidum* (Leyss. ex Fr.) Karst**. (Fig.7-8) Polyporaceae. The sporophore is suberin, the pileus is semicircular or kidney-shaped, it turns yellow to reddish brown after the first birth, has a lacquer-like luster, has ring and radial lines, and the hymenium grows under the pileus inside the tubule. The stipe is nearly cylindrical, lateral, and usually longer than the long diameter of the pileus, its color is same as the pileus and has sheen. Under the microscope,

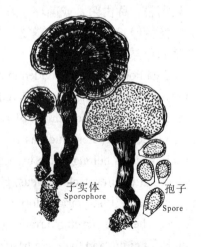

子实体
Sporophore

孢子
Spore

图 7-8　灵芝

Fig. 7-8　*Gandoerma lucidum* (Leyss. ex Fr.) Karst.

the basidiospores are wide oval, with two layers of walls, brown inner walls, numerous small warts on the surface, and smooth, transparent and colorless outer walls. It is distributed in many provinces and regions of the country, mostly grow at the base of broad-leaved trees, it is now cultivated artificially. Sporophore can invigorate Qi and tranquillization, relieve cough and asthma. The dried sporophore of *G. sinense* Zhao Xu et Zhang and *G. lucidum* are used as glossy ganoderma. *G. sinense* are distributed in Zhejiang, Jiangxi, Fujian, Hunan, Guangdong, and Guangxi.

茯苓 *Poria cocos*（Schw.）Wolf.（图 7-9）多孔菌科。菌核呈球形或不规则块状，大小不一，表面粗糙，呈瘤状皱缩，灰棕色或黑褐色，内部白色或淡棕色，粉粒状。子实体平伏，伞形，生于菌核表面成一薄层，幼时白色，老熟干燥后变为淡褐色；菌管单层，孔为多角形至不规则形，孔壁薄，边缘渐变成齿状。担孢子长椭圆形至近圆柱状，有一斜尖，壁表平滑，透明无色。茯苓属于腐生菌，寄生于多种松属植物的根上。全国广布，以安徽、云南、湖北、河南、广东等省分布最多。现已人工栽培。菌核（茯苓）能利水渗湿，健脾宁心。

图 7-9 茯苓
Fig. 7-9 *Poria cocos* (Schw.) Wolf

Poria cocos (Schw.) Wolf. (Fig.7-9) Polyporaceae. Sclerotium are spherical, oblong, oval, or irregular in shape, ranging in size from fists to tens of pounds, softer when fresh, hard after drying, with rough surface, tumor-like shrink, gray-brown or dark brown, white or light brown inside, and powdery granular. The sporophore is flat, umbrella-shaped, born on the surface of the sclerotium into a thin layer, white when young, and light brown after maturing and drying; the tubule is single-layered, the holes are polygonal to irregular, the walls of the holes are thin and the edges gradually toothed. Observed under a microscope, the basidiospores are oblong to nearly cylindrical, with an oblique tip, the surface of wall is smooth, transparent and colorless. Poria is a saprophytic fungus and mostly parasitic on the roots of many pineplants. It is widely distributed throughout the country, especially in Anhui, Yunnan, Hubei, Henan, and Guangdong. There is artificial cultivation now.Sclerotium can remove dampness by promote dieresis and strengthen spleen and calm heart.

子实体
Sporophore

孢子
Spore

菌核
Sclerotium

图 7-10 猪苓
Fig. 7-10 *Polyporus umbellatus* (Pers.) Fr.

猪苓 *Polyporus umbellatus*（Pers.）Fr.（图 7-10）多孔菌科。菌核呈长块状或不规则形块状，表面凹凸不平，皱缩或有瘤状突起，灰色、棕黑色或黑色。子实体常多数合生，菌柄基部相连成一丛菌盖，菌盖伞形或伞状半圆形，表面浅褐色至茶褐色；菌管口微小，呈多角形，担孢子卵圆形。分布于全国多省区，寄生在枫、槭、柞、桦、椴等树的树根上。菌核（猪苓）能利水渗湿，含有的猪苓多糖有抗癌作用。

Polyporus umbellatus (Pers.) Fr. (Fig.7-10) Polyporaceae. The sclerotium are long or irregularly shaped, the surface is uneven, with shrinking or nodular protrusions, gray, brownish black or black. The sorophore are often connate. The base of the stipe is connected to form a cluster of pileus. The pileus is umbrella-shaped or semi-circular, and the surface is light brown to dark brown. Tube openings are small, polygonal, and basidiospores are oval. It is distributed in many provinces and regions of the country, parasitized on the roots of trees such as maple, sassafras, birch, and linden. The sclerotium can eliminate dampness and diuresis, and the polysaccharides contained in it have anti-cancer effects.

脱皮马勃 *Lasiosphaera fenzlii* Reich.（图 7-11）马勃科。子实体近球形或长圆形，成熟时浅褐色。外包被薄，成熟时呈碎片状剥落，内包被纸状，成熟后消失，遗留成团的孢体。孢体紧密，有弹性，灰褐色至淡烟色。孢子球形，褐色，外具小刺，分布于全国多省区。子实体（马勃）能清肺利咽，止血。同科大马勃 *Calvatia gigantean*（Batsch ex Pers.）Lloyd、紫色马勃 *C. lilaciana*（Mont. Et Berk.）Lloyd 的子实体也作马勃入药。

图 7-11　脱皮马勃
Fig. 7-11　*Lasiosphaera fenzlii* Reich.

担子菌亚门入药的还有：多孔菌科彩绒革盖菌 *Coriolus versiolor*（LexFr.）Quel，子实体（云芝）能健脾利湿，清热解毒，其云芝多糖具抗肿瘤、增强免疫力和保肝活性；白蘑科雷丸 *Omphalia lapidescens* Schroet.，菌核（雷丸）能消积，杀虫；木耳科木耳 *Auricularia auricula*（L. ex Hook.）Underw.，子实体（木耳）能补气益血，润肺止血；银耳科银耳 *Tremella fuciformis* Berk.，子实体（银耳）能滋阴养胃、润肺生津，益气和血；齿菌科猴头菌 *Hericium erinaceus*（Bull. ex Fr.）Pers 的子实体（猴头菌）能利五脏，助消化，滋补，抗癌；鬼笔科长裙竹荪 *Dictyophora indusiata*（Vent. Pers.）Fisch 的子实体能补气养阴，润肺止咳，清热利湿；白蘑科蜜环菌 *Armillarialla mellea*（Vahl. ex Fr.）Karst. 是天麻的共生菌，子实体能明目，利肺，益肠胃。

Lasiosphaera fenzlii Reich. Lycoperdaceae. Sporophore nearly spherical or oblong, light brown when mature. The outer cover is thin, flaking in pieces and the inner cover is paper-like, and disappears when mature, leaving clustering spores. The spores are compact, elastic, gray-brown to pale smoky. The spores are spherical, brown, with small spines outside, and are distributed in many provinces and regions of the country. The sporophore can clear the lungs, relieve the throat, and hemostasis. The sporophores *of Calvatia gigantean* (Batsch ex Pers.) Lloyd and *C. lilaciana* (Mont. et Berk.) Lloyd of the same family are also used as *L. fenzlii*.

There are still many medicinal fungi from Basidiomycotina such as *Coriolus versiolor* (LexFr.) Quel, *Omphalia lapidescens* Schroet., *Auricularia auricula* (L. ex Hook.) Underw., etc.

（三）半知菌亚门

2.2.3　Deuteromycotina

半知菌亚门是一类尚未发现有性繁殖阶段，仅以分生孢子进行无性繁殖的真菌，故称半知菌。其营养体大多是有隔的分枝菌丝，有些种类形成假菌丝。一旦发现有性阶段，将其重新归属到相应的亚门。已发现有性阶段的半知菌多属于子囊菌。半知菌分类以应用方便为主，不以亲缘关系为依据，一般根据孢子梗和孢子的形态及产生方式分类。目前有 1880 属，约 26000 种，许多是动、植物的寄生菌。

Deuteromycotina is a type of fungi which a sexual stage has not been observed. Most of them reproduces by means of conidia. Most of the vegetative bodies of the fungus are separated branched hyphae, and some species form pseudo-hyphae. Once a sexual phase is found, it is reassigned to the corresponding sub-division. It has been found that sexually known fungi are mostly ascomycetes. The classification of fungi imperfecti is mainly based on the convenience of application, not on the basis of kinship, and is generally classified according to morphology and production methods of the conidiophore and spore. There are currently 1880 genera and about 26,000 species, many of which are parasites of animals and plants.

【药用真菌 / Representative medicinal plants 】

曲霉属 *Aspergillus*（Micheli）Link.，丛梗孢科，现属子囊菌。菌丝有隔，为多细胞，多分枝。无性生殖发达，分生孢子梗从特化了的壁厚而膨大的菌丝细胞垂直生出，无横隔，顶部膨大成棍棒状、椭圆形、半球形的泡囊。泡囊表面产生很多放射状排列的小梗，小梗单层或双层，分生孢子自小梗顶端相继形成，多呈球形，单细胞，有各种颜色和纹饰。分生孢子呈绿、黑、褐、黄、橙各种颜色。

Aspergillus (Micheli) Link. Moniliaceae belongs to Ascomycetes now. The multicellular and multibranched mycelium has septum. Asexual reproduction is well-developed. Conidial stalks emerge vertically from specialized wall thickness? and swollen hyphae, without septum, and the top swells into stick-shaped, oval, hemispherical vesicles. The surface of the vesicle produces many sterigma arranged in a radial pattern. The sterigma is single-layered or double-layered. Conidia are formed successively from the top of the sterigma. Most of them are spherical, single-celled, and have various colors and patterns. Conidia are green, black, brown, yellow, and orange.

曲霉属真菌种类多，广泛分布于空气、土壤、谷物和各种有机物上，是酿造工业和食品工业的重要菌种。但有的种类对农作物及人类的身体健康有极大危害，如黑曲毒 *Aspergillus niger* Van Tieghen 会引起粮食和中药材霉变，杂色曲霉 *A.versicolor*（Vuill）Tirab. 会引起桃果腐烂和中药材霉变，赭曲霉 *A.ochraceus* Wilhelm 会导致苹果、梨的果实腐烂，烟曲霉 *A. fumigates* Fresen. 可引起人畜和禽类的肺曲霉病。其中杂色曲霉（sterigatocystin）产生的杂色曲霉素可致肝脏受损，特别是黄曲霉常在花生和花生粕上发现，会产生毒性很强的能引起肝癌的黄曲霉素（aflatoxin）。

There are many species of aspergillus, which are widely distributed in the air, soil, grain and various organic matters, and are important strain in the brewing and food industry. However, some species are harmful to crops and humans. For example, *Aspergillus niger* Van Tieghen can cause mildew in food and Chinese materia medica, and *A. versicolor* (Vuill) Tirab. can cause peach fruit rot and mildew of Chinese materia medica, *A. ochraceus* Wilhelm can cause apple and pear fruit rot. *A. fumigatus* Fresen. can cause pulmonary aspergillosis in humans, animals and birds. Among them, *A. versicolor* (sterigatocystin) can cause liver damage. In particular, *A. flavus* is often found on peanuts and peanut butter. *A. flavus*, which grows on moist seeds, secretes aflatoxin, the most potent natural carcinogen known. The toxin causes liver cancer, and no more than 50 parts per billion is allowed in human food.

青霉 *Penicillium* Link.（图 7-12）丛梗孢科，现属子囊菌。菌丝体由多数具有横隔的菌丝组成，无色、淡色或颜色鲜明，常以产生分生孢子的形式进行繁殖。有性生殖极少见。产生孢子时，菌丝体顶端产生多细胞的分生孢子梗，分生孢子梗具横隔，光滑或粗糙，顶端生有扫帚状分枝，称帚状枝。帚状枝由单轮或两轮至多轮分枝系统构成，最顶端的小梗上产生分生孢子。分生孢子呈球、椭圆形或短柱形，多呈蓝绿色，有时无色或呈其他淡色。成熟后随风分散，遇适宜环境，萌发成菌丝。

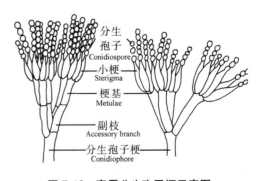

图 7-12　青霉分生孢子梗示意图
Fig. 7-12　Sketch map of conidiophores of *Penicillium* **Link**

Penicillium Link. (Fig.7-12) Moniliaceae. Mycelium is composed of most hyphae with transverse septum, which is colorless, light-colored or brightly colored, and often reproduces in the form of conidia. Sexual reproduction is rare. When spores are produced, multicellular conidiophores are produced at the top of the mycelium. The conidiophores have septum, smooth or rough, and the top has brush branches. The brush branch is composed of a single-round or two-round to multiple-round branching system, and conidia are produced on the apical sterigma. Conidia are spheres, ovals, or short columns, mostly blue-green, and sometimes colorless or other light colors. After maturity, it disperses with the wind, and when the environment is suitable, falls on its substrate to germinate into hyphae.

青霉菌种类多，分布广。常在蔬菜、粮食、肉类、柑橘类水果、皮革和食物上分布。如产黄青霉 *Penicillium chrysogenum* Thom、特异青霉 *P. notatum* Westling 等均可产生青霉素。展青霉 *P. patulum* Bainier 可产生抗真菌的灰黄霉素。黄绿青霉 *P. citreo-viride* Biourge、岛青霉 *P. islandicum* Sopp 可引起大米霉变，产生 "黄变米"，它们产生的霉素如黄绿青霉素对动物神经系统有损害作用，岛青霉产生的环氯素、岛青霉毒素均对肝脏有毒性。柑橘青霉 *P. citrinum* Thom、意大利青霉 *P. italicum* Wehmer 可引起柑橘果实软腐。橘青霉产生的橘青霉素对肾脏有损害作用。

There are many species of penicillium and they are widely distributed on vegetables, grains, meats, citrus fruits, leather, and food. For example, *Penicillium chrysogenum* Thom and *P. notatum* Westling can produce penicillin. *P. patulum* Bainier can produce antifungal griseofulvin. *P. citreo-viride* Biourge and *P. islandicum* Sopp can cause mildew of rice and produce "yellowing rice". The mycotoxins produced by them, such as citreoviridin, can damage the nervous system of animals. The cyclochlorin and penicillin produced by penicillium are toxic to the liver. *P. citrinum* Thom and *P. italicum* Wehmer can cause soft rot of citrus fruits. Citricin produced by *P. citrinum* has damaging effects on the kidneys.

球孢白僵菌 *Beauveria bassiana*（Bals.）Vuill. 寄生于家蚕幼虫体内（可寄生于 60 多种昆虫体上），使家蚕病死，干燥后的尸体称为僵蚕，入药能息风止痉、祛风止痛、化痰散结。

Beauveria bassiana (Bals.) Vuill. Parasitic on silkworm larvae (can be parasitic on more than 60 insects), causing silkworm to die. The dried corpse is called Bombyx Batryticatus. It can relieve spasm by calming endogenous wind, dispel pathogenic wind and relieve pain, dissipate phlegm and resolve masses.

重 点 小 结
Summaries

菌类是一类不含光合色素的异养型原植体植物，也属于低等植物。菌类通常是自然生态系统中的分解者，在地球物质循环中有着重要作用，其营养方式包括寄生、腐生和共生。菌类种类繁多，常分为细菌门、真菌门和黏菌门。细菌门是单细胞原核生物，以裂殖或芽殖方式繁殖，根据形态特征分为球菌、杆菌和螺旋菌，按照细胞壁组成和染色不同分为革兰阳性菌和阴性菌。放线菌是介于细菌和真菌之间的中间类型，是目前抗生素的主要生产菌。

Fungi are usually heterotrophic organisms without photosynthesis. They are decomposers in nature. They play an important role in the earth's material cycle. Their nutritional methods include parasitism, saprophytic and symbiotic. There are many types of fungi, which are often divided into bacterial, fungal, and myxomycota. The bacteriophyta is a single-celled prokaryote that reproduces by fission or budding. It is divided into coccus, bacillus, and spirilla according to its morphological characteristics. It is divided into gram-positive bacteria and negative bacteria according to cell wall composition and staining. Actinomyces are an intermediate type between bacteria and fungi, and are currently the main producers of antibiotics.

Myxomycetes are eukaryotic organisms between fungi and protozoa. Their actions and predation are similar to protozoa, and their reproduction is similar to fungi.

真菌的营养体为无定型的菌丝体，在环境不良或生殖时还会形成多种菌丝组织体。真菌门是中药材来源最多的一类菌类植物，已知真菌植物约12万种，常分为5个亚门。鞭毛菌亚门和接合菌亚门是低等真菌，营养体为单细胞或无隔多核菌丝；子囊菌亚门和担子菌亚门是高等真菌。子囊菌亚门多为有隔菌丝组成，内生子囊孢子；担子菌亚门也为有隔菌丝组成，外生担孢子；半知菌亚门的营养体为有隔菌丝组成，尚未发现其有性阶段。《中国药典》收载的真菌药材有12种，如灵芝、云芝、茯苓、猪苓、马勃、雷丸等。

About 120,000 species of fungi are known, often divided into five subdivisions. Mastigomycotina and Zygomycotina are lower fungi, and vegetative bodies are single-celled or septum-free multinucleated hyphae. Ascomycotina and Basidiomycotina are higher fungi. Ascomycotina is mostly composed of septate hyphae and endophytic ascospores. Basidiomycotina is composed of septate hyphae and exogenous basidiospores. Vegetative bodies of Deuteromycotina are composed of septate hyphae and have no sexual stages. There are twelve kinds of fungal herbs contained in Chinese Pharmacopoeia, such as Ganoderma lucidum, Trametes versicolor, Poria, Grifola, Puffball, and Stone-like Omphalia, etc.

目 标 检 测
Questions

题库

1. 真菌门植物的主要特征有哪些？

What are the main characteristics of fungal plants that distinguish them from other lower plants?

2. 菌类植物有何开发利用价值？

What are the development and utilization values of fungal plants?

3. 生活中常食用的"蘑菇"是担子菌吗？

Is the "mushroom" commonly used in daily life a basidiomycete?

医药大学堂
WWW.YIYAODXT.COM

第八章 地 衣
Chapter 8 Lichens

学习目标 | Learning goals

1. **掌握** 地衣的特征。
2. **熟悉** 代表性药用地衣。
3. **了解** 地衣的分类。

- Know the characteristics of lichens.
- Be familiar with medicinal lichen.
- Understand the classification of lichens.

第一节　地衣的特征

1 Features of the lichens

地衣是一类藻类和真菌高度共生的复合体，属于低等植物。真菌和藻类两类生命长期紧密地联合在一起，在形态、构造、生理和遗传上都形成稳定的共生关系，成为独立的地衣植物门。地衣体中的菌丝缠绕藻细胞，吸收水分和无机盐供藻类利用，而藻类通过光合作用制造有的机物大部分被菌类利用，两者互利互惠，但并不平等，通常真菌在地衣中占据主导地位。

The bodies (thalli) of lichens consist of an alga or cyanobacterium and a fungus in intimate symbiotic relationship. Fungus and algae are closely associated for a long time, forming a single fixed organism in morphology, structure, physiology and heredity, and becoming independent lichens. The algae cells are twined by the hyphae of lichens, and the algae are surrounded from the outside. The fungus absorb water and inorganic salt for the use of algae, while most of the organic matters produced by photosynthesis of algae are used by fungi, forming a special symbiotic relationship between them.

地衣体中的真菌绝大部分属于子囊菌亚门的盘菌纲和核菌纲，少数为担子菌亚门的伞菌目和非褶菌目（多孔菌目），极少数属于半知菌亚门。藻类多为绿藻和蓝藻，其中绿藻门的共球藻属、橘色藻属和蓝藻门的念珠藻属，约占地衣体藻类的90%。地衣进行营养繁殖、无性生殖和有性繁殖。营养繁殖主要是地衣体的断裂，1个地衣体分裂为数个裂片，每个裂片均可发育为新个体。此外，粉芽、珊瑚芽和裂碎片等都可用于繁殖新的个体。无性生殖由地衣体的藻、菌分别进行，菌类产生分生孢子，孢子萌发出菌丝后并遇到适合的藻类即可发育成新的地衣，否则死亡；藻类在地衣体内进行无性生殖，以增加其数量。有性生殖由地衣体中的子囊菌和担子菌进行，产生子囊孢子和担孢子。分别称子囊菌地衣、担子菌地衣，其中子囊菌地衣占地衣种类的绝大部分。

Most of the fungi in lichens belong to Discomycetes and Pyrenomycetes of Ascomycotina, a few belong to Agaricales and Aphyllophorales of Basidiomycotina, and a few belong to Deuteromycotina. Most of the algae are Chlorophyta and Cyanophyta, among which the Trebouxia, Trentepohlia and Nostoc of Chlorophyta account for 90% of algae in lichens. The reproductions of lichens include vegetative propagation, asexual and sexual reproduction. Vegetative propagation is mainly the rupture of lichens. One lichen divides into several lobes, each of which can develop into a new lichen. In addition, soredium, isidium and lobules can develop into a new lichen too. Asexual reproduction is carried out by the alga and fungus of lichens respectively. The fungus produces conidiospore. After the spores sprouting out hyphae and meeting suitable algae, they can develop into new lichens, otherwise they will die. The alga carries out asexual in lichens to increase their numbers. Sexual multiplication happens to ascomycotina and basidiomycotina of lichens, producing ascospores and basidiospores. They are respectively called ascolichens and basidiolichens, of which ascolichens account for the majority of lichens.

地衣种类多并有广泛的经济意义，具有药用、饲用和食用价值。我国药用地衣有 70 多种。中国古代文献《诗经》就有关于"女萝"（即松萝）的记载。甄泉在《药性本草》中有"松萝"、"石蕊"的记载。李时珍在《本草纲目》中也记载了石蕊的药用价值。地衣含有多种药用成分，约 50% 以上的地衣种类含有多种抗菌成分地衣酸，如松萝酸、地衣硬酸等，对革兰阳性菌和结核杆菌有抗菌活性。研究表明多数地衣中所含的地衣多糖和异地衣多糖等具有极高的抗癌活性。地衣具有食用价值，如石耳、石蕊、冰岛衣等。地衣还可作动物饲料，同时可提取淀粉、蔗糖、酒精等。有些地衣可作香料、染料，如扁枝衣属、树花属、石蕊属等含有芳香油。

There are many kinds of lichens, which have a wide range of economic significance. They have medicinal, feeding value and edibility. There are more than 70 kinds of medical lichens in China. Shi Jing (Classic poetry), an ancient Chinese document, has record of *Usnea diffracta* Vain. Zhen Quan has records of *Usnea diffracta* Vain. and *Cladonia rangiferina* (L.) Weber ex F. H. Wigg. in *Yaoxing Bencao*. Li Shizhen recorded the medicinal value of *Cladonia rangiferina* (L.) Weber ex F. H. Wigg. In *Compendium of Materia Medica*. Lichens contain many kinds of medicinal ingredients, which more than 50% of lichens species contain many kinds of antibacterial ingredients lichenic acids, such as usnic acid and lichenic acid etc. They have antibacterial activities against Gram-positive bacteria and *Mycobacterium tuberculosis*. And the lichenin, isolichenin, are contained in most lichens, have high anticancer activities. Some lichens are also edible, such as *Umbilicaria esculenta* (Miyoshi) Minks, *Cladonia rangiferina* (L.) Weber ex F. H. Wigg., *Cetraria islandica* (L.) Ach., etc. Lichens can also be used as animal feed, and starch, sucrose, alcohol, etc. can be extracted from lichens. Some lichens can be used as perfume and dye, which contain aromatic oil, such as evernia, ramalina, cladonia.

地衣大多数是喜光植物，要求空气新鲜，可以作为鉴别大气污染程度的指示植物。地衣一般生长很慢，数年才长几厘米；能忍受长期干旱，干旱时休眠，雨后恢复生长，因此可以在峭壁、岩石、树皮和沙漠生长。地衣耐寒性很强，在高山带、冻土带和南、北极地区也能生长、繁殖。通常认为地衣是自然界的先锋植物。

Lichens are mostly photophilous plants, which require fresh air and are indicators of air pollution. Lichens generally grow very slowly, only a few centimeters long in a few years. Lichens can endure long-term drought, dormant when dry, and grow again after rain, so they can grow in cliff, rock, bark and desert. Lichens have strong cold resistance, and can grow and reproduce in alpine zone, permafrost zone, the Antarctic Pole and Arctic region.

一、地衣的形态
1.1 Morphology of lichens

1. 壳状地衣 地衣体是颜色多样的壳状物，菌丝与基质紧密连接，有的生有假根伸入基质中，不易剥离。壳状地衣约占全部地衣的80%。如生于岩石上的茶渍衣属和生于树皮上的文字衣属等（图8-1）。

(1) Crustose lichen The morphology of crustose lichens is shells with various colors. The hyphae are closely connected with substrate, some of which have rhizine spreading into substrate, which are difficult to peel off. The crustose lichens account for about 80% of all lichens, such as the *Lecanora* and the *Graphis* (Fig.8-1).

2. 叶状地衣 地衣体呈叶片状，四周有瓣状裂片，常由叶状体下部生出一些假根或脐固着在基质上，易与基质剥离。如生在草地上的地卷衣属、生在岩石上的石耳属和生在树皮上的梅衣属等。

(2) Foliose lichen The foliose lichens are foliose, with valviform lobules around. Some rhizine or umbilicus are often produced from the bottom of the foliose lichens and fixed on the substrate, which are easy to peel off from the substrate, such as the *Peltigera*, *Umbilicaria* and *Parmelia*.

壳状地衣（茶渍衣属） Crustose lichen
壳状地衣（文字衣属）
叶状地衣（梅衣属） Foliose lichen

(Left: *Lecanora sp.*; right: *Graphis sp.*)

(*Parmelia sp.*)

枝状地衣（长松萝） 枝状地衣（雪茶）
Fruticose lichen

(Left: *Usnea longissima*; Right: *Thamnolia vermicularis*)

图 8-1 地衣形态
Fig.8-1 Morphology of the lichens

3. 枝状地衣 地衣体呈树枝状，直立或悬垂，仅基部附着于基质上。如石蕊属、石花属直立地上，松萝属悬垂分枝生于云杉、冷杉树枝上。

(3) Fruticose lichen The fruticose lichens are fruticose, erect or obumbrant, with only the base attached to the substrate. For example, the *Cladonia* and the *Ramalina* are erect on the ground, and the *Usnea* is obumbrant on the branches of *Picea asperata* Mast. and *Abies nephrolepis* (Trautv.) Maxim.

三种类型的区别不是绝对的，其中一部分是过渡或中间类型，如标氏衣属由壳状到鳞片状；粉衣科地衣由于横向伸展，壳状结构逐渐消失，呈粉末状。

The differences among the three types are not absolute, and some of them are transitional or intermediate types, such as the *Buellia* from shell to scaly, the *Caliciaceae* from shell to powder shapes due to lateral extent.

二、地衣的构造
1.2 Structure of lichens

不同类型的地衣其内部构造不尽相同。叶状地衣的横切面分为上皮层、藻胞层、髓层和下皮层。上皮层和下皮层由菌丝紧密交织而成，也称假皮层；藻胞层在上皮层之下，由藻类细胞聚集成一层；髓层由疏松排列的菌丝组成。根据藻类细胞在地衣体中的分布情况，通常又将地衣体分为两大类型（图8-2）。

Different types of lichens have different internal structures. The cross section of the foliose lichens can be divided into four layers, that are, upper cortex, algal layer, medulla and lower cortex. The upper and lower cortex is closely intertwined by hyphae, which are also called false cortex. The algal layer (only a layer) is under the upper cortex, which is composed of algal cells. The medulla is composed of loosely arranged hyphae. According to the distribution of algal cells in lichens, the structure of lichens can be divided into two types (Fig.8-2).

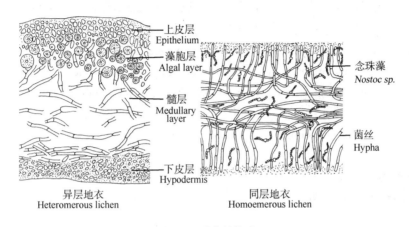

图 8-2 地衣的构造
Fig. 8-2 Structure of lichens

1. **异层地衣** 藻类细胞排列于上皮层和髓层之间，形成明显的一层，即藻胞层。如梅衣属、蜈蚣衣属、地茶属、松萝属等。

(1) Heteromerous lichens The heteromerous lichens' algal layer is an obvious layer, which algal cells are arranged between the upper cortex and medulla, such as *Parmelia*, *Physcia*, *Thamnolia* and *Usnea*, etc.

2. **同层地衣** 藻类细胞分散于上皮层之下的髓层菌丝之间，没有明显的藻胞层、髓层之分，如胶衣属、猫耳衣属，同层地衣类型较少。

(2) Homoenmerous lichens The algae cells are scattered among the hyphae of the medulla. There is no obvious difference between the algae layer and the medulla, such as *Collema* and *Leptogium*. And the homoenmerous lichens are fewer types.

叶状地衣大多数为异层型，从下皮层生出许多假根或脐固着于基物上。壳状地衣多数无皮层，或仅具上皮层，髓层菌丝直接与基物密切紧贴。枝状地衣都是异层型，与异层叶状地衣的构造基本相同，但枝状地衣各层的排列呈圆环状，中央有的有一条中轴，如松萝属，有的中空，如地茶属。

Most of the foliose lichens are heteromerous lichens, which some rhizine or umbilicus are often produced from the bottom of the foliose lichens and fixed on the substrate. Most of the crustose lichens have no cortex or only upper cortex, which the hyphae of the medulla is closely connected with substrate. The fruticose lichens are heteromerous lichens, which are basically the same as the structure of the heteromerous foliose lichens. The arrangement of each layer of fruticose lichens is circular. Some of them have a central axis, such as the *Usnea*. Some of them are hollow, such as the *Thamnolia*.

第二节 地衣的分类

2 Classification of lichens

地衣植物门有 500 余属，25000 余种。它们分布极为广泛，从南、北两极到赤道，从高山到平原，从森林到荒漠都有地衣的存在。我国有 200 属，近 2000 种，其中约 200 种为中国所特有。新疆、贵州、云南等地因其独特的气候和地貌类型，成为我国地衣资源的主要分布区。

There are more than 500 genera and 25000 species of lichens. They are widely distributed from the two poles of the earth to the equator, from mountains to plains, and forests to deserts. Lichens are widely distributed in China, with 200 genera and nearly 2000 species, which about 200 species are endemic to China. Xinjiang, Guizhou, Yunnan and other districts have become the main distribution areas of lichens' resources because of their unique climate and landform types.

一、地衣的分类
2.1　Classification of lichens

通常将地衣分为 3 纲：子囊衣纲、担子衣纲及半知衣纲。

Lichens are usually divided into three classes, which are Ascolichens, Basidiolichens and Deuterolichens.

1. **子囊衣纲**　地衣体中的真菌为子囊菌，本纲地衣的数量占地衣总数量的 99%。包括松萝、文字衣、石蕊等。

(1) **Ascolichens**　The fungus in Ascolichens are ascomycotina, and the number of Ascolichens of this class accounts for 99% of the total number of lichens, including *Usnea diffracta* Vain., *Graphis scripta* (L.) Ach., *Cladonia rangiferina* (L.) Weber ex F. H. Wigg., etc.

2. **担子衣纲**　地衣体菌类多为非褶菌目的伏革菌科菌类，其次为伞目口蘑科的亚脐菇属菌类，还有的属于珊瑚菌科菌类；组成地衣体的藻类为蓝藻，多分布于热带，如扇衣属。

(2) **Basidiolichens**　Most of the Basidiolichens are fungi of Corticiaceae of Aphyllophorales. Secondly are fungi of Omphalina of Tricholomataceae, Clavariaceae. The algae are cyanophyta, which mostly distribute in the tropics, such as Cora.

3. **半知衣纲**　根据半知衣纲地衣体的构造和化学成分分类，其属于子囊菌的某些属，但未见到产生子囊和子囊孢子，是一种无性地衣。如地茶。

(3) **Deuterolichens**　According to the structure and chemical composition of the Deuterolichens, they belong to some genera of ascomycotina. But no asci and ascospore are found, so they are kinds of asexual lichens, such as *Thamnolia vermicularis* (Sw.) Ach. ex Schaer.

二、常见药用地衣
2.2　Important medicinal lichens

松萝 *Usena diffracta* Vain.，菘萝科。植物体悬垂丝状，长 15~30cm，呈多回二叉式分枝，基部较粗分枝少，先端分枝多。表面灰黄绿色，具光泽，有明显的环状裂沟；横断面中央有韧性丝状的中轴，具弹性，可拉长，由菌丝组成，易与皮部分离；其外为藻环，常由环状沟纹分立或呈短筒状。菌层产生少数子囊果，子囊果盘状，褐色，子囊棒状，内生8个椭圆形子囊孢子。分布于全国大部分地区。生于深山老林树干或岩壁上。全草含有松萝酸、去甲环萝酸、松萝多糖等，能止咳平喘，活血通络，清热解毒。

Usnea diffracta **Vain.** Usneaceae, the plant body is obumbrant and filiform, 15-30cm, with multiple dichotomously branched, thick base, few branches of base, many branches at the apex, gray chartreuse of the surface, shiny, with obvious ring-shaped cracks, and the central part of the cross-section has a axis of tough filaments, elastic, elongated, composed of hyphae, easy to be separated from the cortex. The outer part is algae ring, often separated by ring-shaped grooves or in short tube shape. There are a few ascomata in the fungus layer. The ascomata are discoid, brown and clavate, which has 8 ellipsoidal ascospore. It is distributed in most parts of china. Grow in the trunk or rock wall of mountain forest. The plant contains usnic acid, evernic acid, usnea polysaccharides and so on, which can relieve cough and asthma, activate blood circulation and collaterals, clear away heat and detoxify.

药用地衣还有：石蕊 *Cladonia rangiferina*（L.）Web. 全草能祛风，镇痛，凉血止血。石耳 *Umbilicaria esculenta*（Miyoshi）Minks 全草能清热解毒，止咳祛痰，利尿；可食用。冰岛衣 *Cetraria islandica*（L.）Ach. 能调肠胃，助消化。雪茶 *Thamnolia vermicularis*（Sw.）Ach. ex Schaer. 全草能清热解毒，平肝降压，养心明目。

Other medicinal lichens: *Cladonia rangiferina* (L.) Weber ex F. H. Wigg. has the effects of dispelling the wind, ease pain and cooling blood and hemostasis. *Umbilicaria esculenta* (Miyoshi) Minks has the effect of clearing away heat and detoxify, stopping coughing and expectoration, dieresis. The *Cetraria islandica* (L.) Ach. can regulate stomach and intestines, aid digestion. The *Thamnolia vermicularis* (Sw.) Ach. ex Schaer. can clear away heat and detoxify, calm liver and reduce blood pressure, nourish heart and eyesight.

<h1 style="text-align:center">重 点 小 结
Summaries</h1>

地衣是一类特殊的低等植物，是真菌和藻类共生的原植体植物，生存能力强，分布广泛。根据其外部形态可分为壳状地衣、叶状地衣和枝状地衣。按照其内部构造可分为同层地衣和异层地衣。地衣的繁殖方式有营养繁殖、无性生殖和有性生殖。常见的药用地衣有松萝、石耳、雪茶等。

Lichens are very unique plants of the same species, which have strong survival ability and wide distribution. According to the morphology of lichens, they can be divided into crustose lichens, foliose lichens, fruticose lichens. According to the structure of lichens, they can be divided into heteromerous lichens and homoenmerous lichens. The reproductions of lichens include vegetative propagation, asexual reproduction and sexual reproduction. Common medicinal lichens include *Usnea diffracta* Vain., *Umbilicaria esculenta* (Miyoshi) Minks, *Thamnolia vermicularis* (Sw.) Ach. ex Schaer., etc.

题库

目 标 检 测
Questions

1. 地衣可依据什么特征分类？各有哪些类型？

What is the classification basis of lichens? What are the types of lichens?

2. 举例可做中药的地衣名称，并指出属于哪种类型的地衣？

Giving an example of lichens as Traditional Chinese medicine. What kind of lichens do they belong to?

3. 为什么说地衣植物体是复合有机体？

Why are the bodies (thalli) of lichens compound organism?

医药大学堂
WWW.YIYAODXT.COM

第九章 苔 藓
Chapter 9 Bryophyta

学习目标 | Learning goals

1. **掌握** 苔藓的主要特征。
2. **熟悉** 苔纲和藓纲的区别及代表性药用苔藓。
3. **了解** 苔藓的应用现状。

- Know the general characteristics of bryophyta.
- Be familiar with the difference between Hepaticae and Musci and medicinal bryophyta.
- Understand the application of bryophyta.

第一节 苔藓植物的基本特征
1 Features of the bryophyta

　　苔藓植物是一类结构简单、体型矮小，最原始的高等植物。由于苔藓植物的生殖过程依赖水，所以它们虽脱离水生环境进入陆地生活，但大多数仍需生活在潮湿环境中。苔藓植物配子体上产生有性生殖器官以及简单的胚而初步具有了适应陆地生活的基础，但其营养体只是具假根的原始茎叶体或叶状体，而不具备真正的根、茎、叶，也没有典型的维管组织。

　　Bryophyta is a kind of simple, small and the most primitive higher plants. Because the reproductive process of bryophyta depends on water, although they are separated from the aquatic environment and live on land, most of them still need to live in the moist environment. Antheridium, archegonium and simple embryos appear in the life history of bryophyta, which have the basis of preliminary adaptation to the life on land. However, the vegetative body of bryophyta is cormus or thallus with rhizoids, and does not have real roots, stems, leaves and perfect vascular tissues. Therefore, bryophyta is known as an independent phylum by most scholars.

　　苔藓植物生活史中具有明显的世代交替现象。从孢子萌发到形成配子体，配子体产生雌、雄配子，这一阶段为有性世代。从受精卵发育成胚，由胚发育形成孢子体，进而形成孢子囊的阶段称为无性世代。日常描述的苔藓植物是其配子体，在生活史中占优势，体形矮小，构造简单，为具有叶绿体的多细胞自养植物，能独立生活。较原始的苔藓植物常为扁平的叶状体，较进化的则有类似茎、叶分化，茎中尚未分化出维管束构造，无真正的根，仅有假根。苔藓植物的孢子体不发达，不能独立生活，寄生在配子体上，这是苔藓植物的显著特征之一。孢子体由孢蒴、蒴柄和基足 3 部分构成，通过基足从配子体获得营养物质。

In the life history of bryophyta, there is obvious alternation of generation. From spore germination to gametocyte formation, the gametocyte produces male and female gametes, which is a sexual generation. The stage of embryo development from zygote to sporophyte is called asexual generation. The common plant body is gametocyte, which is dominant in life history. They have simple structure and chloroplast, and short. They are small multicellular green plants, autotrophy and independent living. The lower bryophyta are usually flattened thallus, while the higher ones have stem and leaf differentiation. The vascular bundle structure has not been differentiated in the stem. There is no real root, only a single row of cells constitute the rhizoids. The undeveloped sporophyte cannot live independently and parasitize on the gametophyte, which is one of the remarkable characteristics of bryophyta. The sporophyte is composed of capsule, seta and foot, which obtain nutrients from gametocyte by foot.

苔藓植物有性生殖过程产生多细胞的雌、雄生殖器官。雌性生殖器官称颈卵器，呈长颈花瓶状，上部细狭，为颈部，中间有一条沟称颈沟，下部膨大为腹部，成熟时腹部中间有一个大型的细胞称卵细胞。雄性生殖器官称精子器，一般呈棒状或球状，内生多数具 2 条等长鞭毛的精子，以水为媒介游到颈卵器内，与卵结合。卵细胞受精后成为合子，合子在颈卵器内开始分裂，形成的多细胞结构称胚，胚依靠配子体的营养发育成孢子体，借基足依附在配子体上，行寄生生活，故苔藓植物又称有胚植物或颈卵器植物。孢子体最主要部分是孢蒴，孢蒴内的孢原细胞多次分裂后再经减数分裂，形成孢子，孢子散出后，在适宜的环境中萌发成丝状或片状的原丝体，再由原丝体发育成配子体。

The reproductive process produces multicellular female and male reproductive organs. The archegonium of the female reproductive organ is long neck vase shape, with a narrow neck at the upper part, a groove in the middle of the neck, a belly at the lower part, and a large cell in the middle of the belly called the egg. The antheridium of male reproductive organ is generally rodlike or globular, with sperms of two equal length flagella in it. Sperm swim to the archegonium through water, and combine with the egg. After fertilization, the egg becomes a zygote. The zygote develops into an embryo in the archegonium, and the embryo develops into a sporophyte depending on the nutrient substance of the gametophyte. Therefore, bryophyta are also called embryophyta or archegoniatae. The most important part of sporophyte is capsule. The archesporium cells in capsule divide many times and then undergo meiosis to form spores. The spores germinated into filiform or schistose protonema and then developed into gametophyte.

苔藓植物次生代谢产物以萜类和酚类化合物为主，苔类与藓类植物所含成分类别相差较大，如藓类植物中含有双黄酮和三黄酮类化合物，苔类植物中为双联苄、二聚倍半萜、二聚二萜等化合物。许多化合物特别是黄酮、联苄和部分萜类常以二聚体或多聚体形式存在。苔藓植物具有多种生物活性，如细胞毒和细胞生长抑制活性、植物生长调节作用、强心作用、抗炎作用等，其芳香性很强的单萜和倍半萜类化合物，能抵御有害生物的袭击，在农业、食品、医药、化妆品工业、香料中被广泛利用。

The secondary metabolites of bryophyta are mainly terpenes and phenolic compounds. And the components of liverworts and mosses are quite different, such as biflavones and triflavonoids in liverworts, and the compounds of bisbibenzyl, Sesquiterpene dimers and diterpene dimer in mosses. Many compounds, especially flavone, bibenzyl and some terpenoids, often exist as dimers or polymers. bryophyta have a variety of biological activities, such as cytotoxic and cell growth inhibitory activities, plant growth regulation, cardiotonic action, anti inflammatory action and so on. Especially the monoterpene and sesquiterpenoids with strong fragrance in bryophyta can resist the attack of harmful organisms, which can be widely used in agriculture, food, medicine, cosmetics industry and spices.

第二节 苔藓植物的分类
2 Classification of bryophyta and medicinal bryophyta

一、苔藓植物的分类
2.1 Classification of bryophyta

苔藓植物门约有 23000 种，广布世界各地。我国有 2800 多种，药用 50 余种。根据营养体的形态结构，通常将苔藓植物分为苔纲和藓纲，两个纲的植物分别称为苔类和藓类。也有其他方法分类的，如《中国苔藓志》分类编排系统中，将苔藓植物门分为苔纲、角苔纲和藓纲。

There are about 23000 species of bryophyta, which are widely distributed all over the world. There are more than 2800 species in China. More than 50 species are medicinal bryophyta. According to the morphological structure of vegetative body, bryophyta are traditionally divided into hepaticae and musci, which are also called liverworts and mosses respectively. According to the Taxonomic System of Bryophyta of China, bryophyta can be divided into hepaticae, anthocerotae and musci.

苔类植物体多为叶状体，少数有茎和叶的分化，多为两侧对称，有背腹之分；假根为单细胞构造，有茎时通常也不分化出中轴，叶多数只有一层细胞，不具中肋；孢子体的蒴柄柔弱，孢蒴先发育，蒴柄后进行有限的延伸生长，孢蒴成熟后多呈 4 瓣纵裂，孢蒴无蒴齿，内有孢子和弹丝，成熟时顶端不规则开裂。原丝体不发达，每 1 原丝体通常只发育成 1 个植株。

Most of the liverworts are thallus, a few of which have stem and leaf differentiation, most of which are bilateral symmetry. They can be divided into back and belly. The rhizoids are a single cell structure. The stem usually does not differentiate into an axis, and most of the leaves have only one layer of cells, without costa. The seta of sporophyte is weak. The capsule develops before the seta extends and grows. After the capsule maturation, it usually has four longitudinal splits, without peristomal teeth, with spores and elaters. The tip is irregularly cracked at maturity. The protonema is underdeveloped, and each protonema usually develops into one plant.

藓类植物较苔类进化，植物体多为辐射对称、无背腹之分的原始茎叶体；假根由单列多个细胞构成，分枝或不分枝；茎细胞不分化或具分化中轴，叶常具中肋；孢子体一般都有显著而坚挺的蒴柄，孢蒴的发育在蒴柄延伸生长之后；颈卵器壁在孢子体发育初期，随孢子体延长，成熟后上部断裂形成蒴帽，成熟的孢蒴多盖裂，常有蒴齿构造，孢蒴内一般有蒴轴，只形成孢子而不产生弹丝。原丝体通常发达，呈片状、块状或丝状，一个孢子萌发的原丝体上常产生多个植物体（配子体）。

The mosses evolved over liverworts, most of them are actinomorphy. The cormus cannot be divided into back and belly. The rhizoids consist of a single row of cells, which are branched or unbranched. The stem cells usually differentiate into an axis, and the leaf often has a costa. The sporophyte generally has a strong seta. And the capsule develops before the seta extends and grows. At the early stage of sporophyte development, the wall of archegonium extends with the sporophyte. The upper part of the archegonium breaks to form perigynium. The mature capsule often has peristomal teeth, with dehiscent lid. In the capsule, there is usually a capsule axis, which only forms spores without producing elaters. The protonema is usually developed, in the forms of flakes, lumps or filaments. And each protonema usually develops into multiple plants (gametocyte).

二、药用苔藓
2.2　Medicinal bryophyta

地钱 *Marchantia polymorpha* L.，地钱科。植物体（配子体）呈扁平叶状体，暗绿色，宽带状，多回二歧分叉，长 5~10cm，宽 1~2cm，边缘微波状，整齐排列的气室分隔，每室中央具 1 枚烟囱型气孔，孔口边细胞 4 列，呈十字形排列。腹面鳞片紫色；假根平滑或带花纹。雌雄异株。雄托盘状，波状浅裂，盘状体背面的每个小腔含有 1 个精子器；雌托扁平，先端深裂成 9~11 个指状裂瓣；孢蒴生于雌托的指腋腹面。叶状体背面前端常生有杯状的无性胞芽杯，内生胞芽，行无性生殖。生于阴湿的土坡或微湿的岩石及潮湿墙基。全国各地均有分布。植物体含地钱素等有效成分，全株能清热利湿，解毒敛疮。

Marchantia polymorpha L. Marchantiaceae. The gametocyte is thallus, dark green, broad-band, multiple dichotomously branched, 5–10 cm long, 1–2 cm wide, with undulate edge, hexagonal back, and orderly arranged air chambers separation. In the center of each air chamber, there is a pore of stovepipe, with four rows of cells at the edge of the pore, arranged in a cross shape. The palta of belly are purple. Rhizoids are smooth or patterned. Dioecism, the antheridiophore is discal, wavy and shallowly split. Each antheridium is located in the each loculus of back with antheridiophore. The archegoniophore is flat, which has 9–11 fingerlike lobes at the apex. The capsule is living the segmental venter of finger armpit of archegoniophore. There are often cup-shaped asexual gemma cup on the front of the back of the thallus. There are germs in the gemma cup, which is asexual reproduction. *Marchantia polymorpha* L. could be find on the dark and damp slope, slightly damp rock or wet wall base, which has wide distribution in China. The plant contains the mahantin and other effective ingredients. The whole plant has the effects of clearing heat and damp, deintoxication, promoting wound healing.

葫芦藓 *Funaria hygrometrica* Hedw.，葫芦藓科。植物（配子体）矮小直立，淡绿色，高 1~3cm。茎单一或从基部稀疏分枝。叶簇生茎顶，长舌形，叶端渐尖，全缘；叶具中肋，粗壮，消失于叶尖之下，叶细胞近于长方形，壁薄。雌雄同株异苞，雄苞顶生，花蕾状。雌苞生于雄苞下的短侧枝上；蒴柄细长，黄褐色，长 2~5cm，上部弯曲；孢蒴弯梨形，不对称，干时有纵沟槽；蒴齿两层；蒴帽兜形，具长喙，形似葫芦瓢状。生于氮肥丰富的阴湿地上。分布于东北、华北、华东、华中、西南等地区。植物体含苔藓激动素等有效成分，全株能祛风除湿，止痛，止血。

Funaria hygrometrica Hedw. Funariaceae. The gametophyte is short and erect, light green, 1–3 cm height. Stem is simple or sparsely branched from base. The leaves are clustered at the top of the stem, long tongue shaped, tapering at the leaf apex, entire leaves. The leaf often has a costa, strong, disappeared under the leaf opex, and the leaf cells are nearly oblong, with thin cell wall. Autoecious, the perigonium is apical, bud like. And the perichaetium is born on the short lateral branch under the perigonium. The seta is long and thin, yellow-brown, 2–5 cm long, which the upper part is curved. The capsule is pyriform, asymmetric, which has longitudinal groove after drying. The peristomal teeth are two-layer. The perigynium is cucullate, with long beak, like gourd. The *Funaria hygrometrica* Hedw. grows on the wet ground of nitrogen enrichment. It is distributed in northeast, north china, east china, central china, southwest and other regions. The plant contains bryokinin and other effective ingredients. The whole plant has the effects of eliminating Wind and removing Dampness, relieving pain and stopping bleeding.

其他药用苔藓植物：毛地钱 *Dumortiera hirsute*（Sw.）Reinw.，Bl. et. Nees，植物体含倍半萜类化合物，能清热，拔毒，生肌。蛇苔 *Conocephalum conicum*（L.）Dum.，全草能清热解毒，消肿止痛；外用治疗疮，蛇咬伤。暖地大叶藓 *Rhodobryum giganteum*（Sch.）Par.，植物体较大，含挥发油、酚类、黄酮类等成分，全株能养血安神，清肝明目。大叶藓 *R. roseum*（Hedw.）Limpr.，全株能养血安神。匍枝尖叶提灯藓 *Mnium cuspidatum* Hedw.，全株能凉血止血。

Other medicinal bryophyte: *Dumortiera hirsuta* (Sw.) Reinw.Bl. et. Nees contains sesquiterpenes, which has the effects of clearing heat, drawing out poison, and promoting granulation. *Conocephalum conicum* (L) Underw. can clear away heat and toxic material, detumescence, relieve pain. The external application can cure malignant boil and snake bites. The entire body of *Rhodobryum giganteum* (Schwägr.) Paris contains essential oil, phenols, flavones and other components. It has the effects of nourishing the blood and tranquilization, clearing liver and improving vision. The *Rhodobryum roseum* (Hedw.) Limpr. distributed in Heilongjiang, Jilin, Liaoning, Yunnan and other places, which has the effects of nourishing the blood and tranquilization, and the *Mnium cuspidatum* Hedw. has the effects of cooling blood and hemostasis.

重 点 小 结
Summary

苔藓植物是高等植物中最原始的绿色陆生植物，构造简单，体内没有维管组织，所以体态矮小。苔藓植物门通常可分为苔纲和藓纲。苔类植物常为扁平的叶状体，藓类一般有原始的茎叶分化和假根。其生活史中配子体独立生活，占优势地位，孢子体寄生在配子体上。

Bryophyta is a kind of simple, small and the most primitive higher plant, which is divided into hepaticae and music. Most of the liverworts are thallus. Mosses generally have stem, leaf differentiation and rhizoids. And there is no vascular tissue in bryophyta. The undeveloped sporophyte cannot live independently and parasitize on the gametophyte.

	苔纲	藓纲
配子体	叶状体，两侧对称	茎叶分化，多辐射对称
孢子体	孢蒴内有弹丝无蒴齿，纵向开裂；蒴柄柔弱	孢蒴内无弹丝，有蒴齿，盖裂；蒴柄坚挺
原丝体	不发达，只形成一个植株	发达，形成多个植株
药用代表	地钱、蛇苔等	葫芦藓、大叶藓等

	Hepaticae	Music
gametophyte	thallus, bilateral symmetry	Cormus, most actinomorphy
sporophyte	Capsule has elaters, without peristomal teeth, longitudinal splits, seta is weak	Capsule has peristomal teeth, without elaters, with dehiscent lid, strong seta
protonema	underdeveloped, and each protonema usually develops into one plant	developed, each protonema usually develops into multiple plants
medicinal bryophyta	*Marchantia polymorpha* L. *Conocephalum conicum* (L.) Dum.	*Funaria hygrometrica* Hedw. *Rhodobryum roseum* (Hedw.) Limpr.

目 标 检 测
Questions

1. 苔藓植物与低等植物比较，发生了哪些变化?

What changes have taken place between bryophytes and lower plants?

2. 蕨类植物与苔藓植物有哪些主要区别?

What are the main differences between pteridophyta and bryophyte?

3. 列举苔类植物与藓类植物的异同点?

What are the similarities and differences between liverworts and mosses?

4. 苔藓植物门有哪些常用药用植物，主要含有哪些有效成分?

What is the commonly used medicinal bryophyte? What are the main active ingredients?

第十章 蕨 类 植 物
Chapter 10 Pteridophyta

PPT

 学习目标 | Learning goals

1. **掌握** 蕨类植物的形态特征，维管组织和中柱类型以及生殖特点。
2. **熟悉** 蕨类植物的生活史及进化特征和常见药用蕨类。
3. **了解** 蕨类植物的应用概况。

- Know the morphological characteristics, internal strcuture and reproductive characteristics of ferns.
- Be familiar with the life circle, evolutional characteristics and the classification of common medicinal ferns.
- Understand the application of ferns in China.

蕨类植物是地球上最古老的植物类群之一，化石记录可以追溯到中泥盆纪。大多数早期的蕨类植物，如化石记录的蕨类植物最早的祖先 Rhacophytales 类及古树蕨类、假齿目蕨类和短齿目蕨类，以及绝大多数灌木状小型蕨类，都已灭绝。

Ferns are one of the oldest floras on Earth which can date back to the Middle Devonian. According to the fossil record, early ferns have gone extinct like the Rhacophytales, which were possibly some of the earliest progenitors of ferns, the ancient tree-like ferns Pseudosporochnales and Tempskya as well as some small bush-like Stauropterids.

蕨类植物又称羊齿植物，具有根、茎、叶的分化和原始的维管组织构成的输导系统；由于不开花结实，也常称为隐花植物。蕨类植物具有明显的无性世代和有性世代交替现象，配子体和孢子体均能独立生活，孢子体远比配子体发达，占优势地位；其有性生殖过程配子体上产生精子器和颈卵器结构，精子和卵子结合形成的合子萌发形成胚。因此，蕨类植物是介于苔藓植物和种子植物之间的过度类群，它既是高等的孢子植物，又是原始的维管植物。

Pteridophyte, also known as fern plants, have roots, stems, and leaves differentiation and a transmission system composed of primitive vascular tissues; because of the lack of flowering and fruiting, they are also often called Cryptogamia. Ferns have asexual and sexual generation alternation. Both gametophytes and sporophytes can live independently. Sporophytes are far more developed than gametophytes and occupy a dominant position; during the sexual reproduction, archegonium and antheridium are formed on the gametophytes, so the zygote formed by the combination of sperm and egg germinates to form an embryo. Therefore, pteridophytes are an excessive group between bryophytes and seed plants, which are both higher spore plants and the most primitive vascular plants.

蕨类植物广布于世界各地，以热带、亚热带地区为分布中心。地球上现存的蕨类植物约有 12000 种，我国有 2600 多种，多数分布于西南地区和长江流域以南地区，大都喜生于温暖阴湿的森林环境。云南省有 1000 多种蕨类植物，号称蕨类植物的王国。

Ferns are widely distributed all over the world, especially in tropical and subtropical regions, Today, ferns are the second-most diverse group of vascular plants on Earth and with about 12,000 extant species, more than 2600 species in China, most of which are distributed in southwest China and south of the Yangtze river basin. Yunnan province is the kingdom of ferns in China with more than 1000 species of ferns.

第一节　蕨类植物的主要特征

1 Main character of the ferns

一、蕨类植物的形态特征
1.1　Morphological features

（一）孢子体
1.1.1　Sporophyte

蕨类植物的孢子体发达，大多数为多年生草本，通常有根、茎、叶的分化。

The sporophytes of ferns are well developed, most of them are perennial, usually with differentiation of roots, stems and leaves.

1. **根**　主根不发育，常为不定根，须根状，吸收能力较强；少数具假根。

(1) Root　Taproots are not developed, mostly are adventitious roots forming a fibrous root system with strong absorptive ability. A few original species have only rhizoid.

2. **茎**　多为根状茎，少数为高大直立的地上茎，如桫椤。较原始的种类兼具气生茎和根状茎。原始类群无毛茸和鳞片，进化类群常有毛而无鳞片，高等蕨类具鳞片，如真蕨类的石韦、槲蕨等（图 10-1）。

(2) Stem　Stems are mostly rhizomes, a few are upright tree-trunk or other forms of aerial stems,

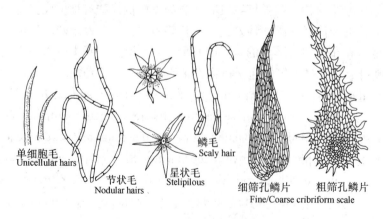

单细胞毛 Unicellular hairs
节状毛 Nodular hairs
星状毛 Stelipilous
鳞毛 Scaly hair
细筛孔鳞片 粗筛孔鳞片 Fine/Coarse cribriform scale

图 10-1　蕨类植物的毛和鳞片
Fig. 10-1　Types of hairs and scales of ferns

the more primitive species have both aerial stems and rhizomes. The stems of primitive taxon are neither hairy nor scaly, such as *Psilotum nudum* (L.) Beauvr. Evolutionary taxon are often hairy without scales. Higher ferns have scales, such as *Pyrrosia lingua* (Thunb.) Farwell and *Drynaria roosii* Nakai. (Fig.10-1).

3. **叶**　按起源和形态，分为小型叶与大型叶。小型叶较原始，由茎表皮突出形成，只具 1 条不分枝的叶脉，无叶隙和叶柄。大型叶属进化类型，由多数顶枝经扁化形成，有叶隙和叶柄，叶脉多分枝，形成各种脉序；真蕨纲植物的叶均为大型叶；大型叶幼时拳卷，成长后分化为叶柄和叶片，叶片全缘或一至多回羽状分裂；叶片中轴称叶轴，第 1 次分裂出的叶片称羽片，羽片的中轴称羽轴，从羽片分裂出的小叶称小羽片，小羽片的中轴称小羽轴，最末次裂片上的中肋称主脉或中脉。

(3) Leaf　The leaves of can be divided into microphyll and macrophyll according to the origin and morphology. The microphyll have small, usually spin-shaped leaves with no veins or only a single vein, no leaf gap and stipe. The most of other evolutionary ferns have macrophyll, leaves with leaf gap, stipes and branched veins. The macrophyll are circinate when young, and often differentiate into stipe and leaf blade. Leaf blade entire or one to many pinnations. The central axis of the leaf blade is termed the leaf axis. The first lobule is called pinna. The center axis of a pinna is called the pinna axis. The lobules that split from the pinna are called pinnule. The central axis of the pinnule is called the pinnule axis, and the central rib on the last lobe is called the principal vein or the midrib.

按叶的功能分为营养叶和孢子叶。营养叶不产生孢子囊和孢子，主要进行光合作用，又称不育叶；孢子叶能产生孢子囊和孢子，又称能育叶。有些蕨类植物无营养叶和孢子之分，形状相同，称同型叶或一型叶；也有孢子叶和营养叶形状完全不同，称异型叶或二型叶（图 10-2）。

Leaves are also divided into foliage leaf and sporophyll types according to their functions. Foliage leaves are mainly used for photosynthesis, but do not produce sporangium or spores, also termed as sterile frond. The leaves that mainly produce sporangia and spores for reproduction are called sporophyll or fertile frond. Some ferns have the homomorphic leaf due to the same type between foliage leaf and sporophyll, and have totally different shape between foliage leaf and sporophyll, it is known as heteromorphic leaf (Fig.10-2).

视频

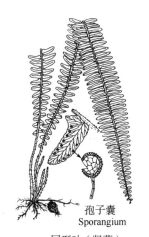

同型叶（肾蕨）
Homomorphic leaf (*Nephrolepis auriculata*)

孢子囊
Sporangium

营养叶
Foliage leaf

孢子叶
Sporophyll

孢子囊
Sporangia

孢子

异型叶（紫萁）
Heteromorphic leaf (*Osmunda japonica*)

图 10-2　蕨类的同型叶和异型叶
Fig. 10-2　The homomorphic leaf and heteromorphic leaf of the ferns

4. 孢子囊和孢子　孢子囊是蕨类植物孢子体上产生孢子的多细胞无性生殖器官。孢子囊群的发育、着生位置、形态和结构等特征是鉴定蕨类植物的重要依据。

(4) Sporangia and spores　A sporangium is a multicellular, asexual reproductive organ that produces spores on the ferns. The development of sporangia, the position of attachment, the morphology and the structure of sporangia are the important basis for the identification of ferns.

孢子囊着生位置：小型叶型蕨类的孢子叶通常聚生在枝的顶端形成球状或穗状的孢子叶穗或孢子叶球，孢子囊单生于孢子叶近轴面的叶腋或叶基部。较进化的真蕨类孢子囊常生于孢子叶背面、边缘或集生于特化的孢子叶上，孢子囊聚集成群，称孢子囊群或孢子囊堆（图 10-3）。水生蕨类的孢子囊群生在特化的孢子果或孢子荚内，如苹、满江红等。原始类群的孢子囊群常裸露，进化类群的由囊群盖覆盖或包被。孢子囊壁由单层或多层细胞构成，有一列细胞的细胞壁不均匀增厚形成环带，有助于孢子囊的开裂和孢子的散布。环带的位置有顶生环带（海金沙属）、横行中部环带（芒萁属）、斜形环带（金毛狗属）、纵行环带（水龙骨属）等（图 10-4）。

Sporangium site: Sporangium occurs singly on the leaf axil of adaxial side or leaf base of the microphyll ferns. A lot of sporophyll usually congregate at the tip of a branch and form a sphere or spike that is called sporophyll spike or strobilus. The sporangia of the more higher ferns usually appear on the back and edge of sporophyll, and concentrate on the top of a specialized sporophyll, which is defined as sporangiorus or sorus that is often clustered by numerous sporangium (Fig.10-3). The sporangia of aquatic ferns are clustered in the specialized sporocarp, such as *Marsilea quadrifolia* L. and *Azolla imbricata* (Roxb.) Nakai. The sporangia are bare in primitive ferns taxa, the evolved group usually has various shapes of indusium that is covered or enveloped, and also degenerated or disappeared.The wall of the sporangium consists of a single or multiple layers of cells. There is a line of cells that is called annulus

   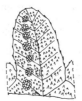 

盖孢子囊群	边生孢子囊群	顶生孢子囊群	有盖孢子囊群	脉背生孢子囊群	脉端生孢子囊群
...less sporangia	Marginal sporangia	Terminal sporangia	Covered sporangia	Sporangia on the back of vein	Sporangia at the apex of vein

图 10-3　孢子囊群着生的位置
Fig.10-3　The location of the sporangia

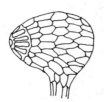  

顶生环带	横行中部环带	斜行环带	纵行环带
（海金沙属）	（芒萁属）	（金毛狗脊属）	（水龙骨属）
Top born annulus	Annulus in the middle	Oblique annulus	Longitudinal annulus
(Genus *Lygodium*)	(Genus *Dicranopteris*)	(Genus *Cibotium*)	(Genus Polypodiodes)

图 10-4　孢子囊群的环带
Fig.10-4　The annulus of the Sporangia

with an inhomogeneously thickened ring on the cell wall. It contributes to the fracture of sporangium and the spread of spore. There are many forms of annulus in sporangium, such as apical annulus (*Lygodium*), transverse central annulus (*Dicranopteris*), oblique annulus (*Cibotium*), and longitudinal annulus (*Polypodiodes*) (Fig. 10-4).

　　孢子是孢子囊内由孢子母细胞经过减数分裂而形成的无性生殖的产物。多数蕨类植物同一植株产生的孢子大小相同，称孢子同型，孢子萌发形成两性配子体。卷柏属植物和少数水生蕨类同一植株有大小孢子囊，分别产生大孢子和小孢子，称孢子异型；大孢子萌发形成雌配子体，小孢子萌发形成雄配子体。无论孢子同型或孢子异型，其形态主要有两类，一类是肾形、单裂缝、两侧对称的两面型孢子；一类是球形或钝三角形，三裂缝，辐射对称的四面型孢子。孢子外壁通常具有不同的突起或纹饰，是蕨类植物重要的分类学特征（图 10-5）。

　　Spores: Spores are the products of asexual reproduction of the sporophyte and are produced in the sporangia. Spores are formed by the meiosis of sporocyte. Most ferns produce spores with the same size that are called isospore. The isospore spores germinate to form an amphoteric Gametophyte. The spores of *Selaginella* plants and a few aquatic ferns are different in size, so it is called heterospore. The megasporangia produce macrospores and the mirosporangia produce microspores.When the megaspores and microspores germinate, they become female gametophytes male gametophytes, respectively. There are mainly two types of spores in morphology, one is the kidney-shaped, monolete, zygomorphic dihedral spore; another is a spherical or obtuse triangular, trilete suture, actinomorphic tetrahedral spore (Fig.10-5). Usually, there is a variety of decorates on the outer surface of the spores, which is an important taxonomic feature for classification of ferns.

两面型孢子　　　　　四面型孢子　　　　球状四面型孢子　　　　丝孢子
（鳞毛蕨属）　　　　（海金沙属）　　　（瓶尔小草属）　　　　（木贼属）
Two sided spores　　Tetrahedral spores　Spherical tetrahedral spore　Filamentous spores
(*Dryopteris*)　　　　(*Lygodium*)　　　(*Ophioglossum*)　　　　(*Equisetum*)

图 10-5　孢子的类型
Fig.10-5　Type of the spore

（二）配子体
1.1.2　Gametophyte

　　蕨类植物的成熟孢子在适宜的环境中直接萌发成配子体，又称原叶体。原叶体形态各异，细小，结构简单，生活期短，无根茎叶的分化，仅具单细胞的假根。绝大多数真蕨类植物的配子体为绿色，有腹背之分，能独立生活。在腹面产生球形的精子器和瓶状的颈卵器，精子器产生有多数鞭毛的精子，颈卵器内有一个卵细胞。精子和卵子成熟后，精子逸出精子器以水为媒介进入颈卵器与卵细胞受精，受精卵发育成胚，胚进一步发育成孢子体，即我们常看到的蕨类植物。孢子体幼时暂时寄生在配子体上，配子体不久死亡，孢子体即行独立生活。

　　The mature spores of the ferns will develop into gametophytes or prothallus in the right environment. The prothallus has only single-celled rhizoid without difference in roots, stems or leaves. Prothallus is different in shape, small and short in life. The gametophytes are green, thallus with ventral and dorsal differentiation, which can live independently. On the abaxial side of the prothallus are the bottle shaped archegonia and globular antheridia. The sperms come from the antheridia and the eggs are in the archegonia. When the sperm and egg get mature, the sperm escapes from the antheridia and enters the archegonia to fertilize the egg with the help of water. The fertilized egg develops into an embryo, which further develops into a sporophyte. The young sporophytes are temporarily parasitic on the gametophytes. When the gametophytes soon die, the sporophytes become independent.

二、维管组织和中柱
1.2 Vascular tissues and Steles

蕨类植物体内维管组织分化、形成各式中柱，主要有原生中柱、管状中柱、网状中柱和散生中柱等（图10-6）。中柱的类型和蕨类植物的演化有关，原生中柱是原始类型，包括单中柱、星状中柱和编织中柱；网状中柱、真中柱和散生中柱是进化的类型。中柱类型是鉴别蕨类植物及研究蕨类系统发育的重要依据之一。如中药绵马贯众的原植物粗茎鳞毛蕨 *Dryopteris cassithcoma* Nakai 叶柄的横切面有 5~13 个大小相似的维管束，排成环状；中药狗脊原植物金毛狗 *Cibotium barometz*（L.）J. Sm. 叶柄横切面肾形维管束 2~4 个，排成半圆形。紫萁贯众的原植物紫萁 *Osmunda japonica* Thunb. 叶柄横切面维管束是 1 个呈 U 字型的维管束（图10-7）。上述特征，可作为鉴别依据。

The vascular tissues of ferns differentiate into various types of steles including protostele, siphonostele, dictyostele, eustele and atactostele (Fig. 10-6). All types of steles are related to the evolution of the ferns. The direction of evolution is from original type of protostele to developed atactostele. The number, type and arrangement of steles can be used to identify medicinal ferns. For example, as the original plants of the herbal medicine 'GuanZhong', there are 5–13 vascular bundles in the cross section of the petiole of *Dryopteris cassithcoma* Nakai., which is similar in size and arranged in a ring. 2–4 reniform, semicircle arranged vascular bundles can be seen in the cross section of the petiole of *Cibotium barometz* (L.) J. Sm., while one U-shaped vascular bundle can be observed in the cross section of the petiole of *Osmunda japonica* Thunb (Fig. 10-7).

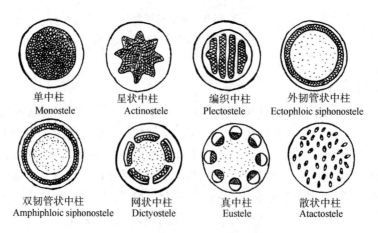

单中柱
Monostele

星状中柱
Actinostele

编织中柱
Plectostele

外韧管状中柱
Ectophloic siphonostele

双韧管状中柱
Amphiphloic siphonostele

网状中柱
Dictyostele

真中柱
Eustele

散状中柱
Atactostele

图 10-6 蕨类植物的中柱类型
Fig.10-6 Vascular bundles and steles of ferns

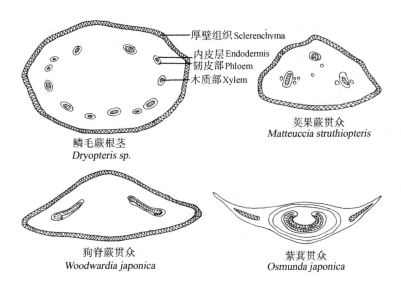

厚壁组织 Sclerenchyma
内皮层 Endodermis
韧皮部 Phloem
木质部 Xylem

荚果蕨贯众
Matteuccia struthiopteris

鳞毛蕨根茎
Dryopteris sp.

狗脊蕨贯众
Woodwardia japonica

紫萁贯众
Osmunda japonica

图 10-7　四种贯众药材叶柄横切面简图

Fig.10-7　Sketch of the cross-section of petiole of four kinds of medicinal Guanzhong

三、生活史
1.3　Life cycle

蕨类植物的生活史中具有世代交替现象。从单倍体的孢子开始至精子和卵结合前的阶段，称配子体世代（有性世代），其细胞染色体单倍（n）。从形成受精卵开始至孢子体上孢子囊中孢子母细胞减数分裂之前，这一阶段称孢子体世代（无性世代），其细胞染色体二倍（2n）。这两个世代有规律地交替完成其生活史（图 10-8）。

There is a generational alternation in the life cycle of the ferns. The period from the haploid spores to the combination between sperm and egg is called gametophytic generation (sexual generation), and the cell has a haploid number of chromosomes at this period. The stage from the formation of the fertilized egg to the meiosis of the sporogonium in the sporangium is called sporophytic generation (asexual generation), the ploidy of a cell is diploid. The two generations regularly alternate in their life cycles. (Fig.10-8).

四、化学成分
1.4　Chemical composition of ferns

1. **黄酮类**　黄酮类成分广泛存在于蕨类植物中，多具生理活性。如问荆含有异槲皮苷，问荆苷，山奈酚等；卷柏、节节草含有芹菜素及木犀草素；槲蕨含橙皮苷，柚皮苷等。在松叶蕨属、卷柏属等小型叶形蕨类中，含有双黄酮类化合物，如穗花杉双黄酮和扁柏双黄酮等。

(1) Flavonoids　Flavonoids are widely existed in ferns and have many physiological activities. e.g. *Equisetum arvense* L. contains isoqurcitrin, equicerin and kaempferol. There were apigenin and luteolin in *Selaginella tamariscina* (P. Beauv.) Spring and *Equisetum ramosissimum* Desf.). Drynaria species contains hesperidin and naringin. *Asplenium ruprechtii* Sa. Kurata contains a variety of kaempferol derivative. Pyrosia contains pyro-sitosterol and mangiferin. There are also some biflavones such as amentoflavone and hinokiflavone in the genus *Psilotum* and *Selaginella*.

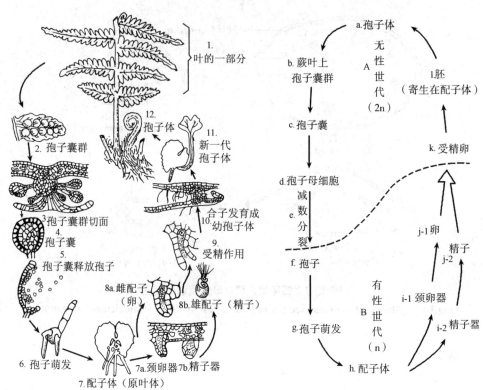

1. Part of the leaf; 2. Sori; 3.Section of sor; 4. Sporangium; 5.Sporangium releasing spore; 6. Spore germination; 7.Gametophyte(prothallus); 7a. Archegonium; 7b.Antheridium; 8a. Megagamete; 8b.Androgamete; 9. Fertilization; 10.Zygote develops into youg sporophyte; 11. New gernaration of porophyte;12. Mature porophyte; A. Asexual generation;B.Sexual generation; a. sporophyte; b. Sporangia on leaves;c. Sporangium; d. Sporoblast; e. Meiosis; f.Spore; g. Spore germinating; h.Gametocyte; i-1. Archegonium; i-2. Antheridium; j-1.Egg; j-2 Sperm; k. Fertilized egg; l. Embryol

图 10-8　蕨类植物生活史图解
Fig. 10-8　Life cycle of ferns

2. 生物碱类　生物碱类广泛存在于小型叶蕨类中。如石松科石松属含石松碱、石松毒碱、石松洛宁、垂穗石松碱等；从石杉科植物中分离的石杉碱甲能治疗阿尔茨海默症。

(2) Alkaloids　Alkaloids are widely found in microphyll ferns, for example Lycopodine, clavatoxine, clavolonine and lycocernuine in *Loeopolin*. Huperzine A can prevent alzheimer's disease from Huperziaceae.

3. 酚类化合物　二元酚类及其衍生物物在真蕨中普遍存在，如咖啡酸、阿魏酸及绿原酸等；这类成分具有抗菌、止血、止痢、利胆和升高白细胞的作用。多元酚类，特别是间苯三酚衍生物常存在于鳞毛蕨属多种植物中，如绵马酚、绵马酸类、东北贯众素等，此类化合物具有较强的驱虫作用和抗病毒活性，有毒。

(3) Phenolics　Diphenols and their derivatives are commonly found in ferns, such as caffeic acid, ferulic acid and chlorogenic acid. These constituents has effects of antisepsis, hemostasis, dysentery control, cholagogic and can increase the number of white blood cells. Polyphenols especially phloroglucinol derivatives, such as aspidinol, filicic acids, and dryocrassin, are found in most species of the genus *Dryopteris*, which have strong anthelmintic action and antiviral activity, some are toxic.

4. 甾体及三萜类化合物　蕨类植物普遍含有三萜类化合物，例如，何帕烷型和羊齿烷型五环三萜。石松含有石松素、石松醇等，蛇足石杉含有千层塔醇，托何宁醇；从紫萁、欧亚多足蕨、乌毛蕨、欧洲蕨等植物中发现的昆虫蜕皮激素，有促进蛋白质合成、排除体内胆固醇、降血脂及抑制血糖上升等活性。

(4) Steroids and triterpenes Triterpenoids are commonly found in ferns. The typical are hoppanes and ferns pentacyclic triterpenes. Lycoclavanin and lycoclavanol were found in *Lycopodium japonicum*. Tohogenol and tohogininol in *Huperzia serrata*. Insect mouting hormones, which can enhance protein synthesis, eliminate cholesterol, lower blood lipid and inhibit the rise of blood sugar, were found in *Osmunda japonica*, *Polypodium vulgare*, *Blechnum orientale* and *Pteridium aquilinum*.

此外，蕨类植物中还含有鞣质。在石松、海金沙等孢子中还含有大量脂肪。鳞毛蕨属植物的地下部分含有微量挥发油。一些蕨类植物还含有香豆素、胡萝卜素及硅化合物。

Ferns also contain tannins and a lot of fat in the spores of fern plants such as *Lycopodium japonicum* and *Lygodium japonicum*. The subterranean parts of the genus *Dryopteris* contain trace volatile oil. Some ferns also contain coumarin, carotene and silicon compounds.

第二节　蕨类植物的分类
2　Classification of ferns

　　蕨类植物的分类研究历史悠久，我国蕨类植物分类学家秦仁昌先生于 1978 年提出了蕨类植物的分类系统，将现代蕨类作为 1 个门，下设 5 个亚门，即松叶蕨亚门（Psilophytina）、石松亚门（Lycophytina）、水韭亚门（Isoephytina）、楔叶蕨亚门（Spheinophytina）和真蕨亚门（Filicophytina）。前 4 个亚门的植物都是小型叶蕨类，是一些较原始而古老的蕨类植物，现存的种类较少。真蕨亚门植物是大型叶蕨类，是最进化的蕨类植物，也是现代最繁盛的蕨类植物。近年来，基于现代分子系统学结果的蕨类植物 PPG I 分类系统（Pteridophyte Phylogeny Group I）是最新被提出的蕨类植物分类系统。本教材采用秦仁昌的分类系统。

The taxonomy of ferns has a long history. In 1978, Chinese pteridophytologist Qin Renchang proposed a classification system for ferns, which take modern ferns as 1 division and 5 subphylums as Psilophytina, Lycophytina, Isoephytina, Spheinophytina and Filicophytina. The first four sub-phytes are all microphyll ferns, which are some of the more primitive and ancient ferns, with fewer existing species. Species in Filicophytina are all macrophyll ferns and are the most evolved ferns. They are also the most prosperous ferns in modern times. With the development of molecular biological techniques and new phylogenetic analysis methods, it not only provides new perspectives and ideas for our understanding of the phylogenetic evolution of ferns, but also brings continuous challenges to some traditional taxonomic views based on morphological and anatomical evidence. In recent years, Pteridophyte Phylogeny Group I (PPG I), based on the results of modern molecular Phylogeny, is the latest proposed classification system of ferns. Qin Renchang's classification system of the ferns are introduced in this textbook.

一、松叶蕨亚门
2.1　Psilophytina

孢子体小至中型，有匍匐根状茎，具毛状假根，具直立气生茎。原生中柱。小型叶。孢子囊

2或3个聚生于叶端或叶腋，孢子同型。配子体柱状，不规则分枝，精子螺旋形，具多数鞭毛。仅存1科2属，约17种；分布于热带到温带地区。我国仅1种。

Plants small to medium-sized. Rhizomes creeping, bearing rhizoids. Stems erect or pendulous, with protostele. Sporangia appearing solitary at or above bases of sporophylls, large, 2- or 3-lobed. Isospores. Gametophytes Columnar, irregular branching. Spiral sperm have many flagellas. Only 1 family, 2 genera (*Psilotum, Tmesipteris*) and about 17 species. From tropics to temperate regions. One species in China.

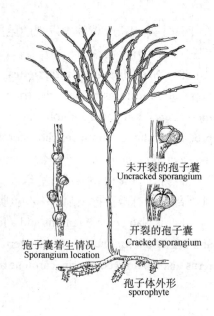

未开裂的孢子囊
Uncracked sporangium

开裂的孢子囊
Cracked sporangium

孢子囊着生情况
Sporangium location

孢子体外形
sporophyte

图 10-9　松叶蕨
Fig. 10-9　*Psilotum nudum* (L.) Beauv

【药用植物】松叶蕨 *Psilotum nudum*（L.）Beauv.（图10-9），松叶蕨科。多年生常绿附生植物。地上茎直立，二叉分枝。叶退化，厚革质。孢子叶顶端2叉；孢子囊生于叶腋，球形，初生时绿色，成熟后金黄色。分布于东南、西南及海南、台湾等地区。全草（松叶蕨）能祛风湿，舒筋活血，化瘀。

Medicinal plants: *Psilotum nudum* (L.) Beauv. (Fig.10-9) Psilotaceae. Perennial evergreen epiphytes. Aerial stems erect to somewhat pendulous, with binary branches. Leaves degenerate and thick leathery. Sporophylls deeply bifid. Sporangia yellow to yellowish brown, obtriangular-globose. Sporangium was located in leaf axilla, spherical, green at birth, golden at maturity. Distributed in southeast, southwest of China, Hainan and Taiwan. The whole grass of *P. nudum* (L.) Beauv. have the effect of expelling wind-damp, relaxing the muscles, stimulating the blood circulation and dispersing blood stasis.

二、石松亚门
2.2　Lycophytina

孢子体有根、茎、叶的分化。茎具二叉分枝，原生中柱或管状中柱。小型叶，常螺旋状排列或对生。孢子叶常聚生枝顶形成孢子叶穗。孢子囊单生于孢子叶基部。孢子同型或异型。配子体块状，精子具两条鞭毛。3科，9属，约1100种；我国各科属均有分布，近140种，药用植物约50种。

The sporophyte has the differentiation of root, stem and leaf. Main stems creeping, with dichotomously branched, with protostele or siphonostele. Leaves as microphylls, spirally arranged or opposite. Sporophyll usually forms sporophyll spike at the top of congregative branch. Strobili terminal on branchlets or main stem. Spores as isospore or heterospory. Gametophytes subterranean or surficial, massive, sperm with two flagella. 3 families, about 9 genera and 1100 species, cosmopolitan, with centers of diversity in the tropics. All families and genera are distributed in China, about 140 species, in which about 50 species are medicinal plants.

【药用植物】石松 *Lycopodium japonicum* Thunb.（图10-10）石松科。多年生常绿草本。匍匐茎细长，蔓生，多分枝；直立茎常2~3回二叉分枝。叶披针形或线状披针形。孢子叶多聚生在分枝顶端，形成孢子叶穗。孢子囊圆肾形，近叶腋生。孢子淡黄色，外壁具网纹。生于稀疏林下阴坡的酸性土壤上。分布于东北、内蒙古、河南、长江流域及以南地区。全草（伸筋草）能祛风散寒，舒筋活络，利尿通经。同属植物玉柏 *L. obscurum* L.、垂穗石松 *L. cernuum* L. 和扁枝石松属

的扁枝石松 *Diphasiastrum complanatum* L. 等的全草有类似功效。

Medicinal plants: *Lycopodium japonicum* Thunb. (Fig.10-10) Lycopodiaceae. Perennial, evergreen, stolons borne on ground, slender and creeping. Leaves lanceolate or linear-lanceolate. Sporophyll accumulate at the top of the branches, forming sporophyll spike. Sporangia round and kidney-shaped, axillary near the leaves. Spores pale yellow, with a reticulated outer wall. Growing in the acidic soil of understory slope of sparse forest. Distributed in the northeast China and Inner Mongolia, Henan, and the Yangtze river basin,etc. The herb (Shenjincao) of *L. japonicum* Thunb. has the effect of dispelling wind and cold, relaxing muscles and activating collaterals, diuresising and clearing channels.

The whole herbs of *L.otscurum L.*, *L.cernuum L.* and *L. complnetm L.* have similar effect as *L. japonicum* Thunb.

蛇足石杉 *Huperzia serrata*（Thunb.ex Murray）Trevis 石杉科。多年生草本，丛生，高 10~30 厘米。叶同型，狭椭圆状披针形。孢子囊肾形，黄色，腋生。我国除华北和西北地区外均有分布。生于林下、灌丛下、路旁。全草入药能止血散瘀、消肿止痛、清热除湿。全草含石杉碱甲等生物碱，用于改善记忆力，治疗重症肌无力和老年性痴呆等疾病。同属植物中华石杉 *H. chinensis*（Christ）Ching 的全株祛风除湿，消肿止痛，清热解毒。

Huperzia serrata (Thunb.ex Murray) Trevis Huperziacea. Perennial herb. Stems erect or ascending. Leaves sparse, homomorphic, narrowly elliptic. Sporangium reniform, yellowish, in axil of sporophyll. Distributed throughout China except for North China and NW China. Growing under orests, shrubs or roadsides. The herb of *H. serata* (Thunb.) Trevis have the effect of hemostasis, removing blood stasis, detumescence, relieving pain, clearing away heat and dampness, detoxification. The herb contains Huperzine A which can be used to improve memory, treat myasthenia gravis and senile dementia. The herb of H. chinensis (Christ) Ching are used for expelling wind and dampness; detumescence and pain; clearing away heat and detoxification.

图 10-10　石松
Fig. 10-10 *Lycopodium japonicum* **Thunb.**

卷柏 *Selaginella tamariscina*（Beauv.）Spring（图 10-11）卷柏科。多年生常绿草本。主茎短，上部分枝丛生。叶鳞片状，二型，覆瓦状近四列排列；侧叶（背叶）斜展，中叶（腹叶）向上斜升。孢子叶穗顶生，四棱形。孢子囊圆肾形，孢子异型。分布于全国各地。全草（卷柏）生用活血通经，炒炭（卷柏炭）能化瘀止血。同属垫状卷柏 *S. pulvinata*（Hook. et Grev.）Maxim. 的全草与卷柏同等入药。本科常用药用植物还有深绿卷柏 *S. doederleinii* Hieron.，翠云草 *S. uncinata*（Desv.）Spring，江南卷柏 *S. moellendorffii* Hieron. 等。

分枝的一段,示中叶及侧叶
Part of a branch, middle and side small pinna are shown

植株全形
Sporophyte

大、小孢子叶
Megasporophyll and microsporophyll

图 10-11　卷柏
Fig. 10-11 *Selaginella tamariscina* **（Beauv.）Spring**

Selaginella tamariscina (P. Beauv.) Spring (Fig.10-11) Selaginellaceae. Perennial evergreen herb. Stem terete, upper dense branched, rosette. Leaves scale-like. Dorsal leaves imbricate, spreading or parallel to axis. Ventral leaves slightly ascending. Sporangium kidney-shaped, with isospores. Distributed all over the country. Dried herb (Juanbai) of *S. tamariscina* (P. Beauv.) Spring have the effect of promoting blood circulation and promoting menstruation. Fried herb can remove blood stasis and stop bleeding. *S. pulvinata* (Hook. et Grev.) is the same herbal medicine as *S. tamariscina. S. doederleinii* Hieron., *S.uncinata* (Desv.) Spring and *S. moellendorffii* Hieron. can also be used medicinally.

三、楔叶蕨亚门
2.3　Spheinophytina

孢子体有根、茎、叶的分化。茎有节和节间，节间中空，表面有纵肋，表皮细胞常硅质化；管状中柱。不育叶小型，无叶绿素，联合成筒状轮生于节上。能育叶盾形，聚生枝顶形成孢子叶球。孢子同型或异型，圆球形，具4条弹丝。

本亚门仅1科，1属，约25种；我国有10种，其中药用8种。

Discription: The sporophyte has the differentiation of root, stem and leaf. Stem branched, with nodes. Aerial stems erect, with nodes and internodes, often with silica tubercles on epidermis, siphonostele. Strobili conelike, terminal on stem or branches, terete or ellipsoid, sometimes stalked. Sporangia saclike, Spores as homospory or heterospore, with 4 elaters. 1 family, 1 genus and about 25 species, cosmopolitan. 10 species in China, in which 8 are medicinal plants.

【药用植物】木贼 *Equisetum hyemale* L.（图 10-12）木贼科。多年生草本。地上茎单一，直立，中空，具纵脊棱 20~30 条。不育叶基部合生成筒状的叶鞘，叶鞘基部和顶端鞘齿呈黑色 2 圈。孢子叶穗顶生，长圆形；孢子同型。分布于东北、华北、西北及四川、重庆。地上部分（木贼）能疏散风热，明目退翳。问荆 *E. arvense* L. 地上茎二型，生殖枝早春先发，孢子散后枯萎。不育枝后发，中部以下多分枝。鞘齿 5~6 个。地上部分能利尿止血，清热止咳。节节草 *E. ramosissima* Desf.、笔管草 *E. ramosissimum* Desf. subsp. *debile*（Roxb. ex Vauch.）Hauke 和木贼功效相似。

Medicinal plants: *Equisetum hyemale* L. (Fig.10-12) Equisetaceae. Perennial. Aerial stem monomorphic, green, hollow, with 20~30-ridged. Leaf sheath blackish brown or both distal portion and base with a blackish brown band. Strobilus terminal ovate with isospores. Distributed in the Northeast, North and NW China. The whole herb (Muzei) of *Equisetum hyemale* L. have the effect of evacuating the wind and heat, clearing the eyes and fading the clouds.

E. arvense L. Aerial stem annual, dimorphic. Fertile stems appearing in spring earlier than sterile branches, yellowish brown. Fertile stems dying back after spores shed, below middle portion of main stem branched. Lateral branches slender, flattened. Sheath teeth 5 or 6. The overground part (Wenjing)

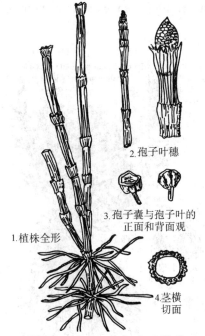

2.孢子叶穗

3.孢子囊与孢子叶的正面和背面观

1.植株全形

4.茎横切面

1.Whole body；2. Sporophyll spike；3. The front view and quilt view of sporangium and sporophyll；4. Cross section of the rhizome

图 10-12　木贼
Fig. 10-12　*Equisetum hyemale* L.

are used for clearing away heat and relieving cough, diuresis and hemostasis. The overground parts of *E. ramosissima* Desf. (Jiejiecao) and *E. ramosissimum* Desf. subsp. *debile* (Roxb. ex Vauch.) Hauke are folk herbal medicines.

四、真蕨亚门
2.4　Filicophytina

孢子体有根、茎、叶的分化。根为须状不定根。除树蕨外，均为根状茎，常被鳞片或毛。中柱各样。幼叶拳卷，叶形多样，具各式脉序。孢子囊常聚生成孢子囊群在叶背部或边缘，有柄或无柄，多具囊群盖。孢子多同型。

Discription: The sporophyte has the various of differentiation of root, stem and leaf. Root fibrous, adventitious. Most are Rhizomes except for tree ferns, often covered with scales or hairs, multiple steles. Young leaves curled, leaves diverse and have various venations. Sporangium morphology is diverse, often clustering in the back or edge of the leaf petiolate or sessile, annulus or not, indusium or not, isospore.

此亚门是现存最多的一个类群，40~58 科，10000~15000 种；我国有约 47 科，2000 余种，其中药用近 400 种。

This subphylum is the largest group of ferns, 40~58 families, about 10000~15000 species, cosmopolitan. About 47 families, 2000 species in China, in which 400 species are medicinal plants.

【药用植物】紫萁 *Osmunda japonica* Thunb. 紫萁科。多年生草本。根状茎粗短，斜升，无鳞片。叶丛生，二型，幼时密被绒毛。营养叶二回羽状；叶脉两面明显，叉状分离。孢子叶卷缩呈线状，沿主脉两侧密生孢子囊，成熟后枯死。分布于秦岭以南温带及亚热带地区。带叶柄残基的根状茎（紫萁贯众）能清热解毒，止血、杀虫；有小毒。

Medicinal plants: *Osmunda japonica* Thunb. Osmundaceae. Perennial herb. Rhizome erect, ascending, or shortly creeping, without scales. Fronds cluster, dimorphic, hairy when young. Sterile fronds 2-pinnate, veins all free, evident on both sides. Fertile fronds 2-pinnate, pinnules linear, extremely narrow, curled linear, on both sides of densely living sporangia along main vein, sporophyll is pulled out between spring and summer, dark brown, falling soon after spore dispersal. Distributed in the south of the Tsinling Mountain, warm temperate and subtropical regions in China. The Rhizome with Petiole residues (Ziqi Guanzhong) have the effect of heat clearing and detoxification, hemostasis and insecticidal. A little apoison.

海金沙 *Lygodium japonicum*（Thunb.）Sw.（图 10-13）海金沙科。多年生攀援草质藤本。根状茎横走，被黑褐色节毛。不育羽片尖三角形，能育羽片卵状三角形。孢子囊穗生于孢子叶羽片的边缘，排列成流苏状，成熟后暗褐色。孢子表面有疣状突起。分布于长江流域及以南各省区。孢子（海金沙），能清利湿热，通淋止痛，并可作丸剂包衣。此外，同属的小叶海金沙 *L. scandens*（L.）Sw.、曲轴海金沙 *L. flexuosum*（L.）Sw. 同做海金沙药用。

Lygodium japonicum (Thunb.) Sw. (Fig.10-13) Lygodiaceae. Perennial climbing vine. Rhizome widely creeping, densely clothed with dark brown hairs. Sterile fronts pointed triangles, fertile secondary branches tripinnate, pinnae smaller than sterile pinnae. Sporophyll fringed at margin, dark brown when mature; often longer than the central sterile portion of a pinnula, sparsely arranged, dark brown, glabrous. Spores finely low tuberculate to verrucose with prominent laesurae. Distributed in the Yangtze River Basin and the provinces and regions south of the Yangtze River. The spores (Haijinsha) have the effect of clearing away dampness and heat, relieving pain through drenching, and can be used as pill coating. Its overground part are used for Clearing away heat and detoxifying, relieving dampness and heat, relieving drenching. Other medicinal plants as *L. scandens* (L.) Sw. and *L. flexuosum* (L.) Sw. are used as folk herbal medicine.

金毛狗 *Cibotium barometz*（Linn.）J. Sm.（图 10-14）蚌壳蕨科。多年生树状草本，高 2~3 米。根状茎粗壮，密被金黄色长柔毛。叶簇生根状茎顶端，叶片三回羽裂。孢子囊群生于末回羽片下部小脉顶端，囊群盖成熟时张开如蚌壳；孢子三角状四面形，透明。分布于华东、华南及西南各省区。根状茎（狗脊）能祛风湿，补肝肾，强腰膝。

Cibotium barometz (L.) J. Sm. (Fig.10-14) Dicksoniaceae. Perennial arborescent herbs, 2~3 m. Rhizome prostrate, densely covered with shiny brown long hairs. Leaves clustered at rhizome apex, triplicate. Sori usually at base of lower pairs of pinnule segments; indusia bivalvate, like a clam shell; spores pale yellowish, triangular tetrahedral, transparent. Distributed in eastern, southern and southwestern provinces of China. The rhrizoma (Gouji) have the effect of dispeling wind dampness, nourishing liver and kidney, strengthening waist and knee.

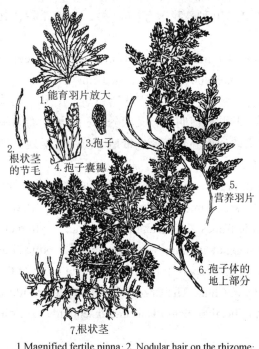

1. 能育羽片放大
2. 根状茎的节毛
3. 孢子
4. 孢子囊穗
5. 营养羽片
6. 孢子体的地上部分
7. 根状茎

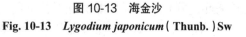

1.Magnified fertile pinna; 2. Nodular hair on the rhizome;
3. Spore; 4. Sporangiate spike; 5.Sterile pinna;
6.Aboveground part of sporophyte; 7. The rhizome

图 10-13　海金沙
Fig. 10-13　*Lygodium japonicum*（Thunb.）Sw

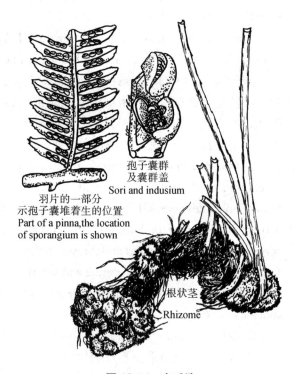

孢子囊群
及囊群盖
Sori and indusium

羽片的一部分
示孢子囊堆着生的位置
Part of a pinna,the location
of sporangium is shown

根状茎
Rhizome

图 10-14　金毛狗
Fig. 10-14　*Cibotium barometz*（L.）J. Sm.

粗茎鳞毛蕨 *Dryopteris crassirhizoma* Nakai（图 10-15）鳞毛蕨科。多年生草本，根状茎粗壮，密生棕色披针形大鳞片。叶簇生，2 回羽状全裂，叶轴密被黄褐色鳞片。孢子囊群生于叶片中部以上的羽片背面。囊群盖圆肾形，棕色，孢子具周壁。分布于东北及河北省东北部。根状茎和叶柄残基（绵马贯众）能清热解毒，止血、杀虫。

Dryopteris crassirhizoma Nakai. (Fig.10-15) Dryopteridaceae. Perennial. Rhizome erect, stout, densely scaly with the fronds. Leaves clustered, 2 pinnately split; the leaf axis densely covered with yellow-brown scales. Sori orbicular-reniform, brown. Spore with peripheral wall. Distributed in the Northeast and northeast of Hebei. The rhrizoma and petiole residues (Mianma Guanzhong) have the effect of heat clearing and detoxification, insect repellent.

石韦 *Pyrrosia lingua*（Thunb.）Farwell（图 10-16）水龙骨科。多年生常绿草本。根状茎长而横走，密被褐色披针形鳞片。叶远生，近二型，革质；叶片披针形，叶背密生淡棕色或砖红色星状毛。孢子囊群近椭圆形，在侧脉间整齐成多行排列，布满整个叶面，或聚生于叶片的大上半部，无囊群盖。分布于长江以南各省区。石韦和庐山石韦 *P. sheareri*（Baker）Ching、有柄石韦 *P. petiolosa*（Christ）Ching 的叶（石韦）能利尿通淋，清肺止咳，凉血止血。光石韦 *P. calvata*（Baker）Ching、粘毛石韦 *P. drakeana*（Franchet）Ching、贴生石韦 *P. adnascens*（Swartz）Ching 的叶也可药用。

Pyrrosia lingua (Thunb.) Farwell (Fig.10-16) Polypodiaceae. Perennial epiphytic and epilithic. Rhizome is long creeping, densely covered with lanceolate scales. Phyllopodia apart. Leaves lanceolate, with densely light brown or brick red stellate hairs on the back. Sori superficial, nearly elliptic, arranged in rows between lateral veins, distributed under the whole frond, exindusiate. Distributed in the provinces south of the Yangtze river. The fronds (Shiwei) have the effect of diuresis, clearing lung, relieving cough, cooling blood and hemostasis. Other species of the genus *Pyrrosia* have similar effects, for example *P. sheareri* (Baker) Ching, *P. petiolosa* (Christ) Ching, *P. calvata* (Baker) Ching, *P. drakeana* (Franchet) Ching and *P. adnascens* (Swartz) Ching.

槲蕨 *Drynaria roosii* Nakaike（图 10-17）槲蕨科。多年生附生草本。根状茎粗壮，密被钻状披针形鳞片。叶 2 型，不育叶黄绿色或枯棕色，厚干膜质；孢子叶绿色，羽状深裂。孢子囊群圆形，生于叶背主脉两侧，各排列成 2~4 行，无囊群盖。分布于长江以南各省区及台湾省。根状茎（骨碎补）能疗伤止痛，补肾强骨；外用消风祛斑。秦岭槲蕨 *D. baronii* Diels、团叶槲蕨 *D. bonii* Christ、石莲姜槲蕨 *D. propinqua*（Wall. ex Mett.）J. Sm. ex Bedd. 等的根状茎在部分地区也作"骨

图 10-15　粗茎鳞毛蕨
Fig. 10-15　*Dryopteris crassirhizoma* Nakai

1. *Pyrrosia lingua* (1a.Amplified stellate hair, 1b.Local amplification of lamina, 1c. Amplified sporangium, 1d. Amplified base secales); 2.*P. sheareri* (2a.Local amplified lamina; 2b.Amplified stellate hair; 2c. Amplified base secales)

图 10-16　石韦
Fig. 10-16　*Pyrrosia lingua*（Thunb.）Farwell

碎补"入药。

Drynaria roosii Nakai. (Fig.10-17) Drynariaceae. Perennial epiphytic. Rhizome shortly creeping, scales peltate densely. Fronds dimorphic; basal foliage leaf yellowish green or dry brown, thickly dry membranous. sporophyll green, pinnately split. Sori in 2-4 rows between costa and margin, exindusiate. Distributed in the south of the Yangtze river provinces, and east toTaiwan province. The rhizome (Gusuibu) have the effect of healing and relieving pain, strengthening kidney and bone, and dispeling the wind freckle by external use. The rhizomes of *D. baronii* Diels, *D. bonii* Christ, *D. propinqua* (Wall. ex Mett.) J. Sm. ex Bedd. have the similar effect with *D. roosii* Nakai..

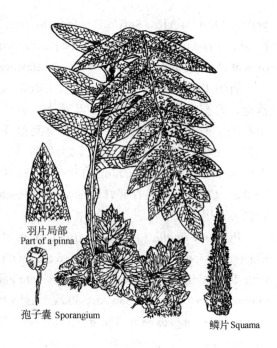

羽片局部
Part of a pinna

孢子囊 Sporangium

鳞片 Squama

图 10-17 槲蕨
Fig. 10-17 *Drynaria roosii* Nakaike

本亚门凤尾蕨科井栏边草 *Pteris multifida* Poir.、铁线蕨 *Adiantum capillus-veneris* Linnaeus、掌叶铁线蕨 *A. pedatum* Linnaeus 等的全草，中国蕨科银粉背蕨 *Aleuritopteris argentea*（Gmel.）Fee 的全草（金牛草），鳞毛蕨科贯众 *Cyrtomium fortunei* J. Sm. 的根状茎及叶柄残基，水龙骨科日本水龙骨 *Polypodiodes niponica*（Mett.）Ching、披针骨牌蕨 *Lemmaphyllum diversum*（Rosenstock）Tagawa、抱石莲 *L. drymoglossoides*（Baker）Ching 的根状茎等亦可药用。

There are also many medicinal plants in Filicophytina. For instance, the whole grass of *Pteris multifida* Poir., *Adiantum capillus-veneris* L., *A. pedatum* L.; the Rhizome and petiole residue of *Aleuritopteris argentea* (Gmel.) Fee, *Cyrtomium fortunei* J. Sm.; the Rhizome of *Polypodiodes niponica* (Mett.) Ching, *Lemmaphyllum diversum* (Rosenstock) Tagawa, *L. drymoglossoides* (Baker) Ching, etc.

重 点 小 结
Summaries

蕨类植物具有发达的孢子体，有根、茎、叶的分化。根多为不定根，须根状；茎常为根茎；叶有小型叶和大型叶以及孢子叶和营养叶之分。蕨类植物的维管系统为多种中柱。其配子体能独立生活，但孢子体在生活史中占优势。常用的中药有紫萁贯众、木贼、海金沙、石韦等。

Pteridophytes have developed sporophytes, which have roots, stems and leaves. Roots of Ferns are mostly adventitious roots, fibrous. The stems are often rhizome. The leaves are divided into microphyll and macrophyll, sporophyll and foliage leaf. The vascular system of ferns is a variety of steles. The gametophytes of ferns can live independently, but sporophytes are dominant in the life cycle. Some species commonly used as Chinese medicines that are *Equisetum hyemale*, *Osmunda japonica*, *Lygodium japonicum*, *Pyrrosia lingua*, etc.

题库

目 标 检 测
Questions

1. 列表比较蕨类植物各亚门的主要形态特征。

Compare the main morphological characteristics of each subphylum of ferns.

2. 真蕨亚门的主要药用植物有哪些？有什么形态特征及药用功效？

What are the main medicinal plants of the Filicophytina? What are the morphological features and medicinal properties?

3. 蕨类植物的生活史有什么特点？

What are the characteristics of the life cycle of ferns?

4. 怎么理解蕨类植物既是最高等的孢子植物，又是最原始的维管植物？

How to understand that ferns are both the highest spore plants and the most primitive vascular plants?

第十一章 裸子植物门
Chapter 11 Gymnospermae

学习目标 | Learning goals

1. **掌握** 裸子植物的基本特征；松科、柏科、红豆杉科、麻黄科的主要特征。
2. **熟悉** 重点科代表性药用植物。
3. **了解** 裸子植物的系统学地位及进化意义。

- Understand the basic characters of the Gymnospermae; the major characters of Pinaceae, Cupressaceae, Taxaceae, Ephedraceae.
- Be familiar with the main medicinal plants in the important families.
- Understand the characteristics and significance of Gymnosperms evolution over Ferns.

第一节 概述

1 Introduction

　　裸子植物大多数具有颈卵器构造，又产生种子，因此既是颈卵器植物，又是种子植物，是介于蕨类植物与被子植物之间的一个类群。裸子植物最早出现在距今约3亿5千万年的古生代泥盆纪，到了古生代二叠纪，银杏、松柏等裸子植物的出现，逐渐取代了古生代盛极一时的蕨类植物。古生代末期的二叠纪到中生代的白垩纪，这长达1亿年的时间是裸子植物的繁盛时期。由于地史和气候经过多次重大变化，古老的种类相继灭绝，新的种类陆续演化出来。现存裸子植物多为第三纪的孑遗植物，被称为活化石。其中银杏、银杉、水杉、巨柏、红豆杉属等是国家一级保护植物，苏铁属、福建柏、台湾杉、金钱松等属种是国家二级保护植物，均具有较高的研究价值、经济价值。

Most gymnosperms plants have the structure of archegonium and produce seeds. Therefore, they are not only an archegomiatae but also a spermatophyta, they are the a group between pteridophytes and angiosperms. The earliest seed plants were gymnosperms (naked seed plants) and appeared about 350 MYA, in the Devonian of the Paleozoic. In the Permian of the Paleozoic, the appearance of gymnosperms such as ginkgo, pine, cypress and so on gradually replaced the ferns of the Paleozoic. The 100 million years, from the Permian of the late Paleozoic to the Cretaceous of the Mesozoic, was the prosperous period of the gymnosperm. Because of many major changes in the geohistory and climate of the earth, ancient species have become extinct successively, and new species have evolved one after another. The most of the living gymnosperms are tertiary relics and are called living fossils. Among them, *Ginkgo biloba* L., *Cathaya argyrophylla* Chun et Kuang, *Metasequoia glyptostroboides* Hu et Cheng, *Cupressus gigantea* Cheng et L. K. Fu, *Taxus* and other genera are our national first-class protected plants, *Cycas*, *Fokienia hodginsii* (Dunn) A. Henry et Thomas, *Taiwania cryptomerioides* Hayata, *Pseudolarix amabilis* (J. Nelson) Rehder and so on are our national second-class protected plants. Most species of the gymnosperms have higher research and economic value.

一、形态特征
1.1　Morphological features of gymnospermae

1. **孢子体发达**　孢子体多为高大常绿乔木，少落叶，极少为亚灌木。枝条常有长、短枝之分；网状中柱，并生型维管束，具形成层和次生生长，且次生构造发达；多数科属木质部为管胞，只有麻黄科和买麻藤科植物具导管，韧皮部为筛胞，无筛管及伴胞。叶针形、条形或鳞片形，稀为扁平的阔叶，在长枝上常螺旋状排列，在短枝上簇生。常有明显的条状排列的浅色气孔带。

(1) Gymnosperms with developed sporophyte　The sporophytes are all woody, mostly evergreen, less deciduous, and rarely subshrubs. The branches are often divided into long and short ones. The vascular bundle in the Gymnosperms are collateral ones. They have the cambium and have the ability for secondary growth, and the secondary structure is well developed. The xylem of the most families and the genera of the gymnosperms are tracheids, only the species in the Ephedraceae and Gnetaceae have the vessels, and the phloem consists of sieve cells, without sieve-ture and companions. The leaves are many needle-shaped, strip-shaped, or scale-shaped, and very few species have flat broad leaves. The leaves often spirally arranged on long branches, and clustered on short ones. There are often obvious, multi-arranged, light-colored stomatal band.

2. **孢子叶球单性，胚珠裸露，形成种子**　孢子叶聚合成球果状，称孢子叶球。孢子叶球单性同株或异株。小孢子叶（雄蕊）聚生成小孢子叶球（雄球花），每个小孢子叶下面生有小孢子囊（花粉囊），囊内贮满小孢子（花粉粒）。大孢子叶（心皮）聚生成大孢子叶球（雌球花），胚珠裸露，不被大孢子叶所形成的心皮所包被，所以称为裸子植物。大孢子叶常变态为羽状大孢子叶（铁树）、珠领或珠座（银杏）、珠鳞（松柏类）、套被（罗汉松）和珠托（红豆杉）等。

(2) The unisexual strobile, exposed ovules and with seeds formation　The sporophyll of the gymnospermae aggregate into strobiliform, this is called strobilus. The unisexual strobiles are homophyletic or heterologous strain. The microsporophyll (stamen) aggregates into staminate strobilus (male cone). The microsporangium (pollen sac) are born under the microsporophyll. The microsporangium is full of microsporum (pollen grain). The macrosporophyll (carpel) aggregate into ovulate strobilus (female cone). Because of the ovule, developing are exposed, not covered by carpels formed by macrosporophyll, this group of plants are called gymnospermae. The macrosporophylls usually have different names because of the different shape, for example the pinniform macrosporophyll, such as the macrosporophyll of that in *Cycas revoluta* Thunb., the collar or the seat of ovule such as that in *Ginkgo biloba* Linn., the ovuliferous scale, such as the macrosporophyll of the conifers, the epimetium, such as the macrosporophyll of the *Podocarpus macrophyllus* (Thunb.) Sweet and salver, such as the macrosporophyll of the *Taxus chinensis* (Pilger) Rehd..

胚珠发育成种子，珠被发育成种皮。种子是携带营养的幼小孢子体。它能以不定期休眠的方式度过不良的环境，待时机合适再萌芽继续生长，种子内的营养（胚乳）保证了幼苗早期生长的需要。

The ovules develop into seeds and the integuments develop into seed coats episperm. The seed is the young sporophyte that carries nutrients and it can withstand the adverse environment by irregular dormancy. When the time is right, it will geminate and continue to grow. The nutrition (endosperm) in the seeds ensures the need for early growth of the seedlings.

3. **配子体退化，寄生在孢子体上** 世代交替中孢子体占优势，配子体退化。雄配子体在多数种类中仅由 2 个退化的原叶细胞、1 个生殖细胞和 1 个管细胞组成。除百岁兰属、买麻藤属外，雌配子体的近珠孔端均产生 2 至多个颈卵器，但结构简单，埋于胚囊中，仅有 2~4 个颈壁细胞露在外面，颈卵器内有 1 个卵细胞和 1 个腹沟细胞，无颈沟细胞，比蕨类植物的颈卵器更为简化。雌、雄配子体均寄生在孢子体上，不能独立生活。

(3) **The gametophyte degenerate, parasitizing on the sporophyte** The sporophytes are dominant in the alternate generations, and the gametophytes are degenerate. In most species, the male gametophytes are composed of only 4 cells, they are 2 degenerate prothallial cells, 1 germ cell and 1 tube cell. Except the genera of Welwitschiaceae and Gnetaceae, most of the female gametophyte of the gymnospermae produce 2 to more archegoniums near the foramen, but the structure is simple. The archegoniums are buried in the blastocyst, only 2 to 4 cervical wall cells are exposed. There are 1 egg cell and 1 ventral canal cell in the archegonium structure, no cervical sulcus cell. The structure is more simpler than the pteridophyte's ones. All the structures are parasitic in sporophytes.

4. **常具多胚现象** 大多数裸子植物出现多胚现象。根据多胚的形成过程可以分为简单多胚现象及裂生多胚现象。由 1 个雌配子体上的若干颈卵器的卵细胞分别受精，各自发育成 1 个胚，形成多胚，称为简单多胚现象；由 1 个受精卵在发育过程中发育成原胚，再由原胚组织分裂为几个胚而形成多胚，称为裂生多胚现象。

(4) **Polyembryony** The polyembryony turns up in the most gymnosperms. According to the formation process of multiple embryos, they can be divided into simple polyembryony and cleavage polyembryony. The polyembryony formation from the fertilization of the egg cells in several archegonium on one female gametophyte, they develop into embryo respectively, this phenomenon is called simple polyembryony. A fertilized egg develops into an embryo initial cell during development, and then the embryo tissue is divided into several embryos to form a polyembryony, this phenomenon is called cleavage polyembryony.

5. **形成花粉管，受精作用不再受水的限制** 裸子植物的珠孔能分泌液体，形成传粉滴，借风力传播的花粉接触时，即被吸附，传粉滴逐渐干涸，花粉经珠孔被吸入胚珠。进入胚珠的花粉先在珠心上方的贮粉室里停留一定的时间后才萌发，产生花粉管伸入胚囊，精子逸出与卵细胞受精。

此外，裸子植物的花粉粒为单沟型，有时具有气囊，无 3 沟、3 孔沟或多孔沟的花粉粒，这也是裸子植物的特征。

(5) **Formation of pollen tubes, not limited by water for the fertilization** The micropyle of gymnosperms can secrete mucus, this is the pollination drop. And the pollens dispersed by wind can be adsorbed. After the pollination drop gradually drying up, the pollens are absorbed in to the ovule through the micropyle. And there are some other characteristics in the gymnosperms, its pollens are the ascon, some of them with the aerocyst. There is no 3 grooves, 3 colporates or many more

colporates pollens.

　　裸子植物是由蕨类植物演化而来，因此其生殖器官结构具有系统发育上的紧密联系（表 11-1 ）。

The gymnosperms evolve from the pteridophytes. So there are close relationship in the reproductive organ structure between the gymnosperms and the pteridophytes (Table 11-1).

表 11-1　裸子植物与蕨类植物生殖器官形态术语的关系
Table 11-1　The Congruent Relationship of the different developmental stages in the spermatophytes and the pteridophytes

蕨类植物	裸子植物
小孢子叶球（microstrobilus）	雄球花（staminate strobilus）
小孢子（microsporophyll）	雄蕊（stamen）
小孢子囊（microsporangium）	花粉囊（pollen sac）
小孢子（microspore）	单核花粉粒（pollen grain）
大孢子叶球（megastrobilus）	雌球花（female cone）
大孢子叶（megasporophyll）	心皮或雌蕊（carpel）
大孢子囊（megasporangium）	珠心（胚珠）［nucellus（ovule）］
大孢子（megaspore）	胚囊（单细胞期）（embryo sac）

二、裸子植物的化学成分
1.2　Mainly chemical composition in the gymnosperms

　　裸子植物的化学成分类型较多，主要有黄酮类、生物碱类、萜类及挥发油、树脂等。

There are many more kinds of chemical composition in the gymnosperms, which mainly include flavonoids, terpenes, volatile oils, resins and so on.

　　1. 黄酮类　黄酮类在裸子植物中普遍存在，常见的有槲皮素、山奈酚、芸香苷等。双黄酮类化合物是裸子植物的特征性成分，如银杏双黄酮、穗花杉双黄酮、扁柏双黄酮等。从银杏叶中提取的总黄酮（含量达 24%）对冠心病具有一定的治疗效果。

　　(1) Flavonoids　Flavonoids are commonly contained in the gymnosperms, the quercetin, kaempferol, rutin and so on are the common ones. The biflavone is the distinctive in the gymnosperms, such as the ginkgetin, amentoflavone, hinokiflavone and so on. The total flavonoids, content up to 24%, extracted from the leaves of *Ginkgo biloba* L. have a good therapeutic effect on coronary heart disease.

　　2. 生物碱类　裸子植物中含有的生物碱结构并不复杂，分布不普遍。多存在于三尖杉科、红豆杉科、罗汉松科、麻黄科及买麻藤科。红豆杉属特有紫杉醇是一种二萜类生物碱，对多种肿瘤具有一定疗效。而麻黄科含麻黄类生物碱（麻黄碱、伪麻黄碱、盐酸麻黄碱等），其中麻黄碱具有镇咳、发汗、兴奋作用。

　　(2) Alkaloids　The structure of alkaloids in gymnosperms are not complicated, and the distribution are not universal. The alkaloids consist in the families of Cephalotaxaceae, Taxaceae, Podocarpaceae, Ephedraceae and Gnetaceae. The taxol in the genus of Taxeae, a kind of diterpenoid alkaloid, has good effect on several tumors. The species in Ephedraceae contain many kinds of alkaloids, the alkaloids in this family are mainly the ephedra alkaloids, such as the ephedrine, pseudoephedrine, ephedrine hydrochloride and so on. And the effect of the ephedrine is the antitussive, sweating, excitement.

3. 萜类、挥发油及树脂 萜类、挥发油和树脂等在裸子植物中普遍存在。如金钱松根皮中的土荆皮酸是一种二萜类化合物；松属植物中多含挥发油（松节油）和树脂（松香）等。

(3) Terpenes, volatile oils and resins The terpenes, volatile oils and resins are ubiquitous in the gymnosperms. The pseudolaric acid in the velamen of the *Pseudolarix amabilis* (J. Nelson) Rehder is a kind of diterpenoid. The volatile oils (Song Jie You) and resins (Song Xiang) are commonly consist in the plants of *Pinus*.

第二节　裸子植物的分类

2 Classification of gymnosperm

现存的裸子植物分为苏铁纲、银杏纲、松柏纲、红豆杉纲、买麻藤纲5纲，9目，12科，71属，近800种。我国是裸子植物种类最多，资源最丰富的的国家之一，有5纲，8目，11科，41属，约240种。已知药用植物100多种。

The existing gymnosperms can be divided into 5 classes, Cycadopsida, Ginkgopsida, Coniferopsida, Taxopsida and Gnetopsida, 9 orders, 12 families, 71 genera and about 800 species. China is one of the countries with the most types of gymnosperms and the most abundant resources. There are 5 classes, 8 orders, 11 families, about 41 genera and approximately 240 species. And more than 100 species have been known with the medicinal effect.

一、苏铁纲
2.1　Cycadopsida

【形态特征】常绿木本，茎粗壮，常不分枝。营养叶大型，羽状深裂，集生于茎干顶部。雌雄异株，雌雄花球均顶生，游动精子具多数鞭毛。种子核果状，具三层种皮，胚乳丰富。本纲始于二叠纪，盛于侏罗纪，现存1科10属，约110种，分布于南北半球的热带、亚热带地区。我国仅有苏铁属（*Cycas*）分布，8种。

Characteristic: Evergreen woody plant with thick, cylindrical stems, often unbranched. The vegetative leaves are large, pinnately split and gathering on the top of the stem. Dioecious, both staminate strobilus and female cone are born on the top of the stem, the swimming sperm has many more flagellum. Seeds drupeous, with three layers of seed coat, endosperm abundant.

The cycadopsida dated from Permian and flourished in the Jurassic.Now existing 1 family, 10 genera, about 110 species, distributing in tropical and subtropical regions of the northern and southern hemispheres. There is only *Cycas,* about 8 species in China.

【药用植物】苏铁 *Cycas revoluta* Thunb.，又称铁树（图 11-1）。常绿木本，柱状主干不分枝，具发达的髓部和皮层，内始式木质部。叶革质，羽状深裂，簇生茎顶，幼时拳卷。雌雄异株，雌雄花球均生于茎顶，雄花球圆柱状，花药通常3个聚生；雌花球球状，每大孢子叶（心皮）着生胚珠2~6枚，有绒毛。种子核果状。分布于我国南方，多栽培。根能祛风活络，补肾；叶能理气止痛，散瘀止血，消肿解毒；种子能平肝降压，镇咳祛痰，收敛固涩。

Medicinal Plant: *Cycas revoluta* Thunb. (Fig.11-1), evergreen woody plant with thick, cylindrical stems, often unbranched, with well-developed medulla and cortex, endarch xylem. Leaves are leathery, large, pinnately split, gathering on the top of the stem tufted, boxing when young. Dioecious, both staminate strobilus and female cone are born on the top of the stem. The staminate strobilu is cylindrical, anthers usually 3 aggregate. The female cone is globular with the tomentum, 2 to 6 ovules in the carpel. Seed obovate or ovoid.The root, leaves, flowers and seeds can be used as herbal medicine.

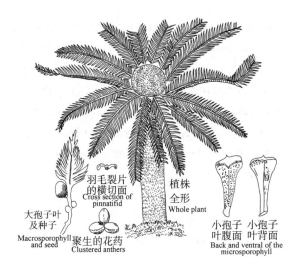

图 11-1 苏铁
Fig.11-1 *Cycas revoluta* **Thunb**.

二、银杏纲
2.2 Ginkgopsida

【形态特征】落叶乔木，枝有长、短之分，次生木质部由管胞组成。单叶，扇形，在长枝上螺旋状散生，在短枝上簇生，叶柄长，有多数叉状并列细脉。雌雄异株；雄球花葇荑花序状，雄蕊多数，各具2花药；雌球花具长梗，梗端有2个杯状珠领，又称珠座，其上各生1直立胚珠。种子核果状，外种皮肉质，中种皮白色，骨质；内种皮淡红色，膜质。胚乳肉质，子叶2枚。本纲现仅存1科1种，原产我国。

Characteristic: Deciduous arbor is with long branches and short branches. Secondary xylem consists of tracheids. Single leaves, fan shaped, clearly spirally arranged on the long branchlets or clustered on the short branches. Dioecious, the staminate strobilus has the short peduncle, like catkin, with many more stamen. The stamen has short stem, with two microsporangia at its tip and the sperm has flagellum. The female cone has two cup shaped carpels (called collar or seat of the ovule) on the top of the long stalk. Every carpel has 1 erect ovule, but just 1 ovule can develop. The seed droops, and with long stalk likes the drupe, with a fleshy, orange-coloured outer layer covered with whiting. The mesosperm is white and harder, sclerotin. The endopleura is reddish and membranous. The endosperm is meat quality. And the seed has 2 pieces of cotyledon.

1 family (Ginkgoaceae) and 1 specie (*G. biloba* Linn.). China is the origin of *G. biloba*.

【药用植物】银杏 *Ginkgo biloba* L.（图11-2），形态特征与纲同。中生代孑遗植物，系我国特产，仅在海拔为500~1000 m 的浙江西天目山有半野生种群，现广泛栽培于世界各地。去掉肉质外种皮的种子（白果）能敛肺定喘，止带缩尿。叶（银杏叶）能活血化瘀，通络止痛，敛肺平喘，化浊降脂。从叶中提取的总黄酮能扩张动脉血管，用于治疗冠心病。

Medicinal Plant: *Ginkgo biloba* L. is a famous Mesozoic relic plant that is original in our country (Fig.11-2). The semi-wild population are only in the western Tianmu Mountain of Zhejiang Province at an altitude of 500m to 1000 m. And now, it is widely cultivated around the world. The seeds (Ginkgo Semen,Baiguo) can astringe the lung qi to relieve wheezing, relieve leukorrhea and reduce urination.

The leaves (Folium ginkgo, Yinxingye) can promote blood circulation for removing blood stasis and obstruction in collaterals for relieving pain, astringing lung for preventing asthma, eliminate turbid pathogen reducing lipid. The total flavonoids extracted from the leaves can dilate arterial blood vessels and are used to treat coronary heart disease.

三、松柏纲
2.3　Coniferopsida

常绿或落叶乔木或灌木。茎有长、短枝之分，次生木质部发达，多具树脂道。叶针形、鳞形、钻形、条形或刺状，单生或成束。雄球花单生或组成花序，雄蕊多数。雌球花的珠鳞两侧对称，生于苞鳞腋部。球果的种鳞鳞片状，扁平或盾形，两侧对称，成熟时张开，稀合生。种子核果状或坚果状，有翅或无翅，胚乳丰富，子叶 2~16 枚。4 科，约

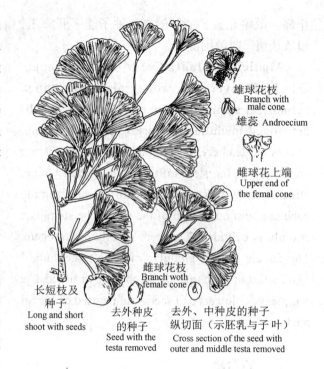

雄球花枝
Branch with male cone

雄蕊 Androecium

雌球花上端
Upper end of the femal cone

雌球花枝
Branch woth female cone

长短枝及种子
Long and short shoot with seeds

去外种皮的种子
Seed with the testa removed

去外、中种皮的种子纵切面（示胚乳与子叶）
Cross section of the seed with outer and middle testa removed

图 11-2　银杏
Fig.11-2　*Ginkgo biloba* L.

400 种，是现代裸子植物中数目最多、分布最广的类群。我国 3 科，23 属，约 150 种。

Evergreen or deciduous arbor or shrub, with long branches and short ones. The secondary xylem is well developed and consist of tracheids and without a catheter. The stem in most of the species has the resin duct. The leaves show acerose, obdeltoid, subulate or linear, solitary or fascicle. The staminate strobilus is solitary or inflorescence. There are many more stamen in one staminate strobilus and no flagellum on the sperm. The ovuliferous scale on the female cone are zygomorphous, growing in the axilla of the bract. The seed-scale on the cone is lepidoid, deplanate or peltate, bilateral symmetry. It will open when the seeds mature, rare ones are concrescent. The seeds have wing or not, liking nut or drupe. The endosperm is abundant, and there are 2 to 16 cotyledon in the seed.

There are more than 400 species in this class, and they can be divided into 4 families: Pinaceae, Taxodiaceae, Cupressaceae and Araucariaceae. And the Coniferopsida is the most numerous and widely distributed taxa in modern gymnosperms. There are 3 families, 23 genera, about 150 species in China.

1. 松科
(1) Pinaceae

【形态特征】常绿乔木，稀落叶。叶针形或条形，常 2~5 针一束，在长枝上螺旋状散生，在短枝上簇生。雄球花穗状，腋生或单生枝顶；雌球花球状，由多数螺旋状排列的珠鳞（心皮）及苞鳞（苞片）组成，每个珠鳞的腹面基部有 2 枚倒生胚珠，背面具 1 苞鳞（苞片），与珠鳞分离。花后珠鳞增大发育成种鳞，聚成木质球果。种子顶端常具单翅；有胚乳，子叶 2~16 枚。10 属，约 230 种。广布于全世界。我国 10 属，约 110 种，分布全国各地。已知药用 8 属，40 余种。

Characteristic: Evergreen trees or rarely deciduous. Leaves acicular or liner, mostly 2 to 5 in one bunch, clearly spirally arranged on the long branchlets or tufted on the top of the short ones. The staminate strobilus axillary or solitary, or with numerous spirally arranged on the top of the branchlets.

The female cone shows globos, consists of carpels which spirally arranged and the bracts. There are 2 upright ovules in the base of the carpels front and with 1 bract in its back. The ovules are free or only basally adnate with bracts. After flowering, the bracts develop into seed scales. And all the seed scales are appressed into globos, exserted or included. Seeds terminally winged (except in some species of *Pinus*) with endosperm. Cotyledons 2~16.

10 genera, about 230 species distributed almost all over the world. 10 genera, about 110 species in China. 8 genera and more than 40 species have been known with medicinal value.

【药用植物】马尾松 *Pinus massoniana* Lamb（图 11-3）常绿乔木。叶 2 针 1 束，细长柔软，长 12~20cm，树脂道 4~8 个，边生。种鳞鳞脐微凹，无刺尖。分布于长江流域及以南各地。树枝的瘤状节（松节）能祛风除湿，通络止痛；花粉（松花粉）能收敛止血，燥湿敛疮。渗出的油树脂，经蒸馏所得的挥发油（松节油）能减轻肌肉痛、关节痛、神经痛及扭伤；油树脂经加工后得到的非挥发性天然树脂（松香）能燥湿祛风，生肌止痛；鲜叶（鲜松叶）、种子、树皮等均可药用。油松 *P. tabuliformis* Carr.、云南松 *P. yunnanensis* Franch. 与马尾松同等入药。

Medicinal Plants: *Pinus massoniana* (Fig.11-3) is evergreen tree. The leaves are needle-like in the short branchlets, with two per fascicle, 12–20 centimetres, 4 to 8 resin canal in the ege of the leaves. The seed scale is apophysis-rhombus or slight hypertrophy, and the scale navel is slightly sunken. The joint or knob (Songjie) can dispel pathogenic wind and remove dampness, dredge channels and relieve pain. The pollen (Pine Pollen,Song Huafen) has the ability of astringing for hemostasis, eliminating dampness and astringing sores. The volatile oil getting from the oleoresin of the plant with the method of distil or other extraction methods (Turpentine) can be used for relieving the pain in the muscle, joint, nerve and sprain. The colophony which is a kind of non-volatile natural resin getting from the oleoresin after processing can dispel pathogenic wind and remove dampness, promote granulation and relieve pain. And the fresh leaves, seeds, bark and so on also can be used as medicine.

常用药用植物还有金钱松 *Pseudolarix amabilis*（J.Nelson）Rehder，根皮（土荆皮）能杀虫，止痒。

Other species as *P.tabuliformis* and *P.yunnanensis* have the same function as *P. massoniana*. In

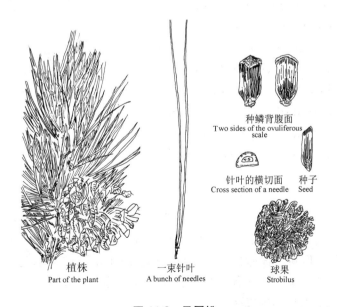

种鳞背腹面
Two sides of the ovuliferous scale

针叶的横切面 种子
Cross section of a needle Seed

植株 一束针叶 球果
Part of the plant A bunch of needles Strobilus

图 11-3 马尾松
Fig.11-3 *Pinus massoniana* Lamb.

addition, the velamen of the ***Pseudolarix amabilis*** named as Golden Larch Bark can kill parasites and relieve itching.

2. 柏科

(2) Cupressaceae

【形态特征】常绿乔木或灌木。叶交互对生或轮生，鳞形或刺状。雄球花顶生，椭圆状卵形，花粉无气囊；雌球花球形，珠鳞交互对生或轮生，珠鳞与苞鳞合生。球果木质或革质，或肉质浆果状。种子无翅或具窄翅，具胚乳，子叶2枚。约22属，150余种，分布与南北两个半球。我国8属，约30种，几遍全国。已知药用20种。

Characteristic: Evergreen trees or shrubs, Leaves decussate or in whorls of 3 to 4, lepidoid or acerose. The oval staminate strobilus are arranged on the top of the branches, each bearing 2 to 6 pollen sacs. The pollen has no saccate. The female strobilus is globose, the cone scales are decussate as 3 to 16 or 3 to 4 whorled. The cones are globos, leathery or woody. Seeds with or no winged has the endosperm. Cotyledons usually 2.

22 genera, about 150 species in Cupressaceae. Distributed in the southern and northern hemispheres. 8 genera, about 30 species in China, in which 20 species are medicinal plants.

【药用植物】侧柏 *Platycladus orientalis*（L.）Franco（图11-4），常绿乔木。小枝扁平，排成一平面，直展。叶鳞形，交互对生，贴生于小枝上。球花单性同株，球果熟时木质，开裂。种子卵形或近椭圆形。除新疆、青海外，分布几遍全国。枝梢和叶（侧柏叶）能凉血止血，化痰止咳；种子（柏子仁）能养心安神，润肠通便。

Medicine plant: *Platycladus orientalis* (Fig.11-4) Evergreen trees. The branchlets flat, arranged in a plane and spread straightly. Leaves lepidoid, apex bluntly pointed, decussate or whorled, adnate on the branchlets. Unisexual, monoecious. The cones are woody, reddish brown, dehiscent after mature. Seeds grayish brown or purplish brown, ovoid or subellipsoid. It is distributed all the country, except Xinjiang Province and Qinghai Province. Twigs and leaves (Cebaiye) can cool blood, stop bleeding, resolve phlegm and relieve coughing. Mature seed (Platycladi Semen, Baiziren) can calm the mind for tranquillization, moisten the large intestine and relieves constipation and arrest sweating.

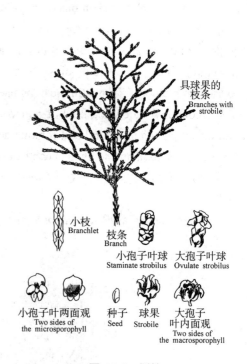

图 11-4　侧柏
Fig.11-4　*Platycladus orientalis*（L.）Franco

3. 杉科

(3) Taxodiaceae

【形态特征】常绿或落叶乔木。叶同型或2型，条形、披针状、鳞片状或钻形，螺旋状排列或交互对生。雄球花单生或簇生枝顶，花粉无气囊；雌球花顶生，由珠鳞与苞鳞组成，二者多为半合生（仅顶端分离），珠鳞腹面基部有2~9枚胚珠。球果木质或者革质，当年成熟，种子小，有窄翅，子叶2~9枚。10属，约16种，主要分布于北温带。我国7属，7种，引入栽培4属7种，分布于长江流域及秦岭以南各省区。

Characteristic: Evergreen or deciduous arbor. Leaves are homomorphic or 2 kinds of shapes, liner, lanceolate, lepidoid or subulate. The staminate strobilus are solitary or acervate on the top of the branches. The strobilus femineu terminal, consisting of cone scale and bract, they are half cleft, just cleft on the top. 2 to 9 ovules on the ventral base of the cone scale. The cones leathery or woody, maturing in the flowering year. Seeds no or winged. Cotyledons 2 to 9.

10 genera, about 16 species. Distributed is northern temperate. 7 genera, about 7 species in China. Importing 4 genera, 7 species for cultivation. Distributed is in the Yangtze River Basin and the south of Qinling Mountain.

【药用植物】杉木 *Cunninghamia lanceolata*（Lamb.）Hook.（图 11-5），常绿乔木。叶条状披针形，沿中脉两侧各有 1 条白气孔带。苞鳞与珠鳞合生，苞鳞大，珠鳞先端 3 裂，腹面具 3 胚珠。种子扁平，两侧具窄翅。分布于秦岭、淮河以南。树皮能祛风止痛、燥湿止血；杉木炭能收敛止血；枝干结节祛风止痛、散瘀止血。

Medicinal plant: *Cunninghamia lanceolata* (Fig.11-5) Evergreen tree. Leaves narrowly linear or lanceolate, 1 white powdery stomatal band on each side of the midrib. The bract scale and cone scale are connate. The bract scale is bigger than the cone scale. There are tilobation on the top of the cone scale and 3 ovule on the venter. The seeds flat and round, two flanks slightly ridged.Distributed in the south of the Qinlin Mountain and Huaihe River. The bark can dispel pathogenic wind for relieving pain, eliminate dampness for stopping bleeding. Its charcoal can astringing for hemostasi. And the nodule of the branches resin can dispel pathogenic wind for relieving pain, eliminate blood stasis and stopping bleeding.

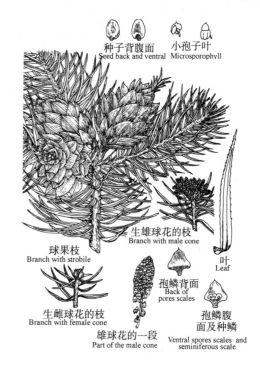

种子背腹面
Seed back and ventral
小孢子叶
Microsporophyll

生雄球花的枝
Branch with male cone

球果枝
Branch with strobile

叶
Leaf

孢鳞背面
Back of pores scales

生雌球花的枝
Branch with female cone

雄球花的一段
Part of the male cone

孢鳞腹面及种鳞
Ventral spores scales and seminiferous scale

图 11-5 杉木
Fig.11-5 *Cunninghamia lanceolata*（Lamb.）Hook.

松柏纲除上述 3 科外，还有分布南半球热带和亚热带的南洋杉科（Araucariaceae），共 2 属，约 40 种。我国引种 2 属，4 种。常见为南洋杉（*Araucaria cunninghamia* Sweet），为 5 大庭院树种之一，其木材可供建筑、器具、家具等用。

Except for those families, there is another family named Araucariaceae distributed in the tropical and subtropical zones of the Southern Hemisphere, with 2 genera and about 40 species. China imported 2 genera, 4 species. *Araucaria cunninghamia* is one of the 5 best known garden trees. Its timber can be used for buildings, appliances, furniture and so on.

四、红豆杉纲
2.4　Taxopsida

常绿乔木或灌木。叶条形、披针形、鳞形等。球花单性异株，稀同株。胚珠生于盘状或漏斗状的珠托上，或包于杯状或囊状的套被中。种子具肉质假种皮或外种皮。3 科，14 属，约 162 种。我国 3 科，7 属，33 种。

Evergreen trees or shrubs. Leaves linear, lanceolate, squamiform and so on. Unisexual and dioecism, rarely monoecious, Ovules develop on the collar which is discoid or funnelform, or in the cupulate or bladderlike bracts. The seeds are with fleshy aril or outer seed coat. 3 families, 14 genera and about 162 species. 3 families, 7 genera and 33 species in China.

1. 三尖杉科

(1) **Cephalotaxaceae**

【形态特征】常绿乔木或灌木，叶条形或条状披针形，交互对生或近对生，基部扭转成2列，背面有白色气孔带2条；球花单性异株，稀同株；雄球花腋生；雌球花有多对交互对生的苞片，基部苞片成珠托并发育成肉质假种皮，包围种子；种子核果状。

本科仅三尖杉属（*Cephalotaxus*）1属，9种。我国产7种，主要分布于秦岭淮河以南各省。本科木材结构细致，材质优良；枝、叶、根、种子可提取多种生物碱，对白血病及淋巴肉瘤等有一定的疗效。

Characteristic: Evergreen trees or shrubs. Leaves linear or linear-lanceolate, decussate or nearly opposite, twisted into 2 lines on lateral branches, with 2 white stomas bands in the lower surface. Unisexual and dioecism. The staminate strobilus caespitose. The strobilus femineu with many pairs of bracts, the basal bracts are the trays of the ovules. The seeds completely coated by the fleshy aril developed from the trays of the ovules. Only 1 genera, as *Cephalotaxus,* 9 species. 7 species China, distributed in the provinces in the south of the Qinling Mountain and Huanghe River. The wood structure and quality of this family are excellent. Many kinds of alkaloid with the effect of curing the leukemia, lymphatic sarcoma and other cancer can be extracted from the branches, leaves, roots and seeds.

【药用植物】 三尖杉 *Cephalotaxus fortunei* Hook. f. 常绿乔木。小枝对生。叶螺旋状排成两列，背面中脉两侧各有1条白色气孔带。雄球花单生叶腋；雌球腋生。种子假种皮成熟时紫色或红紫色，顶端有小尖头。种子能润肺，消积，杀虫；从其枝叶中提取的三尖杉总碱对淋巴肉瘤、肺癌有较好疗效，对胃癌、上颚窦癌、食道癌有一定疗效。粗榧 *Cephalotaxus sinensis* （Rehd. et Wils.）Li.（图11-6）常绿小乔木。叶条形，通常直。雄花有短梗。与三尖杉功用相似。

Medicinal plant: *Cephalotaxus fortunei,* evergreen trees. Leafy branchlets obovate, persistent blastozoid on the base. Leaves borne spirally, 2-ranked, blade linear-lanceolate with 1 white stomatal bands on each side of the midvein abaxially. Staminate strobilus 8–10 capitula globose, solitary. When ripen the aril of the seed turns to purple or reddish violet, and there is a small cusp on its top. Seed can moisten the lung, disperse accumulation and kill worms. Cephalotaxine, the alkaloid extracted from branches and leaves, have certain degree of efficacy on gastric cancer, maxillary sinus carcinoma, and oesophageal

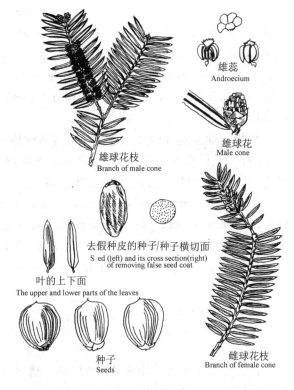

雄蕊
Androecium

雄球花
Male cone

雄球花枝
Branch of male cone

去假种皮的种子/种子横切面
S ed (left) and its cross section(right) of removing false seed coat

叶的上下面
The upper and lower parts of the leaves

种子
Seeds

雌球花枝
Branch of female cone

图 11-6 粗榧
Fig. 11-6 *Cephalotaxus sinensis*（Rehder et Wils）Li

carcinoma, and have better efficacy on lymphatic sarcoma, lung cancer. *C. sinensis* (Fig.11-6) has the similar medicinal value as *C. fortunei*.

2. 红豆杉科

(2) Taxaceae

【形态特征】常绿乔木或灌木。叶条形或披针形，螺旋状排列或交互对生，基部常扭转排成 2 列，上面中脉明显，下面沿中脉两侧各有 1 条气孔带。雌雄异株；雄球花单生叶腋，或成穗状花序集生于枝顶；雌球花单生或成对生腋生，胚珠 1 枚，基部具盘状或漏斗状珠托。种子核果状或坚果状，包于肉质假种皮。5 属，20 余种，主要分布于北半球。我国 4 属，约 12 种，已知药用 10 种（含变种）。

Characteristic: Evergreen trees or shrubs. Leaves spirally arranged or decussate, abaxial surface with 1 stomatal band, resin canal present or absent. Dioecious, rarely monoecious; staminate strobilus solitary in leaf or bract axils, or aggregated into spike-like complexes apically on branches; female cone solitary or paired in axils of leaves or bracts; ovule solitary, erect, with discoid or funnel-shaped ovule tray at base. Seed drupelike, sessile, completely enclosed within fleshy aril, or pedunculate, partially enclosed in a succulent, saccate aril.

5 genera and about 20 species, mainly in northern hemisphere. 4 genera and about 12 species in China, 10 species (including variety) known as medicinal plants.

【药用植物】红豆杉 *Taxus wallichiana* var. *chinensis*（Pilg.）Rehder，常绿乔木。叶条形，微弯或较直，长 1~3cm，宽 2~4mm，叶上面深绿色，下面淡黄色，有 2 条气孔带。种子生于杯状红色肉质假种皮中。我国特有种，分布于甘肃南部至两广和西南等省区。叶治疥癣；种子能消积，驱虫。茎皮中所含紫杉醇（paclitaxel）具有明显的抗肿瘤作用。同属南方红豆杉 *T.chinensis*（Pilg.）Rehder var. *mairei*（Lemee & H. Léveillé）W.C.Cheng et L.K.Fu、西藏红豆杉 *T.wallichiana* Zucc.、东北红豆杉 *T.cuspidata* Sieb. et Zucc. 均可作提取紫杉醇的原料。

Medicinal plants: *Taxus wallichiana* var. *chinensis* Trees everygreen. Leaves 2-ranked, linear, straight to distally falcate, upper part slightly tapered, apex usually with slightly raised pike; adaxial surface deep-green, abaxial surface lightly yellow, 2 stomatal bands. Seed enclosed within red fleshy cupular aril but with apex exposed, sometimes on the fleshy discoid seed tray Endemic in China. Distributed from southern Gansu to Guangxi and Guangdong. Leaves are effective in treating sacies, seeds can disperse accumulation and kill worms. Paclitaxel in barks have obvious anti-tumor effects and better therapeutic efficacy on oophoroma, non-small cell lung cancer, breast carcinoma, gastric carcinoma, metrocarcinoma. *T.chinensis* (Pilg.) Rehder var. *mairei* (Lemee & H. Léveillé) W.C.Cheng et L.K.Fu, *T.wallichiana* Zucc, and *T.cuspidata* Sieb. et Zucc.have resemble with *T. chinensis* (Pilg.) Rehder in medicinal parts and efficacy.

榧树 *Torreya grandis* Fortune ex Lindl. 高大乔木。叶条形，先端凸尖。种子熟时假种皮淡紫褐色，有白粉。种子（榧子）能杀虫消积，润燥止咳，润燥通便。

Torreya grandis: Trees to 25m tall. Leaves linear-lanceolate, usually straight ridged. Seeds (Torreyae Semen, Feizi) can kill worms and disperse accumulation, moisten dryness and relieve cough, moisten dryness and promote defecation.

五、买麻藤纲
2.5 Gnetopsida

灌木或木质藤本。次生木质部具导管，无树脂道。叶对生或轮生，鳞片状或阔叶状。球花单性，有假花被；成熟雌球花球果状、浆果状或穗状。种子包于由盖被发育而成的假种皮中。3 目 3 科 3 属，约 80 种；我国有 2 目，2 科（麻黄科及买麻藤科）2 属，19 种，分布几乎遍布全国。百岁兰科主要分布于非洲西南的海岸沙漠地带，为单种科，仅百岁兰（*Welwitschia bainesii* Hook.f.）一种，我国部分植物园的温室中有栽培。

Usually shrub or ligneous liana. Secondary xylem often have vessels, resin duct absence. Leaves opposite or in whorls, lepidoid or latifolious shape. Cone unisexual.False perianth membranous, leathery, fleshy. Female cone strobiloid, baccate or spicate. Seed concealed in aril that evolved from operculums.

3 orders, 3 families, 3 genus and about 80 species. 2 orders, 2 families, 2 genera and 19 species in China, nearly all over the country. In addition, Welwitschiaceae is mainly in coastal desert region of southwestern Africa and is the family with only one species, *Welwitschia bainesii,* which is cultivated in greenhouse of some Chinese botanical garden.

1. 麻黄科
(1) Ephedraceae

【形态特征】灌木、亚灌木或草本状。小枝对生或轮生，绿色，具节和节间，横断面常有棕红色髓心。叶退化成膜质鳞片状，对生或轮生，2~3 片合生成鞘状。雌雄异株，稀同株。雄球花单生或数个丛生，具 2~8 对交互对生或轮生的苞片，每苞片有 1 雄花，外包膜质假花被，有雄蕊 2~8 个，花丝合成 1~2 束；雌球花具 2~8 对交互对生或轮生的苞片，仅顶端 1~3 枚苞片内有雌花，雌花具顶端开口的囊状革质假花被，胚珠 1，珠被上部延长成珠被管，由假花被开口处伸出，珠被管直或弯曲，假花被发育成革质假种皮，包围种子，成熟时变成肉质，红色或橘红色。1 属，约 40 种，分布于亚洲、美洲及欧洲东部及非洲北部等干旱、荒漠地区。我国约有 12 种，4 变种。已知药用约 10 种。

Characteristic: Shrubs, subshrubs, or herbs. Branchlets opposite or whorled, green, terete, with longitudinally groove between internodes, brownish red pith in cross section usually. Leaves opposite or in whorls of two or three, small, basally connate to membranous sheaths, scalelike. Dioecious, rarely monoecious. Staminate strobilus solitary or clustered at nodes, each cone composed of membranous bracts arranged in 2 to 8 decussate pairs or whorls, each bract subtending a male flower. Stamens 2 to 8, 1 to 2 filaments into bunch; female cone composed of membranous bracts arranged in 2 to 8 decussate pairs or whorls, only the 1 to 3 bracts on the top with one female flower, cystic and leathery false perianth, opening on the top of the female flower, enclosing ovule with a single membranous integument prolonged into a slender, tubular micropyle, exserted from false perianth, integument tube straight or curved, false perianth developed into a leathery aril surrounding the seed, red or orange and fleshy at maturity.

1 genus and about 40 species, in the arid and desert regions of Asia, America, eastern Europe and northern Africa. 12 species and 4 varieties in China. 10 species known as medicinal plants.

【药用植物】草麻黄 *Ephedra sinica* Stapf.（图 11-7），草本状灌木，高 20~40cm；木质茎短或成匍匐状；小枝（草质茎）丛生于基部，直伸或微曲，面细纵槽纹常不明显，具明显的节和节间。鳞片状叶 2 裂，基部鞘状，裂片锐三角形。雄花有雄蕊 7~8；球花单生于幼枝顶端或 2~3 个生于老枝节上，苞片通常 4 对，成熟时变肉质红色。种子常 2 枚，包于肉质的苞片内。分布于辽宁、吉林、内蒙古、陕西、河北、山西等省区。生于沙质干燥地带，适应性强，常组成大面积单纯群落。草质茎（麻黄）能发汗散寒，宣肺平喘，利水消肿；其生物碱含量丰富，作为提取麻黄碱的原料。根和根状茎（麻黄根）能固表止汗。同属中麻黄 *E. intermedia* Schrenk ex C. A.Mey.、木贼麻黄 *E. equisetina* Bunge. 与草麻黄同等入药。

Medicinal plant: *Ephedra sinica*, (Fig.11-7) herbaceous shrub, 20 to 40 cm tall; woody stems short or prostrate; branchlets clustered at base, straight or lightly curved, shallowly furrowed, with evident. Node and the internode. Leaves 2-lobed, scalelike, membranous, base sheath-like, narrowly triangular. Pollen cones usually compound spike, mostly with a basic peduncle, bracts usually in 4 pairs, opposite decussated or whorled, anthers 7 or 8, staminate strobilus on the tip of the young branches solitary, or on the old branches with 2~3 fascicled, bracts usually in 4 pairs, opposite or whorled, red and fleshy at maturity. Seeds usually 2, concealed by fleshy bracts. Distributed in Liaoning, Jilin, NeiMongol, Shaanxi, Hebei, Shanxi. The herbaceous stem (Ephedrae Caulis,Mahuang) can promote sweat and dissipate cold, diffuse the lung and relive panting, promote urination and disperse swell; The root and the rhizome (Radix et Rhizoma Ephedrae,Mahuang Gen) have the efficacy of consolidating the exterior and arrest sweating. In addition, *E. intermedia* and *E. equisetina* are equivalent to *E. sinica*.

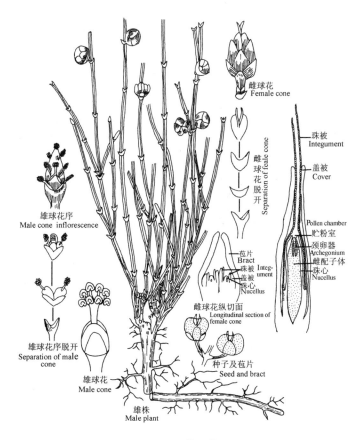

图 11-7 草麻黄
Fig.11-7 *Ephedra sinica* Stapf.

2. 买麻藤科

(2) Gnetaceae

【**形态特征**】常绿木质藤本。单叶对生，无托叶，叶片革质或半革质，具羽状网脉。雌雄异株；雄球花穗单生或数穗组成顶生及腋生聚伞花序，着生在小枝上，雄花具有肉质杯状假花被；

雌球花穗单生或数穗组成聚伞圆锥花序，通常侧生于老枝上，雌花假花被囊状，紧包于胚珠之外，胚珠具两层珠被，内珠被的顶端延长成珠被管，自假花被顶端开口伸出，外珠被分化为肉质外层与骨质内层，肉质外层与假花被合生并发育成假种皮。种子核果状，包于红色或桔红色肉质假种皮中，胚乳丰富，肉质；子叶2枚。1属，30余种，分布于热带及亚热带。我国产7种。

Characteristic: Evergreen woody vines. Simple leaves opposite, without stipules, leaves leathery or half-leathery, pinnately net-vein. Flowers unisexual, dioecious or sometimes monoecious. Staminate strobilus solitary or several in cyme inflorescences, born the axil, usually arranged on old stems, Staminate flower carnose cupulate pseudoperianth, stamens are usually 2, rarely 1, exserting from false perianth; Female spikes solitary or several in thyrses inflorescences, usually cauliflorous clusters on old stems. Female flowers with a false perianth tightly enclosing ovule; ovule with 2 integuments, innermost integument elongated into a micropylar tube exserted from false perianth; outer integument with a fleshy, outer layer connate with false perianth and developing into a false seed coat. Seeds drupelike, enclosed in a red, or orange, fleshy false seed coat. 1 genus and about 30 species, tropical and subtropical. 7 species in China.

【药用植物】小叶买麻藤 *Gnetum parvifolium*（Warb.）C. Y. Cheng ex Chun（图 11-8），常绿木质缠绕藤本，长 10 米以上。叶对生，革质，先端急尖或渐尖而钝，基部楔形或稍圆，边全缘。雄球花序具 5~10 轮环状总苞片，每轮有雄花 40~70；雌球花序一回三出分枝，每轮总苞片内有雌花 5~8 朵。种子核果状，熟时假种皮红色。产于福建、广东、广西及湖南等省区。其根与茎（麻骨风）以及叶入药，能祛风活血，消肿止痛，化痰止咳。同属买麻藤 *Gnetum montanum* Markgr. 亦同等入药。

Medicinal plant: *Gnetum parvifolium* (warb.) C.Y.Cheng ex Chun (Fig.11-8), evergreen volube vines to more than 10m tall. Leaf opposite, leathery, with acute or acumen and blunt tip, base cuneate to subrounded, apex obtuse, margin entire. Male inflorescences simple or once branched, involucral collars 5 to 10, each collar with 40 to 70; female spikes once branched, ternate, with involucral collars, nodes each with 5 to 8 female flowers. Seeds drupelike, red false seed coat at maturity. Distributed in Fujian, Guangdong, Guangxi, Hunan.The roots, stems, and leaves are used as medicine, which can dispel wind and invigorate blood, disperse swell and relieve pain, dissolve phlegm and relieve cough. *Gnetum montanum* Markgr. has the equivalent efficacy.

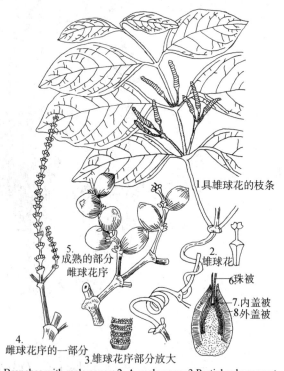

1.具雄球花的枝条
5.成熟的部分雌球花序
2.雄球花
6.珠被
7.内盖被
8.外盖被
4.雌球花序的一部分
3.雄球花序部分放大

1.Branches with male cones; 2. A male cone; 3.Partial enlargement of male inflorescence; 4.Part of female inflorescence; 5.Mature artial female inflorescence; 6.Integument; 7.Inner cover; 8.Outer cover

图 11-8 小叶买麻藤
Fig.11-8 *Gnetum parvifolium*（Warb.）C. Y. Cheng ex Chun

重 点 小 结
Summaries

　　裸子植物的基本特征为孢子体发达，配子体退化，具颈卵器结构，胚珠裸露，多具变态大孢子叶，多胚现象，传粉时精子直达胚珠，产生种子。银杏、红豆杉、草麻黄等为裸子植物中重要的药用植物。

The basic characteristics of the gymnosperms are the developed of sporophytes, the degradation of gametophytes, with the structure of the archegonium, the exposed ovules, most of the plant with abnormal megaspores. Polyembriony, the sperm can reach the ovules with nonstop and produce seeds. *Ginkgo biloba* L., *Taxus wallichiana* var. *chinensis* (Pilger) Florin, and *Ephedra sinica* Stapf are important medicinal plants in gymnosperms.

目 标 检 测
Questions

题库

　　1．怎样理解裸子植物与苔藓植物、蕨类植物相比，其整个生命过程减少了对水的依赖。

How to understand the differences between gymnosperm and bryophyte, pteridophyte about the relying on the water in the life process?

　　2．如何判断某一未知植物是麻黄科植物而不是木贼科植物？

How to judge one unknown plant is *Ephedra* plant but not an *Equisetum* plant?

　　3．本校药植园中存在的裸子植物有多少种？并列出名称和药用部位。

How many kinds of gymnosperm plants exist in our school's medicinal plant garden? And please list the name and section for the medicine.

第十二章 被子植物门
Chapter 12 Angiospermae

被子植物门是目前进化程度最高、种类最多、分布最广、结构最精细的植物类群。自新生代起源以来，被子植物在空间分布和谱系发育系统中均占据着绝对优势。与其他类群相比，被子植物的生态习性和形态结构更加多样化，生殖器官和生殖过程进一步特化，使之对环境的适应能力更强。因被子植物具有典型的花，故又称有花植物。

Angiospermae is a plant group with the highest degree of evolution, the most species, the widest distribution and the most fine structure. Since they are originated in the Cenozoic era, Angiosperms form the most dominant group both in spatial distribution and phylogenetic system. Compared with other plant taxa, the vegetative and reproductive organs of Angiosperms are more complex and diversified, and have stronger adaptability to various environments. Because angiosperms have typical flowers, they are also called flowering plants.

第一节 被子植物的一般特征

1 General characteristics of angiosperms

1.**孢子体高度分化，配子进一步简化** 被子植物的孢子体在形态、结构和生活型方面，更加多样化。有乔木、灌木和草本之分；陆生和水生兼有。木质部出现导管，韧皮部出现筛管和伴胞，输导组织的完善使植物体内水分和营养物质的运输更加快捷有效。被子植物的配子体进一步简化，终生寄生在孢子体上，无独立生活能力。雌配子体即成熟的胚囊，成熟时只有8核（7个细胞）结构，即1个卵细胞、2个助细胞、3个反足细胞、1个中央细胞（2个极核），原始的颈卵器结构完全退化。雄配子体由单核花粉粒发育而来，仅具1个管细胞（营养细胞）和1~2个精核。

(1) Sporophyte highly differentiated and the gametophyte extremely simplified　The sporophytes of angiosperms are more diverse in morphology, histology and habitats. There are arbors, shrubs and herbs; both terrestrial and aquatic. The vesels in the xylem and sieve tubes and companion cells in the phloem enhance the transportation of water and nutrients more efficiently. The gametophyte of angiosperms is further simplified, parasitic on the sporophyte for life, and has no ability to live independently. The female gametophyte evolved into the mature embryo sac. At maturity, it has only 8 nuclei (7 cells), that is, 1 egg, 2 synergids, 3 antipodal cells, and 1 central cell (2 polar nuclei) and the primitive archegonia are completely degraded. The male gametophyte is developed from mononuclear pollen grains, with only one tube cell (vegetative cell) and 1 to 2 sperm nuclei.

2. **具双受精现象和新型胚乳**　雌配子体在受精过程中，2枚精子细胞经花粉管导入8核胚囊后，1枚与卵细胞结合，形成受精卵，将来发育成2n的胚；另1个与两个极核结合，发育成3n胚乳。双受精作用产生的新型三倍体胚乳，为幼胚发育提供营养，增强了幼胚的生命力。

(2) Angiosperms have unique double fertilization and new endosperm　In the process of fertilization, two sperm cells are introduced into the 8-nucleated embryo sac through the pollen tube, and one sperm combines with the egg to form a fertilized egg, which will develop into a biploid embryo. The other sperm combines with two polar nuclei and develop into triploid endosperm.The new kind of triploid endosperm by double fertilization provides nutrition for the development of immature embryo and enhances its vitality.

3. **具有真正的花和果实**　被子植物典型的花由花被（花萼和花冠）、雄蕊群、雌蕊群等组成。花被的出现给异花传粉提供了条件，极大地提高了传粉效率。大孢子叶（心皮）发育形成封闭的子房，胚珠包藏在子房内，得到良好保护。子房受精后发育成果实，果实既保护种子又以各种方式帮助种子散布。

(3) Angiosperms have real flowers and fruit　The typical flowers of Angiosperms are composed of perianth (calyx and corolla), androecium and gynoecium. The appearance of perianth provides the condition for cross-pollination and greatly improves the efficiency of pollination. The megaspore leaf (carpel) develops into a closed ovary in which the ovule is enclosed and well protected. The ovary develops into a fruit after fertilization, which both protects the seed and assists in seed dispersal in various ways.

4. **营养方式和传粉方式多样化**　大多数被子植物为自养型植物，也有寄生、半寄生和腐生等；被子植物还与微生物形成复杂的共生关系，增强了对环境的适应能力。花部结构的高度分化推动了被子植物和传粉媒介的协同进化，使被子植物能适应风媒、虫媒、鸟媒、水媒等多种传粉方式，促进了基因流动和物种分化。

(4) With diversified nutritional and pollination patterns　Most angiosperms are autotrophs, sometimes parasitic, semi-parasitic or saprophyte plants, and have a complex symbiotic relationship with microorganisms, which enhances their adaptability to the environment. The high differentiation of floral structure promoted the co-evolution of angiosperms and pollinators, which enabled angiosperms to adapt to various modes of pollination such as wind, insect, bird and water, and promoted gene flow and species differentiation.

第二节 被子植物的起源与演化规律

2 Origin and evolution of angiosperms

一、被子植物的起源
2.1 Origin of angiosperms

由于相关化石资料的缺乏，目前人们对被子植物花的起源了解有限。系统分类学家们根据现有证据及推断，提出了一些假说，影响比较大的有假花说与真花说。

The origin and early evolution of angiosperms are enigmas that have intrigued botanists for well over a century. Due to the lack of fossil data of flowers, the origin of angiosperms is poorly understood. Based on the available evidence and extrapolation, botanists have proposed several hypotheses, among which the most influential hypothesis are Pseudanthium theory and Euanthial theory.

1. 假花学说 该学说设想被子植物起源于裸子植物麻黄类如弯柄麻黄 *Ephedra campylopoda* C. A. Mey.，认为被子植物的花是由裸子植物的单性孢子叶穗演化而来，即，裸子植物的小孢子叶球（雄花序）的苞片演变成花被，小苞片退化，每一雄花只剩 1 枚雄蕊；大孢子叶球（雌花序）的苞片演变成心皮，其他部分退化后，仅剩下胚珠内藏于子房中。按照此学说，有单性花、无被花、柔荑花序和风媒花是被子植物的原始特征。现在多数学者不赞成此看法。

(1) Pseudanthium theory Commonly associated with the Englerian School, the theory was first proposed by Wettstein, who postulated that angiosperms were derived from the Gnetopsida, represented by Ephedra, Gnetum and Welwitschia (formerly all placed in the same order Gnetales).It is believed that the flowers of angiosperms are derived from the strobilus of gymnosperms, that is, the bracts of the microsporophylls (male inflorescences) of the gymnosperms become perianth, the bracteoles degenerate, each male flower has only one stamen. The bracts of the megasporophyll (female inflorescence) evolve into carpels, and the rest degenerate, leaving only the ovules enclosed in the ovary.

2. 真花学说 该学说设想被子植物来源于已灭绝的原始裸子植物—拟苏铁植物，认为被子植物的花由裸子植物的两性孢子叶球演化而来，裸子植物的大孢子叶演变成被子植物的雌蕊（心

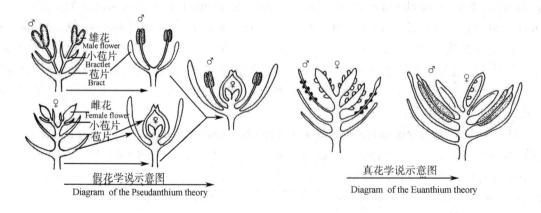

图 12-1 真花学说与假花学说

Fig.12-1 Euanthium theory and pseudoanthium theory

皮），小孢子叶演变成雄蕊。据此，两性花、重被花、离生心皮雌蕊、虫媒花试是被子植物的原始特征。该学说被多数学者赞同。

(2) Euanthial Theory　Also known as Anthostrobilus theory, Euanthial theory was first proposed by Arber and Parkins. According to this theory, the angiosperm flower is interpreted as being derived from an unbranched bisexual strobilus bearing spirally arranged ovulate and pollen organs, similar to the hermaphrodite reproductive structures of some extinct bennettitalean gymnosperms. The carpel is thus regarded as a modified megasporophyll (phyllosporous origin of carpel).The bisexual flower of Magnoliales has been considered to be evolved from such a structure.

二、被子植物的演化规律
2.2　Basic evolutionary trends

被子植物的形态特征，包括营养器官和生殖器官，特别是花和果的形态特征，是研究其演化的主要依据。而原始被子植物花的化石的缺乏，使通过化石研究被子植物演化和亲缘关系相当困难。表 12-1 是一般公认的被子植物形态构造的主要演化规律。

The morphological characteristics of angiosperms, including vegetative and reproductive organs, especially flowers and fruits, are the main basis for studying their evolution. Evolution within angiosperms has proceeded along different lines in different groups. Numerous trends in the evolution of angiosperms have been recognized from comparative studies of extant and fossil plants. The lack of fossil makes it difficult to study the evolution and relationship of angiosperms. Table 12-1 is generally recognized as the main evolution of angiosperm morphology and structure.

表 12-1　被子植物形态特征的主要演化规律

	初生的性状特征	次生的性状特征
根	主根发达（直根系）	主根不发达（须根系）
茎	乔木、灌木	多年生或一、二年生草本
	直立	藤本
	无导管，有管胞	有导管
	环纹、螺纹导管，梯纹穿孔，斜端壁	网纹、孔纹导管，单穿孔，平端壁
叶	单叶	复叶
	互生或螺旋排列	对生或轮生
	常绿	落叶
花	单生	形成花序
	各部螺旋排列	各部轮生
	重被花	单被花或无被花
	各部离生	各部合生
	各部多数而不固	各部有定数（3、4 或 5）
	辐射对称	两侧对称或不对称
	子房上位	子房下位
	两性花	单性花
	雌雄同株	雌雄异株
	花粉粒具单沟	花粉粒具 3 沟或多孔
	虫媒花	风媒花

续表

	初生的性状特征	次生的性状特征
果实	单果、聚合果	聚花果
	菁葖果、蒴果、瘦果	核果、浆果、梨果
种子	种子多	种子少
	胚小、有发达胚	胚大、无胚
	子叶 2 枚	子叶 1 枚
生活型	多年生	一或二年生
	绿色自养植物	寄生、腐生植物

Table 12-1　Basic evolutionary trends of Angiosperms

	Primitive traits	Evolutionary traits
Root	Taproot well developed (taproot system)	Taproot undeveloped (fibrous root system)
Stem	Arbors, shrubs	Herbs Perennial or annual or biennial
	Erect	Vine
	No vessels, having tracheid	Vessels exist
	Annular and spiral vessels, scalariform perforation	Reticulate and pitted vesels, mono-scalariform
Leaf	Simple leaf	Compound leaf
	Alternate or spiral arrangement	Opposite or whorled
	Evergreen	Deciduous
Flower	Flowers solitary	Forming inflorescences
	Spiral arrangement	Whorled
	Double Perianth flower	Simple perianth or achlamydeous flower
	Choripetalous	Synpetalous
	Numerous but not fixed	Numbers fixes (3,4 or 5)
	Actinomorphic flower	Zygomorpic flower or asymmetric flower
	Superior ovary	Inferior ovary
	Bisexual flower	Unisexual flower
	Monoecism	Dioecism
	Pollen with a single groove	Pollen with 3 grooves or multiplepores
	Entomophilous flower	Anemophilous flower
Fruit	Simple fruit or aggregate fruit	Collective fruit
	Follicle,capsule, achene	Drupe, berry nand pome
Seed	Many seeds	Less seeds
	Embryo small, endosperm developed	Embryo large, no endosperm
	Cotyledons 2	Cotyledon 1
Life forms	Perennial	Annual or biennial
	Green autotroph	Parasitic saprophyte

PPT

第三节　被子植物的分类系统
3　Classification systems of angiosperm

自 19 世纪后半期以来，有许多植物分类学工作者依据前述的假说，提出了多种不同的被子植物分类系统，以下对影响较大的系统做简要说明。

Since the second half of the 19th century, many plant taxonomists have proposed many different Angiosperm systems based on the theories above. The followings are brief descriptions.

一、恩格勒分类系统
3.1　Engler's system

恩格勒系统是德国分类学家恩格勒和勃兰特以假花学说为基础合著的《植物自然分科志》中提出的，于 1897 年发布，共 23 卷。该系统是分类学史上第一个比较完整的系统，将植物界分为 13 门，第 13 门为种子植物门，种子植物门又分为裸子植物和被子植物两个亚门。后经多次修订，至 1964 年修订出版第 12 版《植物分科志要》，已接受了很多真花说的论点，如把双子叶植物调整在单子叶植物之前等。这部巨著对植物分类学和被子植物系统学产生了很大影响，因为其内容丰富、全面、系统，除英、法外的许多国家至今仍在使用该系统。《中国植物志》《中国高等植物图鉴》和我国许多地方植物志、本教材以及中国国家标本馆和许多科研教学机构的标本馆，大都采用恩格勒系统。

It was proposed by German taxonomists A. Engler and K. Prantl in " Die Natürichen Pflanzen-familien " based on the *Pseudanthium Theory*, published in 1897 and had 23 volumes. This system is the first complete Angiosperm system in the taxonomic history. The plant kingdom is divided into 13 phylum, the 13th phylum is seed plant which be divided into Gymnosperms and Angiosperms. Later, after several revisions, the 12th edition of Syllabusder Pflanzen-familien was revised and published in 1964. Many viewpoints of Euanthium Theory have been accepted, such as putting Dicotyledons in front of Monocotyledons. This great work has a great influence on plant taxonomy and Angiosperm systematics, because its content is rich, comprehensive and systematic, which is still used in many countries except England and France. The Engler system is widely used in Flora of China, Atlas of Higher Plants of China, many flora of local region, this textbook, china national herbarium, many research institutes and universities.

二、哈钦松分类系统
3.2　Hutchinson system

英国植物学家哈钦松在 1926 年和 1934 年发表了该系统，1973 年做了修订，共 111 目，411 科。该分类系统依据真花学说，认为被子植物单子叶植物和双子叶植物有共同的起源；双子叶植物中多心皮类包括木兰目和毛茛目是最原始的，其他木本植物起源于木兰目，草本植物起源于毛茛目；柔荑花序类群较进化，起源于金缕梅目等。其所持的"草本植物和木本植物两支平行发

医药大学堂
WWW.YIYAODXT.COM

展"的观点，使原本关系较接近的类群过早分支，如唇形目与马鞭草目、五加科与伞形科等，被现代研究结论所抵触，被多数分类学家反对。我国南方很多单位如中科院昆明植物所、华南植物所、广西植物所等植物标本室均采用该系统。

It is published in 1926 and 1934 by British botanist J. Hutchinson, and be revised in 1973, with a total of 111 orders and 411 families. According to Euanthium Theory, the Hutchinson system considers that Monocotyledons and Dicotyledons have the same origin; protopolycarpicae are the most primitive in Dicotyledons, such as Magnolia and Ranunculaceae; the herbaceous plants are derived from Ranunculales, the woody plants are derived from Magnoliales; catkin group, which is relatively evolved, is derived from Hamamelidales, etc. The order of plants is as follows: Dicotyledons is in front of Monocotyledons; plants whith double perianth flowers are in front of those with simple perianth flowers; plants whith bisexual flowers are in front of those with unisexual flowers, etc. Although the system has been widely used in many herbarium of southern of China, such as Kunming Institute of Botany, Huanan Institute of Botany, Guangxi Institute of Botany and so on, the viewpoint of "parallel development of two branches of herbaceous and woody plants" can lead to early branches of groups originally close to each other, such as Lamiales and Verbenales, Araliaceae and Umbelliferae, which has been repudiated by studied in modern times.

三、塔赫他间分类系统
3.3　Takhtajan system

该系统是前苏联植物学家塔赫他间于1954年《被子植物起源》一书中公布的。塔赫他间系统经多次修订，1980年增加了"超目"分类单元，首先打破了传统把双子叶植物分为离瓣花亚纲和合瓣花亚纲的框架，摈弃了人为因素。该系统收录28超目，92目，416科，其谱系体系多数被当今植物解剖学、孢粉学、植物细胞分类学和化学分类学的研究所证实，但其在分类等级上增设的"超目"，因过于繁杂，应用受到一定局限。

This system was published by A. Takhtajan, a former Soviet botanist, in his book the Origin of Angiosperms in 1954. When the system was revised in 1980, the "superorders" taxon was added, which broke the traditional theory that Dicotyledons dividing into Choripetalae and Sympetalae. This system includes 28 superorders, 92 orders and 416 families. Most of its pedigree systems have been confirmed by the research of plant anatomy, palynology, plant cell taxonomy and chemical taxonomy. However, it is too complicated to be applied extensively.

四、克朗奎斯特分类系统
3.4　Cronquist system

美国植物学家克朗奎斯特以真花说为基础理论，于1968年发表了该系统，于1981年修订出版的《有花植物的综合分类系统》中最终完善，包括64目、383科。本系统将被子植物门称为木兰植物门，分木兰纲和百合纲。该系统接近于塔赫他间系统，但取消了"超目"，使系统更为简化，有利于教学和应用，还引用了解剖学、古植物学、植物化学、地理学等证据，所以显得更为成熟。最新的克氏系统由美国分类学家维利尔修订，载于其网络讲义（1998年）中，未正式发表，科数增加至389科。

A. Cronquist, an American botanist, based on Euanthium Theory, published this system in 1968, and finally perfected it in The Evolution and Classification of Flowering Plants which published in 1981, including 64 orders and 383 families. In this system, Angiospermae is named Magnoliophyta, which is divided into Magnoliopsida and Liliopsida. This system is close to Takhtajan system, but it cancels the " superorders ", makes the system more simplified, and is conducive to teaching and application. It seems more mature because of accepting the evidence of anatomy, palaeobotany, phytochemistry, geography etc. Now many botanists still recommend using this classification system. The latest Cronquist system was revised by James L. reweal, an American taxonomist, and was published in his network handout (1998) which including 389 families.

五、APG 系统
3.5 APG system

国际研究组织"被子植物系统发育研究组"以分子系统学理论和成果为依据，于 1998 年首次提出被子植物分类系统 APG Ⅰ；2003 年修订发布了 APG Ⅱ；2009 年修订出版 APG Ⅲ；2016 年修订的 APG Ⅳ 是目前最新的被子植物分类系统，该系统共有 64 个目，416 个科，数目适宜，是目前国际植物分类学界很多人推荐代替传统分类系统的应用和教学的基础工具。

An international research organization, Angiosperm phylogeny group (APG) based on the theory and achievements of molecular systematics, first proposed APG I in 1998; APG II was revised and published in 2003; APG III was revised and published in 2009; APG IV is the latest Angiosperm system published in 2016, which has 64 orders, 416 families. APG IV is a system which be recommended to replace the traditional taxonomic system as an application and basic teaching tools by many taxonomists in the world.

第四节 被子植物的分类
4 Classification of angiosperms

PPT

目前已知的被子植物有 1 万余属，25 万种左右。我国有 3148 属，约 3 万余种，是世界上被子植物最丰富的地区之一，其中药用 213 科、1957 属、10027 种。本书采用恩格勒系统（1964），将被子植物门分为单子叶植物纲和双子叶植物纲进行介绍。

At present, there are more than 10 000 genera and 250 000 species of Angiosperms. China is one of the most abundant angiosperms in the world which has 3,148 genera, about 30,000 species. Among them, there are 213 families, 1 957 genera, 10 027 species are medicinal plants. This textbook follows the Engler's System (1964), which divides angiosperms into Dicotyledoneae and Monocotyledoneae.

一、双子叶纲
4.1 Dicotyledoneae

子叶 2 枚；茎具中央髓部，木本植物具年轮；叶多网状脉；花常 5 或 4 数。分为离瓣花亚纲和合瓣花亚纲。

Cotyledons 2; stems with central pith, woody with rings; leaves with reticulate veins; flowers often 5 or 4. Divided into Choripetalae and Sympetalae.

（一）离瓣花亚纲

4.1.1　Choripetalae

离瓣花亚纲又称原始花被亚纲（Archichlamydeae），花两性或单性；单被、重被或无花被；花瓣分离；雄蕊着生在花托上；种子多少有胚乳；风媒或虫媒传粉。

Also known as Archichlamydeae. Flower bisexual or unisexual; perianth simple, double or achlamydeous; petals separate; stamens grow on the receptacle; seeds have endosperm; pollination by wind or insect.

1. 三白草科　$\male\female * P_0 A_{3-8} \underline{G}_{3-4:1:2-4, (3-4:1:\infty)}$

(1) Saururaceae

【形态特征】多年生草本。单叶互生，托叶与叶柄合生或缺。穗状或总状花序，基部常有总苞片；花小，两性，无花被；雄蕊3~8；心皮3~4，离生或合生，每离生心皮有胚珠2~4，或心皮合生成1室子房，侧膜胎座，胚珠多数。蒴果或浆果。

Characteristic: Perennial herbs. Alternating single leaves whose stipules connate with petiole or absent. Spike or raceme, usually with involucre at base; flowers small, bisexual, without perianth; stamens 3~8; carpels 3~4, free or connate; ovules 2-per free carpels,or carpels connate into 1-roomed ovary; parietal placentation; ovules numerous. Capsule or berry.

【分布】4属，约7种，分布于亚洲东部和北美洲。我国3属，4种，主产中部以南各省区，均药用。

Distribution: 4 genera, about 7 species, distributed in eastern Asia and North America. 3 genera, 4 species in China. All are medicinal plants.

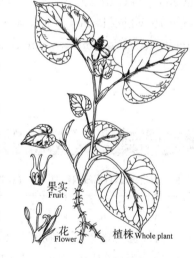

图 12-2　蕺菜
Fig.12-2　*Houttuynia cordata*

【药用植物】蕺菜 *Houttuynia cordata* Thunb. 全草（鱼腥草）能清热解毒，消痈排脓。三白草 *Saururus chinensis*（Lour.）Baill（图12-2）地上部分（三白草）能利尿消肿，清热解毒。

Medicinal plants: *Houttuynia cordata* Thunb.(Fig.12-2), the whole herb (Houttuyniae Herba,Yu xing cao) have the effect of clearing heat and detoxifying, removing carbuncle and eliminating pus. *Saururus chinensis* (Lour.) Baill, the aboveground part (Saururi Herba,San bai cao) have the effect of inducing diuresis to alleviate edema and clearing heat and detoxifying.

2. 胡椒科　$\male P_0 A_{1-10}$；$\female P_0 G_{(2-5:1:1)}$；$\male\female P_0 A_{1\sim10} G_{(2-5:1:1)}$

(2) Piperaceae

【形态特征】藤本、灌木或草本。常有辛辣香气。单叶互生，稀对生或轮生，全缘；托叶与叶柄合生或缺。穗状花序或再排成伞形花序，与叶对生或腋生；花小，无花被，两性、单性异株或间有杂性；雄蕊1~10，花丝离生；心皮2~5，合生，子房上位，1室。浆果小，果皮肉质、薄或干燥。种子具丰富的外胚乳。

Characteristic: Vines, shrubs, or herbs, often spicy. Leaves single, alternate, sparsely opposite or whorled, entire; stipules connate with petiole or absent. Spikes or umbels, opposite or axillary to leaves; flowers small, without perianth, bisexual, dioecism or polygamy; stamens 1–10; filaments free; carpels 2–5, connate;ovary superior, 1-roomed. Berry small; pericarp fleshy, thin or dry. Seeds with abundant perisperm.

【分布】8 或 9 属，约 3100 种；我国 4 属，70 余种；已知 2 属 34 种药用。

Distribution: 8 or 9 genera, about 3100 species; 4 genera in China, more than 70 species; 2 genera and 34 species are known for medicinal use.

【药用植物】胡椒 *Piper nigrum* L.（图 12-3）. 果实（胡椒）能温中散寒，下气，消痰。未成熟果实干后变黑色，称黑胡椒；成熟果实除去果皮后呈白色，称白胡椒。荜茇 *P. longum* L. 干燥近成熟果穗（荜茇）能温中散寒，下气止痛。风藤 *P. kadura*（Choisy）Ohwi 干燥茎藤（海风藤）能祛风湿，通经络，止痹痛。石南藤 *P. wallichii*（Miq.）Hand.-Mazz. 茎（穿壁风）能祛风湿，强腰膝，补肾壮阳。

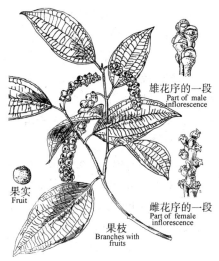

雄花序的一段
Part of male inflorescence

果实
Fruit

雌花序的一段
Part of female inflorescence

果枝
Branches with fruits

图 12-3　胡椒
Fig.12-3　*Piper nigrum*

Medicinal plants: *Piper nigrum* L.(Fig.12-3), the fruit (Piperis Fructus, Hu jiao) have the effect of warming the middle warmer to expelling cold,descending Qi and reducing sputa. The immature fruit turns black after drying, called black pepper; the ripe fruit is white after removing the peel, called white pepper. *P. longum* L., the dried mature infructescence (Piperis Longi Fructus, Bi ba) have the effect of warming the middle warmer to expell cold, descending Qi to relieve pain. *P. kadura* (Choisy) Ohwi, the dry stems (Piperis Kadsurae Caulis, Hai feng teng) have the effect of dispelling rheumatism,dredging meridian and stopping arthralgia. *P. wallichii* (Miq.) Hand.-Mazz., the stem (Piperis Wallichii Caulis, Chuan bi feng) have the effect of dispelling rheumatism, improving waist and knee,and tonifying kidney.

3. **金粟兰科**　$*P_0A_{(1\sim3)}\overline{G}_{(1:1:1)}$

(3) Chloranthaceae

【形态特征】草本或灌木。茎节膨大。单叶对生，叶柄基部常合生；托叶小。花小，单性或两性，排成穗状、头状或圆锥花序。两性花无花被；雄蕊 1~3 枚合生，花丝极短；单雌蕊，子房下位，1 顶生胚珠。单性花：雄花多数，雄蕊 1 枚；雌花少数，具浅杯状 3 齿裂萼管。核果卵形或球形。

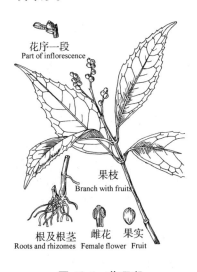

花序一段
Part of inflorescence

果枝
Branch with fruits

根及根茎　雌花　果实
Roots and rhizomes　Female flower　Fruit

图 12-4　草珊瑚
Fig.12-4　*Sarcandra glabra*

Characteristic: Herbs or shrubs. Stem nodes usually inflated. Leaves single, opposite; petiole base usually connate; stipules small. Flowers small, unisexual or bisexual, arranged in spike, capitulum or panicle. Bisexual flowers without perianth; stamens 1~3, connate; filaments extremely short; simple pistil;ovary inferior,1 terminal ovule. Unisexual flowers: male flowers numerous, stamen 1; female flowers few, with shallow cup-like 3-dentate calyx tube. Drupe ovoid or globose.

【药用植物】草珊瑚 *Sarcandra glabra*（Thunb.）Nakai.（图 12-4）全草（肿节风）能清热凉血，活血消斑，祛风通络。宽叶金粟兰 *Chloranthus henryi* Hemsl. 多穗金粟兰 *Ch.multistachys* S.J.Pei 的根及根状茎（四块瓦），及已 *Ch. Serratus*（Thunb.）Roem. et Schult. 的全草等也可药用。

Medicinal plants: *Sarcandra glabra* (Thunb.) Nakai (Fig.12-

4), the whole herb (Sarcandrae Herba, Zhong jie feng) have the effect of removing pathogenic heat from blood, promoting blood circulation to eliminate speckle, dispelling rheumatism and dredging meridian. Other species as *Chloranthus henryi* Hemsl., *Ch.multistachys* S.J.Pei and *Ch. Serratus* (Thunb.) Roem. et Schult. are used as herbal medicine.

4. 桑科 ♂$P_{4-5}A_{4-6}$；♀$P_{4-6}\underline{G}_{(2:1:1)}$

(4) Moraceae

【形态特征】木本，稀草本和藤本；常有乳汁。单叶互生，稀复叶；托叶早落。单性花，雌雄同株或异株；常集成头状、葇荑、穗状或隐头花序；单被花，常4~6片；雄蕊与花被片同数且对生；雌花花被有时肉质，子房上位，2心皮合生，常1室，1胚珠。瘦果或核果，常与花被或花轴等形成聚花果。

Characteristic: Woody, sparsely herbs or vines; usually with laticifer. Leaves simple, alternate, sparsely compound lear; stipules caducous. Flowers unisexual, monoecism or dioecism, usually arranged as capitulum, catkin, spike, or hypanthodium; simple perianth flower, usually 4~6; pistil and perianth identical and opposite; female flowers perianth sometimes fleshy; ovary superior; carpels 2, connate, often 1-roomed; ovule 1. Achenes or drupes, often forming fleshy multiple fruit with perianth or rachis.

【分布】53属，1400种，分布于热带和亚热带。我国12属，153种，全国各省区均有分布，但以长江以南较多。已知药用12属，约80种。

Distribution: 53 genera,1400 species, distributed in the tropical and subtropical. 12 genera,153 species in China, mostly in south of the Yangtze River. 12 genera, about 80 species are medicinal plants.

【药用植物】桑 *Morus alba* L.（图12-5）. 落叶乔木，有乳汁。单叶互生，广卵形至卵形；托叶早落。花单性，腋生，与叶同出；雄花序下垂，花被4，花丝在芽时内折；雌花无梗，无花柱。聚花果卵状椭圆形。分布全国各地，栽培或野生。根皮（桑白皮）能泻肺平喘，利水消肿；嫩枝（桑枝）能祛风湿，利关节；叶（桑叶）能疏风清热，清肝明目；果穗（桑葚）能滋阴养血，生津润燥。

Medicinal plants: *Morus alba* L.(Fig.12-5). Deciduous trees, with milky juice. Alternating single leaves broadly ovate to ovate. Flowers unisexual, axillary or bud-bearing scaly axillary, with leaves; male flowers inflorescence pendulous; perianthes 4; filaments folded in bud; female flowers without stalk and style. Collective fruit ovate-elliptical. Widely distributed in China, cultivated or wild. Root barks (Mori Cortex, Sang bai pi) have the effect of removing heat from lung to relieve asthma, inducing diuresis to alleviate edema;Young shoots (Mori Ramulus, Sang zhi) have the effect of dispelling rheumatism, benefiting joint; leaves (Mori Folium,Sang ye) have the effect of dispelling wind and heat from the body,removing heat from the liver and improving eyesight; infructescence (Mori Fructus,Sang shen) have the effect of nourishing Yin and nourishing blood, promoting the production of body and moisturizing dry.

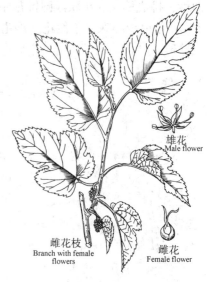

雄花
Male flower

雌花枝
Branch with female flowers

雌花
Female flower

图 12-5 桑
Fig.12-5 *Morus alba*

无花果 *Ficus carica* L.（图12-6）落叶灌木。叶厚纸质，表面粗糙，3~5裂。隐头花序单生叶腋，梨形，成熟时紫黑色。原产地中海沿岸，我国中、南部均有栽培。隐头果（无花果）能润肺止咳，清热润肠；根、叶散瘀消肿，止泻。

Ficus carica L.(Fig.12-6)**,** Deciduous shrub. Leaves thick papery, surface rough,3~5-lobed.Hypanthodium solitary leaf axils, pear-shaped, atropurpureus when ripe. Native to the Mediterranean coast, cultivated in the middle and south of China. Syconia (Fici Caricae Syconium,Wu hua guo) have the effect of moistening lung to relieve cough, clearing heat and loosening the bowel; and its roots, leaves have the effect of eliminating stasis to activate blood circulation, alleviating swelling, stopping diarrhea.

聚花果纵切面
Longitudinal section of aggregate

果枝 Branch with fruit

图 12-6 无花果
Fig.12-6 *Ficus carica*

大麻 *Cannabis sativa* L. 一年生高大草本。叶掌状全裂，裂片披针形。花单性，雌雄异株；雄花花序圆锥状；雌花丛生于叶腋。瘦果扁卵形。原产亚洲西部，我国各地有栽培。种仁（火麻仁）能润燥滑肠、通便。

Cannabis sativa L., Tall annual herbs. Leaves palmately divided, lobes lanceolate. Flowers unisexual, dioecism; male inflorescence panicle; female flowers clustered in leaf axils. Achenes obovate. Native to western Asia, cultivated all over China. Kernels (Cannabis Semen, Huo ma ren) have the effect of moisturizing dryness syndrome and loosening the bowel,relieving constipation.

5. 马兜铃科 $\male\female$ *P $_{(3)}$ A $_{6\text{-}12}$ $\overline{G}$ $_{(4\text{-}6:4\text{-}6:\infty)}$ $\overline{G}$ $_{(4\text{-}6:4\text{-}6:\infty)}$

(5) Aristolochiaceae

【形态特征】草本或木质藤本。单叶互生，叶基常心形，无托叶。总状、聚伞花序，或花单生；花两性，单被，辐射或两侧对称，常具腐肉臭气，下部合生并膨大成管状、球状、钟状或瓶状，顶端3裂或向一侧延伸成舌状；雄蕊6~12，花丝短，分离或与花柱合生；心皮4~6，合生，子房下位或半下位座。蒴果。

Characteristic: Herbs or wooden vines. Leaves simple, alternate; the base of leaf often cordate; stipules absent. Racemes, cymes, or flowers solitary; flowers bisexual, simple perianth, actinomorphic or zygomorphic, usually with a smell of rotting flesh; lower flowers connate and bulging tubular, globose, campanulate, or ampuliform; apical flowers 3-lobed or extending to one side into ligule; stamens 6–12; filaments short, isolated or connate with style; carpels 4–6, connate; ovary inferior or half-inferior. Capsule.

【分布】约8属，600种，分布于热带和亚热带，以南美洲较多。我国4属，71种，6变种，除华北和西北干旱地区外，全国均有分布。已知3属70种药用。

Distribution: Circa 8 genera, 600 species, distributed in the tropics and subtropics, mostly in South America. 4 genera,71 species and 6 varieties in China, except in the arid regions of North and Northwest China. 3 genera, about 70 species are medicinal plants.

【药用植物】马兜铃 *Aristolochia debilis* Sieb. et Zucc.（图12-7）. 草质藤本。根圆柱形，有香气。叶三角状卵形。单花腋生，花被管稍弯曲，基部球形，上部扩大成斜喇叭状。雄蕊6；子房下位。分布于黄河以南各省区。果实（马兜铃）能清肺降气，止咳平喘；地上部分（天仙藤）能行气活血，通络止痛。根（青木香）能平肝止痛，行气消肿；因具肾毒性，已被《中国药典》禁用。同属北马兜铃 *A. contorta* Bge. 叶三角状心形；花数朵簇生于叶腋，花被片顶端具线形长尖尾。其果实、茎与马兜铃同等入药。

Medicinal plants: *Aristolochia debilis*(Fig.12-7), Grassy vines. Roots cylindrical with aroma. Leaves triangular-ovate. Flower single, axillary; perianth tube slightly curved, base spherical, upper enlarged into oblique trumpet-shaped. Stamens 6; inferior ovary, 6-roomed. Distributed in south part of the Yellow River. Fruit (Aristolochiae Fructus, Ma dou ling) have the effect of clearing the lungs and reducing Qi, relieving cough and preventing asthma; aboveground parts (Aristolochiae Herba, Tian xian teng) have the effect of promoting the circulation of Qi and invigorating the circulation of blood, dredging meridian to relieve pain; Roots (Aristolochiae Radix, Qing mu xiang) have the effect of calming the liver to relieve pain, promoting the circulation of Qi to relieve swelling. Due to nephrotoxicity, it has been banned by the Chinese Pharmacopoeia. *A. contorta* Bge., Leaves, triangular- cordate; several flowers clustered in leaf axils; apical perianth with linearly long caudate. Fruits and stems are equally used as herbal medicine as *A.debilis*.

辽细辛（北细辛）*Asarum heterotropoides* Fr. Schmidt var. *mandshuricum*（Maxim.）Kitag.（图 12-8）多年生草本。根状茎横走，根多而细长，气味辛辣。叶基生，常 2 枚，叶片心形至肾状心形，下表面密被短柔毛。单花腋生；花被壶状，紫褐色，顶端 3 裂向下反卷；雄蕊 12；子房半下位。蒴果半球形。分布于吉林、辽宁、黑龙江等省。根及根状茎（细辛）能祛风散寒，通窍止痛，温肺止饮。同属汉城细辛 A. *sieboldii* Miq.var. *seoulense* Nakai 和华细辛 A. *sieboldii* Miq. 的根及根状茎同等入药。

Asarum heterotropoides var. mandshuricum (Fig.12-8)**.** Perennial herb. Rhizome transverse, slender and pungent. Leaves basal, usually 2, cordate to reniform-cordate, densely pubescent below surface. Flowers single, axillary; perianth gyalectiform, puce; apical perianth 3-lobed downwards retrorse; stamens 12; ovary half-inferior. Capsule hemispherical. Distributed in Jilin, Liaoning, Heilongjiang and other provinces. Roots and rhizomes (Asari Radix et Rhizoma, Xi xin) have the effect of dispelling rheumatism and expelling cold, deoppilating cavity to relieve pain, and warming lung and stopping drinking. *A. sieboldii* Miq. *var. seoulense* Nakai and *A. sieboldii* Miq. are equally used as *as A. heterotropoides var. mandshuricum.*.

图 12-7 马兜铃
Fig.12-7 *Aristolochia debilis*

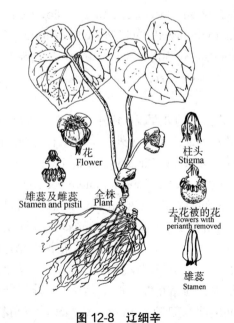

图 12-8 辽细辛
Fig.12-8 *Asarum heterotropoides* var. *mandshuricum*

6. **蓼科**　$\male * P_{3\sim6}, {}_{(3\sim6)}A_{3\sim9} \underline{G}_{(2\sim4:1:1)}$

(6) Polygonaceae

【形态特征】草本。茎节常膨大；单叶互生，有膜质托叶鞘。花常两性，稀单性；花序穗状、总状、头状或圆锥状；花被片3~6，宿存；雄蕊3~9；子房上位，1室，心皮通常3，稀2~4，合生，胚珠1。瘦果，常包于宿存的花被内。种子有胚乳。

Characteristic: Herbs. Stems erect, often with swollen nodes. Leaves simple, alternate, stipules often united to a sheath (ocrea). Inflorescence terminal or axillary, spicate, racemose, capitate, or paniculate. Flowers small, bisexual, rarely unisexual. Perianth 3–6-merous, often enlarged in fruit. Stamens usually (3–) 6–9. Ovary superior, 1-loculed; carpels 3, rarely 2 or 4; ovule 1. Fruit an achene, often included in persistent perianth.

视频

【分布】50属，1150种；我国有13属，约235种，分布于全国，药用10属，约136种。

Distribution: 50 genera and 1150 species. 13 genera, about 235 species distributed throughout China. 10 genera, about 136 species are medicinal plants.

【主要药用属检索表】

1. 瘦果具翅，花被片6，果时不增大 ·················	大黄属 *Rheum*
1. 瘦果不具翅	
2. 花被片6，果时内轮花被片增大 ·················	酸模属 *Rumex*
2. 花被片5或4，果时通常不增大	
3. 瘦果具三棱或双凸镜状，常比宿存的花被短 ···	蓼属 *Polygonum*
3. 瘦果具三棱，明显比宿存的花被长 ···········	荞麦属 *Fagopyrum*

Key to medicinal genera of Polygonaceae

1. Achenes with wings; Tepals 6, not enlarged in fruit ···········	*Rheum*
1. Achenes without wings	
2. Tepals 6, inner 3 enlarged in fruit ···········	*Rumex*
2. Tepals 5 (or 4), not enlarged in fruit	
3. Achenes trigonous or biconvex, shorter than persistent perianth ···········	*Polygonum*
3. Achenes trigonous, much longer than persistent perianth ···········	*Fagopyrum*

【药用植物】掌叶大黄 *Rheum palmatum* L. 多年生高大草本。基生叶大，宽卵形或近圆形，掌状半裂，裂片3~5，裂片有时再羽裂；托叶鞘大，筒状。圆锥花序大型；花小，常为紫红色。瘦果具3棱，有翅。分布于陕西、甘肃、四川、青海和西藏等地。掌叶大黄和同属植物唐古特大黄 *R. tanguticum* Maxim. ex Balf.、药用大黄 *R. officinale* Baill. 的根和根茎（大黄）能泻下攻积，清热泻火，凉血解毒，逐瘀通经，利湿退黄（图12-9）。

Medicinal plants: *Rheum palmatum* (Fig.12-9) Herbs large. Leaf blade large, palmately divided into pinnatisect lobes, apex acuminate or narrowly acute; ocrea large, tubular. Panicle large. Flowers small, purple-red. Achenes trigonous, with wings. Distributed in Shaanxi, Gansu, Sichuan, Qinghai and Tibet. The roots and rhizomes (Rhei Radix et Rhizoma, Da huang) of *Rheum palmatum*, *R. tanguticum*, or *R. officinale* are used for inducing diarrhea, purging accumulation, heat, toxic heat and damp heat, cooling blood, arresting bleeding, activating blood and transforming blood stasis.

何首乌 *Polygonum multiflorum* Thunb.（图12-10）. 多年生缠绕草本。块根肥厚，长椭圆形或不规则形，断面具异型维管束形成的"云锦花纹"。膜质托叶鞘短筒状。圆锥花序顶生或腋生；花小，白色，外侧花被背部有翅。瘦果具3棱。全国各地多有分布。块根（何首乌）能解毒消

花
Flower

掌叶大黄
Rheum palmatum

雌蕊
Pistil

果实
Fruit

药用大黄
R.officinale

唐古特大黄
R. tanguticum

图 12-9　大黄属植物
Fig.12-9　Plants of *Rheum*

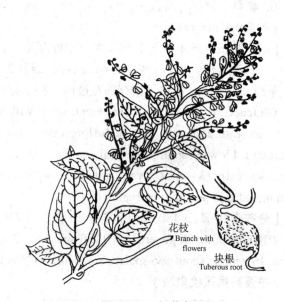

花枝
Branch with flowers

块根
Tuberous root

图 12-10　何首乌
Fig.12-10　*Polygonum multiflorum*

痛，润肠通便；制首乌能补肝肾，益精血，强筋骨，乌须发；茎藤（首乌藤）能养血安神，祛风通络。

Polygonum multiflorum (Fig.12-10) Herbs perennial. Stems twining. Roots hypertrophy, irregularly shaped, with auxillary stele. Leaf blade ovate, ocrea membranous. Inflorescence terminal or axillary. Perianth white. Achenes trigonous. Distributed throughout China. The root tuber (Polygoni Multiflori Radix, He shouwu) are used for decreasing toxins, treating and preventing malaria, and inducing diarrhea; the processed tuber (Polygoni Multiflori Radix Praeparata) is used for tonifying blood and essence of the liver and kidney, and improving hair growth; the stem (Polygoni Multiflori Caulis) is used for nourishing the heart and tranquilizing the mind, expelling wind and unblocking collaterals.

虎杖 *P. cuspidatum* Sieb. et Zucc.（图 12-11）多年生草本。根状茎粗壮。茎中空，散生红色或紫红色斑点。托叶鞘短筒状。单性异株。瘦果卵形。分布于陕西、甘肃及长江流域及以南各省区。根和根状茎（虎杖）能利湿退黄，清热解毒，散瘀止痛，止咳化痰。

P. cuspidatum (Fig.12-11) Herbs perennial. Rhizomes thickened. Stems hollow, often with red or purple spots. Flowers small, dioecism; perianth white, 5-parted. Achenes included in persistent perianth. Distributed in Shaanxi, Gansu and the area south of the Yangtze river basin. Roots and rhizomes (Polygoni Cuspidati Rhizoma et Radix, Hu zhang) are used for clearing dampness and relieving jaundice, purging heat and toxins, transforming blood stasis and alleviating pain, suppressing cough and transforming phlegm.

拳参 *P. bistorta* L. 多年生草本。根状茎肥厚。茎直立。托叶鞘筒状，无缘毛。总状花序穗状，顶生，紧密；花白色或淡红色。分布于东北、华北、华中和华东等地。根状茎（拳参）能清

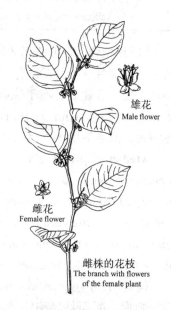

雄花
Male flower

雌花
Female flower

雌株的花枝
The branch with flowers of the female plant

图 12-11　虎杖
Fig.12-11　*Polygonum cuspidatum*

热解毒，消肿止痛。

P. bistorta Herbs perennial. Rhizomes large. Stems erect. Basal leaves long petiolate. Ocrea tubular, glabrous. Inflorescence terminal, spicate, dense. Perianth pinkish or white. Distributed in northeast, north, central and eastern China. The rhizomes (Bistortae Rhizoma, Quan shen) are used for purging heat and toxins, diminishing swelling and alleviating pain.

本科其他药用植物还有：羊蹄 *Rumex japonicus* Houtt.，根（土大黄）能清热解毒，凉血止血，通便。萹蓄 *Polygonum aviculare* L. 地上部分（萹蓄）能利尿通淋，杀虫止痒。蓼蓝 *P. tinctorium* Ait.，叶（蓼大青叶）能清热解毒，凉血消斑。红蓼 *P. orientale* L.，果实（水红花子）能散血消癥，消积止痛，利水消肿。金荞麦 *Fagopyrum dibotrys*（D. Don）Hara，根状茎（金荞麦）能清热解毒，排脓祛瘀。

Other medicinal plants: *Rumex japonicus*, the roots are used for purging heat and toxins, cooling blood, and purging stools. *P. aviculare* L., the above-ground part (Polygoni Avicularis Herba, Bian xu) is used for promoting urination, relieving stranguria, killing parasites and relieving itching. *P. tinctorium*, the leaves (Polygoni Tinctorii Folium, Liao da qing ye) are used for purging heat and toxins, and cooling blood. *P. orientale*, the fruits (Polygoni Orientalis Fructus, Shui hong hua zi) are used for moving blood and relieving pain, inducing perspiration and diminishing swelling. *Fagopyrum dibotrys* (D. Don) Hara, the rhizomes (Fagopyri Dibotryis Rhizoma, Jin qiao mai) are used for purging heat and toxins, and dissipating abscesses.

7. 苋科 $\male$ * $P_{3\sim5} A_{3\sim5} \underline{G}_{(2\sim3:1:1\sim\infty)}$

(7) Amaranthaceae

【形态特征】多草本。叶对生或互生，无托叶。花小，常两性，稀单性；花被片3~5，干膜质，每花下常有1枚干膜质苞片和2枚小苞片。胞果。

Characteristic: Herbs. Leaves alternate or opposite, exstipulate. Flowers small, bisexual or unisexual, subtended by 1 membranous bract and 2 bracteoles. Bracteoles scarious, spiculate. Tepals 3–5, scarious. Utricle.

【分布】约70属，900种；我国有15属，约40种，分布于全国，药用9属，28种。

Distribution: About 70 genera and 900 species. 15 genera and about 40 species in China. Distributed throughout China; 9 genera, about 28 species are medicinal plants.

【药用植物】牛膝 *Achyranthes bidentata* Blume（图12-12）。根（牛膝）能补肝肾，强筋骨，利尿通淋，引血下行。川牛膝 *Cyathula officinalis* Kuan，根（川牛膝）能逐瘀通经，通利关节，利尿通淋。鸡冠花 *Celosia cristata* L.，花序（鸡冠花）能收敛止血，止带，止痢。青葙 *C. argentea* L.，种子（青葙子）能清肝泻火，明目退翳。

Medicinal plants: *Achyranthes bidentata* Blume (Fig.12-12), the roots (Achyranthis Bidentatae Radix, Niu xi) are used for tonifying liver and kidney, strengthening muscles and bones, inducing diuresis and relieving stranguria, leading the blood and fire downwards. *Cyathula officinalis* Kuan, the roots (Cyathulae Radix, Chuan ni xi) are used for dredging channels, leading the blood and fire down and inducing diuresis. *Celosia cristata* L., the inflorescences (Celosiae

小苞片
Bractlet

去掉花
被的花
A flower with
perianth removed

果枝
Branch with fruit

花
Flower

图 12-12 牛膝
Fig. 12-12 *Achyranthes bidentata*

视频

Cristatae Flos,Ji guan hua) are used as astringent for hemostasis, tourniquet and dysentery. ***C. argentea*** **L.**, the seeds (Celosiae Semen,Qing xiang zi) are used for cleaning liver, improving acuity of vision and removing nebula.

8. 石竹科　♀ * $K_{4\sim5,(4\sim5)} C_{4\sim5,0} A_{8,10} \underline{G}_{(2\sim5:1:\infty)}$

(8) **Caryophyllaceae**

【形态特征】草本。节常膨大。单叶对生，全缘，基部多少联合。花两性，辐射对称，聚伞花序；萼片4~5，分离或连合；花瓣4~5，常具爪；雄蕊8或10，二轮列；子房上位，2~5心皮合生，1室，特立中央胎座。蒴果，稀浆果。

Characteristic: Herbs. Stems and branches usually swollen at nodes. Leaves opposite, entire, usually connate at base. Flowers actinomorphic, bisexual. Sepals 4 or 5, free, imbricate, or connate into a tube. Petals 4 or 5, often comprising claw and limb. Stamens 8 or 10, in 2 series. Carpels 2–5, united into a compound ovary; ovary superior; placentation free, central. Fruit usually a capsule, rarely berrylike.

【分布】约75属，2000种；我国有30属，约390种，分布于全国。药用21属，约106种。

Distribution: About 75 genera, 2 000 species. 30 genera, about 390 species in China. Distributed throughout China; medicinal plants 21 genera, about 106 species.

【药用植物】孩儿参 *Pseudostellaria heterophylla* (Miq.) Pax（图12-13）。块根（太子参）能益气健脾，生津润肺。瞿麦 *Dianthus superbus* L. 和石竹 *D. chinensis* L. 地上部分（瞿麦）能利尿通淋，活血通经。麦蓝菜 *Vaccaria segetalis* (Neck.) Garcke，成熟种子（王不留行）能活血通经，下乳消肿，利尿通淋。银柴胡 *Stellaria dichotoma* L. var. *lanceolata* Bge. 根（银柴胡）能清虚热，除疳热。

Medicinal plants: *Pseudostellaria heterophylla* (Miq.) Pax ex Pax et Hoffm. (Fig.12-13), the root tubers (Pseudostellariae Radix, Tai zi shen) are used for nourishing Qi and generating body fluid. *Dianthus superbus* L. or *D. chinensis* L., the above-ground part (Dianthi Herba, Shi zhu) is used for promoting urination, relieving stranguria, breaking up blood stasis and dredging channels. *Vaccaria segetalis* (Neck.) Garcke, the mature seeds (Vaccariae Semen, Wang bu liu xing) are used for moving blood and milk flow, and inducing diuresis. *Stellaria dichotoma* L. var. *lanceolata* Bge., the roots (Stellariae Radix,Yin chai hu) are used for clearing deficiency heat and relieving infantile malnutritional fever.

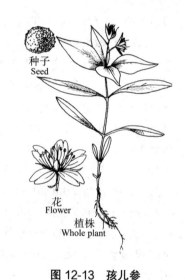

种子
Seed

花
Flower

植株
Whole plant

图 12-13　孩儿参
Fig. 12-13　*Pseudostellaria heterophylla*

9. 毛茛科　♀ * ↑ $K_{3\sim\infty} C_{3\sim\infty,0} A_{\infty} \underline{G}_{1\sim\infty:1:1\sim\infty}$

(9) **Ranunculaceae**

【形态特征】草本，稀木质藤本。叶互生，少对生，单叶或复叶，无托叶。花常两性，单生或排成聚伞花序、总状花序和圆锥花序；萼片3至多数，有时呈花瓣状；花瓣3至多数或缺；雄蕊和心皮多数，离生，螺旋状排列隆起的花托上；子房上位，1室，胚珠1至多数。聚合蓇葖果或聚合瘦果，稀浆果。

Characteristic: Herbs, sometimes woody vines. Leaves alternate, rarely opposite, simple or compound. Inflorescence a simple or compound monochasium, dichasium, simple or compound raceme, or flowers solitary. Flowers bisexual. Sepals 3 or more, free, sometimes petaloid; petals present or absent, 3 or more. Stamens numerous, free. Carpels numerous, free. Ovary superior, 1-loculed, 1 to many ovules.

Fruit follicles or achenes, rarely berries.

【分布】约 50 属，2000 余种；我国有 42 属，800 余种，分布于全国。药用 30 属，近 500 种。

Distribution: 50 genera, 2 000 species. 42 genera and about 800 species in China. Distributed throughout China; 30 genera, about 500 species are medicinal plants.

【主要药用属检索表】

1. 叶互生或基生
　2. 花辐射对称
　　3. 瘦果，每心皮各有 1 胚珠
　　　4. 花序有由 2 枚对生或 3 枚以上轮生苞片形成的总苞；叶均基生
　　　　5. 花柱在果期延长呈羽毛状 ·························· 白头翁属 *Pulsatilla*
　　　　5. 花柱在果期不延长 ····························· 银莲花属 *Anemone*
　　　4. 花序无总苞；叶常基生和茎生
　　　　6. 花有花瓣，花瓣具蜜槽 ························· 毛茛属 *Ranunculus*
　　　　6. 花无花瓣 ································· 唐松草属 *Thalictrum*
　　　4. 蓇葖果，每心皮各有 2 枚以上胚珠
　　　　7. 有退化雄蕊
　　　　　8. 总状或复总状花序；退化雄蕊位于雄蕊外侧；无花瓣 ············ 升麻属 *Cimicifuga*
　　　　　8. 单花或单歧聚伞花序；退化雄蕊位于雄蕊内侧；有花瓣 ············ 天葵属 *Semiaquilegia*
　　　　7. 无退化雄蕊，花序无总苞
　　　　　9. 心皮无细柄；花大，黄色、近白色或淡紫色 ············ 金莲花属 *Trollius*
　　　　　9. 心皮有细柄；花小，黄绿色或白色 ············· 黄连属 *Coptis*
　2. 花两侧对称
　　10. 后萼片船型或盔状，无距；花瓣有长爪，无退化雄蕊 ············ 乌头属 *Aconitum*
　　10. 后萼片平或船型，不呈盔状，有距；花瓣无爪，退化雄蕊 ············ 翠雀属 *Delphinium*
1. 叶对生，常为藤本 ································ 铁线莲属 *Clematis*

Key to important medicinal genera

1. Leaves alternate or basal
　2. Flowers actinomorphic
　　3. Ovary with 1 ovule; fruit an achene
　　　4. Leaves all basal; inflorescence with an involucre
　　　　5. Style strongly elongate and plumose in fruit ················· *Pulsatilla*
　　　　5. Style not as above ···················· *Anemone*
　　　4. Leaves both basal and cauline; inflorescence without an involucre
　　　　6. Petals present, petals with nectaries ··············· *Ranunculus*
　　　　6. Petals absent ······················ *Thalictrum*
　　3. Ovary with several or many ovules; fruit a follicle
　　　7. Staminodes present
　　　　8. Raceme or compound raceme; petals absent ············· *Cimicifuga*
　　　　8. Single flower or monochasium; petals present ············ *Semiaquilegia*
　　　7. Staminodes absent, inflorescence without an involucre
　　　　9. Carpels sessile ······················· *Trollius*
　　　　9. Carpels slender stipitate ·················· *Coptis*
　2. Flowers zygomorphic
　　10. Upper sepal spurless; petals clawed; staminodes absent ············ *Aconitum*
　　10. Upper sepal spurled; petals clawless; staminodes ············ *Delphinium*
1. leaves opposite, often vines ····················· *Clematis*

【药用植物】乌头 *Aconitum carmichaeli* Debx.（图 12-14）. 多年生草本。块根倒圆锥形。叶互生，常 3 全裂。总状花序，萼片 5，蓝紫色，上萼片盔状；花瓣 2，具长爪；雄蕊多数；心皮 3~5，离生；聚合蓇葖果。分布于长江中下游及西南地区。栽培品的主根（川乌）能祛风除湿，温经止痛；子根（附子）能回阳救逆，补火助阳，散寒止痛。

Medicinal plants: *Aconitum carmichaeli*; Debx. (Fig.12-14). Herbs perennial. Root tuber obconical. Leaves alternate, 3-sect or nearly to base. Inflorescence terminal, racemose. Sepals 5, blue-purple, upper sepal high galeate; petals clawed. Stamens numerous. Carpels 3–5, free. Follicles. Distributed in the middle and lower reaches of the Yangtze river and southwest China. The mother roots (Aconiti Radix,Chuan wu) are used for dispelling wind dampness, dissipating cold and alleviating pain. The processed secondary roots (Aconiti Lateralis Radix Praeparata, Fu zi) are used for reviving yang from collapse, tonifying life-gate fire and assisting yang, dispersing cold and relieving pain.

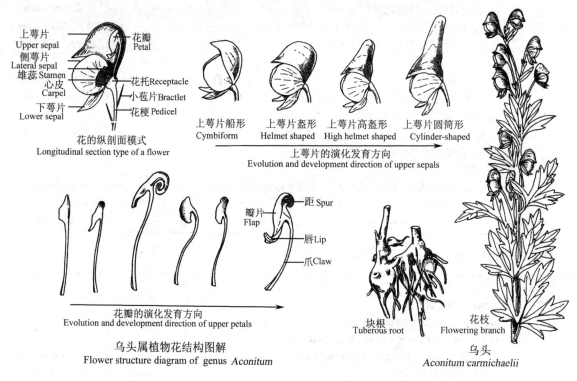

图 12-14　乌头

Fig. 12-14　*Aconitum carmichaeli*

黄连 *Coptis chinensis* Franch.（图 12-15）. 多年生草本。根状茎黄色，分枝成簇。叶基生，叶片 3 全裂。小花黄绿色，萼片 5；花瓣条状披针形，中央有蜜腺；雄蕊多数；心皮 8~12，基部具柄。聚合蓇葖果。分布于四川、贵州、湖南、湖北、陕西等地。黄连和同属植物三角叶黄连 *C. deltoidea* C. Y. Cheng et P. K. Hsiao、云连 *C. teeta* Wall. 的根茎（黄连）能清热燥湿，泻火解毒。

Coptis chinensis Franch. (Fig.12-15). Herbs perennial. Rhizomes yellow, attached to each other at the bottom. Leaf blade ovate-triangular, 3-sect. Sepals 5, greenish yellow, lanceolate; petals linear-lanceolate, with nectaries. Stamens numerous. Carpels 8–12, free, stipitate. Distributed mainly in Sichuan, Guizhou, Hunan, Hubei and Shaanxi. The rhizomes (Coptidis Rhizoma,Huang lian) of *Coptis chinensis* Franch., *C. deltoidea* C. Y. Cheng et P. K. Hsiao, or *C. teeta* Wall. are used for purging heat and toxins, and drying dampness.

花瓣 Petal　萼片 Sepal

植株 Whole plant

图 12-15　黄连

Fig. 12-15　*Coptis chinensis*

花枝
Branches
with flower

根 Roots　果实 Fruit

图 12-16　威灵仙

Fig. 12-16　*Clematis chinensis*

　　威灵仙 *Clematis chinensis* Osbeck（图 12-16）。藤本，干燥后茎叶变黑色。一回羽状复叶对生，小叶 5。圆锥状聚伞花序；萼片 4，白色；无花瓣；雄蕊多数；心皮多数。聚合瘦果，宿存花柱羽毛状。分布于长江中下游及以南地区。威灵仙和同属植物棉团铁线莲 *C. hexapetala* Pall.、东北铁线莲 *C. mandshurica* Rupr. 的根和根状茎（威灵仙）能祛风湿，通经络。

Clematis chinensis Osbeck (Fig.12-16). Vines woody, turning black on drying. Leaves pinnate, usually 5-foliolate. Cymes axillary or terminal, usually paniclelike and many flowered. Sepals 4, white; petals absent. Stamens numerous. Carpels numerous, free. Achenes, persistent style plumose. Distributed in the middle and lower reaches of the Yangtze river and south area. The roots and rhizomes (Clematidis Radix et Rhizoma, Wei ling xian) of *Clematis chinensis* Osbeck, *C. hexapetala* Pall, or *C. mandshurica* Rupr. are used for expelling wind and dampness, activating and dredging channels.

　　白头翁 *Pulsatilla chinensis*（Bunge）Regel 多年生草本。全株密生白色柔毛。叶基生。花单生，萼片 6，蓝紫色；无花瓣。瘦果聚合成头状，宿存花柱羽毛状，下垂如白发。分布于东北、华北、华东和河南、陕西、四川等地。根（白头翁）能清热解毒，凉血止痢。

Pulsatilla chinensis (Bunge) Regel Herbs perennial, densely white pilose. Leaf blade broadly ovate. Flowers solitary, bracts 3. Sepals 6, blue-purple; petals absent. Achenes, with a long plumose beak formed by persistent style. Distributed in northeast, north China, east China and Henan, Shaanxi and Sichuan. The roots (Pulsatillae Radix, Bai tou weng) are used for clearing toxic heat, cooling blood and relieving dysentery.

　　升麻 *Cimicifuga foetida* L. 多年生草本。根状茎粗壮，表面黑色，有多个内陷的圆洞状老茎残基。二至三回羽状复叶。圆锥花序，密被腺毛和短柔毛；萼片 5，白色；无花瓣；心皮 2~5。蓇葖果，有柔毛。分布于青海、甘肃、四川和云南。升麻和同属植物大三叶升麻 *C. heracleifolia* Kom.、兴安升麻 *C. dahurica*（Turcz.）Maxim. 的根状茎（升麻）能发表透疹，清热解毒，升举阳气。

Cimicifuga foetida L. Herbs perennial. Rhizome robust. Leaves 4 or 5, petiolulate. Inflorescence racemose, rachis densely gray glandular pubescent. Sepals 5, white; petals absent. Carpels 2–5, free.

医药大学堂
WWW.YIYAODXT.COM

183

Follicles appressed pubescent. Distributed in Qinghai, Gansu, Sichuan and Yunnan. The rhizomes of *Cimicifuga foetida* L., *C. heracleifolia* Kom., or *C. dahurica* (Turcz.) Maxim. (Cimicifugae Rhizoma, Sheng ma) are used for releasing the exterior, promoting rash eruption, clearing toxic heat and raising yang qi.

小毛茛 *Ranunculus ternatus* Thunb. 一年生小草本。簇生多数肉质小块根。茎铺散，较柔软。单叶或 3 出复叶。花单生，5 数，黄色，带蜡样光泽。分布于华东及广西、湖南、河南等地。块根（猫爪草）能化痰散结，解毒消肿。

Ranunculus ternatus Thunb. Herbs annual. Root tuber ovoid or fusiform. Stems branched, soft. Basal leaves. Flowers solitary, sepals 5; petals 5, yellow, with a waxy sheen. Distributed mainly in east China, Guangxi, Hunan and Henan. The roots (Ranunculi Ternati Radix, Mao zhua cao) are used for reducing phlegm and resolving masses, transforming toxin and diminishing swelling.

常用的药用植物还有：北乌头 *Aconitum kusnezoffii* Reichb. 的块根（草乌）能祛风除湿，温经止痛；叶（草乌叶）能清热，解毒，止痛。天葵 *Semiaquilegia adoxoides*（DC.）Makino 的块根（天葵子）能清热解毒，消肿散结。腺毛黑种草 *Nigella glandulifera* Freyn et Sint. 的种子（黑种草子）能补肾健脑，通经，通乳，利尿。

Other medicinal plants: *Aconitum kusnezoffii* Reichb. roots (Aconiti Kusnezoffii Radix, Cao wu) are used for dispelling wind and eliminating dampness, regulating menstruation and relieving pain; Leaves (Aconiti Kusnezoffii Folium,Cao wu ye) are used for purging heat and toxins, and alleviating pain. *Semiaquilegia adoxoides* (DC.) Makino, roots (Semiaquilegiae Radix,Tian kui zi) are used for purging heat and toxins and resolving masses. *Nigella glandulifera* Freyn et Sint., seeds (Nigellae Semen, Hei zhong cao zi) are used for nourishing brain and kidney, moving blood and milk flow, and inducing diuresis.

10. 小檗科　$\male\female$ * $K_{3+3, \infty} C_{3+3, \infty} A_{3\sim9} \underline{G}_{1:1:1\sim\infty}$

(10) Berberidaceae

【形态特征】灌木或草本。叶互生，单叶或复叶。花两性，辐射对称；萼片与花瓣相似，3 基数，花瓣常具蜜腺；雄蕊 3~9，常与花瓣同数而对生；子房上位，花柱缺或极短，柱头常为盾形。浆果、蒴果或蓇葖果。

Characteristic: Herbs or shrubs. Leaves alternate, simple or compound. Flowers bisexual. Sepals 6–9, often petaloid, in 2 or 3 whorls. Nectary of petals present or absent. Stamens 3–9, opposite petals. Ovary superior, style present or absent. Fruit a berry, capsule, or follicle.

【分布】17 属，约 650 种；我国有 11 属，约 320 种，分布于全国，以四川、云南、西藏种类最多；药用 11 属，140 余种。

Distribution: 17 genera, about 650 species. 11 genera and about 320 species in China. Distributed throughout China, and mostly in Sichuan, Yunnan and Tibet; 11 genera, about 140 species are medicinal plants.

【药用植物】箭叶淫羊藿 *Epimedium sagittatum*（Sieb. et Zucc.）Maxim（图 12-17）. 淫羊藿 *E. brevicornum* Maxim. 、柔毛淫羊藿 *E. pubescens* Maxim. 和朝鲜淫羊藿 *E. koreanum* Nakai 的地上部分（淫羊藿）能补肾阳，强筋骨，祛风湿。八角莲 *Dysosma versipellis*（Hance）M. Cheng ex Ying、六角莲 *D. pleiantha*（Hance）Woodson. 的根状茎能化痰散结，祛瘀止痛，清热解毒。拟豪猪刺 *Berberis soulieana* Schneid.、小黄连刺 *B. wilsonae* Hemsl.、细叶小檗 *B. poirelii* Schneid. 和匙叶小檗 *B. vernae* Schneid. 的根（三颗针）能清热燥湿，泻火解毒。桃儿七 *Sinopodophyllum hexandrum*（Royle）Ying，根及根状茎能祛风除湿，活血止痛、祛痰止咳。

Medicinal plants: Epimedium sagittatum (Fig.12-17), E. brevicornum, E. pubescens, or E. koreanum, above-ground part (Epimedii Folium, Yin yang huo) is used for invigorating the kidney yang, strengthening the muscles and bones, and dispelling wind damp.Rhizomes of *Dysosma versipellis*, or *D. pleiantha* (Hance) Woodson are used for reducing phlegm and resolving masses, removing stasis pain, purging heat and toxins. *Berberis soulieana* Schneid., *B. wilsonae, B. poirelii,* or *B. vernae*, the roots (Berberidis Radix, San ke zhen) are used for heat-clearing and damp-drying, purging fire for removing toxin. Roots and rhizomes of *Sinopodophyllum hexandrum* are used for dispelling wind and eliminating dampness, invigorating blood circulation and alleviating pain, resolving phlegm and reliving cough.

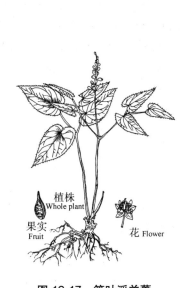

图 12-17 箭叶淫羊藿

Fig. 12-17 *Epimedium sagittatum*

图 12-18 粉防己

Fig. 12-18 *Stephania tetrandra*

11. 防己科 ♂*$K_{3+3}C_{3+3}A_{3\sim6, \infty}$; ♀$K_{3+3}\underline{G}_{3+3}A_{3\sim6:1:1}$

(11) Menispermaceae

【形态特征】多年生草质或木质藤本。单叶互生，无托叶。花单性异株，萼片与花瓣均 6，2 轮，每轮 3 片，雄蕊通常 6 枚，稀 3 或多数，分离或合生，核果。

Characteristic: Perennial grass or woody vine. Leaves alternate, exstipulate. Flowers unisexual. Sepals and petals 6 in 2 whorls, 3 on each whorl. Stamens usually 6, seldom 3 or many, detached or connate. Drupe.

【分布】65 属，350 种，分布于热带和亚热带。我国 19 属，78 种，主要分布于长江流域及其以南各省区，药用 15 属，67 种。

Distribution: 65 genera, 350 species, distributed tropical and subtropical. 19 genera, 78 species in China, distributed the Yangtze river basin.

【药用植物】粉防己 *Stephania tetrandra* S. Moore（图 12-18）的根（防己）能利水消肿，祛风止痛。蝙蝠葛 *Menispermum dauricum* DC. 的根状茎（北豆根）能清热解毒，祛风止痛。木防己 *Cocculus orbiculatus*（L.）DC. 根（木防己）能祛风止痛，利尿消肿。

Medicinal plants: The roots of *Stephania tetrandra* (Fig.12-18) (Stephaniae Tetrandrae Radix,Fen fangji) are used for draining water and alleviating edema, dispelling wind and relieving pain. *Menispermum dauricum* DC., the rhizome (Menispermi Rhizoma, Bei dou gen) are used for clearing heat and removing toxin, dispelling wind and relieving pain. *Cocculus orbiculatus*, the roots (Cocculus Orbiculatus Radix, Mu fang ji) are used for dispelling wind and relieving pain, promoting urination and alleviating edema.

12. 木兰科 ♀*P$_{6\sim12}$A$_\infty$G$_{\infty:1:1\sim2}$

(12) Magnoliaceae

【形态特征】落叶或常绿木本，具有油细胞，芳香。单叶互生，全缘，托叶有或缺；托叶包被幼芽，早落，在节上留下环状托叶痕。花大，单生；多两性，稀单性；辐射对称；花被片6至多数，每轮3，花瓣状；雄蕊和雌蕊均多数，离生，螺旋排列于伸长或隆起的花托上；子房上位，1室。聚合蓇葖果、聚合浆果。

Characteristic: Trees or vines, with oil cell, sweet-smelling. Leaves simple, alternate, entire; stipules present, early caducous and leaving an annular scar around the node. Flowers solitary, usually bisexual; perianth usually petaloid, lobes 6-9- (45) in 2-many whorls, 3 (6) on each whorl; pistils and stamens many, distinct, spirally arranged on elongated torus. Berries or follicles, aggregate fruit.

【分布】18属，330余种。我国14属，约160种，主要分布于东南和西南地区，药用8属，约90种。

Distribution: 18 genera, circa 330 species. 14 genera and about 160 species in China. Mostly in SE to SW China; 8 genera, about 90 species are medicinal plants.

【主要药用属检索表】

1. 乔木或灌木；花两性；聚合蓇葖果
  2. 芽多为托叶包围；小枝上具环状托叶痕；雄蕊和雌蕊螺旋状排列于伸长的花托上
    3. 花顶生；雌蕊群无柄
      4. 每心皮具3~12胚珠 ·· 木莲属 *Manglietia*
      4. 每心皮具2胚珠 ·· 木兰属 *Magnolia*
    3. 花腋生；雌蕊群具明显的柄 ······························· 含笑属 *Michelia*
  2. 芽多具芽鳞；无托叶；雄蕊和雌蕊轮状排列于平顶隆起的花托上 ·········· 八角属 *Illicium*
1. 木质藤本；花单性；聚合浆果
  5. 雌蕊群的花托发育时不伸长；聚合果球状或椭圆状 ························· 南五味子属 *Kadsura*
  5. 雌蕊群的花托发育时明显伸长；聚合果长穗状 ························· 五味子属 *Schisandra*

Key to important medicinal generas

1. Trees or shrubs; Flowers bisexual; aggregate fruit
  2. Buds surrounded by 2 valvate and hooded stipules; branchlets with annular stipule scars; stamens and pistils spirally arranged on elongated receptacles
    3. Flowers terminal; gynoecium sessile
      4. ovules 3–14 in each carpel ··· *Manglietia*
      4. ovules 2 in each carpel ·· *Magnolia*
    3. Flowers axillary, gynoecium long stipitate ···························· *Michelia*
  2. Buds with numerous imbricate bud scales; stipule absent; stamens and pistils arranged on receptacle ······*Illicium*
1. Woody vine; flowers unisexual; matured carpels in fleshy berries
  5. Torus of gynoecium not elongated when fruiting; arranged in dense globose or ellipsoid aggregate ······ *Kadsura*
  5. Tori gradually elongated; sparse or firm long spike-like aggregates ························· *Schisandra*

【药用植物】厚朴 *Magnolia officinalis* Rehd. et Wils.（图12-19）.落叶乔木。单大，生于枝顶；花白色，花被9~12或更多。分布于长江流域、陕西、甘肃南部。现有人工栽培。树皮（厚朴）能燥湿健脾，温中下气，化食消积。花蕾（厚朴花）能理气宽中，化湿。同属植物凹叶厚朴 *M. officinalis* Rehd. et Wils. *var. biloba* Rehd. et Wils. 叶先端凹缺，分布于福建、浙江、江西和湖南等省。功效同厚朴。

Medicinal plants: *Magnolia officinalis* (Fig.12-19)**.** Deciduous tree. Leaves clustered on twig apex, oblong-obovate, nearly leathery. Flowers large, white, solitary at terminal; tepals 9–12 or more. Distributed the yangtze river basin and southernShanxi, Gansu, cultivated. The bark (Magnoliae Officinalis Cortex, Hou po) has the effect of eliminating dampness and strengthening the spleen, digesting food and eliminating accumulation. *M. officinalis* var. *biloba* are also used the same as as *M. officinalis.*

望春花 *M. biondii* Pamp（图 12-20）. 落叶乔木。叶长椭圆状披针形。花先叶开放；萼片 3，近线性，花瓣 6，匙形，白色，基部带紫红色。分布于陕西、甘肃、河南、湖北、四川等地。花蕾（辛夷）有散风寒、通鼻窍之功。同属植物玉兰 *M. denudeata* Desr. 和武当玉兰 *M. sprengeri* Pamp 的花蕾亦做辛夷药用。

M. biondii (Fig.12-20) Deciduous tree. Flowers appearing before leaves; sepal 3, nearly narrowly linear, petal 6, spoon-shaped, white, usually purplish red at base abaxially. Distributed in Shaanxi, Gansu, Henan, Hubei and Sichuan. The flower buds (Magnoliae Flos, Xin yi) have the effect of eliminating wind-cold, deoppilation nasal cavity. The flower buds of *M. denudeata* or *M. sprengeri* are equally used as *M. biondii.*

图 12-19 厚朴

Fig. 12-19 *Magnolia officinalis*

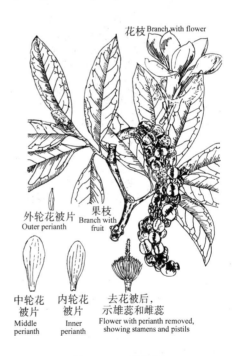

图 12-20 望春花

Fig. 12-20 *M. biondii*

五味子 *Schisandra chinensis*（Turcz.）Baill.（图 12-21）. 木质藤本。单叶互生，边缘有细齿。花单性同株；花被 6~9，白色。雄蕊 5；雌蕊心皮多数，花后花托伸长，聚合果浆果。分布于华北、东北等地。果实（北五味子）能收敛固涩，益气生津，补肾宁心。同属南五味子 *S. sphenanthera* Rehd. et Wils. 果实（南五味子）入药，功效同北五味子。

Schisandra chinensis (Fig.12-21). Woody vines. Leaves broadly elliptic or elliptic, margin callose-serrate. Flower dioecious; tepals 6–9; stamens 5–9; carpels 17–40. Aggregate berries red. Distributed in Northeast, Northern China and Central China. The fruits (Schizandrae Fructus, Wu wei zi) have the effect of astringing lung to treat cough and asthma, nourishing kidney, promoting the production of body fluid, adduction astringent taste.

八角 *Illicium verum* Hook.f.（图 12-22）常绿乔木。单叶互生，螺旋状排列，革质。花单生于叶腋。聚合果放射星芒状，由 8 个蓇葖果组成，顶端呈鸟喙状；每个蓇葖果含 1 粒种子。分布于福建、广西、广东、贵州、云南等地。果实（八角茴香）入药，能温阳散寒，理气止痛。

Illicium verum Hook. f (Fig.12-22). Tree. Leaves simple, alternate; Aggregate consisted of 8 follicles, in one whorl, apex obtuse. Distributed in Guangxi, Yunnan, Guizhou, Guangdong, Fujian etc. The fruits (Anisi Stellati Fructus, Ba jiao hui xiang) have the effect of eliminating cold, regulating qi, and odynolysis.

Other medicinal plants as *Schisandra sphenanthera.* the fruits (Nan Wuweizi) are used for astringe, nourishing qi to generate fluid, invigorating the kidney and calming heart. *Kadsura longipedunculata*, Root and Root bark are used for regulating qi to alleviate pain, expelling wind-evil and removing wetness, and promoting blood flow and detumescence. *Manglietia fordiana*, Fruits (ford manglietia fruit) are used to bowels open, stop coughing. *Michelia alba* DC., Flower (bailan flower) are used for dissipating hygrosis, promoting qi circulation, and stopping coughing.

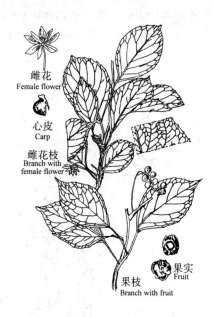

雌花
Female flower

心皮
Carp

雌花枝
Branch with female flower

果枝
Branch with fruit

果实
Fruit

图 12-21　五味子

Fig. 12-21　*Schisandra chinensis*

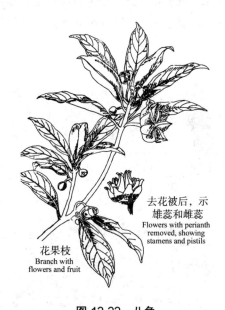

去花被后，示雄蕊和雌蕊
Flowers with perianth removed, showing stamens and pistils

花果枝
Branch with flowers and fruit

图 12-22　八角

Fig. 12-22　*Illicium verum*

13. 樟科　♀*P$_{(6, 9)}$A$_{3, 6, 9, 12}$G$_{(3:1:1)}$

(13) Lauraceae

【形态特征】多常绿乔木，有香气。单叶互生，无托叶。两性单被花，通常 3 基数，2 轮；雄蕊 3~12，通常 9，3 轮。核果或浆果。种子 1 粒。

Characteristic: Mostly evergreen tree, fragrance. Leaves simple, alternate, entire. Exstipulate. Flower bisexual, monochlamydeous flower, usually the basic number of flower is 3, 2 whorl, base connate. Stamens 3 to 12, usually 9 in 3 whorl. Drupe or berry. Seed 1.

【分布】45 属，2000 余种，分布于热带和亚热带。我国 20 属，400 余种，主要分布于长江以南各省区，药用 13 属，110 种。

Distribution: 45 genera, 2 000 species, distributed in tropical and subtropical regions. 20 genera, 400 species in China, distributed in the provinces south of the Yangtze river.

【药用植物】肉桂 *Cinnamomum cassia* Presl.（图12-23）的树皮（肉桂）能温肾壮阳，散寒止痛；嫩枝（桂枝）能解表散寒，温经通络。樟 *C. camphora*（L.）Presl. 全株可提取樟脑。乌药 *Lindera aggregata*（Sims）Kosterm 的根（乌药）能行气止痛，温肾散寒。

Medicinal plants: *Cinnamomum cassia* (Fig.12-23), the bark (Cinnamomi Cortex, Rou gui) are used for warming kidney and assisting yang, dissipating cold and relieving pain; the young branches (Cinnamomi Ramulus,Gui zhi) are used for releasing exterior and dissipating cold, warming meridian and dredging collaterals. *C. Camphora*, the volatile oils are used for relieving stuffy orifices, removing stagnation of Qi, killing worms and relieving itching, decreasing swelling and alleviating pain. *Lindera aggregata*, the roots (Linderae Radix,Wu yao) are used for moving Qi to relieve pain, warming kidney and relieving cold.

图 12-23 肉桂
Fig.12-23 *Cinnamomum cassia*

14. 罂粟科 ♀ *$K_2C_{4\sim6}A_{4\sim6,\ \infty}\underline{G}_{(2\sim\infty:1:\infty)}$

(14) Papaveraceae

【形态特征】草本。常含乳汁或黄色液汁。单叶互生，无托叶。花两性，辐射对称或两侧对称；萼片2，早落；花瓣4~6；雄蕊多数，离生，或6枚合生成2束；子房上位，侧膜胎座，胚珠多数。蒴果。

Characteristic: Herbal. Laticifers or elongated idioblasts present. Leaves simple, alternate, exstipulate. Flowers bisexual, equal. Sepals 2 and petals 4~6, stamens numerous, detached or stamens 6 connate into 2 bundles. Ovary superior, parietal placentation, ovules numerous. Capsule.

【分布】38属，约700种；主要分布于北温带。我国18属，362种；分布于全国，以西南地区为多；药用15属，约130种。

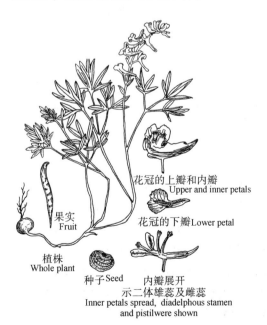

花冠的上瓣和内瓣
Upper and inner petals
花冠的下瓣 Lower petal
果实
Fruit
植株
Whole plant
种子 Seed
内瓣展开
示二体雄蕊及雌蕊
Inner petals spread, diadelphous stamen and pistilwere shown

图 12-24 延胡索
Fig.12-24 *Corydalis yanhusuo*

Distribution: 38 genera, 700 species, mainly distributed in the north temperate zone. 18 genera, 362 species in China, distributed all over the country, mostly in SW China. Medicinal plants 15 genera, about 130 species.

【药用植物】延胡索 *Corydalis yanhusuo* W. T. Wang（图12-24）块茎（延胡索），能行气止痛、活血散瘀。伏生紫堇 *C. decumbens*（Thunb.）Pers. 块茎（夏天无）能舒筋活络、活血止痛。罂粟 *Papaver somniferum* L. 果壳（罂粟壳）能敛肺、涩肠、止痛。果实中的乳汁干燥后（鸦片）能镇痛、止咳、止泻。

Medicinal plants: *Corydalis yanhusuo*, the rhizoma (Corydalis Rhizoma, Yan hu suo) are used for moving Qi to relieve pain, activating blood and resolving stasis (Fig.12-24). *C. decumbens*, the rhizoma (Corydalis Decumbentis Rhizome, Xia tian wu) are used for relaxing sinew and activating col laterals, activating blood and relieving pain. *Papaver somniferum* L., the shell (Papaveris Pericarpium, Yin su ke) are used for astringing lung, astringing intestines and relieving pain; the latex in the fruit (opium) are used for relieving cough and pain, checking diarrhea.

15. 十字花科 $\male *K_{2+2}C_4A_{2+4}\underline{G}_{(2:1\sim2:1\sim\infty)}$

(15) Brassicaceae

【形态特征】草本。单叶互生；无托叶。花两性，辐射对称，总状花序；萼片 4，2 轮；花瓣 4，十字形排列；雄蕊 6，4 长 2 短，为四强雄蕊，常在雄蕊基部有 4 个蜜腺；子房上位，心皮 2，合生，由假隔膜分成 2 室，侧膜胎座。长角果或短角果。

Characteristic: Herbal. Leaves simple, alternate, exstipulate. Flowers bisexual, actinomorphic, usually arranged in racemes. Sepals 4 in 2 whorls. Petals 4, cross arrangement. Stamens 6, 4 long and 2 short, four strong stamens, usually four nectaries at the base of it. Ovary superior, connate by 2 carpels, it is divided into two chambers by central diaphragm (false diaphragm replum). Parietal placentation. Silique or silicle, usually dehiscence 2 bivalve.

【分布】约 350 属，3200 种。我国 96 属，425 种，全国各地均有分布；药用 30 属，103 种。

Distribution: Circa 350 genera, 3 200 species. 96 genera and about 425 species in China, distributed all over the country. Medicinal plants 30 genera, about 103 species.

花
Flower

果实
Fruit

根
Root

花枝
Branches with flowers

图 12-25 菘蓝
Fig. 12-25 *Isatis indigotica*

【药用植物】菘蓝 *Isatis indigotica* Fort.（图 12-25）一年生或二年生草本。主根圆柱形。基生叶有柄；茎生叶半抱茎。短角果扁平，边缘有翅，紫色，不开裂。各地有栽培。根（板蓝根）、叶（大青叶）、叶加工品（青黛）均为清热解毒药，能清热解毒、凉血消斑。

Medicinal plants: *Isatis indigotica* (Fig.12-25)**,** herbs annual or biennial. The taproot is cylindrical. Basal leaves long oval with petioles, cauline leaves semiamplexicaul. Brachycarpus flat, purple, margin winged, indehiscent silicle with a single locule. Cultivated all over the country. The roots (Isatis Radix, Ban lan gen), the leaves (Isatidis Folium, Da qing ye) and the processed products of leaves (Indigo Naturalis, Qing dai) as heat-clearing and detoxicating medicinal, have the effect of clearing heat and removing toxin, cooling blood and eliminating spot.

莱菔 *Raphanus sativus* L. 一年生或二年生草本。直根，肉质。基生叶大头状羽裂，茎生叶长圆形。总状花序；花白色、紫色或粉红色。长角果圆柱形，肉质。各地有栽培。鲜根（莱菔）能消食、下气、化痰、止血、解渴、利尿；开花结实后的老根（地骷髅）能消食理气、清肺利咽、散瘀消肿；种子（莱菔子）为消食药，能消食导气、降气化痰。

Raphanus sativus, herbs annual or biennial. Taproot, fleshy. Basal leaves lyrate, cauline leaves oblong. Raceme, flowers white, purple or pink. Siliqua cylindrical, fleshy. Cultivated all over the country. The fresh roots (Raphani, Lai fu) have the effect of removing food stagnation, lowering Qi, resolving phlegm, stopping bleeding, diuresis; The old roots after flowering and fruiting (Di ku lou) have the effect of promoting digestion and regulating qi, clearing lung and relieving sore throat, eliminating stasis and alleviating edema; the seeds (Raphani Semen, Lai fu zi) as digestant medicinal, have the effect of promoting digestion and removing food stagnation, descending qi and resolving phlegm.

独行菜 *Lepidium apetalum* Willd.：一年生或二年生草本。茎自基部多分枝。基生一回羽状浅裂或深裂。总状花序顶生；花小，雄蕊 2 或 4；短角果。种子椭圆状卵形。分布于全国大部分地区。种子（葶苈子）能祛痰平喘、利水消肿。

Lepidium apetalum, herbs annual or biennial. Stem much branched from base, leaves with long stalks at base of stem. Racemes terminal. Flower small, sepals 4, stamens 2 or 4. Silicle.. Distributed in most parts of the country. The seeds (Lepidii Semen, Ting li zi) as antitussive and antiasthmatic medicinal, have the effect of dispelling phlegm and relieving cough, excreting water and alleviating edema.

白芥 *Sinapis alba* L.：一年生或二年生草本，全株被白色粗毛。基部叶大头羽裂或近全裂。总状花序顶生或腋生；花黄色。长角果，密被白色长毛。种子近球形。原分布于欧洲；我国有栽培。种子（白芥子）为化痰药，能化痰逐饮、散结消肿。

Sinapis alba, herbs annual or biennial, the whole plant is covered with white coarse hair. Leaves with long stalks at base of stem, great head pinnate or subdivided. Racemes terminal or axillary, yellow flowers. Seeds subglobose.Originally distributed in Europe, cultivation in China. The seeds (White Mustard Seed,Bai jie zi) as phlegm-resolving medicine, have the effect of expelling fluid retention by resolving phlegm, dissipating mass and alleviating edema.

常用药用植物还有：荠菜 *Capsella bursa - pastoris*（L.）Medic.，全草（荠菜）能凉肝止血、平肝明目、清热利湿。播娘蒿 *Descurainia sophia*（L.）Webb ex Prantl，种子（葶苈子）为止咳平喘药，能祛痰平喘、利水消肿。蔊菜 *Roiippa indica*（L.）Hiem，全草（蔊菜）能祛痰止咳、解表散寒、活血解毒、利湿退黄。

Other medicinal plants: *Capsella bursa-pastoris*, whole herb (Shepherdspurse Herb, Ji cai) are used for cooling liver and stopping bleeding, pacifying liver and improving vision, clearing heat and draining dampness. *Descurainia sophia*, the seeds (Descurainiae Semen,Ting li zi) as antitussive and antiasthmatic medicinal, are used for dispelling phlegm and relieving asthma, promoting urination and alleviating edema. *Roiippa indica*, whole herb (Indian Rorippa Herb, Han cai) are used for dispelling phlegm and relieving cough, releasing exterior and dissipating cold, activating blood and removing toxin, excreting damp and dispelling jaundice.

16. 虎耳草科 ☿ * ↑ $K_{4\sim5}C_{4\sim5,0} A_{4\sim5,8\sim10}$, $\underline{G}_{(2\sim5:2\sim5:\infty)}$;$\overline{G}_{(2\sim5:2\sim5:\infty)}$

(16) Saxifragaceae

【形态特征】草本或木本。单叶或复叶，互生或对生，常无托叶。聚伞状、圆锥状或总状花序；花两性，辐射对称；花被 4~5 基数；萼片有时花瓣状，花瓣常离生；雄蕊（4~）5~10，或多数；心皮 2，通常多少合生，子房上位、半下位或下位。蒴果或浆果。

Characteristic: Herbs or shrubs.Leaves simple or compound, usually alternate or opposite, usually exstipulate. Flowers usually in cymes, panicles or racemes; flowers usually bisexual,actinomorphic;4~ or 5 (~10) -merous,sepals sometimes petal-like,petals usually free;stamens (4 or) 5~10 or many;carpels 2, usually connate,ovary superior or semi-inferior to inferior.Fruit a capsule or berry.

【分布】80 属，约 1200 种。中国 28 属，约 500 种，全国广布；药用 24 属约 115 种。

Distribution: 80 genera, circa 1200 species. 28 genera and

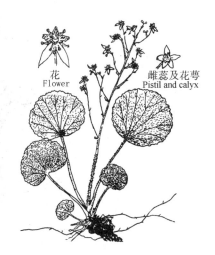

花
Flower

雌蕊及花萼
Pistil and calyx

图 12-26 虎耳草
Fig. 12-26 Saxifragastolonifera

视频

about 500 species in China. Distributed all over the country. 24 genera, about 155 species are medicinal plants.

【药用植物】常山 Dichroa febrifuga Lour. 根（常山）能涌吐痰涎，截疟；虎耳草 Saxifragasto-lonifera Curt.（图 12-26）全草（虎耳草）疏风清热、凉血解毒。

Medicinal plants: *Dichroa febrifuga* (Fig.12-26), the roots (Radix Dichroae, Chan shan) are used for inducing vomiting of phlegm and slobber, and interrupting malaria. Saxifragastolonifera Curt. *Saxifraga stolonifera*, the whole grass (Huercao) winds away heat, cools blood and detoxifies.

17. 蔷薇科 $\male\female * K_5C_5A_\infty \underline{G}_{1\sim\infty:1:1\sim\infty};\overline{G}_{(2\sim5:2\sim5:2)}$

(17) Rosaceae

【形态特征】草本、灌木或乔木。有刺或无刺。单叶或复叶；有明显托叶，离生或与叶柄贴生，稀无托叶。花单生或形成伞形、伞房、总状、聚伞状圆锥花序；花两性，辐射对称；花托扁平、凸起或凹陷，边缘延伸成一碟状、杯状、坛状或壶状的托杯（图 12-27）；萼片 5，花瓣 5，分离；雄蕊多数；心皮 1 至多数，离生或合生，子房下位、半下位、或上位。果实为蓇葖果、瘦果、梨果或核果，稀蒴果。

Characteristic: Trees,shrubs,or herbs,armed or unarmed. Leaves simple or compound; stipules paired, free or adnate to petiole,rarely absent.Inflorescences various,from single flowers to umbellate,corymbose,racemose or cymose-paniculate; flowers bisexual,actinomorphic; receptacle flat, raised, or sunken, hypanthium with the edge extending into a dish, cup, jar or pot shape (Fig.12-27);

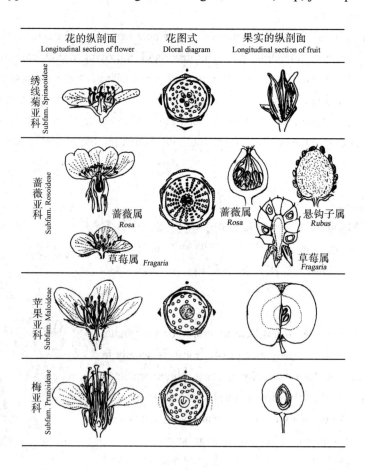

图 12-27　蔷薇科四亚科花、果的比较
Fig. 12-27　The flower and fruit of 4 subfamilies in Rosaceae

sepals 5,petals 5, free; stamens numerous; carpels 1 to many, free, or connate,ovary inferior, semi-inferior, or superior. Fruit a follicle, pome, achene, or drupe, rarely a capsule.

【分布】124 属，3300 余种。我国 51 属，约 1100 种，全国广布；药用 48 属，约 400 种。

Distribution: 124 genera, circa 3300 species. 51 genera and about 1100 species in China, nationwide. Medicinal plants 48 genera, about 400 species.

【亚科及主要药用属检索表】

1. 果开裂；多无托叶 .... 绣线菊亚科 Spiraeoideae 绣线菊属 *Spiraea*
1. 果不开裂；有托叶
　2. 子房上位
　　3. 心皮通常多数，分离；聚合瘦果或聚合小核果；萼宿存；多为复叶 .... 蔷薇亚科 Rosoideae
　　　4. 瘦果，着生在杯状或坛状花托里面
　　　　5. 雌蕊多数；花托成熟时肉质；灌木 .... 蔷薇属 *Rosa*
　　　　5. 雌蕊 1~4；花托成熟时干燥坚硬；草本
　　　　　6. 有花瓣，托杯上部有钩状刺毛 .... 龙芽草属 *Agrimonia*
　　　　　6. 无花瓣，托杯无钩状刺毛 .... 地榆属 *Sanguisorba*
　　　4. 瘦果或小核果，着生在扁平或隆起的花托上
　　　　7. 心皮含两枚胚珠；小核果或聚合果；植株有刺 .... 悬钩子属 *Rubus*
　　　　7. 心皮含 1 枚胚珠；瘦果，分离；植株无刺
　　　　　8. 花柱顶生或近顶生，在果期延长 .... 路边青（蓝布正）属 *Geum*
　　　　　8. 花柱侧生，基生或近基生，在果期不延长
　　　　　　9. 托杯成熟时干燥 .... 委陵菜属 *Potentilla*
　　　　　　9. 托杯成熟膨大成肉质 .... 蛇莓属 *Duchesnea*
　　3. 心皮常 1，稀 2 或 5；核果；萼不宿存；单叶 .... 梅亚科 Prunoideae
　　　10. 果实有沟
　　　　11. 侧芽 3，两侧为花芽，具顶芽；核常有孔穴 .... 桃属 *Amygdalus*
　　　　11. 侧芽 1，顶芽缺；核常光滑
　　　　　12. 子房和果实常被短绒毛；花先叶开 .... 杏属 *Armeniaca*
　　　　　12. 子房和果实均光滑无毛；花叶同开 .... 李属 *Prunus*
　　　10. 果实无沟 .... 樱属 *Cerasus*
　2. 子房下位或半下位 .... 苹果亚科 Maloideae
　　13. 内果皮成熟时骨质，果实含 1~5 小核 .... 山楂属 *Crataegus*
　　13. 内果皮成熟时革质或纸质，每室含 1 至多数种子
　　　14. 伞形或总状花序，有时单生
　　　　15. 心皮含 1~2 枚种子 .... 梨属 *Pyrus*
　　　　15. 心皮含 3 至多枚种子 .... 木瓜属 *Chaenomeles*
　　　14. 复伞花序或圆锥花序
　　　　16. 心皮全部合生，子房下位；常绿 .... 枇杷属 *Eriobotrya*
　　　　16. 心皮部分合生，子房半下位；常绿或落叶 .... 石楠属 *Photinia*

Key to subfamilies and important medicinal generas of Rosaceae

1. Fruit a dehiscent follicle;stipules absent .... Spiraeoideae, *Spiraea*
1. Fruit indehiscent; stipules present
　2. Ovary superior
　　3. Carpels usually numerous,separable;aggregate achenes or aggregate drupes;sepals persistent;mostly leaves compound...Rosoideae
　　　4. Achenes enclosed in cupular or urn-shaped torus
　　　　5. Carpels numerous; torus fleshy at maturity; shrubs .... *Rosa*

continued

    5. Carpels 1~4; torus dry and hard at maturity; herbs
        6. Petals present,hypanthium with hooklike spines abaxially .... *Agrimonia*
        6. Petals absent,hypanthium without hooklike spines abaxially .... *Sanguisorba*
    4. Achenes or drupelets borne on flat or convex torus
        7. Carpel with 2 ovules; drupe or aggregate; plant spiny .... *Rubus*
        7. Carpel with 1 ovule;achenes,separable; stems unarmed
            8. Style terminal or subterminal,elongated at fruiting .... *Geum*
            8. Style basal, lateral, or subterminal, not elongated at fruiting.
                9. Hypanthium dry when ripe .... *Potentilla*
                9. Hypanthium inflated and fleshy when ripe .... *Duchesnea*
    3. Carpel 1 (~5); fruit a drupe; sepals often deciduous; leaves simple .... Prunoideae
        10. Drupe grooved
            11. Axillary buds 3 with 2 lateral buds and 1 central bud; endocarp often pitted .... *Amygdalus*
            11. Axillary bud 1; terminal buds absent; endocarp usually smooth.
                12. Ovary and fruit usually pubescent;flowers first leaves .... *Armeniaca*
                12. Ovary and fruit glabrous;flowers and leaves bloom together .... *Prunus*
        10. Drupe not grooved .... *Cerasus*
2. Ovary inferior or semi-inferior .... Maloideae
    13. carpels bony when mature,fruit with 1~5 pyrenes .... *Crataegus*
    13. carpels leathery or papery when mature,each locule with 1 to many seeds.
        14. Inflorescence umbellate or racemose, sometimes flowers solitary.
            15. Carpels 1- or 2-seeded .... *Pyrus*
            15. Carpels 3- to many seeded .... *Chaenomeles*
        14. Inflorescence compound-corymbose or paniculate.
            16. Carpels wholly connate, ovary inferior;evergreen .... *Eriobotrya*
            16. Carpels partly free; ovary semi-inferior;Evergreen or deciduous .... *Photinia*

【药用植物】蔷薇亚科：掌叶覆盆子 *Rubus chingii* Hu 藤状灌木，有皮刺。叶互生，掌状深裂；托叶线状披针形。花单生于短枝顶端，白色。聚合小核果。分布于江苏、安徽、浙江、江西、福建，现有人工栽培。果实（覆盆子）能益肾固精缩尿，养肝明目。金樱子 *Rosa laevigata* Michx（图 12-28）常绿攀援灌木，有刺。羽状复叶，小叶 3，稀 5。花大，白色。蔷薇果倒卵形，密生直刺，具宿存萼片。分布于华东、华中及华南。果实（金樱子）能固精缩尿，固崩止带，涩肠止泻。

Medicinal plants: Rosoideae: *Rubus chingii*. Shrubs lianoid, with prickles.Leaves simple, margin usually palmately,doubly serrate;stipules linear-lanceolate.Inflorescences terminal on short branchlets, 1-flowered,petals white.Aggregate drupe. Distributed in Jiangsu, Anhui, Zhejiang, Jiangxi, Fujian, mostly cultivated.The fruit (Rubi Fructus,Fu pen zi) have the effect of securing essence, reducing urination and cultivating yin and yang of the kidney. *Rosa laevigata*. (Fig.12-28) Shrubs evergreen,climbing,with prickles. Pinnate compound leaves,leaflets 3, rarely 5,leathery. Flower solitary, petals white.Hip obovoid,densely glandular bristly,with persistent sepals. Distributed in East,central and South China.The fruit (Rosae Laevigatae Fructus,Jin ying zi) have the effect of securing essence,

花、果枝
Flowering and fruiting branch

蔷薇果
Achene

图 12-28 金樱子
Fig.12-28 *Rosa laevigata*

reducing urination and astringing the large intestine to check diarrhea.

　　苹果亚科（梨亚科）：山楂 *Crataegus pinnatifida* Bunge.（图 12-29）落叶乔木。小枝有刺。叶宽卵形或三角状卵形，两侧各有 3~5 羽状深裂片；托叶较大，镰形。伞房花序；花白色。梨果近球形，直径 1~1.5cm，深红色，有浅色斑点。分布于东北、华北及陕西、河南、江苏。果实（山楂）能消食健胃，行气散瘀，化浊降脂。叶（山楂叶）能活血化瘀，理气通脉，化浊降脂。同属植物山里红 *C. pinnatifida* Bge.var.*major* N.E.Br. 的果实亦作山楂药用；叶亦作山楂叶药用。枇杷 *Eriobotrya japonica*（Thunb.）Lindl. 常绿小乔木。小枝密生锈色或灰棕色绒毛。叶片革质，上面光亮，下面密生绒毛。圆锥花序顶生。分布于长江流域及以南地区，多栽培。叶（枇杷叶）能清肺止咳，降逆止呕。

　　Maloideae: *Crataegus pinnatifida*. (Fig.12-29) Trees deciduous.Branchlets with thorns.Leaf broadly ovate or triangular-ovate, with 3-5 pairs of lobes; stipules falcate.Corymb, petals white. Pome dark red,with light spots,subglobose,1-2.5 cm in diam.Distributed in Northeast China, North China, Shaanxi, Henan and Jiangsu.The fruit (Crataegi Fructus, Shan zha) have the effect of promoting digestion to resolve food stagnation and moving Qi and blood to transform stasis.The leaves (Crataegi Folium,Shan zha ye) have the effect of activating blood to resolve stasis.The fruits and leaves of C. pinnatifida.*var*. *major*. are equally used as *C.pinnatifida*. *Eriobotrya japonica* Small evergreen trees. Branchlets densely rusty or grayish rusty tomentose. Leaf leathery, adaxially lustrous,abaxially densely gray rusty tomentose. Panicle terminal. Distributed in the Yangtze River Basin, mostly cultivated. The leaves (Eriobotryae Folium, Pi pa ye) have the effect of clearing lung heat to relieve cough and descending stomach qi to arrest vomiting.

　　梅亚科（李亚科）：杏 *Prunus armeniaca* L.（图 12-30）落叶乔木。单叶互生。先叶开花；花萼、花瓣 5，白色或浅粉红色。核果球形。全国广布；多为栽培。种子（苦杏仁）能降气止咳平喘，润肠通便。同属植物山杏 *P. armeniaca* L. var. *ansu* Maxim.、西伯利亚杏 *P. sibirica* L.、东北杏 *P. mandshurica*（Maxim.）Koehne 的种子亦作苦杏仁药用。桃 *P. persica*（L.）Batsch.；落叶乔木。单叶互生；叶柄常有腺体；核果，表面有短绒毛。种子扁卵状心形。全国广布；多为栽培。种子（桃仁）能活血祛瘀，润肠通便，止咳平喘；枝条（桃枝）能活血通络、解毒杀虫。同属植物山桃 *P. davidiana*（Carr.）Franch. 的种子亦作桃仁药用；枝条亦作桃枝药用。

种子表面观
Seed surface

种子纵剖面
Longitudinal section of the seed

果枝
Fruiting branch

图 12-29　山楂
Fig.12-29　*Crataegus pinnatifida*

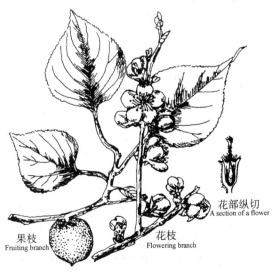

花部纵切
A section of a flower

果枝
Fruiting branch

花枝
Flowering branch

图 12-30　杏
Fig.12-29　*Prunus armeniaca*

Prunoideae: Prunus armeniaca (Fig.12-30) Deciduous trees. Leaves simple,alternate. Flowers solitary,opening before leaves; calyx 5-lobed; petals 5,white or light pink. Drupe globose. Nationwide; mostly cultivated.The seeds (Armeniaca Semen Amarum, Ku xing ren) have the effects of arresting cough and wheezing and lubricating the bowels. The seeds of *P. armeniaca* L. *var. ansu, P. sibirica, P. mandshurica* are equally used as Ku xing ren. *P. persica* Deciduous trees. Leaves simple; petiole with 1 to several nectaries. Drupe,outside pubescent.Seed flat ovate cardioid.Nationwide; mostly cultivated. The seeds (Persicae Semen,Tao ren) have the effect of activating blood and transforming blood stasis, lubricating the large intestine to improve bowel movements, and suppressing cough and wheezing.The twigs (Persicae Ramulus,Tao zhi) have the effect of activating blood to unblock the collaterals,removing toxin and killing the worms. *Prunus davidiana* is equally used as *P. persica* (L.) Batsch.

18. 豆科　$\male\female * \uparrow K_{5,(5)} C_5 A_{(9)+1,10,\infty} \underline{G}_{1:1:1\sim\infty}$

(18) Leguminosae

【形态特征】乔木、灌木或草本。根部常有根瘤。叶常互生，多为复叶，少单叶；有托叶，有叶枕。花序各种；花两性，两侧对称或辐射对称；花萼 5 裂；花瓣 5，常分离；多为蝶形花；雄蕊 10，二体，少数分离或下部合生，稀多数；心皮 1，子房上位，胚珠 1 至多数，边缘胎座。荚果。种子无胚乳。

Characteristic: Trees,shrubs,or herbs,sometimes climbing.Often bearing root-nodules that harbor nitrogen-fixing bacteria.Leaves compound, rarely simple; petiole present,pulvinus.Various inflorescences;flowers bisexual, actinomorphic or zygomorphic; calyx 5-lobed; petals 5, often separated; mostly papilionaceous; stamens mostly 10, diadelphous, a few filaments free or only connate at base, rarely numerous; ovary superior, carpel 1. Legumes.

【分布】650 属，18000 余种。我国 169 属，约 1539 种，全国广布，药用 109 属，约 600 种。

Distribution: 650 genera, circa 18 000 species. 169 genera and about 1 539 species in China, nationwide. Medicinal plants 109 genera, about 600 species.

【亚科及主要药用属检索表】

1. 花辐射对称；花瓣镊合状排列；雄蕊多数或有定数 ………………………… 含羞草亚科 Mimosoideae
  2. 雄蕊多数；荚果不横裂为数节
    3. 花丝下部联合成管状 …………………………………………………………合欢属 Albizia
    3. 花丝分离 …………………………………………………………………金合欢属 Acacia
  2. 雄蕊 5 或 10；荚果成熟时裂为数节 ……………………………………… 含羞草属 Mimosa
1. 花两侧对称；花瓣覆瓦状排列；雄蕊常 10
  4. 花冠假蝶形；雄蕊分离 …………………………………………… 云实亚科 Caesalpinioideae
    5. 单叶，具掌状叶脉 ………………………………………………………… 紫荆属 Cercis
    5. 羽状复叶
      6. 常二回偶数羽状复叶；茎枝或叶轴有刺
        7. 花杂性或单性异株；小叶边缘有齿 …………………………………… 皂荚属 Gleditsia
        7. 花两性；小叶全缘 ………………………………………………… 云实属 Caesalpinia
      6. 一回羽状复叶；植株无刺…………………………………………………… 决明属 Cassia
  4. 花冠蝶形 ………………………………………………… 蝶形花亚科 Papilionoideae
    8. 雄蕊 10，分离或仅基部合生…………………………………………………… 槐属 Sophora

续表

8. 雄蕊 10，合生成单体或二体
  9. 单体雄蕊
    10. 三出复叶；藤本
      11. 花萼钟形；具块根 ··· 葛属 *Pueraria*
      11. 花萼二唇形；不具块根 ······························· 刀豆属 *Canavalia*
    10. 单叶；草本
      12. 荚果不肿胀，常含 1 枚种子，不开裂；单叶，有腺点 ············ 补骨脂属 *Psoralea*
      12. 荚果肿胀，含种子 2 枚以上，开裂；单叶或复叶 ··········· 猪屎豆属 *Crotalaria*
  9. 二体雄蕊
    13. 小叶 1~3 片
      14. 小叶边缘有锯齿；托叶与叶柄连合 ··················· 葫芦巴属 *Trigonella*
      14. 小叶全缘或具裂片；托叶不与叶柄连合
        15. 花轴上的节不增厚··································· 大豆属 *Glycine*
        15. 花轴的节增厚成结
          16. 花柱不具须毛
            17. 旗瓣大于翼瓣和龙骨瓣；枝条有刺 ··········· 刺桐属 *Erythrina*
            17. 所有花瓣长度几相等；枝条无刺 ··········· 密花豆属 *Spatholobus*
          16. 花柱上部具纵列的须毛，或于柱头周围具毛茸
            18. 柱头倾斜 ··· 豇豆属 *Vigna*
            18. 柱头顶生 ··· 扁豆属 *Dolichos*
    13. 小叶 5 至多片
      19. 木质藤本；圆锥花序 ······························ 鸡血藤属 *Millettia*
      19. 草本；总状、穗状或头状花序
        20. 荚果通常肿胀，常因背缝线深延而纵隔为 2 室············ 黄芪属 *Astragalus*
        20. 荚果通常有刺或瘤状突起，1 室 ···················· 甘草属 *Glycyrrhiza*

Key to subfamilies and important medicinal generas of Leguminosae

1. Flowers actinomorphic; petals valvate in bud; stamens numerous, usually more than 10 ·············· Mimosoideae
  2. Stamens numerous; legume not transversely septate
    3. Filaments connate into a tube ·· *Albizia*
    3. Filaments free ··· *Acacia*
  2. Stamens 5 or 10; legume dehiscent in several segments ······························· *Mimosa*
1. Flowers zygomorphic;petals imbricate in bud; stamens 10
  4. Corollas not papilionaceous, uppermost petal overlapped on each side by adjacent lateral petals; stamens
    separated ··· Caesalpinioideae
  5. Leaves simple, palmately nerved·· *Cercis*
  5. Pinnate compound leaves
    6. Leaves usually bipinnate; plants often armed with branched spines
      7. Flowers polygamous or plants dioecious;leafles margin serrulate····················· *Gleditsia*
      7. Flowers bisexual; leaflets margin entire························· *Caesalpinia*
    6. Leaves once pinnate; plant without thorn ···························· *Cassia*
4. Corollas papilionaceous ·· Papilionoideae
  8. Stamens 10, filaments free or only connate at base ······················· *Sophora*
  8. Stamens 10, monadelphous or diadelphous
    9. Monadelphous
      10. Leaves pinnately 3-foliolate;vine

continued

11.　Calyx campanulate; roots sometimes tuberous .... *Pueraria*

11.　Calyx 2-lipped;roots without tuberous .... *Canavalia*

10.　Leaves simple; herbs

12.　Legume not inflated, indehiscent, seed 1; leaves simple,with glands .... *Psoralea*

12.　Legume inflated, dehiscent, 2-to many seeded; leaves simple or compound .... *Crotalaria*

9.　Diadelphous

13.　Leaflets 1~3

14.　Leafles serrulate;stipules adnate to petiole .... *Trigonella*

14.　Leaflets entire or lobed;stipules not adnate to petiole

15.　Rachis without swollen nodes .... *Glycine*

15.　Rachis often with swollen nodes

16.　Style glabrous

17.　Stems with prickles; vexillum is larger than wing and keel .... *Erythrina*

17.　Stems without prickles; petals almost equal .... *Spatholobus*

16.　Style bearded lengthwise on adaxial side, or pilose around stigma

18.　Stigma oblique .... *Vigna*

18.　Stigma terminal .... *Dolichos*

13.　Leaflets 5-many

19.　Lianas, scandent shrubs;panicle .... *Millettia*

19.　Herbs; racemes, spikes, or heads

20.　Legume inflated,mostly keeled ventrally and grooved dor-sally, 1-locular, incompletely or completely 2-locular .... *Astragalus*

20.　Legume prickly or tuberculate, 1-locular .... *Glycyrrhiza*

【药用植物】含羞草亚科：合欢 *Albizia julibrissin* Durazz.（图 12-31）落叶乔木。二回偶数羽状复叶，小叶镰刀状。头状花序排成顶生圆锥花序；雄蕊多数，花丝细长，淡红色。荚果扁条形。全国广布。树皮（合欢皮）能解郁安神，活血消肿；花或花蕾（合欢花）能解郁安神。

Medicinal plants: Mimosoideae. *Albizia julibrissin* (Fig.12-31)**.** Trees, deciduous.Biphasic even-pinnate compound leaves,leaflets falcate.Inflorescences of heads, arranged in terminal panicles; stamens numerous, filaments thin and long, pink. Legume flat.Nationwide.The barks (Albiziae Cortex, He huan pi) have the effect of relieving depression and tranquilizing the mind, activating blood and alleviating swelling.The flowers or buds (Albiziae Flos,He huan hua) have the effect of relieving depression, tranquilizing the mind and regulating the stomach Qi.

云实亚科：决明 *Cassia obtusifolia* L.（图 12-32）一年生亚灌木状草本。偶数羽状复叶；总状花序腋生；花瓣黄色；雄蕊 10，发育雄蕊 7。荚果细长，近四棱形。种子多数，菱柱形，光亮。全国广布。种子（决明子）能清热明目，润肠通便。同属植物小决明 *C. tora L.* 的种子亦作决明子药用。

Caesalpinioideae: *Cassia obtusifolia* (Fig.12-32) Herbs, suffrutescent, annual. Even-pinnate compound leaves. Racemes axillary; petals yellow; stamen 10, fertile stamens 7. Legume slender,subtetragonous. Seeds glossy, rhom-boid. Nationwide.The seeds (Cassiae Semen,Jue ming zi) have the effect of clearing liver heat, improving vision and moistens the bowels to relieve constipation.The seeds of *C. tora* are equally used as "Cassiae Semen".

皂荚 *Gleditsia sinensis* Lam. 乔木。刺粗壮，通常分枝。一回偶数羽状复叶。花杂性；花萼钟状，裂片 4；花瓣 4，白色；雄蕊 6~8；子房条形。荚果条形，黑棕色，被白色粉霜。分布于东北、华北、华东、华南及四川、贵州等地。果实（大皂角）、不育果实（猪牙皂）能祛痰开窍、

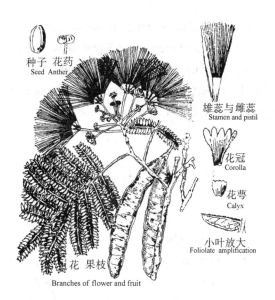

图 12-31　合欢

Fig. 12-31　*Albizia julibrissin*

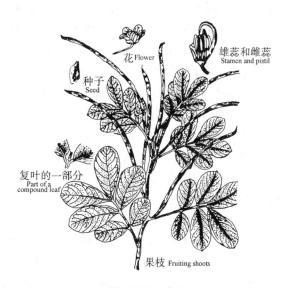

图 12-32　决明

Fig. 12-32　*Cassia obtusifolia*

散结消肿；棘刺（皂角刺）能消肿托毒，排脓，杀虫。

Gleditsia sinensis Trees.Spines robust,often branched. Even-pinnate compound leaves. Flowers polygamous; calyx campanulate,4-lobed; petals 4, white; stamens (6~) 8; ovary strap-shaped.Legume brown,strap-shaped, often farinose. Distributed in Northeast, North, Eastand South China, Sichuan, Guizhou and other places.The fruits (Gleditsiae sinensis Fructus,Da zao jiao), sterile fruit (Gleditsiae Fructus Abnormalis, Zhu ya zao) have the effect of dispelling phlegm to open the orifices,dissipating binds and dispersing swelling;The spine banks (Gleditsiae Spina,Zao jiao ci) have the effect of dispersing swelling and expelling toxin, expelling pus and killing worms.

蝶形花亚科：苦参 *Sophora flavescens* Ait. 草本或亚灌木。根圆柱状，黄白色。奇数羽状复叶。雄蕊 10，分离。果实串珠状。全国广布。根（苦参）能清热燥湿，杀虫，利尿。

Papilionoideae: *Sophora flavescens* Herbs or subshrubs. Roots cylindrical,white, pale yellow.Odd-pinnate compound leaf. Corolla white, pale yellow; stamens 10, free. Legumes slightly constricted between seeds. Nationwide. The roots (Sophorae Flavescentis Radix, Ku shen) have the effect of purging heat, drying dampness, relieving itching, killing worms and promoting urination.

甘草 *Glycyrrhiza uralensis* Fisch.（图 12-33）多年生草本。根和根状茎粗壮，红棕色或暗棕色。羽状复叶。花冠蓝紫色。荚果镰刀状或环状弯曲，密集成球，密生瘤状突起和刺毛状腺体。分布于东北、华北、西北。根和根状茎（甘草）能补脾益气，清热解毒，祛痰止咳，缓急止痛，调和诸药。同属植物胀果甘草 *G. inflata* Batalin 和光果甘草 *G. glabra* L. 的根和根状茎亦作甘草药用。

Glycyrrhiza uralensis (Fig 12-33) Herbs perennial. Roots and rhizomes strong, reddish brown or dark brown. Pinnately

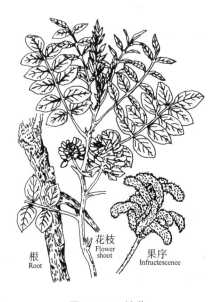

图 12-33　甘草

Fig. 12-33　*Glycyrrhiza uralensis*

compound leaf. Corolla purple. Legume falcate to curved into a ring, densely tuberculate and glandular hairy. Distributed in Northeast, North and Northwest China.The roots and rhizomes (Glycyrrhizae Radix et Rhizoma,Gan cao) have the effect of tonifying qi, clearing heat and toxins, transforming phlegm, arresting coughing, alleviating pain and moderating the harsh properties of other herbs.The roots and rhizomes of *G. inflata* Batalin, *G. glabra* L.are used as'Glycyrrhizae Radix et Rhizoma'.

19. 芸香科　$\varphi * K_{3\sim5} C_{3\sim5} A_{3\sim\infty} \underline{G}_{(2\sim\infty:2\sim\infty:1\sim2)}$

(19) Rutaceae

【形态特征】木本，稀草本。叶、花、果常有透明腺点。多为复叶或单身复叶，少单叶；无托叶。两性花，辐射对称；雄蕊与花瓣同数或为其倍数；花盘明显；子房上位，心皮2至多数，多合生。柑果、蒴果、核果和蓇葖果，稀翅果。

Characteristic: Shrubs、trees, rarely herbs. Oils contained in glands visible at surface of leaves, young branchlets, inflorescences, flower parts, fruit, or cotyledons in seed.Compound leaves or unifoliate compound leaves. Flowers bisexual; actinomorphic; stamens usually as many as or 2 × as many as petals or sometimes more numerous; disk nectariferous, flattened; ovary superior,carpels 2-5 or more, usually connate. Fruit a hesperidium, capsule, drupe or follicle, rarely a samara.

【分布】150余属，1700余种。我国28属，约150种，全国广布。药用23属，约105种。

Distribution: 150 genera,circa 1 700 species.28 genera and about 150 species in China, nationwide; 23 genera, about 105 species are medicinal plants.

【药用植物】橘 *Citrus reticulata* Blanco 常绿小乔木。单身复叶。花单生或2-3朵簇生。柑果。分布于长江流域及以南地区，广泛栽培。成熟果皮（陈皮）能理气健脾，燥湿化痰；未成熟果皮（青皮）能疏肝破气，消积化滞；外层果皮（橘红）能理气宽中，燥湿化痰。

Medicinal plants: *Citrus reticulate*. Small evergreen trees. Unifoliate compound leaves. Flowers solitary to 3 in a fascicle. Fruit oblate to subglobose, sarcocarp with 7~14 segments. Distributed in the Yangtze River Basin and its south area, widely cultivated.The ripe pericarp (Citri Reticulatae Pericarpium, Chen pi) have the effect of regulating qi and strengthening the spleen, drying dampness and resolving phlegm. The immature pericarp (Citri Reticulatae Pericarpium Viride,Qing pi) have the effect of breaking up liver qi and resolving accumulation; ectocarp exocarp (Citri Exocarpium Rubrum, Ju hong) have the effect of moving stomach qi, drying dampness and resolving phlegm.

吴茱萸 *Evodia rutaecarpa*（Juss.）Benth.（图 12-34）小乔木或灌木。幼枝、叶轴及花序均被黄褐色长绒毛。羽状复叶互生。圆锥状聚伞花序顶生；单性异株。蓇葖果紫红色，表面有粗大油腺点。分布于华东、中南、西南等地区，现多栽培。近成熟果实（吴茱萸）能散寒止痛，降逆止呕，助阳止泻。其变种石虎 *E. rutaecarpa*（Juss.）Benth. var. *officinalis*（Dode）Huang 和疏毛吴茱萸 *E. rutaecarpa*（Juss.）Benth. var. *bodinieri*（Dode）Huang 的未成熟果实亦作吴茱萸入药。

果枝
Fruiting branches
雄花　雌花
Male and female flower

图 12-34　吴茱萸
Fig. 12-34　*Evodia rutaecarpa*

Evodia rutaecarpa (Fig.12-34) Shrubs or trees. Twig、rachis and inflorescence covered with yellow brown villi. Pinnate compound leaves. Follicles subglobose, purplish red,with large glands. Distributed in East China, central and South China, southwest and other regions, cultivated. The fruits (Euodiae Fructus, Wu zhu yu) have the effect of dispersing cold to relieving pain, dredging the liver qi downwards, drying dampness and assisting yang. The fruits of *E. rutaecarpa* var. *officinalis* and *E.*

rutaecarpa var. *bodinieri* are used as "Euodiae Fructus".

20. 大戟科 ♂*$K_{0\sim5}C_{0\sim5}A_{1\sim\infty}$；♀*$K_{0\sim5}C_{0\sim5}\underline{G}_{(3:3:1\sim2)}$

(20) Euphorbiaceae

【形态特征】草本、灌木或乔木，常含乳汁。单叶互生，叶基部常有腺体，有托叶。花单性同株或异株，常为聚伞花序，或杯状聚伞花序；子房上位，3 心皮 3 室，中轴胎座，每室 1~2 胚珠。蒴果，稀浆果或核果。

Characteristic: Trees, shrubs, or herbs,latex often present.Leaves simple,alternate,sometimes with glands at base.Flowers unisexual,monoecious or dioecious, cymes or cyathium; ovary superior, carpels 3, placentation axile,with 1~2 ovules per carpel. Fruit a capsule, raely a berry or drupe.

【分布】300 属，8000 余种。我国 66 属，365 种，全国广布。药用 39 属，160 余种。

Distribution: 300 genera, circa 8000 species. 66 genera and 365 species in China, nationwide; medicinal plants 39 genera,about 160 species.

【药用植物】大戟 *Euphorbia pekinensis* Rupr.（图 12-35）多年生草本。根圆柱形。茎被短柔毛。花序基部有 5 枚苞片，伞梗 5；杯状总苞裂片圆形，腺体 4。蒴果球形，表皮有疣状突起。全国广布。根（京大戟）能泻水逐饮，消肿散结。

Medicinal plants: *Euphorbia pekinensis* (Fig.12-35) Herbs, perennial. Roots cylindric.Stems pilose. Inflorescence dichotomous or multibranched,a flowerlike cyathium; involucre cuplike, lobes rounded, glands 4. Capsule globose, sparsely tuberculate.Nationwide.The roots (Euphorbiae Pekinensis Radix,Jing da ji) have the effect of purging to expel retained fluid,dispersing swelling and dissipating binds.

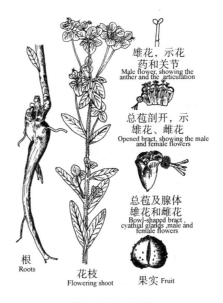

雄花，示花药和关节
Male flower, showing the anther and the articufation

总苞剖开，示雄花、雌花
Opened bract, showing the male and female flowers

总苞及腺体 雄花和雌花
Bowl-shaped bract, cyathial glands ,male and female flowers

根 Roots 花枝 Flowering shoot 果实 Fruit

图 12-35 大戟
Fig.12-35 *Euphorbia pekinensis*

巴豆 *Croton tiglium* L. 小乔木或灌木，幼枝绿色有星状毛。叶互生，近叶柄两侧各有一无柄腺体。总状花序顶生，单性同株；雄花疏生星状毛或几无毛；雌花萼片脱落，子房上位，3 室，密被星状毛。蒴果。分布于长江以南，野生或栽培。种子（巴豆）外用蚀疮。炮制加工品巴豆霜能峻下冷积，逐水退肿，豁痰利咽；外用蚀疮。

Croton tiglium L. Trees or shrubs,young branches green, indumentum of stellate hairs.Leaves alternate, with discoid glands. Racemes terminal, monoecism; male sparsely stellate-hairy or glabrescent; female flowers sepals oblong-lanceolate, ovary superior, 3-loculed,densely stellate-hairy.Capsules. Distributed in the south of the Yangtze River, wild or cultivated.The seeds (Crotonis Fructus, Ba dou) have the effect of healing wounds. Processed seeds have the effect of purging cold accumulation drastically,expelling water to abate edema,sweeping phlegm to soothe the throat.Topical application for healing wounds.

21. 无患子科 ♀*↑$K_{4\sim5}C_{4\sim5,\,0}A_{8\sim10}\underline{G}_{(2\sim4:2\sim4:1\sim2)}$

(21) Sapindaceae

【形态特征】多木本。叶互生，常为羽状复叶，无托叶。花单性、杂性或两性，花小；萼片 4~5；花瓣 4~5，有时无；花盘发达，肉质。浆果、核果、蒴果或翅果。种子常有假种皮。

Characteristic: Trees or shrubs. Leaves alternate, usually estipulate; rarely simple. Flowers unisexual, rarely polygamous or bisexual, actinomorphic or zygomorphic, usually small. Sepals 4 or 5.

Petals 4 or 5, sometimes absent. Disk conspicuous, fleshy. Fruit a loculicidal capsule, berry, or drupe, or consisting of 2 or 3 samaras. Seeds often with a conspicuous fleshy aril or sarcotesta.

【分布】135 属，约 1500 种。国产 21 属 52 种，主要分布在东南；药用 11 属约 19 种。

Distribution: 135 genera and about 1 500 species. 21 genera and 52 species in China, distributed in southeast China, medicinal plants 11 genera, about 19 species.

【药用植物】龙眼 *Dimocarpus longan* Lour.（图 12-36）假种皮（龙眼肉）能补益心脾，养血安神。荔枝 *Litchi chinensis* Sonn. 种子（荔枝核）能行气散结，祛寒止痛。无患子 *Sapindus mukorossi* Gaertn. 果实（无患子果）与根能清热解毒，止咳化痰。

果枝
Shoot with fruit

花枝
Flowering shoots

花
Flower

图 12-36 龙眼
Fig.12-36 *Dimocarpus longan*

Medicinal plants: *Dimocarpus longan* Lour. (Fig.12-36) the fake seed coat (Longan Arillus, Long yan rou) can nourish the heart and spleen, nourish the blood and calm the nerves. *Litchi chinensis* Sonn., seeds (Litchi Semen,Li zhi he) can move Qi and disperse knots, dispel cold and relieve pain. *Sapindus mukorossi* Gaertn., fruit and root can clear heat and detoxify, relieve cough and phlegm.

22. 鼠李科 $\male * K_{(4\sim5)} C_{(4\sim5)} A_{4\sim5} \underline{G}_{(2\sim4:2\sim4:1)}$

(22) Rhamnaceae

【形态特征】乔木或灌木，常多刺。单叶互生或对生，有托叶。两性花小，淡黄色；萼片、花瓣及雄蕊均为 4~5 枚；花盘肉质。多为核果，有时为蒴果或翅果。

Characteristic: Deciduous or evergreen, often thorny trees, shrubs, woody climbers, or lianas, rarely herbs. Leaves simple, alternate or opposite, stipules small, caducous or persistent, sometimes transformed into spines. Flowers yellowish to greenish, small, bisexual. Calyx, Petals, Stamens 4 or 5. Disk intrastaminal, nectariferous, thin to fleshy. Fruit sometimes winged, schizocarpic capsule, or a fleshy drupe with 1~4 indehiscent, rarely dehiscent.

【分布】约 50 属，逾 900 种。国产 13 属 137 种，全国广布；药用 12 属约 77 种。

Distribution: About 50 genera and more than 900 species. 13 genera and 137 species in China, distributed all over the country; medicinal plants 12 genera, about 77 species.

【药用植物】枣 *Ziziphus jujuba* Mill.（图 12-37）果实（大枣）能补中益气，养血安神。酸枣 *Ziziphus jujuba* Mill. *var. spinosa*（Bunge）Hu ex H. F. Chow 种子（酸枣仁）能养心补肝，宁心安神，敛汗，生津。多花勾儿茶 *Berchemia floribunda*（Wall.）Brongn. 全株（黄鳝藤）能清热，凉血，利尿，解毒。

Medicinal plants: *Ziziphus jujuba* (Fig.12-37), fruit (Jujubae Fructus, Da zao) can tonify middle-Jiao and Qi, nourishing the blood and tranquilization. *Z.s jujuba var. spinosa*, seeds (Ziziphi Spinosae Semen, Suan zao ren) can nourish the heart and liver, calm the heart and nerves, constrain sweat, and promote the secretion of saliva or body fluid. *Berchemia floribunda*, the whole plant (Huang shan teng) have the function of clearing heat, cooling blood, diuresis and detoxify.

23. 锦葵科 ♀*$K_{5,(5)}C_5A_{(\infty)}\underline{G}_{(3\sim\infty:3\sim\infty:1\sim\infty)}$

(23) Malvaceae

【形态特征】木本或草本。多具黏液细胞和星状毛。单叶互生，有托叶。花两性，辐射对称；萼片5，常有副萼，萼宿存；花瓣5；单体雄蕊。蒴果。

Characteristic: Herbs, shrubs, or less often trees.Indumentum usually with stellate hairs. Leaves simple, alternate, with stipule. Flowers bisexual, radiant. Sepals 5, valvate, free or connate. Petals 5. Filaments connate into tube. Capsule.

【分布】100属，约1 000种。中国产19属，81种；全国广布。药用12属，约60种。

Distribution: 100 genera and about. 1 000 species. 19 genera and 81 species in China, distributed all over the country; 12 genera, about 60 species are medicinal plants.

【药用植物】木槿 *Hibiscus syriacus* L.（图12-38）根、茎皮（木槿皮）能清热燥湿，杀虫止痒。冬葵 *Malva verticillata* L. 果实（冬葵果）能清热利尿，消肿。苘麻 *Abutilon theophrasti* Medik. 种子（苘麻子）能清热利湿，解毒，退翳。黄蜀葵 *Abelmoschus manihot*（L.）Medik. 花冠（黄蜀葵花）能清利湿热，消肿解毒。

Medicinal plants: *Hibiscus syriacus* L (Fig.12-38). Root and stem bark can clear away heat and dryness, kill insects and relieve itching. *Malva verticillata* fruit (Malvae Fructus, Dong kui guo) can clear away heat and diuresis, reduce swelling. *Abutilon theophrasti*, seeds (Abutili Semen,Qing ma zi) can clear away heat and dampness, detoxify and removing nebula. *Abelmoschus manihot*, corolla (Abelmoschi Corolla, Huang shuikui hua) can clear away dampness and heat, reduce swelling and detoxify.

花
Flower

花枝
Flowering shoots

果实
Fruit

图 12-37 枣
Fig.12-37 *Ziziphus jujube*

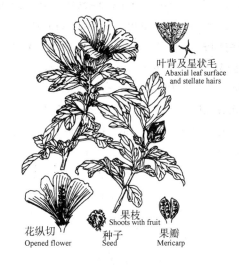

叶背及星状毛
Abaxial leaf surface and stellate hairs

花纵切
Opened flower

果枝
Shoots with fruit

种子
Seed

果瓣
Mericarp

图 12-38 木槿
Fig.12-38 *Hibiscus syriacus*

24. 董菜科 ♀* ↑ $K_5C_5A_5\underline{G}_{(3\sim5:1:1\sim\infty)}$

(24) Violaceae

【形态特征】一年生或多年生草本、灌木。单叶互生，有托叶。花两性，萼片5，花瓣5，下面1枚通常较大，基部囊状或有距，雄蕊5，子房上位。果实为沿室背弹裂的蒴果或为浆果状。

Characteristic: Herbs annual or perennial, shrubs. Leaves simple, usually alternate, with small or leaflike stipules. Flowers bisexual. Sepals 5. Petals 5, anterior one usually larger than others, saccate, gibbous or spurred at base. Stamens 5. Ovary superior. Fruit a loculicidal capsule, usually with elastic and abaxially carinate valves, rarely baccate.

【分布】22属，900~1000种。中国产3属101种；1属，约50种药用。

Distribution: 22 genera, 900–1000 species. 3 genera and 101 species in China; 1 genus, about 50 species are medicinal plants.

【药用植物】紫花地丁 *Viola yedoensis* Makino 全草（紫花地丁）能清热解毒，凉血消肿（图12-39）。董菜 *V. arcuata* Blume 全草能清热解毒。白花地丁 *V. patrinii* DC. ex Ging. 全草能清热解毒，消肿去瘀。

Medicinal plants: *Viola yedoensis* (Fig.12-39) the whole plant (Violae Herba, Zi hua di ding) has the effect of clearing heat and detoxify, cooling blood and reducing swelling. The whole plant of *V. arcuata* and *V. patrinii* also can clear heat and detoxify, eliminate swelling and remove blood stasis.

25. 桃金娘科 ♀ *K $_{(4\sim5)}$ C$_{4\sim5}$ A$_{\infty, (\infty)}$ $\overline{G}_{(2\sim5 : 1\sim5 : 1\sim\infty)}$

(25) Myrtaceae

【形态特征】常绿乔木或灌木，含挥发油。单叶对生。花两性；萼4~5裂，萼筒略与子房合生；花瓣4~5，或与萼片连成一帽状体。浆果、蒴果、稀核果。

Characteristic: Trees or shrubs, evergreen, usually with essential oils-containing cavities. Leaves simple, opposite. Flowers bisexual. Calyx lobes 4 or 5, distinct or connate into a calyptra. Petals 4 or 5, sometimes absent, distinct or connate into a calyptra. Fruit a capsule, berry, drupaceous berry, or drupe.

【分布】130属，4500~5000种。国产10属，121种；主要分布在东南地区。药用10属，约31种。

Distribution: 130 genera and 4500~5000 species. 10 genera and 121 species in China, distributed in southeast China, medicinal plants 10 genera, about 31 species.

【药用植物】丁香 *Eugenia caryophyllata* Thunb.（图12-40）花蕾（丁香）、近成熟果实（母丁香）均能温中降逆，补肾助阳。蓝桉 *Eucalyptus globulus* Labill. 及同属多种植物的挥发油（桉油）能祛风止痛。桃金娘 *Rhodomyrtus tomentosa*（Aiton）Hassk. 根（桃金娘根）能养血、祛风通络、收敛。

Medicinal plants: *Eugenia caryophyllata*. (Fig.12-40) flower bud (Caryophylli Flos, Ding xiang) and near mature fruit (Caryophylli Fructus, Mu ding xiang) can reduce the adverse temperature, tonify the kidney and help Yang. Volatile oil extracted from *Eucalyptus globulus* Labill. and other species of the same genus can dispel wind and relieve pain. *Rhodomyrtus tomentosa*, root (Radix Rhodomyrti) can nourish blood, dispel wind and dredge collaterals, and converge.

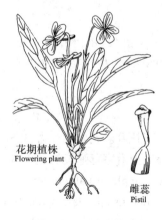

花期植株
Flowering plant

雌蕊
Pistil

图 12-39 紫花地丁
Fig.12-39 *Viola yedoensis*

花枝
Flowering shoot

花蕾纵切
Longitudinal section of flower bud

图 12-40 丁香
Fig.12-40 *Eugenia caryophyllata*

医药大学堂
WWW.YIYAODXT.COM

26. 五加科　♀*$K_5C_{5\sim10}A_{5\sim10}\overline{G}_{(2\sim15:2\sim15:1)}$

(26) Araliaceae

【形态特征】多木本，稀多年生草本；茎常有刺。多具树脂道。叶互生，单叶、掌状复叶或羽状复叶。两性花，伞形花序、头状花序，再排成圆锥状；花瓣5~10；雄蕊5~10，生于花盘边缘；子房下位，2~15室，每室1枚胚珠。浆果或核果。

Characteristic: Trees or shrubs, or woody vines with aerial roots, rarely perennial herbs; secretory canals present in most parts. Leaves alternate, simple, palmately compound, or 1~3-pinnately compound. Inflorescence umbellate, compound-umbellate, ultimate units usually umbels or heads. Flowers bisexual or unisexual. Corolla of 5~10 petals. Stamens usually as many as petalss, inserted at edge of disk. Ovary 2~15 carpellate. Fruit a drupe or berry. Seeds 1 per pyren.

【分布】50属，1350余种。我国23属，180多种，除新疆未发现外，分布于全国各地。药用19属，112种。

Distribution: About 50 genera and 1 350 species. 23 genera and 180 species in China. Medicinal plants 19 genera, about 112 species.

【主要药用属检索表】

---
1. 叶轮生，掌状复叶；草本 ·· 人参属
1. 叶互生，木本，稀多年生草本
  2. 羽状复叶，有托叶；茎常有刺；木本或多年生草本 ······················ 楤木属
  2. 单叶或掌状复叶
    3. 掌状复叶；植物体常有刺 ··· 加属
    3. 单叶，叶片掌状分裂；植物体无刺 ······························ 脱通木属

Key to important medicinal generas of Araliaceae

---
1. Herbs; Leaves verticillate, palmately compound ·························· *Panax*
1. Woody, rarely perennial herbs; leaves alternate
  2. Woody or herbs; Steams with prickles; Pinnately compound, with stipules ·············· *Aralia*
  2. Leaves simple or palmately compound
    3. Plants with prickles; pinnately compound ······················ *Acanthopanax*
    3. Plants without prickles; leaves simple and palmately lobed ············ *Tetrapanax*

【药用植物】人参 *Panax ginseng* C. A. Mey.（图12-41）多年生草本。主根肉质，圆柱形或纺锤形。根状（芦头）短，每年增生1节。掌状复叶，轮生于茎端。顶生伞形花序。浆果成熟时呈红色，扁球状。根和根状茎（人参）能大补元气，复脉固脱，补脾益肺，生津养血。同属三七 *P. notoginseng*（Burkill）F. H. Chen 根和根状茎（三七）能散瘀止血，消肿定痛；花能清热，平肝，降压。

Medicinal plants: *Panax ginseng* (Fig.12-41) Herbs. Rootstock usually with fascicled roots, fusiform or cylindric. Leaves verticillate at apex of stem, palmately compound. Inflorescence a solitary, terminal umbel; Fruit red, compressed-globose. Roots and rhizomes (ginseng) can greatly nourish vital energy, nourish the spleen and the lungs, promote the growth

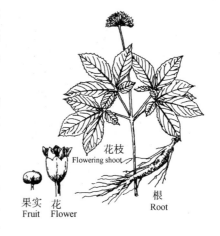

花枝
Flowering shoot

果实　花
Fruit　Flower

根
Root

图 12-41　人参
Fig. 12-41　*Panax ginseng*

of body fluid and blood. Root and rhizome (Panax notoginseng) of *P. notoginseng* can disperse blood stasis, stop bleeding, reduce swelling and pain; flower can clear heat and reduce blood pressure.

刺五加 *Acanthopanax senticosus*（Rupr. & Maxim.）Maxim. 灌木，分枝具刚毛状刺。掌状复叶。伞形花序单生或复生于茎顶。浆果状核果，黑色。根及根状茎或茎（刺五加）能益气健脾，补肾安神。

Acanthopanax senticosus Shrubs. Branches with bristlelike prickles. Leaves palmately compound. florescence a compound umbel, borne on leafy shoots. Fruit ovoid-globose. Root and rhizome or stem (Ci wu jia) can nourish spleen, kidney and anchoring mind.

通脱木 *Tetrapanax papyrifer*（Hook.）K. Koch 灌木或小乔木，密被棕色星状毛。单叶，掌状深裂。伞形花序顶生。茎髓（通草）能清热利尿，通气下乳。

Tetrapanax papyrifer Shrubs or small trees, densely ferruginous or pale brown stellate tomentose. Leaf simple, blade ovate-oblong. Inflorescence terminal. Stem pithes (Tetrapanacis Medulla, Tong tuo mu) can clearing away heat, diuretic, and help lactogenesis in puerpera.

27. 伞形科 $\male \female *K_{(5),0}C_5A_5\overline{G}_{(2:2:1)}$

(27) Apiaceae

【形态特征】草本，常含挥发油。茎中空。叶互生，复叶，少单叶；叶柄基部常为鞘状。两性花小，成复伞形花序；花萼 5，与子房贴生；花瓣 5；雄蕊 5，与花瓣互生；子房下位，花柱 2，具上位花盘（图 12-42）。双悬果，每分果有 5 条主棱（1 条背棱，2 条中棱，2 条侧棱），下有维管束，棱和棱之间的沟槽内和合生面通常有纵走的油管 1 至多数（图 12-43）。

Characteristic: Herbs. Fragrant. Stem hollow or solid. Leaves alternate, compound or sometimes simple, petiole usually sheathing at base. Flowers small, bisexual, in simple or compound umbels. Calyx tube wholly adnate to the ovary. Ovary inferior, 2-celled. Styles 2, usually swollen at the base forming a stylopodium which often secretes nectar (Fig.12-42). Fruit dry, of two mericarps united by their faces (commissure), and usually attached to a central axis (carpophore), from which the mericarps separate at maturity; each mericarp has 5 primary ribs, one down the back (dorsal rib), two on the edges near the commissure (lateral ribs), and two between the dorsal and lateral ribs (intermediate ribs); vittae (oil-tubes)

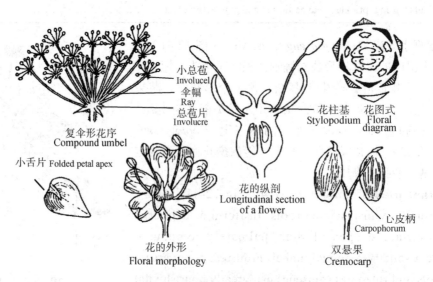

图 12-42　伞形科花、果实模式图
Fig. 12-42　Flower and fruit pattern of Umbelliferae

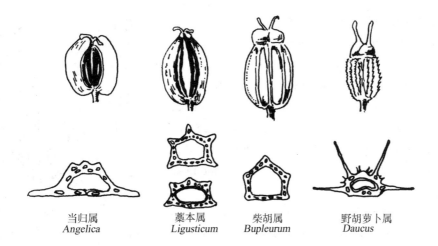

当归属
Angelica　　藁本属
Ligusticum　　柴胡属
Bupleurum　　野胡萝卜属
Daucus

图 12-43　伞形四属植物果实及横切面

Fig. 12-43　Fruit and with its cross section of 4 genera in Umbelliferae

usually present in the furrow (intervals between the ribs sometimes called the valleculae) and on the commissure face (Fig.12-43).

【分布】250~440 属，3300~3700 种。我国 100 属，614 种，分布全国。已知药用 55 属，234 种。

Distribution: Between 250–440 genera and 3 300–3 700 species. 100 genera and 614 species in China. 55 genera, about 234 species are medicinal plants.

【药用植物】当归 *Angelica sinensis*（Oliv.）Diels（图 12-44）多年生草本，根肉质分枝，具强烈芳香味。叶三出式羽状分裂或羽状全裂。双悬果，背棱线形，隆起，侧棱具翅，与果体等宽或略宽。根（当归）能补血活血，调经止痛，润肠通便。同属植物白芷 *A. dahurica*（Fisch. ex Hoffm.）Benth. Et 的根（白芷）能解表散寒，祛风止痛，宜通鼻窍，燥湿止带，消肿排脓。

Medicinal plants: *Angelica sinensis* (Fig.12-44) Perennial, Root cylindric, branched, rootlets many, succulent, strongly aromatic. Blade 2~3-ternate-pinnate, pinnae 3~4 pairs. Fruit ellipsoid or suborbicular, dorsal ribs filiform, prominent, lateral ribs broadly thin-winged, wings as wide as or wider than the body. Root (Radix Angelicae Sinensis, Dang gui) can invigorate blood, regulate menses and relieve pain, moisten intestines and relieve constipation. The root of *A. dahurica* (Radix Angelicae Dahuricae, Bai zhi) can relieve cold, dispel wind and relieve pain. It is suitable to pass through nose and orifices, stop dampness, reduce swelling and discharge pus.

藁本 *Ligusticum sinense* Oliv. 多年生草本。根茎厚，在节上明显膨胀，节间短。叶二回羽状全裂。双悬果宽卵形。根茎（藁本）能祛风散寒，除湿止痛。

Ligusticum sinense Plants perennial. Rootstock thick, apparently swollen at nodes, internodes short. Blade 1- or 2-pinnate. Fruit oblong-ovoid. Rhizome (Radix er Rhizoma Ligustici, Gao ben) can dispel

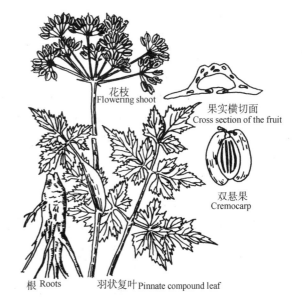

花枝
Flowering shoot

果实横切面
Cross section of the fruit

双悬果
Cremocarp

根 Roots　　羽状复叶 Pinnate compound leaf

图 12-44　当归

Fig. 12-44　*Angelicae sinensis*

wind and cold, dehumidify and relieve pain.

柴胡 *Bupleurum chinense* DC. 多年生草本（图 12-45）主根粗壮，木质，通常有分枝。单叶，具平行脉。花黄色。根（柴胡）能疏散退热，疏肝解郁，升举阳气。

Bupleurum chinense (Fig.12-45) Herbs perennial. Root stout, elongate, brown, woody, usually branched. Leaves parallel veins. Petals bright yellow. Root (Bupleuri Radix, Chai hu) can disperse fever, relieve liver depression and raise Yang Qi.

防风 *Saposhnikovia divaricate*（Turcz.）Schischk. 多年生草本。根粗壮。茎二歧分枝，基部密被纤维状叶鞘。双悬果矩圆状宽卵形。根（防风）能祛风解表，胜湿止痛，止痉。

Saposhnikovia divaricata Herbs perennial. Rootstock stout. Stem dichotomously branched, base densely covered with fibrous leaf sheaths. Root (Saposhnikoviae Radix, Fang feng) can dispel the wind, relieve the external symptoms, relieve pain and spasm.

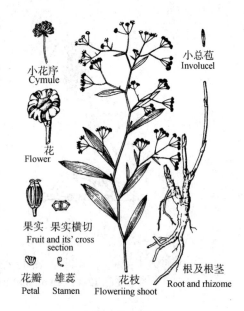

图 12-45　柴胡
Fig. 12-45　*Bupleurum chinense*

视频

（二）合瓣花亚纲
4.1.2　Sympetalae

合瓣花亚纲又称后生花被亚纲，花瓣多少联合，形成各种形状的花冠，利于昆虫传粉，同时雌雄蕊得到更好地保护；花由 5 轮减为 4 轮，各轮数目继续减少；通常无托叶；胚珠具一层珠被。

The Sympetalae also termed as Metachlamydeae. The petals are more or less united, with various types of united corolla adapted to insect pollination and enhanced protection of stamens and pistils. The whorls of petals reduced from 5 to 4, and the number decreased continuously; stipules usually absent; ovule has a layer of integument.

28. 杜鹃花科　$\male\female * K_{(4\sim5)} C_{(4\sim5)} A_{(8\sim10, 4\sim5)} \underline{G}_{(4\sim5:4\sim5:\infty)}$

(28) Ericaceae

【形态特征】常绿木本。单叶互生，对生或轮生；无托叶。花两性，整齐或稍两侧对称；花萼 4~5 裂，宿存；花冠 4~5 裂；雄蕊常为花冠裂片的 2 倍，花药孔裂；子房上位，稀下位，4~5心皮合生。蒴果或浆果。

Characteristic: Shrubs or trees, usually evergreen. Leaves single, alternate. Flowers bisexual, actinomorphic or slightly bilaterally symmetric; calyx 4~5 lobed, persistent; corolla 4~5 lobed; stamens twice as numerous as corolla lobes; anthers pex dehiscing by pores; Ovary superior, sparsely inferior, 4~5 carpels connate. Fruit a capsule or berry.

【分布】约 103 属，3350 种，世界广布。我国有 15 属，约 757 种，全国分布，以西南山区种类最多。药用 12 属 126 种。

Distribution: About 103 genera, 3 350 species, widely distributed. 15 genera, 757 species in China, mostly distributed in the southwest mountains of China. 12 genera,126 species are medicinal plants.

【药用植物】兴安杜鹃 *Rhododendron dauricum* L. 叶（满山红）能止咳祛痰（图 12-46）。羊踯躅 *R. molle*（Blum）G. Don 花（闹羊花）能祛风除湿，散瘀定痛。杜鹃 *R. simsii* Planch. 花（杜鹃花）能活血，调经，止咳，祛风湿，解疮毒。叶（杜鹃花叶）能清热解毒，止血，化痰止咳。南烛 *Vaccinium bracteatum* Thunb. 果实（南烛子）能补肝肾，强筋骨，固精气，止泻痢；叶（南烛叶）益肠胃，养肝肾。

Medicinal plants: *Rhododendron dauricum* L.(Fig.12-46). The leaves (Rhododendri Daurici Folium, Man shan hong) are used for relieving cough and eliminating phlegm. *R. molle* (Blum) G. Don. The flowers (Rhododendri Mollis Flos) are used for expelling wind and dampness, dispersing stasis to settle pain. *R. simsii* Planch., the flowers (Rhododendri Simsii Flos) are used for harmonizing blood, regulating menstruation, relieving cough, expelling wind-dampness, relieving sore poison. The leaves (Rhododendri Simsii Folium) are used for clearing heat and detoxifying, hemostasis, resolving phlegm and relieving cough. *Vaccinium bracteatum* Thunb., the fruits (Vaccinii Bracteati Fructus) can nourish the liver and kidney, strengthen the muscles and bones, strengthen the essence and Qi, and stop diarrhea. The leaves (Vaccinii Folium) are used for benefitting the stomach, liver and kidney.

图 12-46 兴安杜鹃
Fig.12-46 *Rhododendron dauricum*

29. 报春花科 $\male^* K_{(5), 5} C_{(5), 0} A_5 \underline{G}_{(5:1:\infty)}$

(29) Primulaceae

【形态特征】草本。单叶互生或对生，无托叶。花两性，整齐；萼 5 裂，宿存；花冠 5 裂；雄蕊与花冠裂片同数而对生，着生于花冠筒上；子房上位，特立中央胎座；花柱异长；蒴果。

Characteristic: Herbs. Leaves a simple, alternate, opposite, basal or whorled, without stipules. Flowers bisexual, actinomorphic; calyx 5-lobed, persistent; corolla 5-lobed; stamens as many as and opposite corolla lobes, inserted on corolla tube; ovary superior, placentation free central; heterostyle. Fruit a capsule.

【分布】22 属，近 1000 种，世界广布。我国 13 属，近 500 种，全国各地均有分布，以西南地区为多。药用 7 属，100 余种。

Distribution: 22 genera, about 1 000 species. 13 genera, nearly 500 species in China, distributed all over the country, especially in Southwest China. Medicinal plants 7 genera, more than 100 species.

【药用植物】过路黄 *Lysimachia christiniae* Hance 全草（金钱草）能利湿退黄，利尿通淋，解毒消肿（图 12-47）。灵香草 *L. foenum-graecum* Hance 全草（灵香草）能解表，止痛，行气，驱蛔。点地梅 *Androsace umbellata*（Lour.）Merr. 全草（喉咙草）能清热解毒，消肿止痛。

Medicinal plants: *Lysimachia christiniae* Hance (Fig.12-47), the whole grass (Lysimachiae Herba, Jin qian cao) are used for promoting dampness and yellow, diuresis and passing gonorrhea, detoxification and detumescence. *L. foenum-graecum* Hance, the whole grass (Lysimachiae Foenum-graeci Herba) are used for relieving exterior syndrome, relieving pain, promoting qi and driving away Ascaris. *Androsace umbellata* (Lour.) Merr., the whole

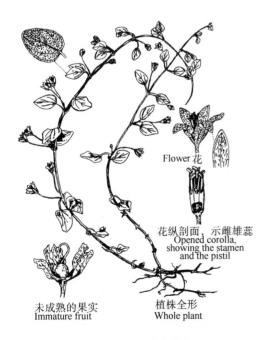

Flower 花

花纵剖面，示雌雄蕊
Opened corolla,
showing the stamen
and the pistil

未成熟的果实
Immature fruit

植株全形
Whole plant

图 12-47 过路黄
Fig.12-47 *Lysimachia christiniae*

grass (Androsaces Umbellatae Herba) are used for clearing away heat, detoxifying, detumescence and relieving pain.

30. 木犀科 ☿ $*K_{(4)} C_{(4), 0} A_2 \underline{G}_{(2:2:2)}$

(30) Oleaceae

【形态特征】灌木或乔木。叶常对生，单叶或复叶，无托叶。聚伞花序圆锥状；花两性，稀单性，整齐；花萼、花冠常4裂，稀无花瓣；雄蕊2；子房上位，2心皮2室。翅果、蒴果、核果、浆果。

Characteristic: Shrubs or trees. Leaves opposite, simple, trifoliolate, or pinnately compound, exstipulate. Inflorescences in panicles, cymes or fascicle; flowers bisexual, rarely unisexual, actinomorphic; Calyx and corolla often 4-lobed, sparsely apetalous; stamens 2; ovary superior, 2 carpels, 2 locules. Fruit a samara, capsule, drupe, berry.

【分布】约27属，400余种。我国12属，178种，各地均有分布。药用8属，80余种。

Distribution: About 27 genera, more than 400 species. 12 genera, 178 species in China, widely distributed. Medicinal plants 8 genera, more than 80 species.

【药用植物】连翘 *Forsythia suspensa*（Thunb.）Vahl 果实（连翘）能清热解毒，消肿散结，疏散风热（图12-48）。女贞 *Ligustrum lucidum* Ait. 果实（女贞子）能滋补肝肾、明目乌发。白蜡树 *Fraxinus chinensis* Roxb. 茎皮（秦皮）能清热燥湿，收涩止痢，止带，明目。

Medicinal plants: *Forsythia suspensa* (Thunb.) Vahl (Fig.12-48), the fruit (Forsythiae Fructus, Lian qiao) can clear away heat and detoxify, eliminate swelling and knots, and disperse wind-heat. *Ligustrum lucidum* Ait., the fruits (Ligustri Lucidi Fructus, Nǚ zhen zi) can nourish the liver and kidney, bright eyes and dark hair. *Fraxinus chinensis* Roxb., the bark (Fraxini Cortex, Qin pi) can clear away heat and dry dampness, stop astringency and dysentery, stop band and make eyes bright.

花冠展开，示雄蕊
Opened corolla, showing the stamen

花萼展开，示雌蕊
Opened calyx, showing the pistil

花枝
Flowering branch

叶枝
Leaf branch

果实
Fruit

图 12-48 连翘
Fig.12-48 *Forsythia suspensa*

31. 龙胆科 ☿ $*K_{(4\sim5)} C_{(4\sim5)} A_{4\sim5} \underline{G}_{(2:1:\infty)}$

(31) Gentianaceae

【形态特征】草本。单叶对生，稀轮生，无托叶。聚伞花序；花两性，整齐，4~5数；萼筒状、钟状或辐状；花冠筒状、漏斗状或辐状；雄蕊与花冠裂片同数而互生；子房上位，2心皮，1室。蒴果2瓣裂。

Characteristic: Herbs.Leaves simple, opposite, sparsely whorled. Cymes; flowers bisexual, actinomorphy, generally 4~5 cardinal; calyx tubular, campanulate or radial; corolla tubular, funnelform or radial, rarely with spur; Stamens and corolla lobes identical but alternate, inserted on corolla tube; Ovary superior, 2 carpels, 1 locule. Capsule, 2-valved.

【分布】约80属700种。我国22属，427种，全国均有分布，以西南山区分布最为集中。药用15属，100余种。

Distribution: About 80 genera, 700 species. 22 genera, 427 species in China, distributed all over the country, especially in the southwest mountainous area. Medicinal plants 15 genera, more than 100 species.

【药用植物】龙胆 *Gentiana scabra* Bunge 和条叶龙胆 *G. manshurica* Kitag. 的根和根茎（龙胆）能清热燥湿，泻肝胆火（图12-49）。秦艽 *G. macrophylla* Pall.，粗茎秦艽 *G. crassicaulis* Duthie ex Burk. 或滇龙胆草 *G. rigescens* Franch. ex Hemsl. 的干燥根（秦艽）能祛风湿，清湿热，止痹痛，退虚热。蒙自獐牙菜 *Swertia leducii* Franch. 全草（青叶胆）能清肝利胆，清热利湿。北方獐牙菜 *S. diluta*（Turcz.）Benth. et Hook. f. 全草（当药）能清湿热，健胃。双蝴蝶 *Tripterospermum chinense*（Migo）H. Smith 全草（肺行草）能清肺止咳，凉血止血，利尿解毒。

Medicinal plants: The roots of *Gentiana scabra* Bunge (Fig.12-49) and *Gentiana. manshurica* Kitag. (Gentianae Radix et Rhizoma, Long dan) can clear heat, dry dampness, diarrhea hepatobiliary fire. Roots of *G. macrophylla*, *G. crassicaulis* and *G.rigescens* (Gentianae Macrophyllae Radix, Qin jiao) are used for dispelling rheumatism, clearing dampness and heat, stopping arthralgia and relieving deficiency-heat. *Swertia leducii* Franch., the whole grass can clear the liver and gallbladder, clear away heat and dampness. *Swertia diluta*, the whole grass (Swertiae Herba) can remove heat and humidity, and strengthen the stomach. *Tripterospermum chinense*, the whole grass (Tripterospermi Chinensis Herba) are used for clearing lung and relieving cough, cooling blood and hemostasis, diuretic and detoxification.

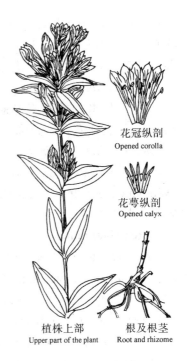

花冠纵剖
Opened corolla

花萼纵剖
Opened calyx

植株上部　　　根及根茎
Upper part of the plant　　Root and rhizome

图 12-49　龙胆
Fig.12-49 *Gentiana scabra*

32. 夹竹桃科　$\male \female \ast K_{(5)} C_{(5)} A_5 \underline{G}_{(2:1\sim2:1\sim\infty)}$

(32) Apocynaceae

【形态特征】灌木、藤本或草本，稀乔木。具白色乳汁或水液。单叶对生或轮生，稀互生，全缘，无托叶或退化成腺体或无。花单生或成聚伞花序；花两性，整齐，5数；萼合生成筒状或钟状，基部内面有腺体；花冠高脚碟状、漏斗状、坛状或钟状，喉部常有副花冠或附属体；雄蕊5，着生于花冠筒上或花冠喉部；子房上位，2心皮离生或合生，1~2室。蓇葖果，稀浆果、核果或蒴果；种子一端常被毛或膜翅。

Characteristic: Trees, shrubs, vines and herbs, with latex or water juice. Leaves simple, opposite or whorled, entire, often exstipulate or reduced to glands. Flowers solitary or in cymes; flowers bisexual, actinomorphy; calyx often 5-lobed, base connate into tubular or campanulate, basal glands usually present; corolla 5-lobed, salverform, funnelform, urceolate or campanulate, corolla throat with corona or scales or membranous or hairy appendages; stamens 5, inserted on corolla tube or corolla throat; ovary superior, 2 carpels distinct or connate, 1~2-locular. Fruit a follicle, rarely berry, drupe, or capsule; seed one end often with hairy or membranous wings.

【分布】约250属，2000余种，分布于热带、亚热带地区，少数在温带地区。我国46属，176种，分布于长江以南各省区。已知药用35属，95种。

Distribution: About 250 genera, more than 2 000 species, mainly distributed in the tropical and subtropical regions, poorly represented in the temperate regions. 46 genera, 176 species in China, distributed in the south of the Yangtze River. Medicinal plants 35 genera, 95 species.

【药用植物】萝芙木 *Rauvolfia verticillata*（Lour.）Baill.（图12-50）全株能镇静，降压，活血止痛，清热解毒，常作提取"降压灵"和"利血平"的原料。同属其他多种植物如蛇根木 *R.*

serpentina（L.）Benth. ex Kurz 等也有相同作用。长春花 *Catharanthus roseus*（L.）G. Don 全株能抗癌，抗病毒，利尿，降血糖；为提取长春碱和长春新碱的原料。罗布麻 *Apocynum venetum* L. 叶（罗布麻叶）能平肝安神，清热利水。黄花夹竹桃 *Thevetia peruviana*（Pers.）K. Schum. 全株有毒，具强心，利尿和消肿作用。络石 *Trachelospermum jasminoides*（Lindl.）Lem. 带叶藤茎（络石藤）能祛风通络，凉血消肿。杜仲藤 *Urceola micrantha*（Wallich ex G. Don）D. J. Middleton 树皮（杜仲藤）能祛风湿，强筋骨。

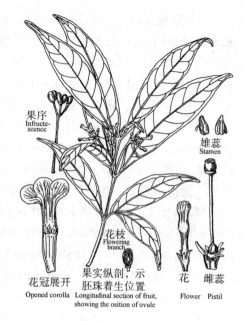

图 12-50　萝芙木

Fig.12-50　*Rauvolfia verticillata*

Medicinal plants: *Rauvolfia verticillata* (Fig.12-50) the whole plant can calm, reduce blood pressure, promote blood circulation and relieve pain, clear heat and detoxify, and is often used as the raw material for extracting " verticilum" and "reserpine". Other plants of the same genus, such as *R. serpentina*, also have the same effect. *Catharanthus roseus* the whole plant is capable of anti-cancer, anti-virus, diuretic and reducing the blood sugar, and often used as the raw material for extracting "vinblastine" and "vincristine". *Apocynum venetum*, the leaves (Apocyni Veneti Folium) are used for calming the liver and mind, clearing away heat and promoting diuresis. *Thevetia peruviana*, the whole plant is toxic and has cardiotonic, diuretic and detumescence effects. *Trachelospermum jasminoides*, vine stem with leaf (Trachelospermi Caulis et Folium) is used for dispelling wind and dredging collaterals, cooling blood and reducing swelling. *Urceola micrantha* the bark (Urceolae Micranthae Cortex) is used for removing rheumatism, strengthening muscles and bones.

33. 萝藦科 ♀ * K$_{(5)}$ C$_{(5)}$ A$_{(5)}$ $\underline{G}_{2:1:\infty}$

(33) Asclepiadaceae

【形态特征】草本、藤本或灌木，具乳汁。单叶对生，稀轮生，叶柄顶端常具腺体；无托叶。聚伞花序；两性花，整齐，5 数；萼筒短，内面基部常有腺体；花冠辐状、坛状，稀高脚碟状，裂片旋转、覆瓦状或镊合状排列；常具副花冠；雄蕊 5，与雌蕊贴生，称合蕊柱；花药连生成一环，贴生于柱头基部的膨大处；花丝合生成为 1 个有蜜腺的筒，称合蕊冠，或花丝离生；花粉粒联合包在 1 层软韧的薄膜内而成块状，称花粉块，通过花粉块柄而系结于着粉腺上，每花药有花粉块 2 个或 4 个；或花粉器匙形，直立，其上部为载粉器，下面为 1 载粉器柄，基部有 1 粘盘，粘于柱头上；无花盘；子房上位，2 心皮，离生；花柱 2，顶部合生，柱头常与花药合生。蓇葖果双生，或因 1 个不育而单生；种子多数，顶端具白色绢丝状毛（图 12-51）。

Characteristic: Herbs, vines, or shrubs, with latex. Leaves simple, opposite, sparsely whorled, entire, apex of petiole with clumped glands, often exstipulate. Cymes; flowers bisexual, actinomorphic, 5-cardinal; calyx tube short, inner base often with glands; corolla rotate, ureolate, sparsely salver-shaped; corona present; stamens 5, with pistil stick to form center column, termed gynostemium; anther connected to form a ring and the abdomen attached to the swollen part of the base of the stigma; Filaments connate into a tube with nectaries, termed gynostegium; pollen grains packed together in a layer of soft and tough thin film to form a block, termed the pollinium, usually attached to the retinaculum by a caudicle, pollinia 2 or 4 per anther;or pollinarium, usually spatulate, upright, the upper part translater containing granular pollen tetrads, the lower part with a translater handle, the handle base with a sticky disk, which sticks to the stigma; flower disc absent; ovary superior; carpels 2, free; styles 2, apically connate, stigma often connate with anthers. Follicles twin, or solitary due to 1 sterile; Seeds numerous, apex with white silky hairs(Fig.12-51).

【分布】约 180 属，2200 种，分布于热带、亚热带地区。我国 44 属，245 种，33 变种，全国分布，以西南、华南种类较多。已知药用 32 属，112 种。

Distribution: About 180 genera, 2 200 species, distributed in tropical, subtropical. 44 genera, 245 species and 33 varieties in China. Mostly in Southwest, South China. Medicinal plants 32 genera, 112 species.

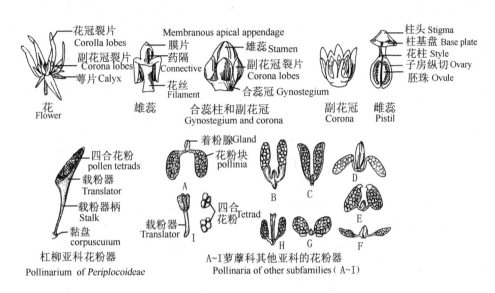

图 12-51　萝藦科花及花粉器的形态和结构

Fig. 12-51　Morphology and structure of flowers and pollinarium in Asclepiadaceae

【药用植物】白薇 *Cynanchum atratum* Bunge（图 12-52）多年生草本，全株被绒毛，根须状，有香气。花冠深紫色。蓇葖果单生，种子一端有长毛。分布全国各地。根及根茎（白薇）能清热凉血，利尿通淋，解毒疗疮。同属植物变色白前 *C. versicolor* Bunge 根及根基亦做“白薇”入药。柳叶白前 *C. stauntonii*（Decne.）Schltr. ex Levl. 根及根茎（白前）能降气，消痰，止咳。牛皮消 *C. auriculatum* Royle ex Wight 块根（白首乌）能补肝肾，强筋骨，益精血，健脾消食，解毒疗疮。

Medicinal plants: *Cynanchum atratum* Bunge (Fig.12-52) Herbs perennial, with milk, whole tomentose, rootlike, aromatic. Corolla dark purple, Follicles solitary, seeds with long hairs on one end. Distributed all over the country. The root and rhizome (Cynanchi Atrati Radix et Rhizoma,Bai wei) have the effect of clearing heat, coolling blood, promoting diuresis, dredging gonorrhea, and detoxifing boils and acne. The root and rhizome of *C. versicolor* is also used as Bai wei. The root and rhizome of *C. stauntonii* (Cynanchi Stauntonii Rhizoma et Radix) have the effect of reducing Qi, eliminating phlegm and cough. The root of *C. auriculatum* (Cynanchi Auriculati Radix) have the effect of tonifying the liver and kidney, strengthening the muscles and bones, benefiting the essence and blood, strengthening the spleen and eliminating food, detoxifying and treating sore.

徐长卿 *Cynanchum paniculatum*（Bunge）Kitagawa 多年生草本，具特异香气。茎直立，常不分枝。花冠黄绿色。蓇葖果单生。全国大部分地区均有分布。根及根茎（徐长卿）能祛风化湿，止痛止痒。

Cynanchum paniculatum (Bunge) Kitagawa Herbs perennial, with specific aroma. Stems erect, often unbranched. Corolla yellow-green. Follicles solitary. Distributed all over the country. Root and rhizome (Cynanchi Paniculati Radix et Rhizoma, Xu chang qin) are used for removing wind and dampness, relieving pain and itching.

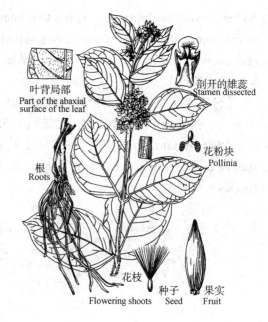

图 12-52　白薇
Fig.12-52　*Cynanchum atratum*

杠柳 *Periploca sepium* Bunge 蔓生灌木。花冠暗紫色，反折。蓇葖果双生。分布于东北、西北、华北、华东以及西南各省。根皮（香加皮）能利水消肿，祛风湿，强筋骨。

Periploca sepium Bunge Vine shrub. Corolla dark purple, reflexed. Follicles twin. Distributed in Northeast, Northwest, North, Eastern China and Southwest provinces. Root bark (Periplocae Cortex, Xiang jia pi) can remove water and reduce swelling, dispel wind and dampness, and strengthen muscles and bones.

常用药用植物还有：萝藦 *Metaplexis japonica*（Thunb.）Makino 果壳（天浆壳）能清肺化痰，散瘀止血。娃儿藤 *Tylophora ovata*（Lindl.）Hook. ex Steud. 根（三十六荡）能祛风湿，化痰止咳，散瘀止痛，解蛇毒。通光散 *Marsdenia tenacissima*（Roxb.）Moon 藤、根或叶（通光散）能清热解毒，止咳平喘，利湿通乳，抗癌。

Other medicinal plants: The pericarp (Metaplexis Japonicae Pericarpium, Tian jiang qiao) of *Metaplexis japonica* (Thunb.) Makino are used for clearing the lung, dissipating phlegm, dispersing blood stasis and stopping bleeding.The roots (Tylophori Ovati Radix, Sanshiliu dang) of *Tylophora ovata* (Lindl.) Hook. ex Steud. can dispel rheumatism, dissipate phlegm and cough, dissipate blood stasis and relieve pain, and relieve snake venom. The vine, root or leaf (Marsdenii Tenacissimi Caulis et Radix) of *Marsdenia tenacissima* (Roxb.) Moon, can clear heat and detoxify, relieve cough and asthma, relieve dampness and promote lactation, and fight against cancer.

34. 旋花科　$\male\female *K_5 C_{(5)} A_5 \underline{G}_{(2:1-4:1-2)}$

(34) Convolvulaceae

【形态特征】草质或木质藤本，有乳汁，具双韧维管束。单叶互生，无托叶。花两性，整齐，5 基数；花萼宿存，外萼常比内萼大；花冠漏斗状、钟状、高脚碟状或坛状，蕾期旋转状。雄蕊 5，冠生；子房上位，2 心皮合生，1~2 室，或因假隔膜成 4 室，每室 1~2 枚胚珠。蒴果，稀浆果。

Characteristic: Herbaceous or woody vines, with milky juice and bicollateral vascular bundle. Leaves alternate, simple. Flowers bisexual, actinomorphic, 5-merous; sepals persistent, outer ones usually larger than inner ones; corolla funnelform, campanulate, salverform, or urceolate. Stamens 5, adnate to corolla. Ovary superior, 2-carpellate connate, or 4-loculed due to false diaphragm, 1~2 anatropous ovules per locule. Fruit a capsule, rarely a berry.

【分布】56 属，1800 余种。我国 22 属，约 125 种，主要分布于西南和华南地区；药用 16 属，56 种。

Distribution: Circa 56 genera, 1 800 species. 22 genera and 125 species in China. Mostly in South and Southwestern China; medicinal plants 16 genera and 56 species.

【药用植物】菟丝子 *Cuscuta chinensis* Lam.（图 12-53）一年生寄生草本。茎缠绕，黄色。叶退化为鳞片状。花簇生成球形；花冠黄白色，壶状。蒴果球形。种子表面粗糙。全国广布。常寄生在豆科、蓼科、菊科、藜科等多种植物上。种子（菟丝子）能补益肝肾，固精缩尿，安胎，明目，止泻，外用消风祛斑。同属植物南方菟丝子 *C. australis* R. Br.、金灯藤（日本菟丝子）*C. japonica* Choisy、欧洲菟丝子 *C. europaea* L. 的种子功效同菟丝子。

Medicinal plants: *Cuscuta chinensis* Lam. (Fig.12-53) annual parasitic herbs. Stems twining, yellow. Leaves scale-like. Inflorescences compact cymose glomerules; corolla yellowish white, urceolate. Capsule enclosed by corolla when maturing, pyxidium. Seeds scabrous. Distributed throughout China, often parasitic on plants of Fabaceae, Polygonaceae, Asteraceae and Zygophyllaceae. Seeds (Cuscutae Semen) can nourish the liver and kidney yin and yang, consolidate essence, relieve enuresis, improve vision and arrest diarrhea. The seeds of *C. australis* R. Br., as well as *C. japonica* Choisy and *C. europaea* L. have the same actions.

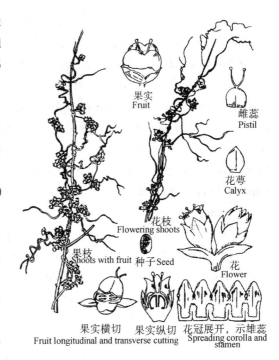

果实 Fruit
雌蕊 Pistil
花萼 Calyx
花枝 Flowering shoots
果枝 Shoots with fruit
种子 Seed
花 Flower
果实横切 果实纵切 Fruit longitudinal and transverse cutting
花冠展开，示雄蕊 Spreading corolla and stamen

图 12-53　菟丝子
Fig.12-53　*Cuscuta chinensis*

牵牛 *Pharbitis nil*（L.）Choisy 一年生缠绕草本。单叶互生，掌状 3 裂。花 1~3 朵腋生；花冠漏斗状，蓝紫色或紫红色；雄蕊 5，不等长。蒴果近球形。种子卵状三棱形。全国广布，野生或栽培。种子（牵牛子）能泻水通便，消痰涤饮，杀虫攻积。同属植物圆叶牵牛 *P. purpurea*（L.）Voigt. 的种子与牵牛同等入药。

Pharbitis nil (L.) Choisy. Annual herbaceous, twining. Leaf simple and alternate, palmately 3- (or 5-) lobed. Corolla funnelform, blue-purple or purplish-red; Stamens 5, unequal. Capsule globose. Seeds ovoid-trigonous. Distributed throughout China, wild or cultivated. Seeds (Pharbitidis Semen, Qian niu zi) have the effects of removing water retention to relieve constipation and diuresis, dissolving phlegm, flushing rheum, killing and expelling parasites to remove accumulation. The seeds of *P. purpurea* (L.) Voigt. have the same effects as *P. nil*.

常用药用植物还有：丁公藤 *Erycibe obtusifolia* Benth. 茎藤（丁公藤）有小毒，能祛风除湿，消肿止痛。同属光叶丁公藤 *E. schmidtii* Craib 干燥茎藤同等入药。

Other medicinal plants: *Erycibe obtusifolia* Benth. the stem (Erycibes Caulis,Ding gong teng) is slightly poisonous, can disperse wind and dampness, reduce swelling and relieve pain. The dried stems of *E. schmidtii* Craib (Erycibes Caulis) have the same effects as *E. obtusifolia*.

35. 紫草科　$\male \female *K_{5,(5)} C_{(5)} A_5 \underline{G}_{(2:2-4:1\sim2)}$

(35) Boraginaceae

【形态特征】多草本，常密被硬毛。单叶互生，常全缘，无托叶。蝎尾状聚伞花序或聚伞花序顶。花两性，整齐，5 数；萼基部至中部合生；花冠筒状、钟状、漏斗状，喉部常有附属物；

雄蕊冠生，与花冠裂片同数而互生；子房上位，2 心皮合生，有时深 4 裂而成假 4 室，每室 1 胚珠；花柱顶生或基生。核果或 4 个小坚果。

Characteristic: Generally herbs, usually bristly or scabrous-pubescent. Leaves simple, alternate, exstipulate, entire at margin. Inflorescences terminal, monochasium or cymes. Flowers bisexual, actinomorphic; calyx connate from base to middle; Corolla tubular, campanulate, funnelform, or salverform, often with appendages at throat; stamens inserted on corolla tube, as many as corolla lobes and alternate with them; Ovary superior, 2-carpellate connate to 2 locules and each with 2 ovules, or 4 locules due to deeply 4-lobed and each with 1 ovule. Style terminal or base. Fruit drupes or 4 nutlets.

【分布】约 100 属，2000 余种。我国 48 属，269 种，主要分布于西南和西北地区，药用 22 属，62 种。

Distribution: 100 genera and 2 000 species. 48 genera and 269 species in China. Mostly in SW and NW China; medicinal plants 22 genera and 62 species.

【药用植物】新疆紫草 *Arnebia euchroma*（Royle）I. M. Johnst. 多年生草本；根粗壮，暗紫色；花冠筒状钟形，深紫色（图 12-54）。分布于新疆及西藏西部。根（紫草）能清热凉血，活血解毒，透疹消斑。内蒙紫草 *A. guttata* Bunge 根与新疆紫草同等入药。

花枝
Flowering shoots

果实
Fruit

植株下部
Lower part of plant

花冠展开
Corolla spreading

子房
Ovary

图 12-54　新疆紫草
Fig. 12-54　*Arnebia euchroma*

Medicinal plants: *Arnebia euchroma* (Royle) I. M. Johnst. (Fig.12-54) Perennial herbs. Roots stout, dark purple. Corolla tubular-campanulate, dark purple. Distributed in Xinjiang and western Tibet. Roots (Arnebiae Radix, Zi cao) can cool and activate blood, and clear toxic heat to relieve skin eruptions. Roots of *A. guttata* Bunge (Arnebiae Radix) is equally used as *A. euchroma*.

常用药用植物还有：紫草 *Lithospermum erythrorhizon* Sieb. Et Zucc. 根（硬紫草）功效同新疆紫草。附地菜 *Trigonotis peduncularis*（Trev.）Benth. ex Baker et Moore 全草（附地菜）能温中健胃，消肿止痛，止血。

Other medicinal plants: *Trigonotis peduncularis* (Trev.) Benth. ex Baker et Moore: Herbs. The whole plants is mainly applied for harmonizing and warming the stomach, reducing swelling and relieving pain, and stopping bleeding. The root of *Lithospermum erythrorhizon* Sieb. Et Zucc. also used as *Arnebia euchroma*.

36. 马鞭草科　$\male \uparrow K_{(4\sim5)} C_{(4\sim5)} A_4 \underline{G}_{(2:4:1\sim2)}$

(36) Verbenaceae

【形态特征】木本，稀草本，常具特殊气味。叶对生。聚伞花序。花两性，不整齐；花萼杯状、钟状或管状，常于果熟后增大而宿存；花冠二唇形或不等的 4~5 裂；雄蕊 2 强，着生花冠管上；子房上位，2 心皮，2~4 室，每室胚珠 1~2。花柱顶生。核果，蒴果，浆果状核果或裂为 4 枚小坚果。

Characteristic: Woody plants, rarely herbs, usually with special odour. Leaves opposite. Inflorescences cymose. Flowers bisexual, often zygomorphic; Calyx cyathiform, campanulate, or tubular, persistent, enlarge after mature fruit; Corolla bilabiate; stamens 4, often didymous, inserted on corolla tube; Ovary superior, 2-carpellate connate to 2~4 locules, ovules 1 or 2 per locule. Drupe, capsule or berry drup, sometimes breaking up into 4 nutlets.

【分布】80 属，3000 余种。我国 21 属，175 种，分布于长江以南各省；药用 15 属，100 余种。

Distribution: Circa 80 genera, more than 3 000 species. 21 genera, 175 species in China. Primarily in south of Yangtze River Basin. Medicinal plants 15 genera, more than 100 species.

【药用植物】马鞭草 *Verbena officinalis* L.（图 12-55）多年生草本。茎四棱形。茎生叶常 3 深裂。穗状花序细长如马鞭。花冠略二唇形。蒴果长圆形，熟时 4 瓣裂成四小坚果。全国广布。全草（马鞭草）能活血散瘀，解毒，利水，退黄，截疟。

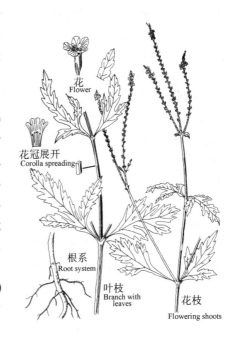

图 12-55 马鞭草
Fig.12-55 *Verbena officinalis*

Medicinal plants: *Verbena officinalis* L. (Fig.12-55) Perennial herbs. Stems quadrate. stem leaves often deeply 3-lobed. Spikes long and slender, like a horsewhip; corolla slightly bilabiate. Capsule, oblong, breaking up into 4 nutlets when mature. Distributed throughout China. The whole plants (Verbenae Herba, Ma bian cao) shows the effects of promoting blood circulation to removing stasis, detoxication, diuresis, removing jaundice and preventing further attack of malaria.

牡 荆 *Vitex negundo* L. var. *cannabifolia*（Sieb. et Zucc.）Hand.-Mazz.：落叶灌木或小乔木，小枝方形。掌状复叶对生。圆锥花序顶生。核果球形，黑色，具膨大宿萼。分布于黄河以南各省区。果实（牡荆子）能祛痰下气，止咳平喘，理气止痛。叶（牡荆叶）能祛痰，止咳，平喘用于咳嗽痰多。其原变种黄荆 *V. negundo* L. 全国广布；叶（黄荆叶）能解表散热，化湿和中，杀虫止痒。

Vitex negundo L. var. *cannabifolia* (Sieb. et Zucc.) Hand.-Mazz.: Shrubs or small trees, deciduous, branchlets quadrate. Leaves palmately compound, opposite. Fruit a drupe, globose, black. Distributed in provinces of south of the Yellow River. The fruit (Viticis Negundo Fructus, Mu jing zi) shows effects of removing phlegm to depress qi, relieving cough and asthma and regulating qi-flowing for relieving pain. The leaves (Viticis Negundo Folium, Mu jing ye) can remove phlegm, relieve cough, and relieve asthma due to coughing and phlegm. *V. negundo* L., leaves (Viticis Negundo Folium, Huang jing ye) have the effects of relieving exterior syndrome and heat dissipation, removing dampness for regulating stomach and are also for insecticidal, anti-itch use.

其他药用植物还有：单叶蔓荆 *Vitex trifolia* L.var. *simplicifolia* Cham. 和蔓荆 *V. trifolia* L. 的果实（蔓荆子）能疏散风热，清利头目。杜虹花 *Callicarpa formosana* Rolfe，叶（紫珠叶）能凉血、收敛、止血，散瘀解毒、消肿。广东紫珠 *C. kwangtungensis* Chun 的干燥茎枝叶（广东紫珠）能收敛止血，散瘀，清热解毒。海州常山 *Clerodendrum trichotomum* Thunb. 嫩枝和叶（臭梧桐）能祛风除湿，平肝降压，解毒杀虫；同属植物大青 *C. cyrtophyllum* Turcz. 叶（臭大青叶）能清热解毒，消肿止痛。

Other medicinal plants: The fruit of *Vitex trifolia* L.var. *simplicifolia* Cham. and *V. trifolia* L. (Viticis Fructus, Man jing zi) is used to dispersing wind and heat to refresh the head and eyes.

The fruit of is equally used as Viticis Fructus.***Callicarpa formosana* Rolfe** The leaves (Callicarpae Formosanae Folium) are used for blood coolding and arresting, scattered stasis detumescence and detoxification. The dried stems, branchlets and leaves of **C. kwangtungensis Chun** (Callicarpae Caulis et Folium) have the acctions as *Callicarpa formosana* Rolfe. The young branchlets and leaves **of *Clerodendrum trichotomum* Thunb.** have the effects of dispelling wind and eliminating dampness, calming the liver and lowering blood pressure, detoxification and insecticide. ***C. cyrtophyllum*** Turcz. is used medicinally for heat-clearing and detoxifying, decreasing swellinging to relieving pain.

37. 唇形科 $\male\female\uparrow K_{(5)} C_{(5)} A_{4,2} \underline{G}_{(2:4:1)}$

(37) Lamiaceae(Labiatae)

【形态特征】芳香草本，多含挥发油。茎四棱形。叶对生。轮伞花序或再集成穗状、总状、圆锥状或头状复合花序。花两性，不整齐；花萼合生，宿存；花冠二唇形，少单唇形（如香科科属 Teucrium）或假单唇形（如筋骨草属 Ajuga）（图 12-56）；雄蕊二强，或仅 2 枚发育（如鼠尾草属 Salvia）；子房上位，2 心皮，常 4 深裂成假 4 室，每室 1 胚珠，花柱基生。4 枚小坚果。

Characteristic: Aromatic herbs, mostly containing volatile oil. Stems and branches 4-angled. Leaves opposite. Flowers bisexual, zygomorphic. Calyx connate, persistent. Corolla limb usually 2-lipped, rarely one-lipped (lower limb 5-lobed, e.g. *Teucrium*) or false pseudo-1-lipped (upper lip entire and lower 4-lobed, e.g. *Ajuga*)(Fig.12-56). Stamens didynamous, or 2 (e.g. *Salvia*). Ovary superior, 2-carpeled deeply 4-parted and each lobe 1-ovuled, style usually gynobasic from bases of ovary lobes. Fruit usually 4 dry nutlets.

【分布】220 属，3500 余种。我国 99 属，800 余种，全国广布；药用 75 属，436 种。

Distribution: 220 genera and ca. 3500 species. 99 genera, 800 species in China. Throughout the country; medicinal plants 75 genera and 436 species.

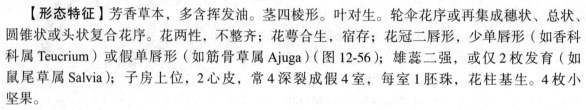

花冠单唇形　　　　　　假单唇形　　　　　　花冠2/3式
Corolla simple lipped　Corolla pseudolabiate　Corolla bilabiate

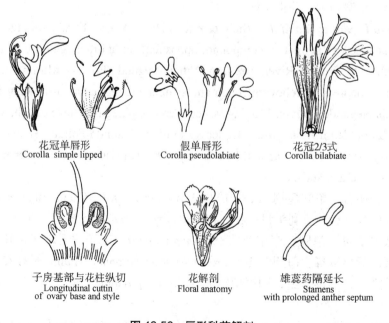

子房基部与花柱纵切　　　花解剖　　　　雄蕊药隔延长
Longitudinal cuttin　　Floral anatomy　　Stamens
of ovary base and style　　　　　　with prolonged anther septum

图 12-56　唇形科花解剖
Fig. 12-56　Flower anatomy of Labiatae

【主要药用属检索表】

1. 花冠单唇形或假单唇形
　　2. 花冠单唇形（上唇缺，下唇 5 裂）；花序常腋生 ·························· 香科科属 *Teucrium*
　　2. 花冠假单唇形（上唇短 2 裂，下唇 3 裂）；花序顶生 ·················· 筋骨草属 *Ajuga*
1. 花冠二唇形或整齐
　　3. 花萼二唇形，上萼有盾状附属物 ·································· 黄芩属 *Scutellaria*
　　3. 花萼整齐，或二唇形，但无附属物
　　　　4. 花冠下唇内凹成船形，上唇外反 ··························· 香茶菜属 *Robdosia*
　　　　4. 花冠下唇不为船形，上唇不外反
　　　　　　5. 花冠管包于萼内；果时宿萼明显增大，后中齿明显反卷·········· 罗勒属 *Ocimum*
　　　　　　5. 花冠管不包于萼内；花萼不为上述情形
　　　　　　　　6. 花药球形，药室叉开；花粉散出后药室平展
　　　　　　　　　　7. 花冠二唇，上唇直立，微缺或全缘，下唇 3 裂；花丝无毛 ······ 香薷属 *Elsholtzia*
　　　　　　　　　　7. 花冠近辐状，上唇 3 裂，下唇全缘；花丝有毛 ········刺蕊草属 *Pogostemon*
　　　　　　　　6. 花药非球形，药室平行或叉开；花粉散出后药室不平展
　　　　　　　　　　8. 花冠近辐射对称；有上唇则扁平或稍穹隆
　　　　　　　　　　　　9. 雄蕊 4，几相等，非 2 强雄蕊
　　　　　　　　　　　　　　10. 能育雄蕊 2 枚，药室略叉开 ············· 地笋属 *Lycopus*
　　　　　　　　　　　　　　10. 能育雄蕊 4 枚，药室平行 ············· 薄荷属 *Mentha*
　　　　　　　　　　　　9. 雄蕊 2，若 4 则为 2 强雄蕊
　　　　　　　　　　　　　　11. 能育雄蕊 4 枚 ····················· 紫苏属 *Perilla*
　　　　　　　　　　　　　　11. 能育雄蕊 2 枚 ···················· 石荠苎属 *Mosla*
　　　　　　　　　　8. 花冠明显二唇形；上唇盔瓣状、镰刀形或弧形
　　　　　　　　　　　　12. 雄蕊 2 枚，花药线形，药隔延长 ··········· 鼠尾草属 *Salvia*
　　　　　　　　　　　　12. 雄蕊 4 枚，花药卵形，药隔不延长
　　　　　　　　　　　　　　13. 后对（上侧）雄蕊比前对（下侧）雄蕊长
　　　　　　　　　　　　　　　　14. 两对雄蕊不平行，后对雄蕊下倾，前对雄蕊上升········· 藿香属 *Agastache*
　　　　　　　　　　　　　　　　14. 两对雄蕊平行，皆向花冠上唇内面弧状上升。
　　　　　　　　　　　　　　　　　　15. 直立草本，顶生花序，花萼先端平截 ··············荆芥属 *Nepeta*
　　　　　　　　　　　　　　　　　　15. 匍匐草本，腋生花序，花萼二唇形 ·······活血丹属 *Glechoma*
　　　　　　　　　　　　　　13. 后对（上侧）雄蕊比前对（下侧）雄蕊短
　　　　　　　　　　　　　　　　16. 花萼二唇形，果熟时闭合；轮伞花序排成假穗状花序········ 夏枯草属 *Prunella*
　　　　　　　　　　　　　　　　16. 花萼不为二唇形，果熟时张开；轮伞花序不排成假穗状花序
　　　　　　　　　　　　　　　　　　17. 小坚果倒卵形，顶端钝圆 ················ 水苏属 *Stachys*
　　　　　　　　　　　　　　　　　　17. 小坚果多少呈三角形，顶端平截
　　　　　　　　　　　　　　　　　　　　18. 花冠上唇呈盔状，萼齿顶端无刺 ·········· 野芝麻属 *Lamium*
　　　　　　　　　　　　　　　　　　　　18. 花冠上唇直立，萼齿顶端有刺 ··········· 益母草属 Leonurus

Key to important medicinal generas of Lamiaceae

1. Corolla 1-lipped (lower limb 5-lobed) or pseudo-1-lipped (upper lip entire and lower 4-lobed)
　　2. Corolla 1-lipped (lower limb 5-lobed), inflorescence usually axillary .... *Teucrium*
　　2. Corolla pseudo-1-lipped (upper lip entire and lower 4-lobed), inflorescence terminal .... *Ajuga*

continued

1. Corolla 2-lipped or actinomorphic
    3. Corolla 2-lipped, calyx with upper lips grow obviouslly into scalelike scutellum (shield) .... *Scutellaria*
    3. Corolla actinomorphic, or 2-lipped, but calyx without scalelike scutellum (shield)
        4. Corolla with lower lip concave and navicular, upper lip recurved .... *Robdosia*
        4. Corolla without lower lip concave and navicular, upper lip not recurved
            5. Corolla tube slightly shorter than calyx, included; upper lip of calyx obviouslly recurved and enlarged at fruiting... *Ocimum*
            5. Corolla tube longer than calyx, exserted; calyx not the same as the above
                6. Anthers globose, cell divaricate, flattened after pollination
                    7. Corolla 2-lipped, upper lip erect, emarginate or entire, lower lip 3-lobed; filaments glabrous.. *Elsholtzia*
                    7. Corolla nearly actinomorphic, upper lip 3-lobed, lower lip entire; filaments hairy .... *Pogostemon*
                6. Anthers not globose, cell parallel or divergent, not flattened after pollination
                    8. Corolla nearly actinomorphic, or 2-lipped with upper lip fat or vaulted
                        9. Stamens 4, subequal, not didynamous
                            10. Two anterior stamens fertile, cell divaricate .... *Lycopus*
                            10. Four stamens fertile, cell parallel .... *Mentha*
                        9. Stamens 2,or 4 and didynamous
                            11. Four stamens fertile .... *Perilla*
                            11. Two stamens fertile .... *Mosla*
                    8. Corolla obviouslly 2-lipped with upper lip galeate, falcate or arcuate
                      12. Stamens 2, anther linear, attached to filaments by dolabriform joints
                      12. Stamens 4, anther ovate, not attached to filaments by dolabriform joints
                          13. Posterior stamens longer than anterior ones
                            14. Stamen pairs not parallel to each other, not all arcuate ascending to upper lip of corolla ....*Agastache*
                            14. Stamen pairs parallel to each other, all arcuate ascending to upper lip of corolla
                                15. Herb erect, inflorescence terminal, calyx apex truncatus .... *Nepeta*
                                15. Herb creeping, inflorescence axillary, calyx 2-liped........................*Glechoma*
                          13. Posterior stamens shorter than anterior ones
                            16. Calyx with very dissimilar teeth, throat closed in fruit; verticillasters in terminal spikes .... *Prunella*
                            16. Calyx with nearly similar teeth, throat open in fruit; verticillasters not in terminal spikes.
                                17. Nutlets obovate, apex obtuse to rounded ....*Stachys*
                                17. Nutlets ± triangular, apex truncatus
                                  18. Corolla with upper lip galeate, calyx apex not spined .... *Lamium*
                                  18. Corolla with upper lip erect, calyx apex spined......................... *Leonurus*

　　【**药用植物**】丹参 *Salvia miltiorrhiza* Bge.（图 12-57）多年生草本，全株密被柔毛和腺毛。根粗大，砖红色。奇数羽状复叶对生。轮伞花序排成总状花序。雄蕊 2。全国广布。根（丹参）能活血祛瘀，通经止痛，清心除烦，凉血消痈。同属植物甘西鼠尾草 *S. przewalskii* Maxim. 分布于甘肃西部、四川西部、云南西北部和西藏；根在云南丽江作丹参代用品。南丹参 *S. bowleyana* Dunn. 分布于长江以南地区，功效同丹参。

　　Medicinal plants: *Salvia miltiorrhiza* Bge. (Fig.12-57) Perennial herb, densely villous and glandular villous. Taproot thickened. Leaves simple to odd-pinnate, opposite. Stamens 2. Distributed troughout China. Root (Salviae Miltiorrhizae Radix et Rhizoma, Dan shen) can activate blood to transform blood stasis, dredge channels to relieve pain, tranquilize and resuscitate the mind and cool blood to relieve abscesses. ***S. przewalskii* Maxim** and ***S. bowleyana* Dunn.** are also used medicinally as Dan shen.

　　黄芩 *Scutellaria baicalensis* Georgi（图 12-58）多年生草本。根肥厚，圆锥形。茎丛生。总状花序顶生，花显著偏向一侧。花萼二唇形，上萼背上具 1 竖起的盾片，果时显著增大。分布于华北、东北及西南地区。根（黄芩）能清热燥湿，泻火解毒，止血安胎。同属植物滇黄芩 *S. amoena* C. H. Wright、粘毛黄芩 *S. viscidula* Bge.、甘肃黄芩 *S. rehderiana* Diels 功效同黄芩。

　　Scutellaria baicalensis Georgi: Herbs perennial(Fig.12-58). Rhizomes fleshy, conical. Stems ascending, much branched. Racemes terminal, flowers obviously sideways. Calyx 2-lipped with scutellum enlarged in fruit. Distributed in N, NE and SW China. The roots (Scutellariae Radix, Huang qin) can drie dampness, purge fire and toxic heat and cool blood to prevent abortion. *S. amoena* C. H. Wright, *S. viscidula* Bge. and *S. rehderiana* Diels. have the same medicinal actions as *S. baicalensis* Ceorgi.

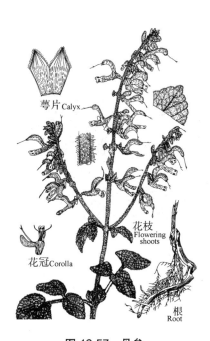

图 12-57　丹参

Fig.12-57　*Salvia miltiorrhiza*

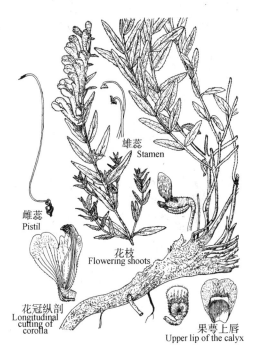

图 12-58　黄芩

Fig.12-58　*Scutellaria baicalensis*

　　益母草 *Leonurus japonicus* Houttuyn（图 12-59）草本。茎四棱形。基生叶近圆形，浅裂，有长柄；茎生叶菱形，深裂，几无柄。全国广布。全草（益母草）能活血调经，利尿消肿，清热解毒；果实（茺蔚子）能活血调经，清肝明目。同属细叶益母草 *L. sibiricus* L. 分布于东北、内蒙古、河北、山西、陕西等省区，在当地亦作益母草入药。

　　Leonurus japonicus Houttuyn: Herbs(Fig.12-59). Stems and branches 4-angled. Lower stem leaf narrowly lobed; mid stem leaf blade palmatipartite, petiole nearly absent. The whole herb (Leonuri Herba, Yi mu cao) is used for activating blood, transforming blood stasis, draining fluid through urination and clearing toxic heat. The fruits (Leonuri Fructus) can promote blood circulation for regulating menstruation and remove liver-fire for improving eyesight. *L. sibiricus* L. is also used medicinally as Leonuri Herba.

薄荷 *Mentha haplocalyx* Briq.（图 12-60）多年生草本，具浓烈清凉香气，被柔毛。萼 5 裂，具腺点；小坚果黄褐色，具小腺窝。分布于南北各省。全草（薄荷）能疏散风热，清利头目，利咽，透疹，疏肝行气。同属植物留兰香 *M. spicata* L.、辣薄荷 *M. piperita* L.、水薄荷 *M. aquatica* L.、欧薄荷 *M. longifolia*（L.）Hudson、长叶薄荷 *M. lavandulacea* L.、唇萼薄荷 *M. pulegium* L. 等在有些地区亦作薄荷药用。

Mentha haplocalyx Briq. (Fig.12-60)**:** Herbs perennial, with strong cool fragrance, villose. Calyx 5-lobed, glandular. Nutlets 4, yellow-brown, small pitted. Throughout China. The whole plants (Menthae Haplocalycis Herba, Bo he) has the effects of dispersing wind and heat, clearing heat and easing the throat, promoting eruption of rashes, soothing the liver and relieving stagnation. *M. spicata* L., *M. piperita* L., *M. aquatica* L., *M. longifolia* (L.) Hudson, *M. lavandulacea* L., *M. pulegium* L., are also used as folk medicine.

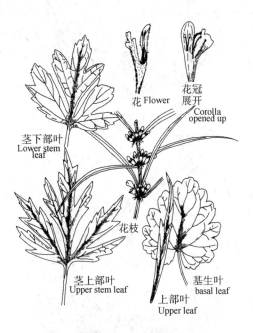

花 Flower
花冠展开 Corolla opened up
茎下部叶 Lower stem leaf
花枝
茎上部叶 Upper stem leaf
基生叶 basal leaf
上部叶 Upper leaf

图 12-59　益母草
Fig.12-59　*Leonurus japonicus*

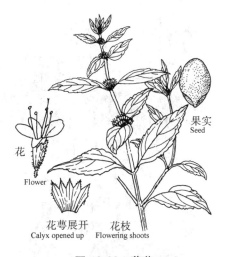

果实 Seed
花 Flower
花萼展开 Calyx opened up
花枝 Flowering shoots

图 12-60　薄荷
Fig.12-60　*Ment haplocalyx*

其他药用植物：广藿香 *Pogostemon cablin*（Blanco）Benth.，茎、叶（广藿香）能芳香化浊，和中止呕，发表解暑。紫苏 *Perilla frulescens*（L.）Britt. 叶（紫苏叶）能解表散寒，行气和胃；茎（紫苏梗）能理气宽中，止痛，安胎；果实（紫苏子）能降气化痰，止咳平喘，润肠通便。石香薷 *Mosla chinensis* Maxim. 全草（香薷）能发汗解表，化湿和中。

Other medicinal plants: *Pogostemon cablin* (Blanco) Benth. Stem and leaves (Pogostemonis Herba,Guang huo xiang) are used for transforming dampness, stopping vomiting and releasing the exterior to relieve summer pathogens. *Perilla frulescens* (L.) Britt. The leaves (Perillae Folium, Zi su) have the effects of resolving the exterior and dispelling cold, descending the lung qi and harmonizing the stomach. The stem (Perillae Caulis, Zi su geng) is used medically for regulating flow of qi and relieving the pain, and anti-abortion. The fruits (Perillae Fructus, Zi su zi) can resolve phlegm, relieve cough and wheezing, and lubricate the bowels to relieve constipation. *Mosla chinensis* Maxim. Whole plants (Moslae Herba, Xiang ru) specialize in inducing perspiration to release the exterior, transforming dampness and harmonizing the middle, and inducing diuresis to alleviate edema.

38. 茄科 ⚥ *K$_{(5)}$ C$_{(5)}$ A$_5$ G$_{(2:2:\infty)}$

(38) Solanaceae

【形态特征】草本或灌木。单叶互生，无托叶。花两性，整齐，5基数；萼宿存，果时增大；花冠联合成辐状、钟状、漏斗状或高脚碟状；雄蕊与花冠裂片同数而互生；子房上位，中轴胎座，2心皮，2室，有时因假隔膜而成4室。浆果或蒴果。

Characteristic: Herbs or shrubs. Leaves simple, alternate, stipules ansent. Flowers bisexual, actinomorphous, 5-merous; calyx persistent, usually enlarged when fruiting; corolla united into wheel-shape, campanulate, funnel-form, or salverform; stamens as many as corolla lobes and alternate with them, inserted within corolla, ovary superior, placentation axile, 2-carpellate, 2 locular, sometimes 4 locular due to false diaphragm. Fruit a berry or capsule.

【分布】30属，3000余种。我国24属，105种，分布于全国，药用25属，84种。

Distribution: About 30 genera, 3 000 species. 24 genera, 105 species in China.Widely distributd. Medicinal plants 25 genera and 84 species.

【药用植物】白花曼陀罗 *Datura metel* L.（图 12-61）一年生草木半灌木状草本，近无毛。花单生叶腋；花冠白色，漏斗状，大型，具5棱。蒴果疏生短刺，4瓣裂。分布华南和华东，栽培或野生。花（洋金花）能平喘止咳，解痉定痛。同属植物毛曼陀罗 *D. innoxia* Mill.、曼陀罗 *D. stramonium* L. 与白花曼陀罗同用。

Medicinal plants: *Datura metel* L. (Fig.12-61) Herbs annual, shrubby, glabrescent. Flowers axillary; corolla white, funnelform, large, 5-angulate. Capsules, short spines, 4-valved. Distributed in South and East China, cultivated or wild. Flowers (Daturae Flos) can relieve asthma, arrest cough, relieve spasmolysis and pain. The flowers of *D. innoxia* Mill. and *D. stramonium* L. have the same effects as Daturae Flos.

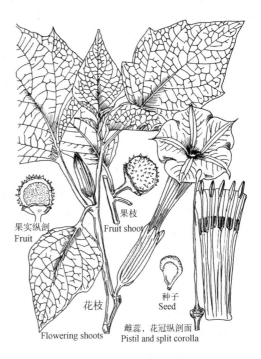

果实纵剖 / Fruit
果枝 / Fruit shoot
种子 / Seed
花枝 / Flowering shoots
雌蕊，花冠纵剖面 / Pistil and split corolla

图 12-61 白花曼陀罗
Fig.12-61 *Datura metel*

宁夏枸杞 *Lycium barbarum* L. 灌木，枝端具刺。叶互生或短枝簇生。花簇生；花冠粉红色或淡紫色。浆果，熟时红色。分布于宁夏、甘肃、青海、新疆、内蒙古、河北等省区，多栽培。果实（枸杞子）能滋补肝肾，益精明目；根皮（地骨皮）可凉血除蒸，清肺降火。同属植物枸杞 *L. chinensis* Miller 同等入药。

Lycium barbarum L.: Shrubs, branches thorny. Leaves alternate or fasciculate. Inflorescences clustered flowers. Corolla pink or pale purple. Berry red when mature. Distributed in Ningxia, Gansu, Qinghai, Xinjiang, Inner Mongolia, Hebei, etc., mostly cultivated. Fruits (Lycii Fructus, Gou qi) can nourish the liver and kidney, consolidate essence, and improve vision. Root bark (Lycii Cortex, Di gu pi) not only cool blood but also clear lung heat. The fruits and bark of *L. chinensis* Miller have similar effects as *L. barbarum*.

其他药用植物还有：颠茄 *Atropa belladonna* L. 全草能解痉止痛，抑制腺体分泌。酸浆 *Physalis alkekengi* L. 干燥宿萼（锦灯笼）能清热解毒，利咽化痰，利尿通淋。

Other medicinal plants: *Atropa belladonna* L. The whole dry plants is effective for spasmolysis, relieving pain, and inhibiting secretory. *Physalis alkekengi* L. Dried persistent calyx (Physalis Calyx seu Fructus) can clear toxic heat, relieve throat and phlegm, promot diuresis and relieve stranguria.

39. 玄参科 $\text{⚥} \uparrow K_{(4\sim5)} C_{(4-5)} A_{4,\,2} \underline{G}_{(2:2:\infty)}$

(39) Scrophulariaceae

【形态特征】草本。单叶对生，轮生或互生，无托叶。总状或聚伞花序。花萼4~5裂，宿存；花冠多少呈2唇形；雄蕊2强，生于花冠管上；子房上位，2心皮，2室，中轴胎座；花柱顶生。蒴果，稀浆果。

Characteristic: Herbs. Leaves simple and opposite, rarely whorled or alternate, stipules absent. Inflorescences racemes or cymes. Calyx 4~5-lobed, persistent; corolla 4~5-lobed, 2-lipped; stamens 4, didynamous, inserted on corolla tube; disk ringlike, copular; ovary superior, 2-carpellate, 2 locules, axile placentas, each locule with numerous ovules; styles terminal. Fruit a capsule, rarely a berry. Seeds numerous and minute.

【分布】200属，3000余种。我国56属，634种，全国广布，主产西南，药用233种。

Distribution: About 200 genera, 3 000 species. 56 genera, 634 species throughout China, with a greater concentration in Southwestern China. Medicinal plants 233 species.

【药用植物】玄参 *Scrophularia ningpoensis* Hemsl.（图12-62）高大草本。根肥大，纺锤形，干后变黑。茎下部叶对生，上部叶有时互生；圆锥聚伞花序。花冠二唇形，紫褐色；雄蕊二强，花丝肥厚。蒴果。分布于华东、中南及西南地区。根（玄参）能清热凉血，滋阴降火，解毒散结。同属植物北玄参 *S. buergeriana* Miq. 功效同玄参。

Medicinal plants: *Scrophularia ningpoensis* Hemsl. (Fig.12-62) Tall herbs. Roots fleshy, fusiform, turning into black when dried. Leaves below opposite, upper ones sometimes alternate. Inflorescences thyrsoid panicles; corolla 2-lipped, purple-brown; stamens didynamous, filaments thickened. Capsule. Distributed in East Middle-south and Soutwest China. The roots (Scrophulariae Radix, Xuan shen) has

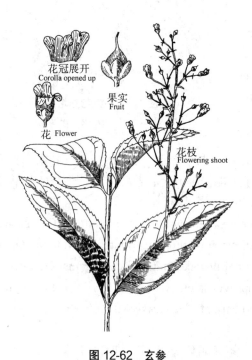

图 12-62 玄参
Fig.12-62 *Scrophularia ningpoensis*

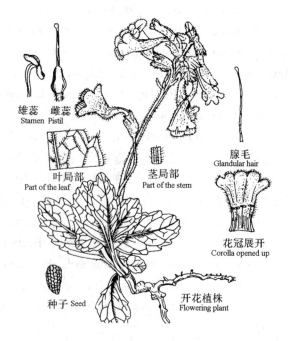

图 12-63 地黄
Fig.12-63 *Rehmannia glutinosa*

the effect of clearing toxic heat, cooling blood, nourishing yin, suppressing fire and dissipating lumps. *S. buergeriana* Miq. is equally used as *S. ningpoensis*.

地黄 *Rehmannia glutinosa*（Gaert.）Libosch. ex Fisch. et Mey.（图 12-63）多年生草本，全株密被灰白色长柔毛和腺毛。根状茎肥大呈块状。叶莲座状，茎生叶互生。总状花序顶生。常栽培，主产河南等地。新鲜块根（鲜地黄）能清热生津，凉血，止血；干燥块根（生地黄）能清热凉血，养阴生津。

Rehmannia glutinosa (Gaert.) Libosch. ex Fisch. et Mey. (Fig.12-63) Perennial herbs, densely grey-white villous with glandular hairs. Rhizomes fleshy, tuber. Leaves rosulate, stem leaves alternate. Inflorescences racemes. Mostly cultivated, mainly in Henan. The fresh root tuber (Rehmanniae Radix Recens, Xian di huang) has the effect of clearing heat, promoting fluid production, cooling blood and stanching bleeding, while the dry root tuber (Rehmanniae Radix, Sheng di huang) is used for clearing heat, cooling blood, nourishing yin and promoting fluid production.

常用药用植物还有：胡黄连 *Neopicrorhiza scrophulariiflora*（Pennell）D. Y. Hong：矮小草本。分布于西藏东部、云南西北、四川西部。根状茎（胡黄连）能退虚热，除疳热，清湿热。阴行草 *Siphonostegia chinensis* Benth. 全草（北刘寄奴）能活血祛瘀，通经止痛，凉血止血，清热利湿。

Other medicinal plants: *Neopicrorhiza scrophulariiflora* (Pennell) D. Y. Hong: Small herbs, distributed in E Tibet, NW Yunnan, W Sichuan. Rhizomes (Picrorhizae Rhizoma, Hu huang lian) are mainly applicable for clearing deficiency heat, relieving infantile malnutritional fever and clearing damp heat. The whole plants of *Siphonostegia chinensis* Benth. (Siphonosegiae Herba) is used for promoting blood circulation, removing blood stasis, inducing menstruation to relieve menalgia, cooling blood and stanching bleeding, clearing heat and promoting diuresis.

40. 爵床科　$\diameter \uparrow K_{(4\text{-}5)} C_{(4\text{-}5)} A_{4, 2} \underline{G}_{(2:2:1\sim\infty)}$

(40) Acanthaceae

【形态特征】草本或灌木，具钟乳体。茎节膨大。单叶对生，无托叶。聚伞花序构成多种复合花序。每花具 1 苞片和 2 小苞片；花萼 4~5 裂；花冠二唇形；雄蕊 2 强或仅 2 枚，贴生于花冠筒上；子房上位，2 心皮 2 室，中轴胎座。蒴果，常具珠柄钩。

Characteristic: Herbs or shrubs, usually with strip cystoliths on the blade, little shoot and calyx. Leaves simple, opposite, simple, stipules absent. Verticillasters arranged into various compound inflorescences. Flowers bisexual, bracts 1 and bracteoles 2 per flower. Calyx 4–5 lobulate; Corolla 2-lobulate. Stamens didynamous or 2. Capsule. Seeds usually borne on hooklike retinacula.

【分布】250 属，3450 种。我国 68 属，311 种，分布于长江以南各省，药用 32 属，70 余种。

Distribution: 250 genera, 3 450 species. 68 genera, 311 species in China. Distributed in south of Yangtze River; medicinal plants 32 genera and about 70 species.

【药用植物】穿心莲 *Andrographis paniculata*（Burm. F.）Nees（图 12-64）一年生草本。茎四棱。叶卵状长圆形至披针形。总状花序集成大型圆锥花序；花冠二唇形，白色；雄蕊 2。蒴果。原产东南亚，我国南方大量栽培。全草（穿心莲）能清热解毒，凉血，消肿。

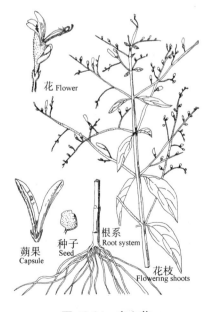

图 12-64　穿心莲
Fig.12-64　*Andrographis paniculata*

Medicinal plants: *Andrographis paniculata* (Burm. F.) Nees (Fig.12-64) Herbs annual. Stems 4-angled. Leaves simple, leaf blade ovate-oblong to lanceolate. Racemes arranged into large panicle; Corolla white, 2-lobulate. Stamens 2. Capsule. Native to Southeast Asia, cultivated in South China. The whole plants (Andrographis Herba, Chuan xin lian) is commonly used for clearing heat and toxins, cooling blood, reducing swelling and dry dampness.

常用药用植物还有：马蓝 *Baphicacanthus cusia*（Nees）Bremek. 草本，多年生一次性结实。分布于华南、西南及台湾。根（南板蓝根）能清热解毒，凉血消斑。叶在产地习作"大青叶"入药，也是生产"青黛"的原料药材。九头狮子草 *Peristrophe japonica*（Thunb.）Bremek. 全草（九头狮子草）能清热解毒，发汗解表。

Other medicinal plants: *Baphicacanthus cusia* (Nees) Bremek. Herbs pliestesial (living for several years then dying after flowering and fruiting). Roots (Baphicacanthis Cusiae Rhizoma et Radix, Nan banlangen) have the effects of heat-clearing and detoxifying, cooling blood and anti-freckle. The leaves are used as Da qing ye for heat-clearing and detoxifying, cooling blood, anti-freckle, discharging fire and calming the frightened. The whole plants of *Peristrophe japonica* (Thunb.) Bremek. shows the effects of heat-clearing and detoxifying, and relieving exterior syndrome by diaphoresis.

41. 茜草科 $\male\female *K_{(4-5)} C_{(4-5)} A_{4-5} \overline{G}_{(2:2:1\sim\infty)}$

(41) Rubiaceae

【形态特征】草本，灌木或乔木，或攀援状。单叶对生或轮生，全缘；托叶2，生叶柄间或叶腋内。二歧聚伞花序排成圆锥状或头状；花两性，整齐；花萼4~5裂；花冠4~5裂，稀6裂；雄蕊与花冠裂片同数且互生；2心皮合生，子房下位，常2室。蒴果、浆果或核果。

Characteristic: Herbs, shrubs, trees and lianas. Leaves simple, opposite or wholed, usually entire; 2 stipules, interpetiolar or connate; Diachasiums usually aggregated into panicle or capitulum; flowers bisexual; calyx lobes 4 or 5; corolla lobes 4–5, rarely 6; the number of the stamens equal to that of the corolla lobes; Carpels usually 2, connate, ovary inferior, 2 locules. Capsule, berry, drupe.

【分布】约637属，11150种；我国98属，约676种。主要分布于西南至东南部。已知药用59属，210余种。

Distribution: 637 genera, circa 11 150 species; 98 genera, about 676 species in China. Mainly in Southwest to Southeast China; medicinal plants 59 genera, about 210 species.

【药用植物】栀子 *Gardenia jasminoides* Ellis（图 12-65）灌木，叶对生，少3枚轮生，革质。花大，白色，芳香，单生枝顶，花冠高脚碟状。子房下位。成熟果实有翅状纵棱5~9条。分布于我国南部和中部。果实（栀子）能泻火除烦，清热利湿，凉血解毒。

Medicinal plants: *Gardenia jasminoides* Ellis. (Fig.12-65) Shrubs. Leaves opposite or rarely ternate, blade leathery. Flower solitary, terminal, fragrant.Corolla salverform. at pilose; Ovary inferior. Mature fruit with 5~9 wing-shaped longitudinal ridges. Mainly in middle and south China. Fruit (Gardeniae Fructus, Zhi zi) can purges fire, calms the mind, drains heat and dampness, cools the blood and clears toxic

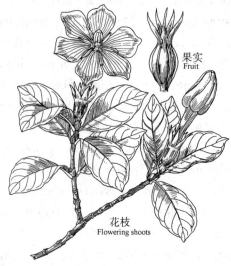

果实
Fruit

花枝
Flowering shoots

图 12-65 栀子
Fig. 12-65 *Gardenia jasminoides*

heat.

钩藤 *Uncaria rhynchophylla*（Miq.）Miq. ex Havil. 藤本；嫩枝无毛。叶腋有钩。托叶狭三角形，深 2 裂。头状花序单生叶腋；蒴果。分布于湖南、江西、福建、广东、广西及西南地区。带钩的茎枝（钩藤）能息风定惊，清热平肝。同属植物大叶钩藤 *U. macrophylla* Wall.、毛钩藤 *U. hirsuta* Havil.、华钩藤 *U. sinensis*（Oliv.）Havil 和无柄果钩藤 *U. sessilifructus* Roxb. 同等入药。

Uncaria rhynchophylla (Miq.) Miq. ex Havil. Lianas.Young stems slender, glabrous, hook in leaf axil; stipules, 2-lobed, lobes triangular-lanceolat. Inflorescences axillary and terminal, solitary; Capsule. Distributed in Hunan, Jiangxi, Fujian, Guangdong, Guangxi and southwest China. Dried vine stems and thorny branches (Uncariae Ramulus Cum Uncis, Gou teng) can extinguishes wind and arrests convulsions, clears heat and calms the liver. *U. macrophylla* Wall., *U. hirsuta Havil.*, *U. sinensis* (Oliv.) Havil and *U. sessilifructus* Roxb. Law are equally used as *U. rhynchophylla*.

常用药用植物还有：茜草 *Rubia cordifolia* L.（图 12-66）全国广布。根（茜草）能凉血，祛瘀，止痛，通经。巴戟天 *Morinda officinalis* F. C. How，根（巴戟天）能补肾壮阳，强筋骨，祛风湿。鸡矢藤 *Paederia scandens*（Lour.）Merr 的全草（鸡矢藤）能清热解毒，镇痛，止咳。白花蛇舌草 *Oldenlandia diffusa* Willd. 的全草（白花蛇舌草）能清热解毒。金鸡纳树 *Cinchona calisaya* Weddell 的树皮含奎宁，能抗疟，退热。

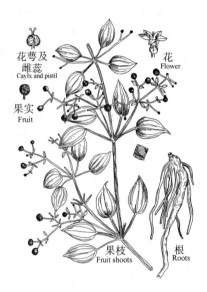

花萼及雌蕊
Caylx and pistil

花
Flower

果实
Fruit

果枝
Fruit shoots

根
Roots

图 12-66　茜草
Fig. 12-66 *Rubia cordifolia*

Other medicinal plants: *Rubia cordifolia L.* (Fig.12-66) Widely distributed. Root (Rubiae Radix et Rhizoma,Qian cao) stops bleeding, cools blood and invigorates blood circulation.The root of *Morinda officinalis* F. C. How (Morindae Officinalis Radix,Ba ji tian) tonifies kidney yang, strengthens the muscles and bones, and dispels wind damp. Whole plant of *Oldenlandia diffusa* Willd. (Hedyotis Diffusae Herba) clears heat and toxins, transforms abscess swelling and drains dampness. Whole plant of *Paederia scandens* (Lour.) Merr (Paederiae Herba, Ji shi teng) Clearing away heat and detoxication, relieves pain. Bark of *Cinchona calisaya* Weddell contains Quinine, which is antimalarial and antipyretic.antipyretic.

42. 忍冬科 $\male\female * \uparrow K_{(4-5)} C_{(4-5)} A_{4-5} \overline{G}_{(2-5:1-5:1-\infty)}$

(42) Caprifoliaceae

【形态特征】灌木或木质藤本。叶对生，单叶，少复叶；常无托叶。聚伞花序；花两性，辐射对称或两侧对称；花萼 4~5 裂；花冠管状，多 5 裂，有时二唇形；雄蕊冠生，与花冠裂片同数且互生；心皮 2~5，子房下位，1~5 室。浆果、核果或蒴果。

Characteristic: Shrubs or woody climbers. Leaves simple, opposite, stipules absent. Cymes. Flowers bisexual, actinomorphic or zygomorphic. Calyx 4- or 5-lobed. Corolla gamopetalous; lobes 5~Stamens alternating with corolla lobes. Ovary inferior, carpels 2~5, fused; locules 1 to 5 Berry, drupe or achene.

【分布】约 13 属，500 种。我国 12 属，200 余种，广布全国。药用 9 属，100 余种。

Distribution: Circa 13 genera, 500 species; 12 genera, about 200 species in China. Widely distributed; medicinal plants 9 genera, about 100 species.

【药用植物】忍冬 *Lonicera japonica* Thunb.（图 12-67）半常绿藤本；密被黄色硬毛。小枝上部叶两面密被短糙毛，下部叶常无毛。花冠白色，后变黄色，唇形。浆果，熟时蓝黑色。全国各省均有分布。花蕾或带初开的花（金银花）能清热解毒，疏散风热；茎枝（忍冬藤）能清热解毒，疏风通络。同属植物灰毡毛忍冬 *L.macranthoides* Hand. - Mazz.、红腺忍冬 *L.hypoglauca* Miq.、华南忍冬 *L.confusa*（Sweet）DC. 或黄褐毛忍冬 *L. fulvotomentosa* P. S. Hsu et S. C. Cheng 的花蕾或带初开的花（山银花）清热解毒，疏散风热。

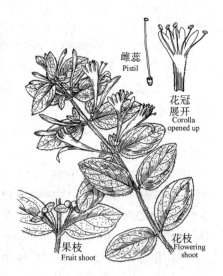

图 12-67　忍冬
Fig. 12-67　*Lonicera japonica*

Medicinal plants: *Lonicera japonica* Thunb. (Fig.12-67) Climbers, semievergreen. Branches, petioles, and peduncles with dense, yellow-brown spreading stiff hairs. Corolla white, then yellow, lip shaped. Fruit a berry, black when ripe. Widely distributed. Bud or early flower (Lonicerae Japonicae Flos, Jin yin hua) can clear away heat and detoxify, and disperses wind and heat. Buds or early flowers of *L. macranthoides* Hand. - Mazz., *L. hypoglauca* Miq., *L. confusa* (Sweet) DC. or *L. fulvotomentosa* P. S. Hsu et S. C. Cheng (Lonicerae Flos, Shan yin hua) have the same function as Jin yinhua.

常用药用植物还有：接骨木 *Sambucus racemosa* L. 的带叶茎枝（接骨木）消肿止痛，祛风通络。荚蒾 *Viburnum dilatatum* Thunb. 枝、叶能清热解毒、疏风解表，根消肿祛瘀。

Other medicinal plants: *Sambucus racemosa* L. stem with leaves can detumescence and pain relief, dispel wind and dredge collaterals; *Viburnum dilatatum* Thunb. Stem and leaf can can clear away heat and detoxify, and disperses win, root can relieves swelling.

43. 败酱科　$\male\female\uparrow K_{(5\sim15)} C_{(3\sim5)} A_{3\sim4} \overline{G}_{(3:3:1)}$

(43) Valerianaceae

【形态特征】二年生或多年生草本，根及根茎常有特殊气味。叶对生或基生，无托叶。聚伞花序组成各种花序。花小，两性；萼宿存；花冠钟状或狭漏斗形；瘦果。

Characteristic: Biennial or perennial herbs, with a characteristic fetid odor. Leaves opposite or basal, lacking stipules. Cymes aggregated into various inflorescences. Flowers small, bisexual; calyx present; corolla campanulate or narrowly funnelform; achenes.

【分布】本科 13 属，400 种。我国 3 属，40 余种，南北均有分布。药用 3 属，24 余种。

Distribution: 13 genera, 400 species; 3 genera, about 40 species in China. Spread throughout the country; medicinal plants 3 genera, 24 species.

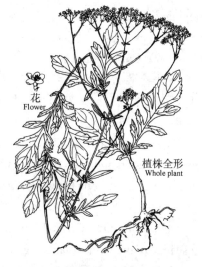

图 12-68　败酱
Fig. 12-68　*Patrinia scabiosifolia*

【药用植物】败酱 *Patrinia scabiosifolia* Fisch. ex Trevir.（图12-68）和白花败酱 *P. villosa*（Thunb.）Juss 全草（败酱草）能清热解毒，祛瘀止痛，消肿排脓。缬草 *Valeriana officinalis* L. 根及根状茎（缬草）能安神、理气、止痛。甘松 *Nardostachys jatamansi*（D. Don）DC. 根及根状茎（甘松）能理气止痛。蜘蛛香 *Valeriana jatamansi* Jones 根状茎和根（蜘蛛香）理气止痛，消食止泻，祛风除湿，镇惊安神。

Medicinal plants: whole plants of *Patrinia scabiosifolia* Fisch. ex Trevir. (Fig.12-68) and *P. villosa* (Thunb.) Juss can clear away heat and detoxify. *Valeriana officinalis* L. root and rhizome can regulate Qi and relieve pain, transforms stasis and reduces swelling and expels pus. *V. jatamansi* Jones root and rhizome can regulate Qi and relieve pain, promotes digestion to resolve food stagnation, dispels wind dampness, calms fright and tranquilizes the mind.

44. 葫芦科 $\male *K_{(5)} C_{(5)} A_{5, (3-5)}$; $\female *K_{(5)} C_{(5)} \overline{G}_{(3:1:\infty)}$

(44) Cucurbitaceae

【形态特征】草质或木质藤本，具卷须；茎具双韧维管束。单叶互生，常为掌状裂，无托叶。花单性，同株或异株；花5基数，整齐；雄蕊5，花丝两两结合，1条分离，形似3雄蕊，花药弯曲成S形；子房下位，3心皮1室，侧膜胎座。瓠果。

Characteristic: Herbs and woody climbers, tendrils present; stem with bicollateral vascular bundle. Leaves simple, alternate, often palmately divided, stipules absent. Plants monoecious or dioecious; flowers unisexual; 5-merous corolla,, actinomorphic; stamens usually 5 or 3, two combination of filaments, one separation; anthers curved into an S shape.Ovary inferior, mostly composed of 3 carpels, 3-locular, parietal placenta. Pepo.

【分布】113属，800种。我国32属，154种，全国分布。药用25属，92种。

Distribution:113 genera, 800 species. 32 genera, 154 species in China. Distributed throughout the country; medicinal plants 25 genera, 92 species.

【药用植物】栝楼 *Trichosanthes kirilowii* Maxim.（图12-69）块根圆柱状。花冠中部以上细裂成流苏状。成熟果实（瓜蒌）能清热涤痰，宽胸散结，润燥滑肠；根（天花粉）能清热泻火，消肿排脓，生津止渴。罗汉果 *Mormordica grosvenorii* Swingle 果实（罗汉果）能清热润肺，利咽开音，滑肠通便。绞股蓝 *Gynostemma pentaphllam*（Thunb.）Makino 根茎或全草（绞股蓝）能益气健脾，清热解毒，化痰止咳。丝瓜 *Luffa cylindrical*（L.）Roem. 干燥成熟果实的维管束（丝瓜络）能祛风、通络、活血、下乳。

Medicinal plants: *Trichosanthes kirilowii* Maxim. (Fig.12-69) Tuberous roots cylindric. Corolla segments 5, usually long fimb. Ripe fruit (Trichosanthis Fructus, Gua lou) can clear heat and phlegm, moves chest Qi, moistens dryness and lubricates the bowels. Root (Trichosanthis Radix, Tian hua

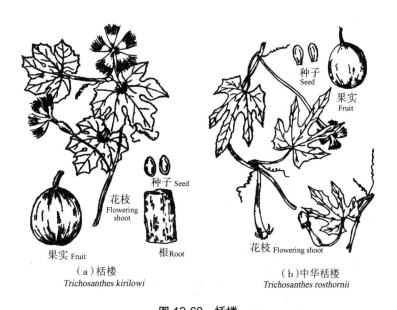

（a）栝楼
Trichosanthes kirilowi

（b）中华栝楼
Trichosanthes rosthornii

图 12-69 栝楼
Fig. 12-69 *Trichosanthes kirilowii*

fen) clear away heat and purges heat, help produce saliva and slake thirst, detumescence and abscess elimination. *Mormordica grosvenorii* Swingle fruit (Siraitiae Fructus, Luo han guo) can clear the lungs and promote the swallowing, dissipate phlegm and relieve cough, moisten the intestines and relieve constipation. *Gynostemma pentaphllam* (Thunb.) Makino rhizome or whole grass can nourish qi and spleen, clear away heat and detoxification, dissipate phlegm and stop cough. *Luffa cylindrical* (L.) Roem. The vascular bundle of the dried and mature fruit can dispel wind, dredge collaterals and activate blood circulation, promote lactation.

45. 桔梗科 $\male * \uparrow K_{(5)} C_{(5)} A_5 \overline{G}_{(2\sim5:2\sim5:\infty)}; \overline{\underline{G}}_{(2\sim5:2\sim5:\infty)}$

(45) Campanulaceae

【形态特征】草本，少木本。常具乳汁。单叶互生或对生，少轮生，无托叶。花两性，5 基数，辐射对称或两侧对称；萼常宿存；花冠钟状或管状；雄蕊与花冠裂片互生；子房下位或半下位，3 室，中轴胎座，胚珠多数。蒴果，稀浆果。

Characteristic: Herbs, rarely shrub, or tree, often with latex. Leaves simple, alternate or opposite, rarely whorled, without stipule. Flowers bisexual, 5-merous corolla, actinomorphic or zygomorphic; calyx often persistent; corolla campanulate or tubular; stamens 5, alternate with corolla lobes; inferior ovary or half-inferior ovary, 3-locular, axile placenta, ovules numerous. Capsule, rarely berry.

【分布】60 余属，约 2000 种；我国 16 属，约 170 种，全国均有分布，以西南地区最为丰富。药用 13 属，111 种。

Distribution: Circa 60 genera, about 2 000 species; 16 genera, about 170 species in China. Widely distributed, abundant in southwest China; medicinal plants 13 genera, 111 species.

【主要药用属检索表】

1. 花两侧对称；雄蕊多合生 ··	半边莲属 *Lobelia*
1. 花辐射对称；雄蕊分离	
2. 果基部孔裂；花柱基部有圆筒状花盘；花冠 5 浅裂 ··············	沙参属 *Adenophora*
2. 果顶端孔裂	
3. 茎缠绕；柱头 3 裂，裂片卵形或椭圆形 ······················	党参属 *Codonopsis*
3. 直立茎；柱头 5 裂，裂片条形 ·······························	桔梗属 *Platycodon*

Key to important medicinal genera

1. Flower zygomorphic, stamens often connate ....Lobelia
1. Flower actinomorphic, stamens separated
  2. Poricidal at the base of the fruit; the base of the style has a cylindrical disk; the Corolla is 5-lobed ... Adenophora
  2. Poricidal on the top of the fruit
    3. Stem twining; stigma 3-lobed, lobes ovate or elliptic .... Codonopsis
    3. Erect stem; stigma 5-lobed, lobes strip .... Platycodon

【药用植物】党参 *Codonopsis pilosula*（Franch.）Nannf.（图 12-70）多年生草质藤本，有乳汁。根肥大呈纺锤形。花单生枝顶，花冠淡黄绿色、红紫色、蓝白色等。分布于东北、华北等省。根（党参）能健脾益肺，养血生津。同属植物素花党参 *C. Pilosula* Nannf. *var. modesta*（Nannf.）L. T. Shen 或川党参 *C. tangshen* Oliv. 的根同等入药。

Medicinal plants: *Codonopsis pilosula* (Franch.) Nannf. (Fig.12-70) Herbs, climbing vine, with latex. Roots spindle shaped. Flower solitary, 5-merous, light yellow green, red purple, blue white and so on. Distributed in Northeast, North China. Root (Codonopsis Radix, Dang shen) can nourish general Qi and blood and generates body fluid. The root of *C. Pilosula* Nannf. *var. modesta* (Nannf.) L. T. Shen and *C. tangshen* Oliv. are also used as "Codonopsis Radix".

桔梗 *Platycodon grandiflorus*（Jacq.）A. DC.（图 12-71）多年生草本。花冠阔钟状，蓝色或蓝紫色。蒴果倒卵形。全国各地均有分布。根（桔梗）能宣肺，利咽，祛痰，排脓。

Platycodon grandiflorus (Jacq.) A. DC. (Fig.12-71) Perennial Herbs. Broad campanulate corolla; Flower colour blue or blue purple. Capsule. Distributed all over the country. Root (Platycodonis Radix, Jie geng) disperses the lung Qi, expels phlegm, eases the throat and improves purulent discharge.

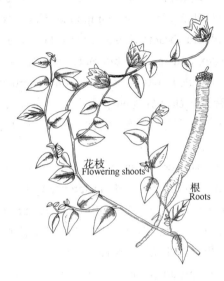

图 12-70 党参

Fig.12-70 *Codonopsis pilosula*

图 12-71 桔梗

Fig.12-71 *Platycodon grandiflorus*

轮叶沙参 *Adenophora tetraphylla*（Thunb.）Fisch. 多年生草本。3~6 叶轮生。花蓝紫色。主产江苏、安徽、浙江等地。根（南沙参）能养阴清肺，清胃生津，化痰益气。同属植物沙参 *A. Stricta* Miq. 的根同等入药。

Adenophora tetraphylla (Thunb.) Fisch. Perennial Herbs. 3-6 leaves whorled. Flowers blue purple. Distributed in Jiangsu, Anhui, Zhejiang and other places. Root (Adenophorae Radix, Nan sha shen) can nourish yin, cools the lung and stomach, generates fluid, resolves phlegm and replenishes qi. The root of *A. Stricta* Miq. is also used as "Adenophorae Radix".

常用药用植物还有：半边莲 *Lobelia chinensis* Lour. 多年生小草本。花冠裂片偏向一侧。分布于长江中下游及以南地区。全草（半边莲）能清热解毒，利尿消肿。羊乳 *Codonopsis lanceolata*（Sieb. et Zucc.）Trautv. 根滋阴润肺排脓解毒。铜锤玉带草 *Lobelia nummularia* Lam. 全草能活血祛瘀，除风利湿。蓝花参 *Wahlenbergia marginata*（Thunb.）A. DC. 根或全草，能益气，祛痰截疟。

Other medicinal plants: *Lobelia chinensis* Lour. Small perennial herbs. Corolla lobes to one side. Distributed in the middle and lower reaches of the Yangtze River. Whole plants (Lobeliae Chinensis Herba, Ban bian lian) can clear heat and relieve toxicity, drain dampness to relieve edema. *Codonopsis lanceolata* (Sieb. et Zucc.) Trautv. Root nourishing yin, moistening lung, expelling pus and relieve toxicity. *Lobelia nummularia* Lam. Whole plant can invigorate blood and transform stasis, eliminate wind and dampness. *Wahlenbergia marginata* (Thunb.) A. DC. Root or whole grass canbenefit qi, dispel phlegm and intercepting malaria.

46. 菊科 ♀*↑$K_{0, \infty} C_{(3-5)} A_{(4-5)} \overline{G}_{(2:2:1)}$

(46) Asteraceae (Compositae)

【形态特征】草本、亚灌木或灌木，稀为乔木。具乳汁管或树脂道。叶互生，稀对生或轮生，单叶或复叶。头状花序单生或排成各式花序，有总苞。花小，两性；花萼退化成冠毛状、刺状、鳞片状或缺；花冠管状、舌状或假舌装、漏斗状；头状花序的小花异型（外围舌状花、假舌状花或漏斗状花，称缘花；中央管状花，称盘花）或同型（全为管状花或舌状花）。聚药雄蕊；子房下位，2 心皮合生，1 室，胚珠 1。子房与萼筒共同形成连萼瘦果，有冠毛。种子无胚乳。

Characteristic: Herbs, subshrubs, or shrubs, rarely trees; many with laticiferous cells or canals and/or resinous ducts. Leaves alternate, rarely opposite or whorled; simple or compound. Capitula solitary or arranged in variously formed synflorescences, involucral bracts present; Corolla gamopetalous, homogamous or heterogamous, tubular or tubular-filiform, and regular, bilabiate, radiate, or ligulate. Synantherous stamen; inferior ovary, carpels 2, connate; 1-loculed; ovule 1; sypsela. Endosperm scanty.

本科植物根据头状花序花冠类型不同、乳汁的有无，通常分为两个亚科：Ⅰ 为管状花亚科：头状花序至少有管状花，植物体不含乳汁。管状花亚科包括菊科大部分的种、属。Ⅱ 为舌状花亚科：花序全为舌状花，植物体含乳汁。

According to the types of corolla, latex present or not, Asteraceae was divided into two subfamilies. I. Tubuliflorae: all florets are tubular, or both tubular and liguliform, lacking latex; most species and genera of Compositae belong to Tubuliflorae; II. Liguliflorae: all florets are liguliform, latex present.

【分布】本科是为被子植物第一大科，占有花植物的 1/10。全世界约 1000 属，25000~30000 种。我国 230 余属，2300 余种，全国各地均有分布。药用 155 属，778 种。

Distribution: Compositae is the largest family of angiosperms, accounting for 1/10 of flowering plants. About 1 000 genera, circa 25 000–30 000 species in the world; about 230 genera, circa 700 species in China, distributed all over the country. Medicinal plants 155 genera, 778 species.

【主要药用属检索表】

1. 头状花序全为管状花，或管状花和舌状花兼有，植物体无乳汁 ┈┈┈┈┈┈┈┈ 管状花亚科 Tubuliflorae
　2. 头状花序全为两性或单性的管状花
　　3. 叶对生；三全裂或单叶 ┈┈┈┈┈┈┈┈┈┈┈┈┈┈┈┈┈┈┈┈ 泽兰属 *Eupatorium*
　　3. 叶互生
　　　4. 无冠毛
　　　　5. 头状花序单性；总苞多钩刺 ┈┈┈┈┈┈┈┈┈┈┈┈┈┈┈ 苍耳属 *Xanthium*
　　　　5. 头状花序内层两性花，外层雌性花；总苞无钩刺 ┈┈┈┈┈ 蒿属 *Artemisia*
　　　4. 具冠毛
　　　　6. 叶缘具刺
　　　　　7. 冠毛基部联合，羽状
　　　　　　8. 花序基部有叶状苞；果多丝状柔毛 ┈┈┈┈┈┈┈┈ 苍术属 *Atractylodes*
　　　　　　8. 花序基部无叶状苞；果无毛 ┈┈┈┈┈┈┈┈┈┈┈ 蓟属 *Cirsium*
　　　　　7. 冠毛鳞片状或缺；总苞外轮叶状，有刺 ┈┈┈┈┈┈┈ 红花属 *Carthamus*
　　　　6. 叶缘不具刺
　　　　　9. 根有香气
　　　　　　10. 多年生高大草本；冠毛羽毛状 ┈┈┈┈┈┈┈┈┈ 风毛菊属 *Saussurea*
　　　　　　10. 多年生低矮草本；冠毛刚毛状 ┈┈┈┈┈┈┈┈┈ 川木香属 *Vladimiria*
　　　　　9. 根不具香气
　　　　　　11. 总苞有钩状刺毛；冠毛短，易脱落 ┈┈┈┈┈┈┈ 牛蒡属 *Arctium*
　　　　　　11. 总苞无钩状刺毛；冠毛长，不易脱落 ┈┈┈┈┈┈ 漏芦属 *Rhaponticum*

续表

2. 头状花序管状花和舌状花兼有
　12. 冠毛较果实长
　　　13. 管状花和舌状花均为黄色，冠毛1轮 ························ 旋覆花属 *Inula*
　　　13. 舌状花白色或蓝紫花，管状花黄色，冠毛1~2轮 ··············· 紫菀属 *Aster*
　12. 冠毛较果实短，或缺
　　　14. 叶对生
　　　　　15. 舌状花1层，先端3裂；瘦果为内层苞片所包裹 ············· 豨莶属 *Siegesbeckia*
　　　　　15. 舌状花2层，先端全缘或2裂；瘦果不为内层苞片所包裹 ········· 鳢肠属 *Eclipta*
　　　14. 叶互生，总苞边缘干膜质；头状花序大，单生茎顶或在枝顶排成伞房或复伞房花序
　　　　　··· 苍耳属 *Xanthium*
1. 头状花序仅有舌状花，植物体具乳汁 ························· 舌状花亚科 Liguliflorae
　16. 冠毛有细毛，瘦果粗糙或平滑，有喙或无喙部
　　　17. 头状花序单生花葶上，瘦果有长喙 ····················· 蒲公英属 *Taraxacum*
　　　17. 头状花序排成伞房状，瘦果无喙部 ····················· 苦苣菜属 *Sonchus*
　16. 冠毛有糙毛，瘦果扁或近圆柱形
　　　18. 瘦果极扁平或较扁，两面具细纵肋 ····················· 莴苣属 *Luctuca*
　　　18. 瘦果近圆形，腹背稍扁，具10翅 ····················· 苦荬菜属 *Ixeris*

Key to important medicinal genera of Asteraceae

1. All florets are tubular, or both tubular and liguliform, lacking latex ························· **Tubuliflorae**
　2. All florets are tubular, bisexual and unisexual
　　3. Leaves opposite, three divided or simple ···························· *Eupatorium*
　　3. Leaves alternate
　　　4. Lacking pappus
　　　　5. Unisexual; involucre spines ······························· *Uncaria*
　　　　5. Central florets bisexual, outer female; involucre spines absent ··············· *Artemisia*
　　　4. Pappus present
　　　　6. Avantha along the margin
　　　　　7. Pappus connate at the base, pinnate
　　　　　　8. Bract phylloid at the base of the inflorescence; fruit covered with fine, silky hairs ······ *Atractylodes*
　　　　　　8. No bract phylloid at the base of the inflorescence; fruit with no hairs ················ *Cirsium*
　　　　　7. Pappus scale-like or absent, outer phyllary leaf-like, spine ················· *Carthamus*
　　　　6. Unarmed along the margin
　　　　　9. Root aroma
　　　　　　10. Tall perennial herbs; pappus pinnate ······················· *Saussurea*
　　　　　　10. Low perennial herbs; pappus bristlelike ····················· *Vladimiria*
　　　　　9. Root with no aroma.
　　　　　　11. Phyllary with barbeIlate hairs; pappus short, falling off easily ··············· *Arctium*
　　　　　　11. Phyllary with no barbeIlate hairs; pappus long, falling off difficulty ············· *Rhaponticum*
　2. Both tubular and liguliform
　　12. Pappus longer than fruit
　　　13. Both tubulars and liguliforms are yellow, one series of pappus ····················· *Inula*
　　　13. Liguliforms are white and bluish, tubulars are yellow, two series of pappus ·············· *Aster*
　　12. Pappus shorter than fruit, or lacking
　　　14. Leaves opposite
　　　　15. Ligulate flowers 1 layer, apex 3-lobed; achenes enclosed by inner bracts ············· *Siegesbeckia*
　　　　15. Ligulate flowers 2-layered, apex entire or 2-lobed; achenes not enclosed by inner bracts ······ *Eclipta*

continued

14. Leaves alternate, margin of periclinium dry membranous. A large, solitary, capitulum or arranged to corymbose at the top of a branch .... *Xanthium*

1. All florets are liguliform, latex present .... **Liguliflorae**

16. Pappus with fine hairs, achenes coarse or smooth; beak present or lacking

17. Capitulum, solitary, at the top of the scape .... *Taraxacum*

17. Capitulum arranged to corymbose at the top of a scape .... *Sonchus*

16. Pappus with coarse hairs, achenes flat or nearly cylindrical

18. Achenes are very flat or flattened, with thin longitudinal ribs on both sides .... *Luctuca*

18. Achenes suborbicular, slightly flat ventral back, with 10 wings .... *Ixeris*

【药用植物】
Medicinal plants

（1）管状花亚科　红花 *Carthamus tinctorius* L.（图 12-72）一年生草本。叶质硬，边缘不规则浅裂，裂片先端成锐齿或无锐齿，齿间有刺。头状花序伞房状；总苞边缘有针刺。花序全为两性管状花，初开时黄色，后转为橘红色。瘦果，无冠毛。全国各地有栽培。管状花冠（红花）能活血通经、散瘀止痛。

Tubuliflorae *Carthamus tinctorius* L. (Fig.12-72) Herbs, annual. Leaves rigid, margin spinosely toothed; teeth apically with spine. Capitula aggregated into corymbose, margin spinosely. All flowers are tubular, bisexual, corolla yellow at first, turn orange red later. Achene without pappus. Cultivated all over the country. Dried flower (Carthami Flos, Hong hua) activates blood, unblocks channels, transforms stasis and relieves pain.

菊花 *Chrysanthemum morifolium* Ramat.（图 12-73）多年生草本，被柔毛。头状花序大小不一。总苞多层。舌状花颜色各异，管状花黄色。瘦果无冠毛。全国各地有栽培。花序（菊花）能散风清热，平肝明目，清热解毒。按品种、产地和加工方法的不同，分为"亳菊"、"滁菊"、"杭菊"、"贡菊"、"怀菊"等。

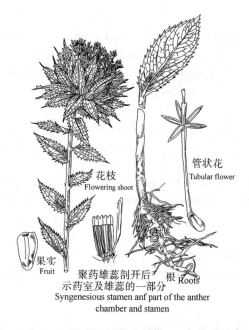

图 12-72　红花
Fig.12-72 *Carthamus tinctorius*

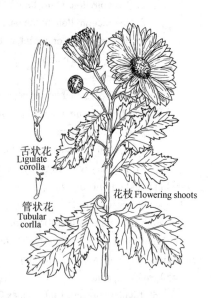

图 12-73　菊花
Fig.12-73 *Chrysanthemum morifolium*

Chrysanthemum morifolium Ramat. (Fig.12-73) Herbs perennial, pilose. Capitula vary in size. Involucre multilayer. Ligulate corolla different colors, tubular corolla yellow. Achene without pappus. Cultivated all over China. Capitulum (Chrysanthemi Flos, Ju hua) can disperses wind and heat, pacifies the liver yang to improve eyesight and clears toxic heat. According the difference of producing areas and processing methods, the medicinal Ju hua are divided into "Boju", "Chuju", "Hangju", "Gongju", etc..

白术 *Atractylodes macrocephala* Koidz.（图 12-74）多年生草本。根茎结节状膨大。头状花序单生茎顶，全为管状花，紫红色。多栽培。根状茎（白术）能健脾益气，燥湿利尿，止汗，安胎。同属植物苍术 *A. lancea*（Thunb.）DC. 根状茎平卧，结节状。管状花白色；华中、华东地区有分布；根状茎（苍术）能燥湿健脾，祛风散寒，明目。北苍术 *A. chinensis*（DC.）Koidz. 根状茎也作苍术入药。

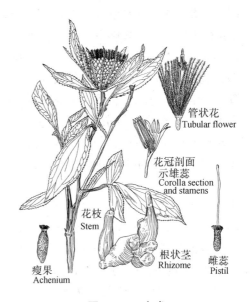

管状花
Tubular flower

花冠剖面
示雄蕊
Corolla section
and stamens

花枝
Stem

根状茎
Rhizome

瘦果
Achenium

雌蕊
Pistil

图 12-74 白术
Fig.12-74 *Atractylodes macrocephala*

Atractylodes macrocephala Koidz. (Fig.12-74) Herbs perennial. Rhizome nodular. Capitulum solitary apical. Corolla tubular, purplish red. Mostly cultivated. Rhizome (Atractylodis Macrocephalae Rhizoma, Bai zhu) can nourishe Qi, strengthen the spleen, dry and drain dampness, arrest sweating and prevent abortion. *A. lancea* (Thunb.) DC. Rhizome recumbent, nodular. Tubulars flower white. Distributed in central and east China. Rhizome (Atractylodis Rhizoma, Cang zhu) can dry dampness, tonify the spleen, expel wind dampness and induce perspiration to release the exterior, improve eyesight. Rhizome of **A. chinensis** (DC.) Koidz. has the same function as "Atractylodis Rhizoma".

茵陈蒿 *Artemisia capillaris* Thunb. 多年生草本。幼苗被白色柔毛。叶一至三回羽状分裂。全国各地均有分布。春季幼苗地上部分（茵陈）能清利湿热，利胆退黄。同属植物滨蒿 *A. scoparia* Waldst. et Kit. 幼苗地上部分也作茵陈入药。黄花蒿 *A. annua* L. 的地上部分（青蒿）能清虚热，除骨蒸，解暑热，截疟，退黄。艾 *A. argyi* Levl. et Vent. 的叶（艾叶）能温经止血，温经止血，散寒止痛，调经。

Artemisia capillaris Thunb. Herbs perennial. The seedlings white pilose. Leaves 1-3 pinnate lobed. Widely distributed. In spring, the above ground part of the seedlings (Artemisiae Scopariae Herba,Yin chen hao) can clear and drain dampness and heat, improve secretion and discharge of bile to relieve jaundice. *A. scoparia* Waldst. et Kit. is equally used as *A. capillaris*. The above ground part of *Artemisia annua* L. (Artemisiae Annuae Herba,Qing hao) is used for suppressing deficiency heat, cooling blood, clearing summerheat and counteracting malaria. *A. argyi* Levl. et Vent. Leaves (Artemisiae Argyi Folium, Ai ye) can warm the channels, arrest bleeding, dissipates coldness to relieve pain, and regulates menses.

木香 *Aucklandia lappa* Decne.（图 12-75）多年生高大草本。主根肥大。叶基部下延成翅。原产印度，我国引种栽培成功。产于印度、巴基斯坦、缅甸者，称为广木香，主产于云南、广西者，称为云木香。根（木香）能行气止痛，健脾消食。

Aucklandia lappa Decne. (Fig.12-75) Tall perennial herbs. Main root hypertrophy. Leaf base extended into wings. Native to India; Successfully introduced and cultivated in China. Those produced in

India, Pakistan and Myanmar are called "Guang mu xiang", and those mainly produced in Yunnan and Guangxi are called "yun mu xiang". Root (Aucklandiae Radix,Mu xiang) can move qi, relieve pain, invigorate the spleen and promote digestion.

常用药用植物还有：野菊 *Chrysanthemum indicum* L. 的花序（野菊花）能清热解毒，泻火平肝；紫菀 *Aster tataricus* L. f. 的根（紫菀）能润肺下气，消痰止咳；牛蒡 *Arctium lappa* L. 的果实（牛蒡子）能疏散风热，宣肺透疹，解毒消肿，利咽；旋覆花 *Inula japonica* Thunb. 的头状花序（旋覆花）能降气，消痰，行水，止呕；豨莶 *Siegesbeckia orientalis* L. 地上部分（豨莶草）能祛风湿，利关节，解毒；鳢肠 *Eclipta prostrata* L. 地上部分（墨旱莲）能滋补肝肾，凉血止血；漏芦 *Rhaponticum uniflorum*（L.）DC. 根（漏芦）能清热解毒，消痈下乳；刺儿菜 *Cirsium setosum*（Willd.）MB. 地上部分或根（小蓟）能凉血止血，散瘀解毒，消痈；蓟

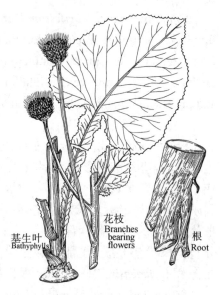

图 12-75　木香
Fig.12-75　*Aucklandia lappa*

基生叶 Bathyphylls　花枝 Branches bearing flowers　根 Root

Cirsium japonicum DC. 地上部分或根（大蓟）能散瘀解毒消痈；苍耳 *Xanthium sibiricum* Patr. 干燥成熟带总苞的果实（苍耳子）能散风寒，通鼻窍，祛风湿；天名精 *Carpesium abrotanoides* L. 干燥成熟果实（鹤虱）能杀虫消积。千里光 *Senecio scandens* Buch.-Ham. 全草（千里光）能清热解毒，明目，利湿。款冬 *Tussilago farlara* L. 干燥头状花序（款冬花），能润肺下气，止咳化痰。

Other medicinal plants: Inflorescence of *Chrysanthemum indicum* L. can clear heat and toxins, purging fire to calm the liver. Root of *Aster tataricus* L. f. (Asteris Radix Et Rhizoma, Zi wan) can moisten the lung, resolve phlegm and arrest cough. Fruit of *Arctium lappa* L. (Arctii Fructus, Niu bang zi) can disperse wind and heat, disperse lung qi, ease the throat and dissipate masses, clear toxic heat and relieve swelling, disperse lung and promoting rash. *Inula japonica* Thunb. Inflorescence (Inulae Flos,Xuan fu hua) can descend the adverse Qi flow, resolve phlegm and arrest vomiting. *Siegesbeckia orientalis* L. Aboveground part (Siegesbeckiae Herba,Xi xian cao) can dispel wind and dampness, dredge channels, activate collaterals and clear toxic heat. *Eclipta prostrata* L. Abovegroud part (Ecliptae Herba) can nsourish the kidney yin, cool blood and arrestsbleeding. *Rhaponticum uniflorum* (L.) DC. Root can clear away heat and detoxify, eliminate carbuncle and loose knots, relax tendons and channels. Aboveground part of *Cirsium setosum* (Willd.) MB. and *C. japonicum* DC. can cool blood, stop bleeding, dissolve stasis. *Xanthium sibiricum* Patr. Fruit can dispels wind and releases the exterior, ventilates and dredges the nasal channels. *Centipeda minima* (L.) A. Braun et Aschers. Whole plant can dispel the wind and cold, clear the nose and orifices, relieve cough and detoxify. *Carpesium abrotanoides* L. Fruit can insecticidal and depopulating. *Senecio scandens* Buch. -Ham. Whole plant can clear away heat and toxin, nourishing liver and eyesight. The inflorescence of *Tussilago farlara* L. (Farfarae Flos) can moisten the lung, descends lung qi, resolves phlegm and relieves cough.

（2）舌状花亚科　蒲公英 *Taraxacum mongolicum* Hand.–Mazz. 多年生草本，具乳汁。叶基生。头状花序，全为舌状花，花黄色。蒲公英与本属等数种植物的干燥全草（蒲公英）能清热解毒，消肿散结，利尿通淋。本属植物广布全国各地。

Liguliflorae: *Taraxacum mongolicum* Hand.–Mazz. Herbs perennial, with latex. Leaves basal.

Capitulumis all glossy, yellow. Flower yellow. This genus is widely distributed all over the country. Dandelion and The whole grass of taraxacum mongolicum and many other plants in this genus are used for clearing away heat and detoxifying, swelling and dispel knots, diuresis and drench.

常用药用植物还有苦荬菜 *Ixeris denticulata*（Houtt.）Stebb. 全草能清热解毒，消痈散结。苦苣菜 *Sonchus oleraceus* L. 全草能清热解毒凉血。

Other medicinal plants: Whole plant of *Ixeris denticulata* (Houtt.) Stebb. can clear away heat and detoxify, eliminate carbuncle and disperse knots. Whole plant of *Sonchus oleraceus* L. can clear away heat and detoxify.

二、单子叶植物纲
4.2 Monotyledoneae

多为草本，少木本。须根系；茎内维管束散生，无形成层；叶多具平行脉或弧形脉；花 3 基数，花被通常相似。种子的胚一般只有 1 枚子叶。

Mostly herbaceous, less woody. Fibrous roots; vascular bundles scattered within stems, cambium-free; leaves many with parallel or arcuate veins; flowers 3-merous, perianth usually similar. The embryo of a seed usually has only one cotyledon.

47. 禾本科 ♀ * P$_{2\sim3}$A$_{3\sim6}$$\underline{G}$$_{(2\sim3:1:1)}$
(47) Poaceae

【形态特征】草本或木本（图 12-76）。常具根状茎，地上茎习称为秆，常中空，节明显。叶排成 2 列，由叶片、叶鞘和叶舌组成，有时有叶耳；叶鞘抱秆，一侧开裂。花序多种，由小穗集成，每小穗有小花 1~ 数朵排列与轴上；小穗轴基部有 2 苞片成为颖片，下面的为外颖，上面的为内颖；花小，常两性；花被退化，而为 2 苞片所包，2 苞片称稃片，分别叫外稃和内稃；每朵小花下有退化的花被 2~3 片，称为浆片；雄蕊常 3，有时 1~6 枚；雌蕊 1，子房上位，2~3 心皮组成 1 室，1 胚珠，花柱 2，柱头羽毛状；颖果。

视频

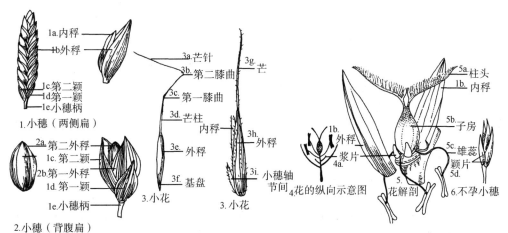

1. Spikelet(flat on two sides)(1a.palea,1b.lemma,1c.the second glume,1d. the first glume,1e. pedicel);2.Spikelet (flat back and belly)(2a.the second lemma,2b. the first lemma); 3. Floret (3a.awn needle,3b.the second geniculation,3c.the first geniculation,3d.awn column,3e.lema, 3f.basal template,3g.awn needle,3h.lemma,3i. rachis internode of spikelet);4. Longitudinal cutting diagram of flowers(4a.lodicule),5.Flower anatomy(5a.stigma,5b.ovary,5c.stamen,5d.glume); 6. Sterile spikelet

图 12-76 禾本科植物小穗、小花及花的构造
Figure 12-76 Spikelets，florets and flower structure of Gramineae

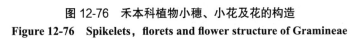

Characteristic: Herbs or woody (Fig.12-76), with rhizomes; above-ground stems (culms) often hollow, nodes obvious. Leaves arranged alternately in 2 ranks, differentiated into sheath, blade, and blade junction (ligule), sometimes with auricle. Inflorescence terminal or axillary. Spikelets composed of distichous bracts arranged along a slender axis (rachilla); typically 2 lowest bracts (glumes) subtending 1 to many florets; Florets composed of 2 opposing bracts enclosing a single small flower, outer bract (lemma) clasping the more delicate, usually 2-keeled inner bract (palea); base of floret often with thickened prolongation articulated with rachilla (callus). Flowers bisexual or unisexual; lodicules 2, rarely 3 or absent; stamens 3 rarely 1, 2, 6, or more in some bamboos; ovary 1-celled, styles free or united at base, topped by feathery stigmas. Fruit normally a dry indehiscent.

【分布】本科植物约有 700 属，10000 余种，广布全球。我国约 200 多属，1800 种，全国分布。已知药用 85 属，173 种。

Distribution: 700 genera, about 10000 species. 200 genera, 1800 species in China. Distributed all over the country. Medicinal plants 85 genera and 173 species.

【主要药用属检索表】

1. 乔木或灌木；叶二型，有箨叶与营养叶之分 ·················· 竹亚科 *Bambusoideae*
　2. 竿箨早落；竿每节分 2 枝，中部各仅具 1 竿芽；竿髓为膜质，呈两端封闭的囊状，紧贴空腔内壁
　　·················· 刚竹属 *Phyllostachys*
1. 草本；叶单型，无箨叶 ·················· 禾亚科 *Agrostidoideae*
　2. 叶片为狭长披针形或线形，中脉明显，通常无叶柄，叶鞘明显，叶片与叶鞘连接处无节
　　3. 雌小穗包于骨质总苞内 ·················· 薏苡属 *Coix*
　　3. 雌小穗外无骨质总苞
　　　4. 须根中下部膨大呈纺锤形。小穗无柄，脱节于颖之下；外稃具芒尖，不育外稃紧密包卷，先端具短芒；雄蕊 2 枚 ·················· 淡竹叶属 *Lophatherum*
　　　4. 雄蕊 3 枚或 6 枚
　　　　5. 多年生，具发达根状茎的苇状沼泽草本；外稃基盘之两侧密生等长或长于其稃体之丝状柔毛；雄蕊 3 枚 ·················· 芦苇属 *Phragmites*
　　　　5. 成熟花下有 2 枚不孕花外稃；颖退化，在小穗柄顶端呈二半月形之痕迹；雄蕊 6 枚 ··· 稻属 *Oryza*

Key to medicinal generas of Poaceae

1. Trees or shrubs; two types of leaf, sheath leaf and nutrition leaf ·················· *Bambusoideae*
　2. The culm sheath falls early; node of the culm is divided into two branches, with one bud in the middle of culm; The rod pulp is membranous, which is a closed capsule, which close to the inner wall of the cavity
　　·················· *Phyllostachys*
1. Herbs; one type of leaf, without sheath leaf ·················· *Agrostidoideae*
　2. Narrow lanceolate or linear leaf, obvious midrib, without petiole, obvious leaf sheath, there is no node at the junction leaf and leaf sheath
　　3. There is osseous involucre outside the female spikelet ·················· *Coix*
　　3. There is not osseous involucre outside the female spikelet
　　　4. The middle and lower parts of fibrous roots were fusiform. Spikelets sessile, disjointed below glume; lemma with aristate, sterile lemma tightly convoluted, apex with short awn; 2 stamens ······ *Lophatherum*
　　　4. 3 or 6 stamens
　　　　5. Perennial reed-like marsh herb with developed rhizome; there is equal or longer than lemma filiform pubescence at two sides of lemma basal; 3 stamens ·················· *Phragmites*
　　　　5. There are 2 infertile lemmas under the mature flowers; glumes degenerate, there is the trace of two half moons at the top of spikelet stalk; 6 stamens ·················· *Oryza*

【药用植物】
Medicinal plants

（1）竹亚科 *Bambusoideae*　灌木或乔木状。叶分为主干叶和普通叶；主干叶由箨鞘、箨叶组成；普通叶具短柄，叶片常披针形，具明显的中脉，无叶鞘，叶片和叶柄连接处有关节，叶片易从关节处脱落。

Bambusoideae: Shrub or arbor. Two types of leaf, sheath leaf and nutrition leaf. The nutrition leaf has short petioles, lanceolate blade, with obvious midrib, without sheath, and there is node between the junction of the leaf and petiole, the leaf is easy to fall off from the node.

淡竹 *Phyllostachys glauca* McClure（图 12-77）秆高 18m，新秆密被白粉，老杆绿或灰黄绿色。秆箨淡红褐或绿褐色，有多数紫色脉纹，无毛，被紫褐色斑点；箨叶带状披针形，有紫色脉纹。分布于黄河流域至长江流域各地，系常见栽培竹种之一。秆的中层（竹茹）能清热化痰、除烦止呕；竹沥（秆加热流出的汁）能清热豁痰。青竿竹 *Bambusa tuldoides* Munro 和大头典竹 *B. beecheyana* Munro var. *pubescens*（P. F. Li）W. C. Lin 茎秆的中层也做竹茹药用。

常用药用植物还有：青皮竹 *Bambusa textilis* McClure 和华思劳竹 *Schizostachyum chinese* Rendle 的秆内分泌物（天竺黄）能清热豁痰，凉心定惊。

Phyllostachys glauca McClure. (Fig.12-77) Culm 18 meters hight, new culm covered with white powder, old culm green or chartreuse. Culm-sheath pale red brown or greenish brown, with many purple veins, glabrous, purple brown spots. Lanceolate sheath leaf, green, with many purple veins. Distributed from the

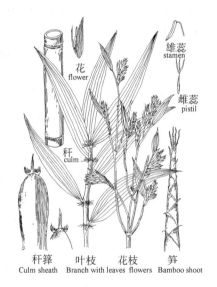

花 flower
雄蕊 stamen
雌蕊 pistil
秆 culm
秆箨 Culm sheath　叶枝 Branch with leaves　花枝 flowers　笋 Bamboo shoot

图 12-77　淡竹
Fig. 12-77　*Phyllostachys glauca*

Yellow River basin to the Yangtze River Basin in China, which is one of the common cultivated species. The middle layer of the culm (Bambusae in Taenias Caulis, Zhu ru) can clear away heat and phlegm, eliminate boredom and stop vomiting. The middle layer of *Bambusa tuldoides* Munro and *B. beecheyana* Munro var. *pubescens* (P. F. Li) W. C. Lin are also used as Zhu ru.

（2）禾亚科 *Agrostidoideae*　草本。只有普通叶，叶片狭长披针形或线形，中脉明显，通常无叶柄，叶鞘明显，叶片与叶鞘连接处无关节。

Agrostidoideae: Herb. leaves narrow lanceolate or linear; midrib obvious; petiole absent. No joint between the blade and the leaf sheath.

薏苡 *Coix lacryma-jobi* L.（图 12-78）一年生草本。秆直立丛生。总状花序腋生成束，雄蕊常退化，雌蕊具细长柱头，伸出，颖果小。分布于我国辽宁以南、西藏以西的湿润及半湿润区。种仁（薏苡仁）能健脾利湿、除痹止泻、排脓、解毒散结。

Coix lacryma-jobi L. (Fig.12-78) Annual sturdy herb. Culms erect, grow thickly. Fasciculate raceme in axillary, stamen degenerate, pistils with slender stigmas, protruding, small caryopsis. Distributed in the moist and sub-humid areas south of Liaoning and west of Tibet. Kernel (Coicis Semen,Yi yi ren) can invigorate the spleen and removing dampness, eliminate arthralgia and relieving diarrhea, clearing away heat and discharging pus.

淡竹叶 *Lophatherum gracile* Brongn. 多年生草本；须根中部膨大呈纺锤形小块根。圆锥花序，雄蕊 2 枚，颖果长椭圆形。分布于华东、华南、西南等地区。茎叶（淡竹叶）能清热泻火、除烦止渴、利尿通淋。

Lophatherum gracile Brongn. Perennial herb; the middle part of fibrous root is expanded and appear fusiform root tuber. Panicle, 2 stamens, long elliptic caryopsis. Distributed in East China, South China, southwest and other regions. Stem and leaf (Lophatheri Herba,Dan zhu ye) can clear away heat and annoyance, diuresis, promote body fluid and quench thirst.

常用药用植物还有：芦苇 *Phragmites australis*（Cav.）Trin. ex Steud. 根状茎（芦根）能清热泻火，生津止渴，除烦，止呕，利尿。大麦 *Hordeum vulgare* L. 发芽果实（麦芽）能行气消食，健脾开胃，回乳消胀。白茅 *Imperata cylindrica*（L.）Beauv. 根状茎（白茅根）能凉血止血，清热利尿。稻 *Oryza sativa* L. 和粟 *Setaria italica*（L.）P. Beauv. 成熟果实经发芽干燥的炮制加工品（稻芽和谷芽）能消食和中，健脾开胃。

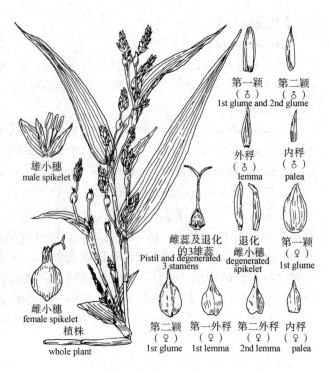

图 12-78　薏苡
Fig. 12-78　*Coix lacryma-jobi*

Other medicinal plants: *Phragmites australis* (Cav.) Trin. ex Steud. Rhizome (Phragmitis Rhizoma, Lu gen) can clear away heat and relieve fire, generate fluid and thirst, eliminate boredom, stop vomiting and diuresis. *Hordeum vulgare* L. Germinating fruit (Hordei Germinatus Fructus,Mai ya) can promote Qi and digest food, invigorate the spleen and appetizer, return breast milk and eliminate swelling. *Imperata cylindrica* (L.) Beauv. Rhizome (Imperatae Rhizoma,Bai mao gen) can cool blood, stop bleeding, clear heat and diuresis. *Oryza sativa* L. and *Setaria italica* (L.) P. Beauv. The mature fruit by germinating and drying (Oryzae Germinatus Fructus, Dao ya; Setariae Germinatus Fructus,Gu ya) can eliminate food consumption and promote spleen and appetizer.

48. 莎草科　$\female * P_0 A_3 \underline{G}_{(2-3:1:1)}$；　$\male * P_0 A_3$；　$\female * P_0 \underline{G}_{(2-3:1:1)}$

(48) Cyperaceae

【形态特征】草本。具根状茎，秆实心，常三棱形，无节。叶常 3 列，叶片狭长，有封闭的叶鞘。花小，两性或单性，小穗复排成穗状花序、总状花序、圆锥状花序、头状花序或聚伞花序等；雄蕊 1~3，子房上位，1 室，有直立的胚珠 1 枚。果为小坚果。

Characteristic: Perennial or annual herbs. Most have rhizomes, culm solid, prism, node absent. Leaves usually arranged in three rows, long and narrow, with closed leaf sheaths. Flowers small, bisexual or unisexual, the spikelets arranged into spikes, racemes, panicles, capitulum or cymes; 1-3 stamens, superior ovary, 1 locular, with 1 atropous ovule. Nut.

【分布】80 属，约 4000 种。国产 28 属，500 种；药用 16 属 110 种。

Distribution: 80 genera, about 4000 species. 28 genera, 500 species in China. Medicinal plants 16 genera and 110 species.

【药用植物】香附 *Cyperus rotundus* L.（图 12-79）分布于浙江、福建、湖南。块茎（香附）能疏肝理气、调经止痛。荸荠 *Heleocharis dulcis*（Burm. F.）Trin. ex Henschel 全国各地都有栽

培；球茎淀粉（荸荠粉）能润喉去燥，清热除烦。荆三棱 *Scirpus yagara* Ohwi 分布于东北、西南及陕西、甘肃、新疆等地；块茎（荆三棱）能祛瘀通经，破血消症。短叶水蜈蚣 *Kyllinga brevifolia* Rottb. 分布湖北、贵州、安徽、浙江以至广东、海南岛；全草能清热利湿，止咳化痰，祛瘀消肿。

Medicinal plants: *Cyperus rotundus* L. (Fig.12-79) Distributed in Zhejiang, Fujian and Hunan. Tuber (Cyperi Rhizoma,Xiang fu zi) can soothe the liver and regulate Qi, regulate meridians and relieve pain. *Heleocharis dulcis* (Burm. F.) Trin. ex Henschel. Cultivated all over the country. The corm starch can moisten the throat and remove dryness, clear away heat and annoyance. *Scirpus yagara* Ohwi. Distributed in Northeast, southwest and Shanxi, Gansu, Xinjiang, etc. Tuber can remove blood stasis and meridian, breaking blood and eliminating disease. *Kyllinga brevifolia* Rottb. is distributed in Hubei, Guizhou, Anhui, Zhejiang, Guangdong and Hainan. The whole grass can clear away heat and dampness, relieve cough and phlegm, remove blood stasis and swelling.

49. 棕榈科 ♀ * $P_{3+3} A_{3+3} \underline{G}_{(3:1\sim3:1)}$ ； ♂ * $P_{3+3} A_{3+3}$ ； ♀ * $P_{3+3} \underline{G}_{(3:1\sim3:1)}$

(49) Arecaceae

【形态特征】乔木、灌木或藤本。茎不分枝。叶互生，聚生茎顶；羽状或掌状分裂。肉穗花序，多分枝；花单性或两性；花被6，2轮；雄蕊6，2轮；心皮3，离生或于合生。核果或浆果；外果皮长纤维质。

Characteristic: Arbors, shrubs, or vines; stems usually unbranched. Leaves aggregated top of stem, alternate, pinnate or palmately divided. Spadix, branched. Flowers bisexual or unisexual. Perith 6, 2 whorled; stamens 6, 2 whorled. Ovary is 1~3-locular or 3 carpels. Drupes or berries.

【分布】210属，约2800种。国产28属，约100种；产东南至西南部。药用16属25种。

Distribution: 210 genera, about 2 800 species. 28 genera, more than 100 species in China, Mostly from southwest to Southeast. Medicinal plants 16 genera and 25 species.

【药用植物】槟榔 *Areca catechu* L.（图12-80）分布在云南、海南及台湾等热带地区。果皮（大腹皮）能行气宽中，行水消肿；成熟种子（槟榔）能杀虫，消积，行气，利水，截疟。棕榈 *Trachycarpus fortunei*（Hook.）H. Wendl. 我国黄河以南地区均有分布。叶柄（煅后称"棕榈炭"）

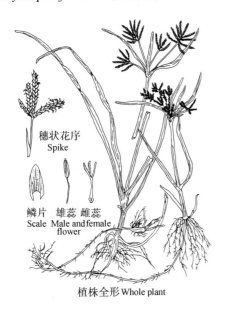

穗状花序
Spike

鳞片 雄蕊 雌蕊
Scale Male and female flower

植株全形 Whole plant

图 12-79 香附
Fig.12-79 *Cyperus rotundus*

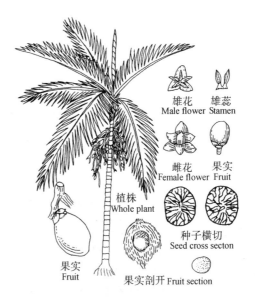

雄花 雄蕊
Male flower Stamen

雌花 果实
Female flower Fruit

植株
Whole plant

种子横切
Seed cross secton

果实
Fruit

果实剖开 Fruit section

图 12-80 槟榔
Fig.12-80 *Areca catechu*

能收涩止血。麒麟竭 *Daemonorops draco*（Willd.）Blume. 我国广东、台湾等地亦有种植。茎或果实渗出的树脂（血竭）能活血定痛，化瘀止血，生肌敛疮。

Medicinal plants: *Areca catechu* L. (Fig.12-80) Distributed in Yunnan, Hainan and Taiwan. The pericarp (Arecae Pericarpium, Da fu pi) is used for promoting qi circulation to alleviate middle energizer, move the water to reduce swelling; the mature seed (Arecae Semen, Binglang) can kill insects, eliminate accumulation, move the Qi, benefit the liquild and stop malaria. *Trachycarpus fortunei* (Hook.) H. Wendl. is distributed in the south of the Yellow River. The petiole (Trachycarpi Petiolus,Zong lv tan) can astringent and stop bleeding. *Daemonorops draco* (Willd.) Blume. Distributed in Guangdong, Taiwan and other places in China. The resin from the stem or fruit (Draconis Sanguis, Xue jie) can activate the blood and relieve pain, remove blood stasis and stop bleeding, and generate muscle to collect sore.

50. **天南星科**　$\female *P_{4\sim6} A_{4\sim6} \underline{G}_{(1\sim\infty:1\sim\infty:1\sim\infty)}$；　$\male *P_0 A_{(1\sim8),(\infty),1\sim8,\infty}$；　$\female *P_0 \underline{G}_{(1\sim\infty:1\sim\infty:1\sim\infty)}$

(50) Araceae

【形态特征】多年生草本。具块茎或根状茎。植物体含水汁、乳汁和草酸钙针晶。单叶或复叶，常基生，常具膜质叶鞘，网状脉。肉穗花序，具佛焰苞；花小，两性或单性；单性花常无花被；两性花具花被片 4~6；雄蕊 4 或 6；子房上位，心皮 1 至多数。浆果密集生于花序轴上。

Characteristic: Perennial herbs, with tubers or rhizomes; contains aqueous juice, latex, and calcium oxalate needle crystals. Leaves simple or compound, basal, with membranaceous sheath at base of leaf; reticulate venation. Spadix with a large spathe at the base; flowers small, bisexual or unisexual; the unisexual flowers have not perianth; the bisexual flowers perianth 4~6; stamens 4 or 6; superior ovary, carples 1 to many. Berries, densely in the rachis.

【分布】约 115 属，3 500 多种；分布于热带和亚热带。我国 35 属，205 种，主要分布于西南、华南地区。已知药用 22 属，106 种。

Distribution: 115 genera, about 3 500 species. 35 genera and 205 species in China; Mainly distributed in southwest and South China. Medicinal plants 22 genera and 106 species.

【主要药用属检索表】

1. 花两性；花被片 6，佛焰苞和叶片同色、同形 ·······················菖蒲属 *Acorus*
1. 花单性；花被通常不存在
　2. 草本，具块茎；肉穗花序有顶生附属器
　　3. 佛焰苞管喉部闭合；肉穗花序下部雌花序与上部雄花序间有不育部分，附属器超出佛焰苞很长
　　　··· 半夏属 *Pinellia*
　　3. 佛焰苞管喉部张开
　　　4. 雌雄同株；叶片常箭形或箭状戟形；佛焰苞常紫红色 ················· 犁头尖属 *Typhonium*
　　　4. 雌雄异株；叶片常掌状或鸟趾状分裂；佛焰苞常绿白色 ·················天南星属 *Arisaema*
　2. 亚灌木状草本，具地上茎；肉穗花序上部无附属器；佛焰苞檐部展开为舟状，先端内弯，果期脱落；雄蕊分离；胚珠多数 ··· 千年健属 *Homalomena*

Key to important medicinal generas of Araceae

1. Bisexual flower; 6 perianths, spathe have the same color and shape as leaf ························· *Acorus*
1. Unisexual flower; without perianth
　2. Herb, tuber; spadix with terminal appendage
　　3. The throat of the spathe is close; there is a sterile part between the lower female inflorescence and the upper male inflorescence, and the appendage is much longer than the spathe ························· *Pinellia*
　　3. The throat of the spathe is open

视频

continued

4. Monoecism; arrow or halberd shaped leaf; amaranth spathe ……………………………… *Typhonium*

4. Dioecism; palmate or bird toe splitting leaf; green-white spathe …………………………… *Arisaema*

2. Subshrub herb, overground stem; spadix withou terminal appendage; the eaves of the spathe are scaphoid, the apex is curved inward, which falls off during fruit period; the stamens are separated; many ovules …………………………………………………………………………………………………… *Homalomena*

【药用植物】天南星 *Arisaema erubescens*（Wall.）Schott.（图 12-81）块茎扁球形。叶 1 枚，放射状、鸟趾状全裂。肉穗花序；花雌雄异株。浆果红色。除西北、西藏外，大部分省区都有分布。块茎（天南星）能散结消肿。异叶天南星 *A. heterophyllum* Blume、东北天南星 *A.amurense* Maxim. 二者的块茎同等入药。

Medicinal plants: *Arisaema erubescens* (Wall.) Schott. (Fig.12-81) Tubers oblate. Leaf 1, radial or bird-toe like divided. Spadix. Flower dioeciou. Berry, red. Cultivated in most provinces and regions in China except for the northwest and Tibet. Tuber (Arisaematis Rhizoma, Tian nan xing) can loose knot and detumescence.The tubers of *A. heterophyllum* Blumev and *A.amurense* Maxim. are equally used as Tian nan xing.

半夏 *Pinellia ternata*（Thunb.）Breit.（图 12-82）块茎圆球形。叶 3 深裂至 3 全裂。花雌雄同株。全国各地广布。块茎（半夏）能燥湿化痰、降逆止呕、消痞散结。同属植物掌叶半夏 *P. pedatisecta* Schott. 块茎（虎掌南星）习作"天南星"入药。

Pinellia ternata (Thunb.) Breit. (Fig.12-82) Tubers bulbous. Leaves 3-parted to 3-sect, with tubercles at the base. Flower monoecious. Widely distributed all over the country. Tuber (Pinelliae Rhizoma, Ban xia) can dry and damp phlegm, reduce adverse reactions and stop vomiting, and eliminate pimple and loose knots. The tuber of *P. pedatisecta* Schott. is customarily used as Tian nan xing.

石菖蒲 *Acorus tatarinowii* Schott（图 12-83）多年生草本。根茎芳香、肉质。叶无柄，叶片线形。佛焰苞叶状，肉穗花序圆柱状。分布于黄河以南各省区。根状茎（石菖蒲）能开窍豁痰、醒神益智、化湿和胃。同属植物菖蒲 *A.calamus* L. 的根状茎（水菖蒲）功效与石菖蒲近似。

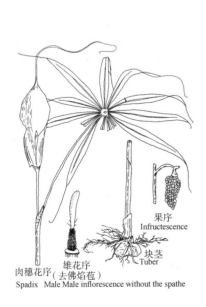

图 12-81　天南星

Fig.12-81　*Arisaema erubescens*

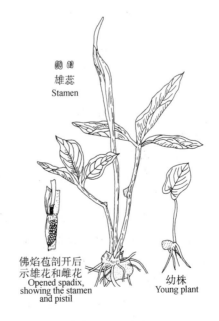

图 12-82　半夏

Fig.12-82　*Pinellia ternata*

Acorus tatarinowii Schott (Fig.12-83) Perennial herbs. Rhizome fragrant, fleshy. Petiole absent, blade linear-lanceolate. Spathe foliated, cylindrical. Distributed in the provinces south of the Yellow River. Rhizome (Acori Tatarinowii Rhizoma, Shi chang pu) can open the orifices and turn away phlegm, awaken the spirits and improve the intelligence, and remove dampness and stomach. *A.calamus* L.is equally used as *A. tatarinowii*.

常用药用植物还有：千年健 *Homalomena occula*（Lour.）Schott 根状茎（千年健）能祛风湿、强筋骨。独角莲 *Typhonium giganteum* Engl.（图 12-84）块茎（禹白附）能燥湿化痰、驱风解痉、解毒散结。鞭檐犁头尖 *T. flagelliforme*（Lodd.）Bl. 块茎（水半夏）能燥湿化痰、止咳，为半夏的地区习用品。

Other medicinal plants: Rhizome of *Homalomena occula* (Lour.) Schott (Homalomenae Rhizoma, Qian nian jian) can dispel rheumatism and strengthen muscles and bones. Tuber of *Typhonium giganteum* Engl. (Fig.12-84) can dry dampness and dissipate phlegm, expel wind and spasm, detoxify and disperse knots. Tuber of *T. flagelliforme* (Lodd.) Bl. can dry dampness, dissipate phlegm and stop coughing, which a regional staple for *Pinellia ternate*.

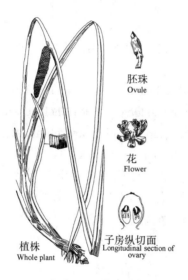

图 12-83　石菖蒲
Fig.12-83　*Acorus tatarinowii*

图 12-84　独角莲
Fig.12-84　*Typhonium giganteum*

51. 百合科　$\female *P_{3+3,(3+3)}A_{3+3}\underline{G}_{(3:3:1\sim\infty)}$

(51) Liliaceae

【形态特征】多年生草本，稀木本。有鳞茎、根状茎或块茎。单叶互生、对生、轮生或退化成鳞片状。花常两性，辐射对称；花被 6，分离或结合，2 轮；雄蕊 6；子房上位，3 心皮，3 室，中轴胎座。蒴果或浆果。

Characteristic: Herbaceous, sparse woody; with bulbs, rhizomes, or tuber. Leaves single, alternate, opposite, whorled or degenerate into scales. Flower bisexual, perianth 6, separated or united; stamens 6; ovary superior, 3 carpels. axial placenta. Capsule or berry.

【分布】250 属，约 3500 种，广布全球，以温带和亚热带地区为多。我国约 60 属，560 多种，全国分布，主要分布于西南地区。已知药用 52 属，374 种。

Distribution: 250 genera, about 3 500 species. 60 genera, 560 species in China, mainly in the southwest. Medicinal plants 52 genera and 374 species.

【主要药用属检索表】

1. 植株无鳞茎
　2. 叶轮生茎顶端；花单朵顶生，外轮花被片叶状，绿色 ················· 重楼属 *Paris*
　2. 叶和花非上述情况
　　3. 叶退化为鳞片，而具簇生的针状、扁圆柱状或近条形的叶状枝 ·············· 天门冬属 *Asparagus*
　　3. 叶较大，多枚基生，或互生、对生、轮生于茎或枝条上；无叶状枝
　　　4. 成熟种子小核果状
　　　　5. 子房上位 ················· 山麦冬属 *Liriope*
　　　　5. 子房半下位 ················· 沿阶草属 *Ophiopogon*
　　　4. 浆果或蒴果
　　　　6. 叶肉质肥厚，边缘常具刺状小齿 ················· 芦荟属 *Aloe*
　　　　6. 叶草质，边缘不具刺状小齿
　　　　　7. 花单性 ················· 菝葜属 *Smilax*
　　　　　7. 花两性
　　　　　　8. 雄蕊 3 枚 ················· 知母属 *Anemarrhena*
　　　　　　8. 雄蕊 6 枚
　　　　　　　9. 蒴果 ················· 萱草属 *Hemerocallis*
　　　　　　　9. 浆果 ················· 黄精属 *Polygonatum*
1. 植株具鳞茎
　10. 伞形花序；植株具有葱蒜味 ················· 葱属 *Alliu*
　10. 花序非伞形花序；植物一般无葱蒜味
　　11. 花被片基部有蜜腺窝 ················· 贝母属 *Fritillaria*
　　11. 花被片基部无蜜腺窝 ················· 百合属 *Lilium*

Key to important medicinal generas of Liliaceae

1. Plant without bulbs
　2. Whorled leaves are at the top of the stem; flowers are borne on apex of stem, outer Whorled perianth leaf-shapped, green ················· *Paris*
　2. Leaves and flowers do not match the above
　　3. Leaves degenerate into scales, fascicular leafy shoots are acicular, oblate columnar, or strip ······ *Asparagus*
　　3. Leaves larger, mostly basal leaf, or alternate, opposite, worled, which is located on the stems or branches; No leafy shoots
　　　4. Mature seeds like drupe
　　　　5. Superior ovary ················· *Liriope*
　　　　5. Half-inferior ovary ················· *Ophiopogon*
　　　4. Capsule or berry
　　　　6. The leave are succulent and thick, leaf margin with spiny teeth ················· *Aloe*
　　　　6. The leaves are herbaceous, leaf margin without spiny teeth
　　　　　7. Unisexual flower ················· *Smilax*
　　　　　7. Bisexual flower
　　　　　　8. 3 stamens ················· *Anemarrhena*
　　　　　　8. 6 stamens
　　　　　　　9. Capsule ················· *Hemerocallis*
　　　　　　　9. Berry ················· *Polygonatum*
1. Plant with bulbs
　10. Umbel; plant has the flavor of onion and garlic ················· *Allium*
　10. Inflorescences are non-umbelliferous; plant has not the flavor of onion and garlic
　　11. The base of perianth has nectary fossa ················· *Fritillaria*
　　11. The base of perianth has not nectary fossa ················· *Lilium*

【药用植物】百合 *Lilium brownii* F. E. Brown var.*viridulum* Baker（图 12-85）多年生草本。鳞茎球形，白色。花喇叭形，白色；柱头 3 裂；蒴果。全国各地均有栽培，主产于河北、陕西、湖南、江西和浙江等地。鳞叶（百合）能养阴润肺，清心安神。同属植物卷丹 *L. tigrinum* 和细叶百合（山丹）*L. pumilum* DC. 肉质鳞叶同等入药。

Medicinal plants: *Lilium brownii* F. E. Brown var.*viridulum Baker* (Fig.12-85) Perennial herbs. Bulbs spherical, white. Flower white, funnel-shaped. Stigma 3 lobed. Capsule. Cultivated all over the country, mainly in Hebei, Shanxi, Hunan,Jiangxi and Zhejiang.Scaly leaf of bulb (Lilii Bulbus, Bai he) can raise Yin embellish lung, clear the heart and calm the mind. *L. tigrinum* and *L. pumilum* DC.are equally used.

黄精 *Polygonatum sibiricum* Delar. ex Red.（图 12-86）根状茎结节状膨大。叶 4~6 枚轮生。花 2~4 朵腋生。分布黑龙江、河北、山西、安徽、浙江等地。根状茎（黄精）能润肺滋阴、补脾气、益肾。同属多花黄精 *P. cyrtonema* Hua 和滇黄精 *P. kingianum* Collett et Hemsl. 的根状茎同等入药；玉竹 *P. odoratum*（Mill.）Druce 根状茎（玉竹）能滋阴润肺、生津养胃。

P. sibiricum Delar. ex Red. (Fig.12-86) Rhizome tuberous enlargement. Leaves 4~6, whorled. Flowers 2~4 axillary. Distributed in Heilongjiang, Hebei, Shanxi, Anhui, Zhejiang, etc. The rhizomes (Polygonati Rhizoma,Huang jing) can raise Yin embellish lung, nourish stomach and promote fluid production. *P. sibiricum* Delar. ex Red, *P. cyrtonema* Hua and *P. kingianum* Collett et Hemsl. are equally used as *P. sibiricum*. The rhizome of *P. odoratum* (Mill.) Druce (Polygonati Odorati Rhizoma,Yu zhu) can raise Yin embellish lung, nourish stomach and promote fluid production.

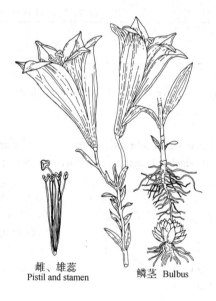

图 12-85 百合
Fig.12-85 *Lilium brownii* var.*viridulum*

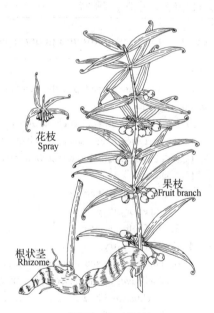

图 12-86 黄精
Fig.12-86 *Polygonatum sibiricum*

川贝母 *Fritillaria cirrhosa* D. Don 叶常对生，先端卷曲。花常 1~3 朵，紫色至淡黄绿色；叶状苞片 3。主要分布在西藏、四川、云南接壤地区，有栽培。鳞茎（川贝母）能清热化痰、润肺止咳。同属暗紫贝母 *F. unibracteata* Hsiao et K.C.Hsia、甘肃贝母 *F. przewalskii* Maxim.、梭砂贝母 *F. delavayi* Franch.、太白贝母 *F. taipaiensis* P. Y. Li 和瓦布贝母 *F. unibracteata* Hsiao et K.C.Hsia *var. wabuensis*（S.Y.Tang et S.C.Yue）Z.D. Liu，S. Wang et S.C.Chen 的鳞茎与川贝母同等入药。

Fritillaria cirrhosa D. Don. Leaves often opposite, apex curly. Flowers 1~3, purple to yellowish green; Bracts leaflike 3. Mainly distributed in Tibet, Sichuan, Yunnan border areas. Bulbus (Fritillariae Cirrhosae Bubus, Chuan bei mu) can clear away heat and phlegm, moisten lung and stop cough. The bulbs of *F. unibracteata* Hsiao et K.C.Hsia, *F. przewalskii* Maxim., *F. delavayi* Franch., *F. taipaiensis* P. Y. Li and *F. unibracteata* Hsiao et K.C.Hsia *var. wabuensis* (S.Y.Tang et S.C.Yue) Z.D. Liu,S. Wang et S.C.Chen are equally used as Chuan bei mu.

浙贝母 *Fritillaria thunbergii* Miq. 鳞茎大，由2~3枚鳞片组成。花1~6朵，淡黄色。分布在江苏（南部）、浙江（北部）和湖南等地。鳞茎（浙贝母）能清热化痰止咳。

F. thunbergii Miq. Bulbus large, consist of 2~3 scaly leaves. Flowers 1~6, light yellow. Dstributed in Jiangsu (South), Zhejiang (North) and Hunan. Bulbus (Fritillariae Thunbergii Bubus, Zhe bei mu) can clear heat and phlegm, stop cough.

常用药用植物还有：七叶一枝花 *Paris polyphylla* Sm. 叶7~10枚轮生；外轮花被绿色，4~6枚，雄蕊8~12枚；蒴果紫色；分布于西藏（东南部）、云南、四川和贵州；根状茎（重楼）能清热解毒、消肿止痛、凉肝定惊。知母 *Anemarrhena asphodeloides* Bunge. 具根状茎。叶具多条平行脉；分布于河北、山西、山东、陕西、内蒙古、辽宁、吉林等地；根状茎（知母）能清热泻火、滋阴润燥。麦冬 *Ophiopogon japonicus*（L.f.）Ker-Gawl.（图12-87）草本，叶基生成丛，具小块根；中国南方等地均有栽培；块根（麦冬）能润肺养阴、益胃生津、清心除烦、润肠。天门 *Asparagus cochinchinensis*（Lour.）Merr. 分布于华北、西北的南部至华东、中南、西南各省区。块根（天冬）能养阴润燥，清肺生津；光叶菝葜 *Smilax glabra* Roxb.（图12-88）分布于甘肃（南部）和长江流域以南各省区；根状茎（土茯苓）能清热解毒、通利关节、除湿。

Other medicinal plants: *Paris polyphylla* Sm. Leaves 7-10, whorled; stamens 8~12. Capsule purple. Distributed in Tibet (southeast), Yunnan, Sichuan and Guizhou. The rhizome (Paridis Rhizoma,Chong lou) can clear away heat, detoxify, detumescence, relieve pain, and calm down the wind. *Anemarrhena asphodeloides* Bunge. Rhizome. Parallel venation. Distributed in Hebei, Shanxi, Shandong, Shanxi, Inner Mongolia, Liaoning, Jilin, etc. Rhizome (Anemarrhenae Rhizoma, Zhi mu) can clear away heat, reduce fire, nourish yin and moisten dryness.

花
Flower

雄蕊
Stamen

花纵剖
示子房
Vertical section of flower,showing the ovary

图 12-87　麦冬
Fig.12-87 *Ophiopogon japonicus*

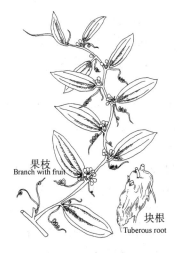

果枝
Branch with fruit

块根
Tuberous root

图 12-88　光叶菝葜
Fig.12-88 *Smilax glabra*

Ophiopogon japonicus (L.f.) Ker-Gawl. (Fig. 12-87) Perennial herb. Leaves are clustered at the base, lanceolate. Cultivated in southern China. Root tuber (Ophiopogonis Radix, Mai dong) can moisten lung, nourish yin, benefit stomach, promote body fluid, clear heart, remove troubles and moisten intestines. *Asparagus cochinchinensis* (Lour.) Merr. root tuber (Asparagi Radix, Tian men dong) can nourish yin, moisten dryness, clear lung and promote fluid production. *Smilax glabra* Roxb. (Fig.12-88) rhizome (Smilacis Glabrae Rhizoma, Tu fu ling) can clear away heat, detoxify, clear joints and dehumidify.

52. 石蒜科 $\male\female *P_{(3+3),\ 3+3}A_{(3+3),\ 3+3}\overline{G}_{(3:3:\infty)}$

(52) Amaryllidaceae

【形态特征】多草本。具鳞茎、根状茎或块茎。叶基生，条形。花单生或成伞形花序；花两性；花被6，2轮；雄蕊6，少数花丝合生成副花冠。3心皮，子房下位。蒴果。

Characteristic: Herbs. Bulbs, rhizomes or tubers. Leaves basal, strip. Flowers solitary or inflorescence umbel, Flowers bisexual; Perianth 6, in 2 whorls; Stamens 6 with or without a corona. Ovary inferior.Capsule.

【分布】100余属，约1200种。分布于热带、亚热带和温带地区。中国产10属，44种；全国分布。药用10属，20种。

Distribution: More than 100 genera, 1 200 species; tropical, subtropical, and temperate regions. 10 genera, 44 species in China, widely distributed. Medicinal plants 10 genera, 20 species.

【药用植物】仙茅 *Curculigo orchioides* Gaertn.（图12-89）根状茎（仙茅）能补肾阳、强筋骨、祛寒湿。石蒜 *Lycoris radiate*（L'Her.）Herb. 鳞茎能解毒、祛痰、利尿、消肿、催吐、杀虫；有毒。

Medicinal plants: *Curculigo orchioides* Gaertn. (Fig.12-89) The rhizomes (Curculiginis Rhizoma, Xian mao) are used for tonifying kidney Yang,strengthening muscles and bones, dispelling cold and dampness. Bulbs of *Lycoris radiata* (L'Her.) Herb. used for detoxification, expectorant, diuretic, vomiting and insecticide.

花
A flower

根状茎
Rhizoma

种子
Seed

雄蕊
Stamen

图 12-89 仙茅
Fig.12-89 *Curculigo orchioides*

53. 薯蓣科 $\male * P_{3+3,\ (3+3)}A_6$；$\female * P_{3+3,\ (3+3)}\overline{G}_{(3:3:2)}$

(53) Dioscoreaceae

【形态特征】缠绕草质或木质藤本。具根状茎或块茎。叶多互生，少对生或轮生，单叶或掌状复叶，基出脉掌状，侧脉网状。花小，单性异株或同株。花被6，2轮，离生或基部合生；雄蕊6，或3枚退化；3心皮合生，子房下位；花柱3。蒴果具三棱形的翅。

Characteristic: Herbs twining or woody vines. Rhizome or tuber. Leaves alternate or opposite, simple or palmately compound, basal veins palmately, lateral veins reticulate. Flowers small, unisexual, dioecious or monoecious. Perith 6, 2 whorled. Stamens 6, sometimes 3 reduced to staminodes; Ovary inferior, 3-loculed Fruit a capsule, with triangular wings.

【分布】9属，650多种，分布于全球的热带和温带地区。我国1属，约50种，主要分布于西南至东南各省区。已知1属37种药用。

Distribution: 9 genera, 650 species, in tropical and temperate regions of the globe. 1 genera and about 50 species in China. Mostly in SW to SE China. Medicinal plants 1 genera, about 37 species.

【药用植物】薯蓣 *Dioscorea opposita* Thunb.（图12-90）缠绕草质藤本；根状茎常圆柱形；茎

右旋。单叶，茎下部互生，中部以上对生。主产河南、湖南、江西、贵州等。根状茎（山药）能补脾养胃，生津益肺，补肾涩精。

同属植物穿龙薯蓣 *D. nipponica* Makino 的根状茎（穿山龙）能祛风除湿，舒筋通络，活血止痛，止咳平喘。粉背薯蓣 *D. collettii* Hook. f. *var. hypoglauca*（Palibin）C. T. Ting et al. 根状茎（粉萆薢）能利湿去浊，祛风除痹。

Medicinal plants: *Dioscorea opposita Thunb.* (Fig.12-90) Herbs twining, rhizomatous oblong cylinder; Stem twining to right. Leaves simple, alternate in lower stems part, opposite above middle. Main produced in Henan, Hunan, Jiangxi, Guizhou. Rhizomes (Dioscoreae Rhizoma, Shan yao) have the effect of tonifying spleen nourishing the stomach, fluiding profit lung, tonifying kidney and arresting seminal emission. The rhizomes (Dioscoreae Nipponicae Rhizoma) of *D. nip-*

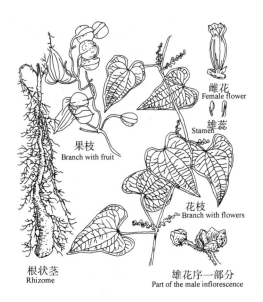

雌花
Female flower

雄蕊
Stamen

果枝
Branch with fruit

花枝
Branch with flowers

根状茎
Rhizome

雄花序一部分
Part of the male inflorescence

图 12-90　薯蓣
Fig.12-90 *Dioscorea polystachya*

ponica Makino have the effect of dispelling wind and dehumidify, relieving muscle and collaterals, promoting blood circulation and relieving pain, relieving cough and relieving asthma. The rhizomes (Dioscoreae Hypoglaucae Rhizoma) of *D. collettii* Hook. f. var. *hypoglauca* (Palibin) C. T. Ting have the effect of removing dampness and turbidity, dispelling wind and removing arthralgia.

54. 鸢尾科　$\female *\uparrow P_{(3+3)} A_3 \overline{G}_{(3:3:\infty)}$

(54) Iridaceae

【形态特征】多年生草本。根状茎、球茎或鳞茎。叶多基生，剑形或条形，基部鞘状，互相套迭。花两性，单生；花被片6，2轮；雄蕊3；花柱3裂。蒴果。

Characteristic: Mostly herbs, perennial, with rhizomes, bulbs, or corms. Leaves usually basal, sword-shaped to linear, base sheathing, overlap with each other. Inflorescence or solitary flower; perianth 6, in 2 whorls; stamens 3; style with 3 branches. Capsule.

【分布】约60属，800种，世界广布。中国11属，约71种；药用8属，39种。

Distribution: About 60 genera, 800 species, nearly worldwide, especially Africa, Asia, and Europe. 11 genera, more than 71 species in China. 8 genera, 39 species for medicinal plants.

【药用植物】番红花 *Crocus sativus* L.（图 12-91）柱头（西红花）能活血化瘀、凉血解毒，解郁安神。射干 Belamcanda chinensis（L.）Redouté（图 12-92）根状茎（射干）能清热解毒，消痰，利咽。鸢尾 *Iris tectorum* Maxim. 根状茎（川射干）能清热解毒，祛痰，利咽。

Medicinal plants: *Crocus sativus* L. (Fig.12-91) Stigmas (Croci stigma, Xi hong hua) are used for invigorating the circulation of blood, cooling blood detoxification, resolving depression nerves. Rhizomes of *Belamcanda chinensis* (L.) Redouté (Fig.12-92) (Belamcandae Rhizoma, She gan) and *Iris tectorum* Maxim. (Iridis Tectori Rhizoma, Chuan she gan) are used for clearing heat detoxify, eliminating phlegm, benefiting pharynx.

55. 姜科　$\female \uparrow P_{3+3} A_1 \overline{G}_{(3:3:\infty)}$

(55) Zingiberaceae

【形态特征】多年生草本，具根状茎或块根，芳香。叶2列，具叶鞘、叶片和叶舌。花两性，不整齐；花被片6，2轮；侧生退化雄蕊2或4，外轮2枚瓣状，内轮2枚合成唇瓣，能育雄蕊1

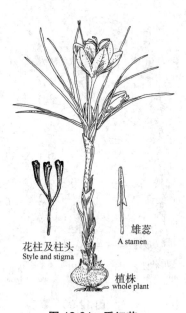

图 12-91　番红花

Fig.12-91 *Crocus sativus*

图 12-92　射干

Fig.12-92 *Belamcanda chinensis*

（图 12-93）。蒴果或浆果状。种子有假种皮。

Characteristic: Herbs perennial, with rhizomes or tuber, aromatic. Leaves distichous, with leaf sheath or ligule. Flowers bisexual; perianth 6, in 2 whorls; staminodes 6, lateral staminodes 2 or 4, outer 2, petaloid, inner 2 into synthetic lip; fertile stamen 1(Fig.12-93). Capsule or berrylike. Seeds with aril.

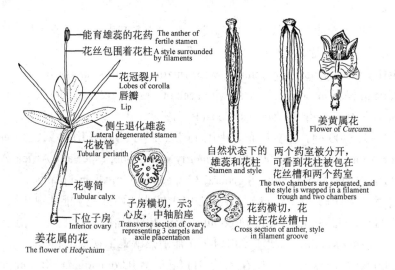

图 12-93　姜黄属和姜花属花的结构

Fig.12-93 **The structure of Curcuma and Zingiber flowers**

【分布】50 属，约 1500 种，分布于热带、亚热带地区，主产于热带亚洲。我国 19 属，150 余种，分布于西南、华南至东南；已知药用 15 属 100 余种。

Distribution: 50 genera, about 1 500 species, distributed in tropical and subtropical areas, mainly produced in Tropical Asia. More than 150 species in 19 Genera in China, distributed from southwest to southeast of China; 15 genera, 100 species are medicinal plants.

【药用植物】姜 *Zingiber officinale* Rose. 根状茎指状分枝，具辛辣味（图 12-94）。干燥根茎（干姜）能温中散寒，回阳通脉，温肺化饮；新鲜根状茎（生姜）能解表散寒，温中止呕，化痰止咳，解鱼蟹毒。

Medicinal plants: *Zingiber officinale* Rose. (Fig.12-94) Rhizome digitate branched, with pungent taste. Dried rhizomes (Zingiberis rhizome, Gan jiang) have the effect of warming in dispersing cold, restoring yang and rescuing patient from collapse, warming lung to drink. Fesh rhizomes (Zingiberis Recens Rhizoma, Sheng jiang) can relieve cold, warm stop vomiting, expectorant cough, ditoxify fish and crab poisoning.

姜黄 *Curcuma longa* L. 根状茎粗短，肉质芳香，须根末端常膨大成块根。分布于我国东南部至西南部。根状茎（姜黄）能破血行气，通经止痛；块根（黄丝郁金）能破血行气、清心解郁、凉血止血、利胆退黄。同属温郁金 *C. wenyujin* Y.H. Chen et C.Ling（图 12-95）块根（温郁金）能活血止痛、行气解郁、清心凉血、利胆退黄；根状茎（莪术）能行气破血、消积止痛。广西莪术 *C. kwangsinensis* S.G.Lee et C.F.Liang、蓬莪术 *C.phaeocaulis* 的块根、根茎与温郁金同等入药，三者块根分别称为"桂郁金"、"温郁金"和"绿丝郁金"。

Curcuma longa L. Rhizomes stout and short, fleshy and aromatic, fibrous root tips often swollen into tuberous roots. Distributed from southeast to southwest of China. The rhizomes (Curcumae longae Rhizoma,Jiang huang) can reak blood circulation, relieve pain. Root Tuber (Curcumae longae Radix, Huang si yu jin) can break the blood, clear heart depression, cool blood to stop bleeding, promoting cholagogue. The root tuber of C. wenyujin Y.H. Chen et C.Ling (Fig.12-95), C. kwangsinensis S.G.Lee et C.F. Liang and C.phaeocaulis Val have the same effect as Huang si yu jin, while the rhizomes of those three species are equally used as Curcumae Rhizoma,E zhu, which can exsanguinate Qi, eliminate accumulation and relieve pain.

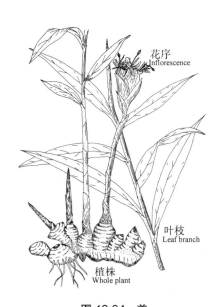

图 12-94　姜
Fig. 12-94　*Zingiber officinale*

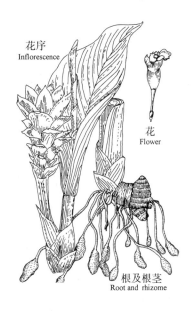

图 12-95　温郁金
Fig. 12-95　*C. wenyujin*

常用药用植物还有：草豆蔻 *Alpinia katsumadai* Hayata 近成熟种子（草豆蔻）能燥湿行气，温中止呕。高良姜 *A. officinarum* Hance 根茎（高良姜）能温胃止呕，散寒止痛。砂仁（阳春砂仁）*Amomum villosum* Lour. 果实（砂仁）能化湿开胃，温脾止泻，理气安胎。

Other medicinal plants: *Alpinia katsumadai* Hayata, the submature seed (Alpiniae katsumadai

Semen,Cao dou kou) can dry wets and stop vomiting. *A. officinarum* Hance, the rhizome (Alpiniae officinarum Rhizoma,Gao liang jiang) can warm stomach to stop vomiting, dispel cold pain. *Amomum villosum* Lour. Fruit (Amomi Fructus, Sha ren) can change wet appetizers, warm spleen and diarrhea, regulate Qi and prevent abortion.

56. 兰科 ♀↑P$_{3+3}$A$_{1~2}$$\overline{G}$$_{(3:1:\infty)}$

(56) Orchidaceae

【形态特征】陆生、附生或腐生草本。单叶互生，常2列，具叶鞘，有时退化为鳞片状。花两性；花被6，2轮；萼片3；花瓣3，中央1枚特化为唇瓣（由于子房180扭转，常位于下方）；花柱、柱头与雄蕊合生成合蕊柱，与唇瓣对生；能育雄蕊1，生合蕊柱顶端；柱头与花药之间有1舌状物，称蕊喙；花粉常粘合成团块状，并进一步特化成花粉块；蕊柱基部有时延伸成足状，称蕊柱足；子房下位，3心皮，1室，胚珠多数（图12-96）。蒴果。种子极小而多。

Characteristic: Terrestrial, epiphytic, or rarely mycotrophic herbs. Leaves alternate, distichous, with sheathing, or sometimes reduced to scales. Flowers bisexual; Perianth segments 6, in 2 whorls; Sepal 3; corolla 3, the middle one is lager and specialized labellum, often down below (ovary twisted through 180°). Style, stigma and stamen connate to form a gynoecium column, opposite labellum; Fertile stamens often 1, on the top of the gynoecium column; Between stigma and anthers there is a tongue-shaped organ called the rostellum; pollen usually forming distinct pollinia; The base of the gynoecium column sometimes extends into a foot-like shape, termed the column foot; Ovary inferior, 3 carpels, 1-loculed, ovules numerous. Capsule. The seeds are tiny and numerous(Fig.12-96).

【分布】700余属，20000余种。我国171属，约1250种，以云南、台湾及海南最为丰富；已知药用76属，287种。

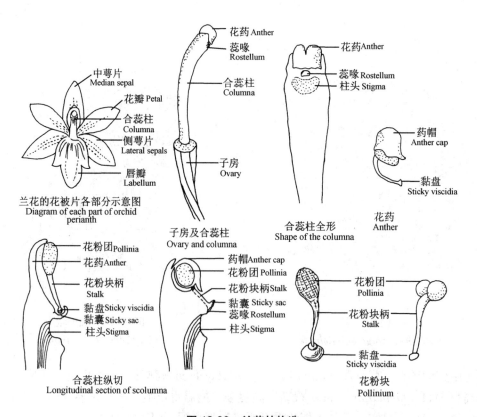

图 12-96　兰花的构造
Fig.12-96　Structure of orchis flower

Distribution: About 700 genera, 20 000 species. 171 genera, about 1 250 species in China. 76 genera, 287 species are medicinal plants.

【药用植物】天麻 *Gastrodia elata* Bl.（图 12-97）腐生草本，与蜜环菌共生。块茎肉质，具较密的节。茎直立。花橙红色或淡黄绿色。分布于全国大部分地区，多人工栽培。块茎（天麻）能息风止痉，平抑肝阳，祛风通络。

Medicinal plants: *Gastrodia elata* Bl. (Fig.12-97) Saprophytic herbs, symbiosis with *Armillaria mellea*. Tubers fleshy, with denser nodes. Stem erect. Flower orange red or yellowish green. Distributed in most parts of the country, mostly cultivated. Tuber (Gastrodiae Rhizoma, Tianma) can calm the wind and spasmodic, calm the liver yang, dispel wind and remove obstruction in the meridians.

石斛（金钗石斛）*Dendrobium nobile* Lindl.（图 12-98）附生草本。茎直立，肉质肥厚，干后金黄色。唇瓣中央具 1 紫红色大斑块。分布长江以南各省及西藏。茎（石斛）能益胃生津，滋阴清热。同属植物铁皮石斛 *D. officinale* Kimura et Migo、鼓槌石斛 *D. chrysotoxum* Lindl.、流苏石斛 *D. fimbriatum* Hook. 的茎同等入药。

Dendrobium nobile Lindl. (Fig.12-98) Epiphytic herbs. Stems erect, fleshy, golden yellow when dry. Labellum broadly ovate, with 1 large purplish red patch in the center Distributed Distributed south of the Yangtze River and Tibet.. Stem (Dendrobii Caulis, Shihu) can tonify stomach and promote fluid, benefit the stomach, nourish Yin and clearing heat. The stems of *D. officinale* Kimura et Migo, *D. chrysotoxum* Lindl. and *D. fimbriatum* Hook. are equally used as "Dendrobii Caulis".

植株
Whole plant

花
Flower

图 12-97　天麻
Fig.12-97　*Gastrodia elata*

唇瓣
Labellum

合蕊柱背面
The Back of columna

合蕊柱正面
（示雄蕊）
Adaxial surface
of columna

合蕊柱剖面
Longitudinal section of
columna

图 12-98　石斛
Fig.12-98　*Dendrobium nobile*

白及 *Bletilla striata*（Thunb.）Rchb. f. 陆生草本。块茎肥厚，多分枝，表面有环纹。花被紫红色。分布于华南、华中、西南以及陕西南部和甘肃东南部。块茎（白及）能收敛止血，消肿生肌。

Bletilla striata (Thunb.) Rchb. f. Terrestrial herbs. Tubers thick, multi-branched, and having ring patterns on the surface. Perianth fuchsia. Distributed in South, Central, and Southwest China, Southern Shaanxi and Southeast Gansu. Tuber (Bletillae Rhizoma,Baiji) can stop bleeding, reduce swelling and build muscle.

重点小结
Summary

被子植物具有具有真正的花和果实、双受精现象和新型胚乳、孢子体高度发达和分化、配子体进一步简化、营养和传粉方式多样化等特征，使其具有优于其他类群的适应能力。被子植物的自然分类系统里面，恩格勒系统支持二元起源和假花学说，哈钦松系统、塔赫他间系统、克朗奎斯特系统等支持单起源和真花学说。本教材采用恩格勒系统介绍了具有重要药用价值和分类地位的56个科。

Angiospermae are characterized by true flowers and fruits, double fertilization, highly developed and differentiated endosperm and sporophytes, simplified gametophytes, and diversified nutrition and pollination patterns, which make them more adaptable than any other groups. Within the natural classification system of angiosperms, the Engler's system supports the dualistic origin and Pseudanthium Theory, while the system of Hutchinson, Takhtajan, and Cronquist system support the monophyletic and anthostrobilus theory. 56 families with important medicinal value and classification status were introduced in this textbook according to Engler's system.

题库

目 标 检 测
Question

1. 试解释说明被子植物是地球上最进化的植物。

Please explain that angiosperms are the most evolved plants on the planet.

2. 为什么说毛茛科与木兰科是被子植物中的原始类群？

Why are Ranunculaceae and Magnoliaceae primitive groups in angiosperms?

3. 为什么说禾本科是风媒传粉中最特化的类群？

Why is Gramineae the most specialized group of wind pollinators?

4. 为什么说兰科是被子植物中最进化的类群？

Why is Orchidaceae the most evolved group of angiosperms?

5. 哪些科具有叶对生、唇形花、二强雄蕊、二心皮等特征？如何区别这些科？

Which families have those features as leaves opposite, corolla labiatae, stamens didynamous and carpels 2? How to distinguish these families?

6. 蔷薇科、豆科、菊科、禾本科分别分为几个亚科？比较它们亚科的区别点。

How to distinguish Rosaceae, Leguminosae, Compositae and Gramineae into several subfamilies? Compare the differences between their subfamilies.

7. 比较毛茛科与木兰科、五加科与伞形科，唇形科与玄参科、马鞭草科、爵床科，茜草科和忍冬科，百合科与石蒜科、鸢尾科的异同点？

Compare the similarities and differences between Ranunculaceae and Magnoliaceae, Acanthaceae

医药大学堂
WWW.YIYAODXT.COM

and Umbelliferae, Labiatae, Scrophulariaceae, Verbenaceae and Acanthaceae, Rubiaceae and Caprifoliaceae, Liliaceae, Lycoridaceae and Iridaceae?

8. 利用所学的植物分类学知识，调查校园药用植物，进行鉴定和编目。

Please investigate, identify and catalog medicinal plants on campus with the knowledge of plant taxonomy that you've learned.

第三篇　药用植物的组织结构

Part III　Anatomy of Medicinal Plants

第十三章　植物的细胞
Chapter 13　Plant Cell

学习目标｜Learning goals

1．**掌握**　植物细胞的一般构造、细胞壁的结构和特化类型；植物细胞后含物的种类及特点。

2．**熟悉**　细胞生理活性物质的种类及特点。

3．**了解**　植物细胞分裂与增殖的方式。

● Master the general structure, cell wall structure and specialized types of plant cells; the types and characteristics of ergastic substances.

● Be familiar with the types and characteristics of physiologically active substances.

● Understand the cell division and proliferation.

第一节　植物细胞的结构
1　Cell structure

PPT

植物细胞是植物体形态结构与功能的基本单位。多数植物细胞较小，肉眼难以分辨，故需借助显微镜观察。在光学显微镜下观察到的细胞结构，称为显微结构，计量单位为微米（μm）；在电子显微镜下观察到的细胞结构更细微，称为亚显微结构或超微结构，其放大倍数是光学显微镜的近 1000 倍。

Cells are the basic structural and functional unit of the plants. Most plant cells are small and difficult to be distinguished by the naked eye. Therefore, to study the morphology and structure of plant cells, we need to observe them under a microscopes. The cell structures observed under an optical microscope are named as micro structure with the unit of measurement micrometer (μm); the cell structures observed under an electron microscope are more subtle, called submicroscopic or ultrastructural, and are magnified nearly a thousand times more than under an optical microscope.

各种植物细胞形状与构造各不相同，同一细胞在不同的发育阶段其结构也有变化，所以通常不可能在 1 个细胞中同时观察到细胞的全部构造。为了便于学习和掌握细胞的构造，将各种细胞结构都集中在 1 个细胞内加以说明，此细胞被称为典型细胞或模式细胞（图 13-1）。植物的各种细胞在形态、结构和功能上各不相同，但基本构造是相同的，均由原生质体、细胞壁和后含物 3

医药大学堂
WWW.YIYAODXT.COM

部分构成。

The shape and structure of different kinds of plant cells are different, even if the same cell in different stages of development, its structure will also change, as a result that is almost impossible to observe all the cell structures in a cell at the same time. To make it easier to describe the structure of cells, the book explains how the various cellular structures are incorporated into one plant cell which is called "model plant cell" (Fig. 13-1).

The microstructure of a typical plant cell consists of protoplast and cell wall. Protoplasts are the living materials in cells, including cytoplasm, nucleus and organelles. Protoplasm is surrounded by a tough layer of cell wall outside. In addition, there are also a variety of inanimate substances produced by protoplast metabolism in plant cells, which are collectively referred to as ergastic substance.

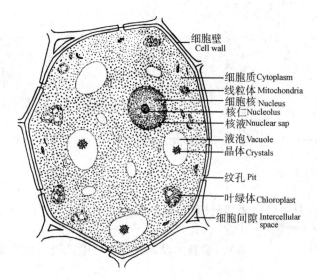

图 13-1　植物模式细胞图
Fig.13-1　A model plant cell

一、原生质体
1.1　Protoplast

原生质体是细胞内有生命物质的总称，是细胞内各种代谢活动的主要场所。在生活的植物细胞中，原生质体由于不断地进行代谢活动，进而分化形成形态、功能及组分不同的细胞质和细胞器等结构。

Protoplasts are a general term of living substances in a cell. All life activities of cells are performed by protoplasts which perform their metabolic activities constantly and further differentiate to form a variety of complex structure, including the cytoplasm and nucleus, plasmids, mitochondria, Golgi apparatus, the ribonucleoprotein bodies (hereinafter referred to as the ribosomes), lysosomes.

（一）细胞质
1.1.1　Cytoplasm

是充满在细胞壁和细胞核之间的半透明、半流动、无固定结构的基质，主要由细胞基质与质膜组成。在细胞质内分散着细胞核、质体、线粒体等细胞器和后含物。

Cytoplasm is a translucent, semi-flowing and non-fixed matrix between cell wall and nucleus. It is the basic component of protoplast, mainly composed of cell matrix and plasma membrane. In the cytoplasm, the nucleus, plastids, mitochondria are also dispersed.

1. 质膜　又称细胞膜，是包围在原生质体表面，由脂质、蛋白质和糖类组成的生物膜。在电子显微镜下，质膜显示出"暗 - 明 - 暗"的三层结构，中间明带主要成分是类脂，两侧暗带主要成分是蛋白质，这三层结构组成一个单位膜。磷脂双分子层构成质膜的骨架，磷脂分子的尾部相对藏在内面组成磷脂双分子层，使膜两侧的水溶性物质不能自由通过。蛋白质分布在膜内，外表面或嵌入磷脂双分子层内部，执行载体、受体、酶等功能。此外，质膜表面存在寡糖类分子构成的膜糖，大多数膜糖和蛋白质结合成糖蛋白或与脂质结合成糖脂，糖蛋白与细胞识别有关。

(1) The plasma membrane also known as the cell membrane, is a biofilm consisting of lipids, proteins and sugars that surrounds the protoplasts. Under an electron microscope, the plasma membrane shows a three-layer structure of "dark-bright-dark", with the middle bright-belt mainly composed of lipids and the two sides of the dark belt mainly composed of proteins. These three layers constitute a unit membrane. The phospholipid bilayer forms the skeleton of the plasma membranes, and the tails of the phospholipid molecules are relatively hidden in the inner surface to form the phospholipid bilayer, so that the water-soluble substances on both sides of the membrane cannot pass through freely. Proteins are distributed in the membrane, on the outer surface or embedded in the phospholipid bilayer, and perform many important functions of the membrane, including carrier, receptor, enzyme, etc. In addition, membrane carbohydrates composed of oligosaccharides exist on the surface of plasma membrane. Most membrane carbohydrates combine with proteins to form glycoproteins or with lipids to form glycolipids, which are related to cell recognition.

质膜主要有两种特性，一是半透性，是控制细胞内外物质运输交换的关键所在；二是选择性，通过由蛋白质或多肽形成的载体有选择性地转运某些物质出入细胞。因而其既能阻止细胞内的许多有机物由细胞内渗出，同时又能调节水分、无机物及其他营养物质进入细胞，并使废物排出。此外，质膜在外界信号的接受和传递、细胞生命活动的调节、抵御病菌的侵害等过程中具有重要的作用。

The plasma membrane has two main characteristics. One is semi-permeability, which is the key to control the transport and exchange of substances inside and outside the cell. The other is selectivity, the selective transport of certain substances into and out of cells by carriers formed by proteins or peptides. Thus, it cannot only prevent many organic matters inside the cell from exudating outside of the cell, but also regulate the water, inorganic matter and other nutrients into the cells, and make waste discharge. In addition, plasma membranes play an important role in the reception and transmission of external signals, the regulation of cells life activities, and the protection against pathogens.

2. 细胞基质 又称细胞溶胶，是除细胞器和后含物以外呈均质、半透明的胶状物质，是活细胞进行新陈代谢的主要场所，并为各类细胞器完成其功能活动提供了必要的原料。随着细胞的新陈代谢，细胞基质常常按照一定的方向，规则地持续地流动，称胞质运动，胞质运动可以促进细胞内营养物质的流动。

(2) The cell matrix also known as cytosol, is a homogeneous, translucent colloidal substance except for organelles and after-containers. It is the main place where living cells metabolize and provides the necessary raw materials for various organelles to complete their functional activities. With the metabolism of the cell, the cytomatrix often flows continuously and regularly in a certain direction, called cytoplasmic movement, which can promote the flow of nutrients in the cell.

（二）细胞器
1.1.2 Organelle
细胞器是细胞质内具有特定形态结构和功能的微器官，也称拟器官或亚结构。植物的细胞器一般包括细胞核、液泡、质体、线粒体、内质网、核糖体、高尔基体、溶酶体等。其中细胞核、液泡、质体、线粒体可以在光学显微镜下观察到，其余需要借助电子显微镜才能看到。

Organelle is a kind of micro-organ with special structure and function in cytoplasm, also called "organoid" or "sub-structure". The organelles of a plant generally include nuclei, vacuoles, plastids, mitochondria, endoplasmic reticula, ribosomes, Golgi apparatuses, lysosomes, microbodies, etc. The nuclei, vacuoles, plastids and mitochondria can be observed under optical microscope, while others which need to be seen by electron microscope.

1. 细胞核 除蓝藻和细菌外，大多数植物细胞（真核细胞）通常只有 1 个细胞核，但在乳汁管等一些特殊的细胞中也具有双核或多核。在幼小的细胞中，细胞核位于细胞中央，随着细胞的长大和中央液泡的形成，细胞核也随之被中央液泡挤压到细胞的一侧，呈扁圆形。

(1) Nucleus In the plant kingdom, prokaryotes such as cyanobacteria and bacteria are usually excluded. All other living cells in plants are eukaryotic cells with nuclei. Higher plants usually have only one cell nucleus, but in the milk tube and some other special cells also have binuclear or multinucleate, lower plant cells such as algae also have binucleate or multinucleate. In a young cell, the nucleus is located in the center of the cell, and as the cell grows and the central vacuole forms, the nucleus is pushed to the side of the cell by the central vacuole, which is oblate.

细胞核是遗传物质贮藏、复制和转录的场所，是细胞生命活动的控制中心。细胞核主要由核膜、核液、核仁、染色质（染色体）四部分构成。

Nuclei are the carriers of genetic information DNA storage, replication and transcription place, and their main function is to control the heredity and growth of cells, control and regulate other physiological activities of cells. The nucleus is mainly composed of nuclear membrane, nuclear fluid, nucleoli and chromatin (chromosome).

核膜是分隔细胞质与细胞核的薄膜。核膜为双层结构，由外膜和内膜组成。膜上具有许多可以开启和关闭的圆形小孔，为核孔，是细胞质与细胞核进行物质交换的通道；核孔的开闭控制细胞核与细胞质之间的物质交换，并调节细胞的代谢。

Nuclear membrane: The membrane on the surface of the nucleus that separates the cytoplasm from the nucleus. The nuclear membrane observed under the electron microscope is a double-layer structure composed of the outer membrane and the inner membrane. There are many round pores in the membrane that can be opened and closed. These are the nuclear pores, which are the channels for material exchange between cytoplasm and nucleus.

核液是细胞核内由纤维蛋白构成的网架体系，网孔中充满液体，其中分散有核仁和染色质。核液主要由蛋白质、RNA 以及多种酶组成，这些物质保证了 DNA 的复制和 RNA 的转录。

Nuclear sap: Liquid substance without obvious structure in the nucleus. Its main components are proteins, RNAs and a variety of enzymes. It guarantees the replication of DNAs and the transcription of RNAs.

核仁是细胞核中折光率更强的近圆形小球体，通常有一个或几个，主要由蛋白质和 RNA 组成，也有少量的类脂和 DNA；其大小随细胞生理状态不同而变化。核仁是核内 RNA 和蛋白质合成的主要场所，与核糖体的形成密切相关。

Nucleoli: A nearly circular globule with a higher refractive index in the nucleus, usually one or more, consisting mainly of proteins and RNAs, but also of small amounts of lipids and DNAs. Its size varies with the physiological state of the cell. Nucleoli are the main sites of RNAs and protein synthesis in the nucleus and are closely related to the formation of ribosomes.

染色质是细胞核内一种易被碱性染料着色的物质，主要由 DNA、蛋白质、少量 RNA 组成。在细胞分裂间期染色质不明显，或呈现交织成网状的细丝，即染色质网。当细胞核分裂时，染色质成为一些螺旋状的染色质丝，进而形成棒状的染色体。染色体是贮存、复制和传递遗传信息的主要物质基础，与植物的遗传密切相关。不同种植物染色体的数目、形状和大小各不相同，对某一物种来说则是相对稳定的，因此染色体是植物分类鉴定和亲缘关系研究的重要依据之一。

Chromatin: Basophilic substance in the nucleus that is easily colored by an alkaline dye (methyl green). It is granular, filamentous, or meshed in the nuclear fluid and consists mainly of DNAs, proteins, and a small amount of RNAs. During intermitotic periods, chromatin is inconspicuous or presents meshed filaments and chromatin reticulations. When the nucleus divides, chromatin forms spiral chromatin filaments that form rods of chromosomes. Chromatin and chromosomes are two forms of the same substance in cells at different times. Chromosomes are the main material basis for storing, duplicating and transmitting genetic information, which are closely related to the heredity of plants. The number, shape and size of plant chromosomes vary from species to species, which are relatively stable for a certain plant. Therefore, chromosomes are one of the important evidences for plant classification, identification and relationship research.

2. **质体** 质体是植物细胞特有的细胞器，与碳水化合物的合成和贮藏有密切关系。根据质体所含色素与功能的不同，分为叶绿体、有色体和白色体（图 13-2）。

(2) Plastid Plastids are organelles peculiar to plant cells and closely related to carbohydrate synthesis and storage. According to the pigment and function of plastids, plastids can be divided into pigment-containing type: chloroplasts, chromoplasts, and non-pigment-containing type: leucoplast(Fig.13-2).

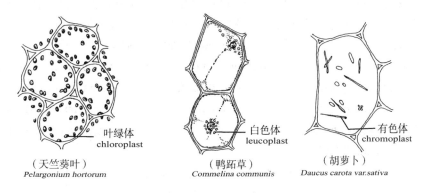

叶绿体 chloroplast	白色体 leucoplast	有色体 chromoplast
（天竺葵叶） *Pelargonium hortorum*	（鸭跖草） *Commelina communis*	（胡萝卜） *Daucus carota var.sativa*

图 13-2　质体类型

Fig.13-2　Three types of plastids

叶绿体是植物进行光合作用和合成淀粉的场所，常存在于植物的叶、幼茎、幼果中。光学显微镜下，高等植物的叶绿体多呈扁球形颗粒状。在电子显微镜下，叶绿体由双层膜围成，其内部是无色的溶胶状蛋白质基质，悬浮着复杂的膜系统；其中有扁平的囊，称类囊体；一些类囊体有规律地垛叠在一起，称为基粒，在基粒之间有基质片层相联系，各类囊体的腔彼此相通。类囊体的腔是 CO_2 转化成糖的场所，光合作用的色素和电子传递系统均位于类囊体膜上。叶绿体主要由蛋白质、类脂、RNA 和色素组成，含有叶绿素 a、叶绿素 b、胡萝卜素和叶黄素。叶绿素是主要的光合色素，能吸收并利用光能直接参与光合作用，其他两类色素不能直接参与光合作用，只能把吸收的光能传递给叶绿素，起辅助作用。

Chloroplast: A place for plants to photosynthesize and synthesize starch. Chloroplasts often exist in leaves, young stems and young fruits. Under the microscope, chloroplasts in higher plants are usually oblate and granular, and there can be dozens and dozens of chloroplasts in a cell. In electron microscopy, chloroplasts are surrounded by a double-layer membrane with a colorless sol-like protein matrix and a complex membrane system suspended inside. There are flat sacs called thylakoids; Some thylakoids are regularly stacked together, called granum. There are matrix lamellae between the granule, and the lumens of various types of thylakoids are connected with each other. The lumen of thylakoid is the place where

CO_2 is converted into carbohydrates, and the pigment and electron transfer system of photosynthesis are located on thylakoid membrane. Chloroplasts are mainly composed of proteins, lipids, RNAs and pigments. There are mainly four kinds of pigments, namely chlorophyll a, chlorophyll b, carotene and lutein. Chlorophyll is the main photosynthetic pigment, which can absorb and utilize light energy and directly participate in photosynthesis. The other two kinds of pigments cannot directly participate in photosynthesis, but can only transfer the absorbed light energy to chlorophyll and play an auxiliary role.

有色体常呈杆状、圆形、多角形，存在于花、果实和根中。主要含胡萝卜素和叶黄素等。由于色素在植物体中比例不同，使植物呈现黄色、橙色或红色。如在胡萝卜的根、金莲花、蒲公英的花瓣、番茄、红辣椒的果肉细胞中均可看到有色体。

Chromoplast: Often rod-shaped, round, polygonal, found in flowers, fruits, and roots. The pigments of chromoplasts are mainly carotene and lutein, etc. Due to the different proportion of pigments in the plant body, the plants appear yellow, orange or red. For example, chromoplasts can be found in the roots of carrot (*Daucus carota* var. sativa), flower of golden lotus (*Trollius chinensis*), petals of dandelion (*Taraxacum mongolicum*), pulp cells of tomato (*Lycopersicon esculentum*) and red pepper (*Capsicum annuum*).

白色体是不含色素的一类微小质体，呈圆形、椭圆形或颗粒状。主要存在于植物的分生组织、种子的幼胚及多数器官的无色部分。白色体与物质的积累贮藏有关，包括合成贮藏淀粉的造粉体、合成贮藏蛋白质的蛋白质体以及合成脂肪及脂肪油的造油体。

Leucoplast: A kind of micro plasmid without pigment, which is round, oval or granular. Leucoplasts mainly found in meristem of plants, young embryos of seeds and colorless parts of most organs. They are related to the accumulation of storage reserves, including the amyloplasts related with the synthesis of storage starch, the protein plasts related with the synthesis of storage protein, and the elaioplasts related with the synthesis of fat and fat oil.

叶绿体、有色体和白色体均由幼小细胞中的前质体分化而来，它们在一定条件下可以转化，如胡萝卜根经长期光照白色体转化为叶绿体；番茄幼果由绿变红，叶绿体逐渐转变成有色体。

The chloroplast, chromoplast and leucoplast are differentiated from the proplastids in the young cells in origin, and can be transformed under certain conditions. For instance, with the young tomato fruit turning from green to red, chloroplast can be gradually changing into chromoplast.

3. **线粒体** 线粒体是细胞内进行呼吸作用的场所，主要与细胞内的能量转换有关。在光学显微镜下，常为球状、棒状或细丝状颗粒。在电子显微镜下，为外膜、内膜、基质以及内膜在不同的部位向内折叠的形成许多管状或隔板状突起，称为嵴。嵴表面和基质中有多种与呼吸作用有关的酶和电子传递系统，其数量变化常可作为判断线粒体的活性和细胞活力的标志。此外，线粒体中还具有环状的双链 DNA 和核糖体，DNA 编码自身的蛋白质，称线粒体基因组。

(3) Mitochondria Mitochondria are the sites for respiration in cells and are mainly related to energy conversion in cells. Under an optical microscope, they are usually spherical, rod-shaped, or filamentous. Under the electron microscope, the mitochondria can be divided into outer membrane, inner membrane, matrix, and the inner membrane folds inward at different places to form many tubular or clap-like protuberances, namely cristae. There are a variety of enzymes and electron transfer systems related to respiration in the cristae surface and matrix. In addition, mitochondria have circular double-stranded DNA and ribosomes, which codes for its own proteins, called mitochondrial genomes.

4. **液泡** 是植物细胞特有的结构。在幼小的细胞中液泡不明显、体积小且数量多，随着细胞生长，液泡逐渐增大，彼此合并成几个大液泡，细胞成熟时往往形成一个中央大液泡，并将细胞质的其余部分连同细胞核挤向细胞的周边。液泡膜具有特殊的选择透性，同时还维持细胞的渗透压和膨胀压。液泡内的细胞液中除含有大量水分外，还有各种代谢物，如糖类、盐类、生物碱、苷类、有机酸等，这些成分往往是有效成分，具有重要的药用价值。

(4) Vacuole A unique structure of plant cells. In the young cells, vacuoles are not obvious, small in size and large in number. With the growth of cells, vacuoles gradually increase and merge into several large vacuoles. When the cells mature, they often form a central vacuole and push the rest of the cytoplasm and the nucleus to the periphery of the cells.

The vacuolar membrane separates the cell fluid within the membrane from the cytoplasm and has a special selective permeability, while maintaining the osmotic and expansion pressures of the cells. In addition, to containing a large amount of water, the cell fluid in vacuoles contains a variety of metabolites produced by cell metabolism as well, such as carbohydrates, salts, alkaloids, glycosides, organic acids, which are usually to be the effective components of plant drugs and have important medicinal value.

5. **内质网** 是细胞质内由膜组成的一系列管状和扁平囊状的腔彼此相通形成的管道系统，在电子显微镜下为两层平行的单位膜结构。内质网有两种类型，一种是粗面内质网，主要功能是合成、运输蛋白质，产生构成新膜的脂蛋白和初级溶酶体所含的酸性磷酸酶；另一种光面内质网，是脂类合成的场所。两种内质网可互相转化，也可同时存在于一个细胞内。

(5) Endoplasmic reticulum A series of tubular and flat saclike cavities connected by membranes in the cytoplasm. There are two types of endoplasmic reticula, one is a kind of rough endoplasmic reticulum, the main function of which is about protein synthesis and transport, producing new membrane lipoprotein and primary lysosome contains acid phosphatase; The other one is a smooth surfaced endoplasmic reticulum that is the site for lipid synthesis. The two endoplasmic reticula can be transformed to each other and can exist simultaneously in a cell as well.

6. **核糖核蛋白体** 核蛋白体简称为核糖体，近球形，游离在细胞质中或排列在粗糙型内质网上，是蛋白质合成的中心。

(6) Ribosome A subglobose body that is free in the cytoplasm or arranged on the rough endoplasmic reticulum and the synthesis center of protein.

7. **高尔基体** 是由一叠单位膜构成的扁平型的泡囊所组成，囊的边缘或多或少出现穿孔，当穿孔扩大时，整体呈现网状结构。在网状部分的外侧，形成小泡，小泡从高尔基体脱离后，游离到细胞基质中。高尔基体主要功能与合成多糖与运输多糖密切有关，并参与细胞壁的形成和生长。

(7) Golgi apparatus Composed of a stack of flat vesicles made up of unit membranes. The edges of the vesicles are more or less perforated. When the perforations expand, the whole network structure is presented. On the outer side of the network, vesicles form, which break off from the Golgi apparatus and migrate into the cell matrix. The main function of Golgi apparatus is closely related to the synthesis and transport of polysaccharides and is involved in the formation and growth of cell walls.

8. **微管** 是分布在细胞质中靠近膜的位置、中空而直的细管。微管除在细胞中起支架作用，参与细胞形态的维持外，也参与细胞壁的形成和生长，也与细胞的运动和细胞内细胞器的运动有密切关系，如构成植物游动细胞的纤毛或鞭毛，细胞分裂时使染色体运动的纺锤丝等。

(8) Microtubule A hollow, straight tube in the cytoplasm near the membrane. The main functions of microtubules are as follows: first, they may act as scaffolds in cells and participate in the maintenance of cell shape; second, they participate in the formation and growth of cell wall; third, they are closely related to the movement of cells and organelles within cells, such as the cilia or flagella that make up the swimming cells of plants, and the spindles that make chromosomes move during cell division.

9. 溶酶体 单层膜构成的圆球状小体，呈颗粒状分散在细胞质中；膜内充满多种水解酶，如蛋白酶、核糖核酸酶、磷酸酶、糖苷酶等。它们能催化蛋白质、多糖、脂质、DNA 和 RNA 等大分子分解，消化贮藏物质，分解细胞受损或失去功能的碎片。

(9) Lysosome A spherical body composed of a single-layer membrane, dispersed in the cytoplasm in a granular form; the membrane is filled with a variety of hydrolases, such as protease, ribonuclease, phosphatase, glycosidase, etc. They can catalyze the breakdown of macromolecules such as proteins, polysaccharides, lipids, DNA and RNA, digest storage materials, and break down fragments of damaged or lost cells.

二、细胞后含物和生理活性物质
1.2　Ergastic substances and physiologically active substances

（一）后含物
1.2.1　Ergastic substance

植物细胞在新陈代谢过程中产生的所有非生命物质统称为后含物。后含物有些是具有营养价值的贮藏物，如淀粉、蛋白质、脂肪油；有些是细胞代谢的废弃物质，如草酸钙结晶。后含物常以液体、晶体和非结晶固体形态存在于细胞质或液泡中，其种类、形态和性质往往与植物种类相关，因此在药用植物、药材鉴定中常作为显微鉴定的重要指标。常见的有以下几类。

All nonliving substances produced by plant cells during metabolism are collectively referred to as ergastic substances. Some of the ergastic substances are storage substance with nutritional value, such as starch, protein, and fatty oils. Some are wastes produced by cell metabolism, such as calcium oxalate crystals. The ergastic substance usually exists in the cytoplasm or vacuoles in the form of liquid, crystal and amorphous solid, and its species, morphology and properties are often related to plant species, so it is often used as an important index for microscopic identification of medicinal plants and Chinese herbal medicine. The ergastic substance of plant cells mainly includes starch, inulin, protein, fat or fat oil, crystal, etc.

1. 淀粉 淀粉是葡萄糖分子以 α-1,4 糖苷键聚合而成的长链化合物。通常绿色植物经光合作用所产生的葡萄糖，暂时在叶绿体内转变成同化淀粉，再被水解为葡萄糖转运至贮藏器官中，再在造粉体内重新形成贮藏淀粉。贮藏淀粉是以淀粉粒的形式贮藏在植物根、茎和种子等器官的薄壁细胞中。淀粉累积时，先从一处开始，这个形成淀粉粒的核心称为脐点，直链淀粉与支链淀粉相互交替，环绕脐点由内向外分层累积。由于两种淀粉在水中膨胀度的不同造成折光上的差异，在显微镜下可观察到围绕脐点有许多亮暗相间的层纹。如果用乙醇处理，使淀粉脱水，这种层纹即随之消失。

(1) Starch A long-chain compound in which glucose molecules are polymerized with α-1,4 glycosidic bonds.

Generally, the glucoses produced by photosynthesis in green plants is temporarily converted into assimilated starches in the chloroplasts, and then hydrolyzed to glucoses to be transported to the storage organs, and the storage starches are reformed in the powder body. Storage starches are stored in the form of starch granules in the thin-walled cells of plant roots, stems, and seeds. When starches accumulate, they start from one place. The core of the starch granules in this process is called "umbilical point". Then, amylose and amylopectin alternately surround the umbilical point and accumulate from the inside to the outside. Due to the difference in refractive index between the two starches in water, there are many light and dark striations around the umbilical point under the microscope. If treated with ethanol to dehydrate the starches, the striations disappear.

淀粉粒多呈圆球形、椭圆球形或多面体，脐点的形状有颗粒状、裂隙状、分叉状等。淀粉粒有单粒、复粒、半复粒三种。只有一个脐点的淀粉粒称为单粒淀粉；具有两个或以上脐点，每个脐点有各自层纹的称为复粒淀粉；具有两个或以上脐点，每个脐点除有本身的层纹环绕外，在外面另被有共同层纹的称为半复粒淀粉（图 13-3）。

Starch granules are mostly spherical, ellipsoidal or polyhedral, and the shape of the umbilical points is granular, fissured, or bifurcated. There are three types of starch granules, including single granules, compound granules, and semi-compound granules. The starch granule with only one umbilical point is called "single starch"; that with two or more umbilical points and each umbilical point have its own striae is called "compound starches"; with two or more umbilical points, every point has its own layered grains and surround all points on the outside, there are a common layered grain. This type is called "semi-compound starch" (Fig.13-3).

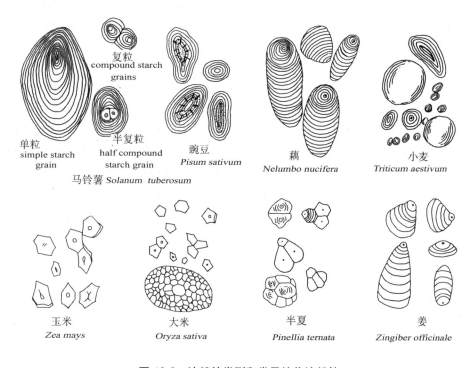

图 13-3　淀粉的类型和常见植物淀粉粒
Fig.13-3　Types of starch and common plant starch grains

淀粉粒不溶于水，在热水中膨胀而糊化，与酸或碱共煮则分解为葡萄糖，遇稀碘液显蓝紫色（直链淀粉显蓝色，支链淀粉则显紫红色）。

Starch granules, being insoluble in water, can swell and gelatinize in hot water, and decompose into glucoses while co-boiling with acid or alkali. Generally, the starch granules of plants contain two kinds of starches at the same time, and they appear blue-purple when dilute iodine solution (amylose is blue, amylopectin is purple-red).

2. 菊糖　菊糖是一种由果糖分子聚合而成的低聚糖，多分布于菊科、桔梗科、龙胆科部分植物根的薄壁细胞中以及山茱萸的果皮细胞中。菊糖能溶于水，不溶于乙醇。在光学显微镜下观察，菊糖常呈扇形、半圆形或圆形的结晶，加 10%α-萘酚的乙醇溶液，再加 80% 硫酸菊糖显紫红色，并很快溶解（图 13-4）。

(2) Inulin The oligosaccharide formed by the polymerization of fructose molecules.

Inulin is mostly distributed in parenchymal cells of the roots of some plants in Asteraceae, Platycodonaceae, and Gentianaceae, as well as in the peel cells of *Cornus officinalis*. Inulin is soluble in water and insoluble in ethanol. It can be loaded with absolute ethanol and observed under an optical microscope that inulin is often fan-shaped, semi-circular, or circular in crystals. When add a 10% solution of α-naphthol in ethanol firstly then add 80% inulin sulfate, the color of it turns purple-red, and quickly dissolves(Fig.13-4).

图 13-4　菊糖
Fig. 13-4　Inulin

3. 蛋白质　植物贮藏的蛋白质与构成原生质体的活性蛋白质完全不同，它是非活性的、较稳定的无生命物质，以结晶体或无定形的小颗粒存在于细胞质、液泡、细胞核和质体中。结晶的蛋白质因具有晶体和胶体二重性，称拟晶体。无定形蛋白质常被一层膜包裹成无定形小颗粒，称为糊粉粒。糊粉粒多存在于种子胚乳或子叶细胞中，在玉米、小麦等果实的胚乳最外一层或几层细胞中含有大量的蛋白质，这层细胞被称为糊粉层（图 13-5）。蛋白遇碘显棕色或黄棕色，加硫酸铜和苛性碱的水溶液显紫红色；加硝酸汞液显砖红色。

(3) Protein The storage proteins in plant cells are completely different from the active proteins that constitute the protoplasts. They are an inactive and relatively stable inanimate substance. They exist as crystalline or amorphous small particles in the cytoplasm, vacuoles, nucleus, and plastids. Crystallized proteins are called pseudo-crystals due to the duality of crystals and colloids to distinguish them from true crystals. Proteins are often wrapped into a small layer of amorphous particles, called aleurone grain. Some aleurone grains contain both pseudo-crystals and aleurone grains, called complex forms.

Aleurone grains are mostly distributed in the endosperm or cotyledon cells of the seed. The outermost layer or layers of cells in the endosperm of corns, wheats, and other fruits contain a large amount of proteins. This layer of cells is called "aleurone layer" (Fig.13-5).

In addition, aleurone and starch granules are often mixed with each other in a same cell.

When the proteins meet iodine, the color of them turn brown or yellow-brown, and the aqueous solution of copper sulfate and caustic is purple-red; the mercury nitrate solution is red-brick.

4. 脂肪和脂肪油　脂肪和脂肪油是由脂肪酸和甘油结合而成的脂。在常温下呈固态或半固态者为脂肪，如可可脂；呈液态者为脂肪油，如大豆油。脂肪和脂肪油在细胞质中常呈小滴状分

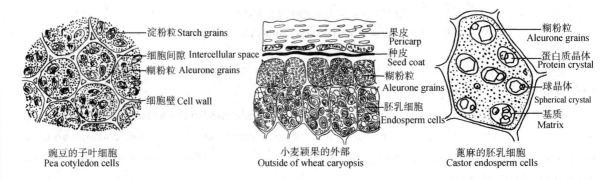

图 13-5　糊粉粒的类型
Fig.13-5　Types of the aleurone grain

散，比重小，折光率强，在植物的种子中含量极其丰富。脂肪和脂肪油遇苏丹Ⅲ溶液显橙红色、红色或紫红色；加紫草试液显紫红色。有的脂肪油可供药用，如油茶、月见草种子的脂肪油，牛蒡属果实的脂肪油等。

(4) Fat and fatty oil They are both fats made from a combination of fatty acid and glycerol. The solid or semi-solid ones at room temperature are fats, such as cocoa butter; the liquid ones are called fatty oils, such as soybean oil, peanut oil, and sesame oil. Fats and fatty oils are often dispersed in droplets in the cytoplasm, have a small specific gravity, high refractive index, and are extremely abundant in plant seeds, such as castor and rape. When added with Sudan Ⅲ, their color turn orange-red, red or purple-red and turn purple-red while adding solution of purple grass.

Fats are the most economical form of storage nutrients, and they are the storage materials with the highest energy and the smallest volume. Moreover, they can release more energy during oxidation. The starches stored in the thin-walled cells of some trunks can be converted into fats in winter, and then converted into starches in the following spring to store more energy. Some kinds of fatty oils are available for medicinal use, such as the seed oil of camellia (*Camellia oleifera*) or *Oenothera biennis*, and the fruit oil of Arctium.

5. 晶体 晶体主要存在于液泡中，常见草酸钙结晶和碳酸钙结晶。

(5) Crystal A waste deposit produced during the physiological metabolism of plant cells.

The crystals in the cells mainly accumulate in the vacuoles, and most of them are calcium salt crystals. There are two common types.

草酸钙结晶是植物体代谢过程中产生的草酸与钙盐结合而成的晶体，是植物细胞中最常见的晶体形式，常无色半透明或暗灰色。它的形成可以减少过量的草酸对细胞的毒害。其形状、大小在不同植物或同一植物的不同部位存在差别，也是药材的鉴定依据之一。它不溶于醋酸，可溶于10%~20%硫酸并形成硫酸钙针状晶体析出。常见有下几类（图13-6）。

Calcium oxalate crystal: A crystal formed by the combination of oxalic acid and calcium salts produced during the metabolism of plants(Fig.13-6).

Calcium oxalate crystals are the most common crystal form in plant cells and is usually colorless, translucent or dark gray. The shape and size of the calcium oxalate crystals are different in different plants or different parts of the consenting plant, and can usually be used as one of the identification evidences for medicinal plants and herbs. They are insoluble in acetic acid and soluble in 10% ~ 20% sulfuric acid and forms as calcium sulfate needle crystals. These crystals mainly come in the following forms.

方晶：又称单晶或块晶，通常呈正方形、菱形、长方形等单独存在于细胞内，如甘草根及根状茎、黄檗树皮。有的单晶交叉形成双晶，如莨菪叶。

Solitary crystal: Also called the single crystal or block crystal.

Solitary crystals are usually square, rhombic, rectangular, etc.. They exist in the cell alone, such as root and rhizome of *Glycyrrhiza uralensis*, bark of *Phellodendron amurense*. Some single crystals cross each other to form to double crystals, such as the ones in the leaves of *Hyoscyamus niger*.

针晶：晶体呈两端尖锐的针状，在细胞中多成束存在，称针晶束。常存在于黏液细胞中，如半夏块茎、黄精根状茎。有的针晶不规则地分散在细胞中，如苍术根状茎。

Acicular crystal: A crystal structure with needle-shaped at both ends.

Acicular crystals mostly exist in bundles in the cell. They are called raphides. They are often found in mucus cells, such as rhizomes of *Pinellia ternata* or *Polygonatum sibiricum*. Some needle crystals are irregularly dispersed in the cells, such as the rhizomes of *Atractylodes lancea*.

1.方晶（甘草根）　2.针晶（半夏块茎）　3.簇晶（人参根）　4.砂晶（牛膝根）5.柱晶（射干根）
1.Solitary crystals in the root of *Glycyrrihza sp.* 2. Acicular crystals in the tuber of *Pinellia ternata;*3. Cluster crystal in the root of *Panax ginseng;*4. Micro-crystals in the root of *Achyranthes bidentata;* 5. Columnar crystals in the root of *Belamcanda chinensis*

图 13-6　常见的草酸钙晶体
Fig.13-6　Types of calcium oxalate crystals

簇晶：由许多八面体、三棱形单晶联合成的复式结构，呈多角形，如大黄、人参根状茎。

Cluster crystal: A compound structure composed of many octahedral and triangular prisms, which is spherical, such as rhizomes of *Rheum palmatum* or *Panax ginseng*.

砂晶：呈细小的三角形、箭头状或不规则形，通常密集地分布在细胞腔中，如颠茄叶、牛膝根、枸杞根皮。

Crystal sand: A small triangle, arrow-shaped or irregular, usually densely distributed in the cell cavity. The cells with sand crystals are darker in color and can be easily distinguished from other cells, such as leaves of *Atropa belladonna*, roots of *Achyranthes bidentata*, and root barks of wolfberry (*Lycium chinense*).

柱晶：呈长柱形，长度通常为直径的 4 倍以上。如射干根状茎、淫羊藿叶片等。

Columnar crystal: With a long column shape, and the length is usually more than 4 times the diameter. Such as shoot dry rhizomes, Epimedium leaves.

碳酸钙结晶又称钟乳体，是细胞壁连接聚集大量的碳酸钙或少量的硅酸钙而形成，一端连接细胞壁，一端垂悬于细胞腔内，形状如一串悬垂的葡萄（图 13-7）。碳酸钙结晶多见于爵床科、桑科、荨麻科等植物的叶表皮细胞中，如穿心莲叶、无花果叶、大麻叶。碳酸钙结晶加醋酸或稀盐酸溶解，并产生二氧化碳气泡，据此可区别于草酸钙结晶。

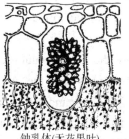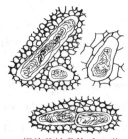

钟乳体(无花果叶)　　　　螺旋状钟乳体(穿心莲)
Cysyolith in the leaf of *Ficus carica*　Spiral cysyolith in *Andrographis paniculata*

图 13-7　碳酸钙结晶示意图
Fig.13-7　Calcium carbonate crystals

Calcium carbonate crystal: Also known as stalactite cystolith, it is formed by the cell wall connected with a large amount of calcium carbonates or a small amount of calcium silicates.

One end of it is connected to the cell wall and other end is suspended in the cell cavity, looking like a bunch of hanging grape sand often shows as a stalactite(Fig.13-7). Calcium carbonate crystals are mostly found in the epidermal cells of plants such as the family Acanthaceae, Moraceae, and Urticaceae, such as and leaves of *Andrographis paniculate*, *Ficus carica* and *Cannabis sativa*. Calcium carbonate crystals can be dissolved with acetic acid or dilute hydrochloric acid. Meanwhile, carbon dioxide bubbles are generated, which can be distinguished from calcium oxalate crystals.

除上述两种结晶外，一些植物还存在其他类型结晶，如柽柳叶含硫酸钙结晶，薄荷叶含橙皮苷结晶，菘蓝叶含靛蓝结晶，槐花含芸香苷结晶等。

Additionally, besides calcium oxalate crystals and calcium carbonate crystals in plants, there are other types of crystals for instance, calcium sulfate crystals in leaf cells of *Tamarix chinensis*, hesperidin crystals in leaf cells of mint (*Mentha haplocalyx*), indigo crystals in leaf cells of *Isatis indigotica*, and rutin crystals in cells of sophora flowers (*Sophora japonica*).

（二）生理活性物质

1.2.2　Physiologically active substances

生理活性物质是对植物细胞内的生化反应和生理活动起调节作用的一类活性成分的总称，包括酶、维生素、植物激素、抗生素等。其对植物的生长发育具有重要作用。

Physiologically active substances are a general term for a class of active ingredients that regulate the biochemical response and physiological activities in plant cells. Including enzymes, vitamins, plant hormones and antibiotics. They play an important role in the growth, development, metabolism and other life activities of plants.

1. **酶**　是具有催化功能的高分子物质，主要为复杂的蛋白质结构。酶具有高度专一性与高催化活性。生物体的新陈代谢、营养和能量转换等过程几乎都是在酶的催化作用下进行。

(1) Enzyme　A high-molecular substance with biocatalytic functions, and a complicated protein structure.

Enzymes have a high degree of specificity and only catalyze specific reactions or produce specific configurations. Almost all processes of the organism's metabolism, nutrition and energy conversion are carried out under the catalytic action of enzymes.

2. **维生素**　是一类复杂的有机化合物，常参与酶的形成，对植物的生长发育和物质代谢有调节作用，分为脂溶性和水溶性两大类。脂溶性维生素包括维生素 A、维生素 D、维生素 E、维生素 K 等；水溶性维生素包括维生素 B、维生素 C 等。维生素在植物中广泛分布，如酸枣、山楂的果实。维生素对人类疾病的治疗和预防有重要的作用，如夜盲症、皮肤类疾病等。

(2) Vitamin　A complex class of organic compounds that often participate in the formation of enzymes.

Vitamins have important regulatory effects on plant growth and material metabolism. There are dozens of vitamins that have been discovered, which can be divided into two categories, the fat-soluble and the water-soluble. Fat-soluble vitamins include vitamins A, D, E, K, etc.; water-soluble vitamins include vitamins B, C, and so on. Vitamins are widely distributed in plants, such as fruits of jujube (*Ziziphus jujuba* var. *spinosa*) and hawthorn (*Crataegus pinnatifida*) contain a large amount of vitamin C, and safflower (*Carthamus tinctorius*) oil is rich in vitamin E. Vitamins play an vital role in the treatment and prevention of some human diseases, like night blindness and skin diseases.

3. **植物激素**　又名植物内源激素，是指植物细胞接受特定环境信号诱导产生的，并从产生部位转移到作用部位，低浓度时可调节植物生理反应的活性小分子有机化合物。虽然含量微小，但在调控植物生长、发育与分化，影响植物发芽、生根、开花、结实、休眠和脱落等方面起着重要辅助作用。如生长素、细胞分裂素、脱落酸、乙烯等。

(3) Phytohormone　Also known as plant endogenous hormones who are is a molecule organic compound that are induced by plant cells by receiving specific environmental signals and transferred from the site of production to the site of action, and can regulate the activity of plant physiological responses at low concentrations.

Although the content phytohormones of is low, they play an important auxiliary role in regulating plant growth, development and differentiation, affecting plant life like germination, rooting, flowering, fruiting, dormancy and shedding. Currently known that plant hormones include auxins, cytokinins, abscisic acid (ABA), ethylene (ethylene, ETH), and the like.

三、细胞壁
1.3　Cell wall

细胞壁是植物细胞特有的结构，由原生质体分泌的非生活物质所构成，具有一定的硬度和弹性，对细胞起保护作用。细胞生长发育的不同时期和不同植物组织中，细胞壁的成分和结构也不尽相同。

Cell walls are a unique structure of plant cells, which are together with vacuoles and plastids constitute the structural features of plant cells that can be distinguished from animal cells. Cell walls can keep plant cells in a certain shape and protect the cells. Plant cells have inconsistent components and structures due to their ages and executive functions.

（一）细胞壁的分层
1.3.1　Stratification of cell walls

根据细胞壁形成的先后和化学成分的不同，分为胞间层，初生壁和次生壁（图 13-8）。

The structure of the cell wall is usually divided into three layers according to the order of formation and chemical composition: intercellular layer, primary wall and secondary wall (Fig.13-8).

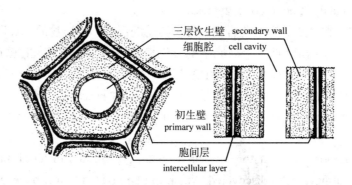

图 13-8　细胞壁结构示意图
Fig. 13-8　The structure of the cell wall

1. **胞间层**　又称中层，为细胞壁的最外面、连接相邻两个细胞的共用薄层。主要成分是果胶类物质，使相邻细胞粘连在一起。果胶易被果胶酶、酸等溶解，使细胞壁彼此分离而形成细胞间隙。实验室常用硝酸和氯酸钾混合液，将植物药材解离后进行观察鉴定。

(1) Intercellular layer　Also known as the middle lamellae which is the outermost layer of the cell wall and a common thin layer connecting two adjacent cells.

The main components intercellular layers are hydrophilic pectin-type substances, which can make adjacent cells stick together. Pectins are easily dissolved by pectinase, acid, etc., so that the cells are separated from each other. Tissue dissociation methods commonly used in the microscopic identification of medicinal materials and agricultural ramie processes are based on this principle. The former is leached with a mixed solution of nitric acid and potassium chlorate, and the latter is used to break down hemp fiber cells by using pectinase produced by microbe intercellular layer.

2. **初生壁**　是胞间层内侧，由原生质体分泌的纤维素、半纤维和果胶质构成。初生壁通常薄而柔软，并富有弹性，能不断地填充新的原生质体分泌物，使初生壁持续增长，这称为填充生长。许多植物细胞终生只具有初生壁。原生质体分泌物也可同时增加在已形成的初生壁的内侧，称为附加生长。

(2) Primary wall They are inner the intracellular layer side and composed of cellulose, hemicellulose and pectin secreted by protoplasts. The primary walls are usually thin, soft, and elastic, and can expand as the cells growing. With the extension of the primary walls, some new protoplast secretions are continuously filled to promote their growth. This phenomenon is referred to as "filling growth". Many plant cells only have one primary wall throughout its existence. Protoplast secretions can also increase on the inside of the formed primary walls at the same time, of which phenomenon is called "additional growth".

3. **次生壁** 是某些植物细胞壁停止生长后，在初生壁的内侧继续积累的细胞层。主要成分为纤维素、少量的半纤维素，常添加木质素或木栓质。次生壁较厚、质地较坚韧，有较强的机械支持能力。不是所有的细胞都有次生壁，大部分具有次生壁的细胞在成熟时，原生质体都已死亡。

(3) Secondary wall After some cell walls stop growing, some substances have been accumulating layer by layer on the inner wall of the primary wall that thicken the cell walls additionally to form a series of concentric layers which are named as "secondary walls". The main component of the secondary walls is cellulose, a small amount of hemicellulose, often accompanied with a small amount of lignin or auxin. At the same time, because the secondary walls are full of lignin and lacks pectin, the secondary walls are thicker, harder in texture, and have stronger mechanical support ability. It should be noted that not all cells have secondary walls, and most of the cells with secondary walls have dead protoplasts when they mature.

（二）纹孔和胞间连丝
1.3.2 Pit and plasmodesmata

次生壁在不均匀增厚过程中，留有没有增厚的孔状区域，称纹孔；相邻两细胞的纹孔常成对存在，称纹孔对。常见纹孔对有单纹孔、具缘纹孔和半缘纹孔三种（图 13-9）。

Pits are a pore-like space without thickening left by the secondary walls of the cell during the uneven thickening, usually in the form of a small nest or a thin tube. The pits of adjacent cell walls are often interconnected in pairs, called pit pairs. There are three types of common pits, namely single pits, bordered pits and half-edge pits (Fig.13-9).

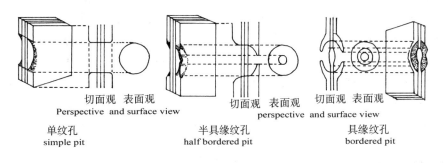

切面观 表面观
Perspective and surface view

切面观 表面观
perspective and surface view

切面观 表面观

单纹孔
simple pit

半具缘纹孔
half bordered pit

具缘纹孔
bordered pit

图 13-9 纹孔结构示意图
Fig. 13-9 The structure of a pit

1. **单纹孔** 单纹孔的细胞壁上未增厚的部分呈圆孔形或圆筒形，结构简单。纹孔对中间有初生壁和胞间层组成的纹孔膜。多存在于薄壁细胞、韧皮纤维和石细胞中。

(1) Simple pit The unthickened part of cell walls of the single pit has a round hole shape or a cylindrical shape, the structure is simple, and it looks like a circle in front view under an optical microscope. The pit pairs of the single pit have a pit membrane consisting of a primary wall and an intercellular layer in the middle. Single pits are mostly found in parenchyma cells, phloem fibers, and stone cells.

2. 具缘纹孔 纹孔边缘的次生壁向细胞腔内呈架拱状隆起，形成一个扁圆的纹孔腔，纹孔腔有一圆形或扁圆形的纹孔口。具缘纹孔正面观，呈 2 个同心圆，外圈是纹孔膜的边缘，中圈是纹孔口的边缘。松柏类裸子植物的管胞上，具缘纹的纹孔膜增厚形成纹孔塞，正面观呈 3 个同心圆。纹孔塞有活塞作用，当水流速度快时，水流压力会把纹孔塞推向一面，堵塞纹孔口起来，使上升水流减缓，可以调节胞间液流。

(2) Bordered pit The secondary walls at the edge of bordered pits bulge into the cell cavity to form an oblate pit cavity which has a round or oblate pit opening. The frontal view of the marginal pits of plants and angiosperms under the light microscope shows two concentric circles.

However, in the tracheids of coniferous gymnosperms, disc-shaped thickened structures are often formed on pit membranes, those are, pit plugs. The front view of the edged pit under the optical microscope is three concentric circles, the outer ring is the edge of the pit cavity, the middle ring is the edge of the pit plug, and the inner ring is the edge of the pit opening. Piston plugs have a piston effect on the edged pits. When the water flows quickly, the pressure of the water flow will push the pit plugs to one side, and the pit plugs will block the pit openings, making the rising water flow slow, so it can be adjusted Intercellular fluid flow rate.

3. 半缘纹孔 是管胞或导管与薄壁细胞间形成的纹孔。纹孔对的一边有架拱状隆起的纹孔缘，而另一边形似单纹孔，中间无纹孔塞。正面观为两个同心圆，外圈是纹孔腔的边缘，内圈是纹孔口的边缘。

(3) Half bordered pit A pit formed between tracheids or ducts and parenchymal cells. One side of a half-margin pit has an arch-shaped raised ridge edge, while the other is shaped like a single pit with no pit plug in the middle. The front view of the light microscope is two concentric circles, the outer ring is the edge of the pit cavity, and the inner ring is the edge of the pit opening.

4. 胞间连丝 是穿过胞间层和初生壁沟通相邻细胞的原生质丝。在电子显微镜下可见胞间连丝中有内质网连接相邻细胞的内质网系统，有利于细胞间特定物质运输和信号传递。有的植物胞间连丝明显，经染色后可在光学显微镜下观察，如柿、马钱子的胚乳细胞。

(4) Plasmodesma Plasmodesmata are microscopic channels that are connected with each other through the fine holes in the primary walls. Observed under an electron microscope, endoplasmic reticulum system can be found that is in the plasmodesmata and consist of endoplasmic reticula connecting with adjacent cells. This system is conducive to the transportation of specific substances between cells and signal transmission. Some plants have obvious plasmodesmata which can be observed under light microscope after staining, such as persimmon (*Diospyros kaki*) kernel, endosperm cells of *Strychnos nux-vomica*.

（三）胞壁的特化

1.3.3 Cell wall specialization

细胞壁的主要化学成分为纤维素。植物细胞壁由于环境的影响和生理机能的不同，细胞壁中会填充他物质，使其理化性质发生变化，称为细胞壁的特化。常见的特化方式有以下几类。

The main chemical component of cell walls is cellulose. However, due to the influence of the environment and different physiological functions, other substances are often deposited in cell walls to complete different physiological functions, so that their physical and chemical properties have undergone obvious special changes, which is called "cell wall specialization". Common specialization forms include lignification, suberization, cutinization, mucusification and mineralization of cell walls.

1. 木质化　细胞壁内填充和附加了木质素，使细胞壁硬度增强，细胞群的机械力增加，如管胞、导管、木纤维、石细胞等细胞的细胞壁。木质化细胞壁过于增厚，其细胞趋于衰老死亡。木质化的细胞壁加间苯三酚溶液和浓盐酸显红色。

(1) Lignification　A special phenomenon of cell walls that the protoplasts secrete more lignin into the cell walls during the process of cell metabolism.

The precipitation of lignin makes the cell walls firmer that greatly increases the supporting capacity of cells. Such as cell walls of tracheids, ducts, wood fibers, stone cells and so on When add one drop of resorcinol solution to the lignified cell walls, and add another drop of concentrated hydrochloric acid for a while, the color of cell walls will turn red.

2. 木栓化　细胞壁中填充了脂肪性化合物木栓质，使细胞壁不透水、不透气，最后细胞内原生质体消失，细胞死亡。木栓化细胞壁具有保护作用，如树干外面的粗皮就是由木栓化细胞组成的木栓组织。木栓化细胞壁遇苏丹Ⅲ试液染成橘红色。

(2) Suberization　A special phenomenon in which cell walls are filled with lipophilic xanthones, making the cell walls impermeable to water and air, and the protoplasts in the cells are gradually necrotic due to isolation from the surrounding environment.

The suberized cells have a protective effect on the internal tissues of the plant. For example, the rough skin outside the bark is composed of suberized cells. Suberized cell walls can be dyed in orange-red when encountering Sudan III test solution.

3. 角质化　是细胞次生壁中填充的脂肪性角质的特化现象，常在茎、叶或果实的表皮外侧形成一层薄角质层。角质化细胞可防止水分过度蒸发和病虫害的侵害。角质遇苏丹Ⅲ试液可被染成橘红色。

(3) Cutinization　A special phenomenon that the fatty keratin fills in the secondary walls, and often forms a thin cuticle on the outside of the epidermis of stems, leaves or fruits. Cutinized cells prevent excessive evaporation of water and pests and diseases. When cutin meets Sudan III test solution, it can be dyed in orange-red.

4. 黏液质化　是细胞壁中的纤维素和果胶质等变成黏液的特化现象。黏液质化所形成的黏液在细胞表面常呈固体状态，吸水则膨胀。如车前、亚麻种子的表皮。黏液质化的细胞壁遇玫红酸钠醇溶液染成玫瑰红色；遇钌红试剂可染成红色。

(4) Mucilagization　A special phenomenon in which the components such as cellulose and pectin in the cell wall change to form the mucus. The mucus formed by mucilagization is usually solid on the surface of cells, and becomes sticky after absorbing water. For example, epidermal cells of seeds of *Plantago asiatica*, *Linum usitatissimum*, and *Nicandra physalodes* have mucoid cells. The mucoid cell walls can be dyed rose red with sodium rosate alcohol solution; dyed red with ruthenium red reagent.

5. 矿质化　是细胞次生壁中渗入硅质或钙质的特化现象，其中以含硅质为最常见，如禾本科植物薏苡的茎、叶以及木贼茎均含大量硅酸盐。的矿质化能增强植物茎、叶的机械强度，提高抗倒伏和抗病虫害的能力。

PPT

(5) Mineralization A special phenomenon of infiltration of siliceous or calcareous deposits in the secondary walls of cells. For example, the stems, leaves and equisetum stems of *Equisetum hyemale* contain large amounts of silicate. The mineralization of the cell walls can enhance the mechanical strength of plant stems and leaves, and improve the ability to resist lodging and pests. Silicon is soluble in hydrogen fluoride, but has no change in the presence of sulfuric acid or acetic acid.

第二节　细胞的分裂与增殖
2 Division and proliferation of plant cells

单细胞植物生长到一定阶段，细胞一分为二，以此方式增殖。多细胞植物生长发育中，细胞分裂增加了细胞数目。植物的生长发育、生殖与细胞分裂密切相关。

When a single-celled plant grows to a certain stage, the cell is divided into two and proliferates in this way. Cell division increases the number of cells in the growth and development of multicellular plants. The growth, development and reproduction of plants are closely related to cell division.

一、细胞周期
2.1　Cell cycle

有丝分裂是真核细胞分裂的主要方式。连续分裂的细胞从一次有丝分裂结束到下一次有丝分裂结束所经历的整个过程叫做细胞周期。细胞周期常分为 4 个时期：DNA 合成前期（G₁ 期）、DNA 合成期（S 期）与 DNA 合成后期（G₂ 期）、有丝分裂期（M 期），其中前三个时期又合称分裂间期。

Mitotic (mitosis) is the main way of eukaryotic cell division. When cells divide, they go through an orderly series of events known as the cell cycle. The process that a continuously dividing cell undergoes from the end of one mitosis to the end of the next is called the cell cycle. This cycle is usually divided into interphase and mitosis, mitosis itself being subdivided into four phases. Interphase living cells that are not dividing are said to be in interphase, a period during which chromosomes are not visible with light microscopes. These intervals are designated as pre-DNA synthesis phase (or growth 1), DNA synthesis phase (S phase), and late DNA synthesis phase (growth 2), usually referred to as G1, S, and G2, respectively.

1. G₁ 期　从上一次有丝分裂结束到 DNA 复制前的一段时期，主要合成 RNA、蛋白质和酶。此外，还进行合成和复制膜系统和细胞器。

(1) The G₁ period The G₁ period is relatively lengthy and begins immediately after a nucleus has divided. During this period, the cell increases in size. The ribosomes, RNA, proteins and enzymes are produced. In addition, membrane systems and organelles are synthesized and replicated.

2. S 期　即核 DNA 复制开始到复制结束的时期，在此期，除了合成 DNA 外，同时还要合成组蛋白并转入细胞核，进行中心粒复制。

(2) The S period During the S period, the unique process of DNA replication (duplication) takes place. In this period, in addition to synthesizing DNA, histones are also synthesized and transferred into the nucleus for centrosomal replication.

3. **G_2期** G_2期为DNA复制结束到有丝分裂开始前的时期。中心粒复制完毕，形成两个中心体，还合成RNA和微管蛋白等。G_2期末两条染色单体已经形成。是有丝分裂的准备期。在这一时期，DNA合成终止，大量合成RNA及蛋白质，包括微管蛋白和促成熟因子等。

(3) The G_2 period In the G_2 period, mitochondria and other organelles divide, and microtubules and other substances directly involved in mitosis are produced. Coiling and condensation of chromosomes also begin during G_2. Two chromatids have been formed. It is the preparatory period for mitosis.

4. **M期** 即分裂期，细胞经过间期后进入分裂期，细胞中出现染色体、纺锤丝，已复制的DNA将以染色体的形式平均分配到子细胞中。细胞周期的运转十分有序，按照G_1-S-G_2-M的次序进行，这是细胞周期有关基因顺序表达的结果。

(4) M stage M stage: division stage, when the cells enter the division stage after interphase, chromosomes and spindle filaments appear in the cells, and the replicated DNA will be evenly distributed to the daughter cells in the form of chromosomes. The operation of the cell cycle is very orderly, according to the order of G_1-S-G_2-M, which is the result of the sequential expression of genes related to the cell cycle.

视频

二、细胞的分裂与增殖
2.2 Cell division and proliferation

细胞分裂是植物个体生长、发育和繁殖的基础，也是生命延续的前提。植物细胞的分裂主要有两方面的作用：一是增加细胞的数量，使植物生长；二是形成生殖细胞，用以繁衍后代。细胞的增殖是细胞分裂的结果。植物细胞的分裂通常有三种方式：有丝分裂、无丝分裂和减数分裂。

Cell division is not only the basis of plant individual growth, development and reproduction, but also the premise of the continuation of life. The division of plant cells mainly plays a role in two aspects: one is to increase the number of cells to make plants grow; the other is to form germ cells for reproduction. Cell proliferation is the result of cell division. Plant cells usually divide in three ways: mitosis, amitosis and meiosis.

1. **有丝分裂** 有丝分裂又称间接分裂，是高等植物和多数低等植物营养细胞的分裂方式，是细胞分裂中最普遍的一种方式，通过细胞分裂使植物生长。有丝分裂所产生的两个子细胞的染色体数目与体细胞的染色体数目一致，具有与母细胞相同的遗传性，保持了细胞遗传的稳定性。植物根尖和茎尖等生长特别旺盛的部位的分生区细胞、根和茎的形成层细胞的分裂就是有丝分裂。有丝分裂是一个连续而复杂的过程，包括细胞核分裂和细胞质分裂，通常人为地将有丝分裂分成分裂间期、前期、中期、后期和末期5个时期。

(1) Mitotic (mitosis) Mitotic, also known as indirect division, is the mode of division of vegetative cells in higher plants and most lower plants, and is the most common way of cell division, which makes plants grow through cell division. The chromosome numbers of the two daughter cells produced by mitosis are the same as those of somatic cells, have the same heredity as mother cells, and maintain the stability of cell inheritance. The division of meristematic zone cells and cambium cells of root and stem is mitosis in particularly vigorous parts such as root tip and stem tip. Mitosis is a continuous and complex process, including nuclear division and cytoplasmic division. Mitosis is usually artificially divided into interphase, Prophase, metaphase, anaphase and telophase.

2. 无丝分裂 无丝分裂又称直接分裂，细胞分裂过程较简单，分裂时细胞核不出现染色体和纺锤丝等一系列复杂的变化。无丝分裂的形式多种多样，有横缢式、芽生式、碎裂式、劈裂式等。最普通的形式是横缢式，细胞分裂时核仁先分裂为二，细胞核引长，中部内陷成"8"字形，状如哑铃，最后缢缩成两个核，在子核间又产生出新的细胞壁，将一个细胞的细胞核和细胞质分成两个部分。无丝分裂速度快，消耗能量小，但不能保证母细胞的遗传物质平均地分配到两个子细胞中去，从而影响了遗传的稳定性。

无丝分裂在低等植物中普遍存在，在高等植物中也较为常见，尤其是生长迅速的部位，如愈伤组织、薄壁组织、生长点、胚乳、花药的绒毡层细胞、表皮、不定芽、不定根、叶柄等处可见到细胞的无丝分裂。因此，对无丝分裂的生物学意义还有待深入研究。

(2) Amitosis Amitosis, also known as direct division, the process of cell division is relatively simple, and the nucleus does not have a series of complex changes such as chromosomes and spindles during division. There are many forms of amitosis, such as transverse constriction, budding, fragmentation, cleavage and so on. The most common form is transverse constriction. During cell division, the nucleolus first divides into two, the nucleus lengthens, the middle part sinks into a "8" shape, shaped like a dumbbell, and finally constricts into two nuclei, producing a new cell wall between the daughter nucleies. the nucleus and cytoplasm of a cell are divided into two parts. The speed of amitosis is fast and the energy consumption is small, but it can not guarantee that the genetic material of the mother cell is evenly distributed between the two daughter cells, thus affecting the genetic stability. Amitosis is common in lower plants and more common in higher plants, especially in fast-growing parts, such as callus, parenchyma, growth point, endosperm, tapetum of anther, epidermis, adventitious buds, adventitious roots, petioles and so on. Therefore, the biological significance of amitosis remains to be further studied.

3. 减数分裂 减数分裂与植物的有性生殖密切相关，只发生于植物的有性生殖产生配子的过程中。减数分裂包括两次连续进行的细胞分裂。在减数分裂中，细胞核进行染色体的复制和分裂，出现纺锤丝等，最终分裂形成4个子细胞，每个子细胞的染色体数只有母细胞的一半，成为单倍染色体（n），故称减数分裂。种子植物的精子和卵细胞由减数分裂形成，均为单倍体（n）。精子和卵细胞结合，恢复成为二倍体（2n），使得子代的染色体与亲代的染色体相同，不仅保证了遗传的稳定性，而且还保留父母双方的遗传物质而扩大变异，增强了适应性。在栽培育种上常利用减数分裂特性进行品种间杂交，以培育新品种。

(3) Meiosis Meiosis are closely related to plant sexual reproduction, which only occurs in the process of gamete production. Meiosis consists of two consecutive cell divisions. In meiosis, the nucleus replicates and divides into spindle filaments, and finally divides into four daughter cells. The number of chromosomes in each daughter cell is only half that of the mother cell, which becomes haploid chromosome (n), so it is called meiosis. The spermatozoa and eggs of seed plants are formed by meiosis and are haploid (n). The sperm combines with the egg cell and returns to diploid (2n), which makes the chromosome of the offspring the same as that of the parents, which not only ensures the genetic stability, but also preserves the genetic material of both parents and expands the variation and enhances the adaptability. In cultivation and breeding, the characteristics of meiosis are often used for inter-variety hybridization in order to cultivate new varieties.

4. 染色体、单倍体、二倍体、多倍体 染色体是在细胞进行有丝分裂和减数分裂时细胞核中出现的一种包含基因的伸长结构，由DNA和组蛋白组成。同种植物含有同样的染色体数，不同种植物所含有的染色体数和形态不同。染色体基数通常以X表示。因此，观察染色体的数目及形态特征，可为植物物种的鉴定和进化提供重要依据。

(4) Chromosomes, haploids, diploids and polyploids　Chromosome: Chromosome is an elongated structure containing genes in the nucleus during mitosis and meiosis, which is composed of DNA and histones. The same kind of plant contains the same chromosome number, but different species contain different chromosome number and morphology, and the chromosome base number is usually expressed by X. Therefore, the observation of the number and morphological characteristics of chromosomes could provide an important basis for the identification and evolution of plant species. The characteristics of centromere, chromosome arm, primary constriction and secondary constriction were observed. In a certain part of the chromosome there is an area called the centromere, which is connected to the traction filaments of the spindle. The centromere position of each chromosome is fixed. Centromeres located in the middle of chromosomes are called metacentric chromosomes (median, m); those near the middle are called submetacentric chromosomes (submedian, sm); those near one end are called submetacentric chromosomes (subterminal, st); and those located at one end are called telocentric chromosomes (terminal, t). The location of the centromere is an important sign to identify the types of chromosomes. With the centromere as the boundary, the chromosome is divided into two parts, which are called chromosome arms, and those with equal length are called equal arm chromosomes; those with different lengths are called long arms and short arms, respectively, and the part with lighter coloring and constriction between the two arms is called primary constriction; the other lighter colored constriction is called secondary constriction. Satellite means that the chromosome has a spherical or rod-shaped projection at the end of the short arm, and it is also an important feature to identify chromosomes. The study of the morphological structure of all chromosomes of a species, including the sum of the number, size, shape, primary constriction and secondary constriction of chromosomes, is called karyotype analysis or karyotype analysis. The absolute size of each chromosome in the karyotype is a fairly stable feature of the species. The absolute length of chromosomes and the relative length of two arms are the main methods to identify specific chromosomes in cells. The application of karyotype analysis in plant species classification is more important than the characteristics of chromosome number.

　　观察细胞分裂中期染色体，可见着丝点、染色体臂、主缢痕、次缢痕等特征。在染色体的一定部位有一个称为着丝点的区域，这个区域就是和纺锤体的牵引丝相连的部位。每种染色体的着丝点位置是一定的。着丝点位于染色体中部的，称中部着丝点染色体（median, m）；近于中部的，称亚中着丝点染色体（submedian, sm）；近于一端的称亚端着丝点染色体（subterminal, st）；位于一端的，称端部着丝点染色体（terminal, t）。着丝点的位置是识别染色体种类的一个重要标志。染色体以着丝点为界分成两个部分，称为染色体臂，两臂长度相等的称等臂染色体；长度不等的则分别称长臂和短臂，两臂间着色较浅而缢缩的部分，称主缢痕；另一着色较浅的缢缩部分，称次缢痕。随体是指染色体在短臂的末端还有一个球形或棒形的突出物，随体也是识别染色体的一个重要特征。

　　研究一个物种的全部染色体的形态结构，包括染色体的数目、大小、形状、主缢痕和次缢痕等特征的总和，称为染色体组型分析或染色体核型分析。染色体组型中的各染色体的绝对大小是物种的一个相当稳定的特征。染色体的绝对长度和两臂的相对长度是识别细胞中特定染色体的主要方法。染色体组型分析应用于植物种级分类，要比染色体数目的特征更为重要。

　　细胞内的染色体成组存在，一组的染色体之间的形状各不相同，不能配对。细胞内仅含一组染色体的个体称为单倍体（用 n 表示）。经过减数分裂产生的精子和卵细胞的染色体数均为单倍。如菘蓝 *Isatis indigotica* Fort. 单倍体植株体细胞中的染色体是 7 个，即 n=X=7。

Haploid: The chromosomes in haploid cells exist in groups, and one group of chromosomes have different shapes and cannot be paired. An individual with only one set of chromosomes in a cell is called a haploid (represented by n). The chromosome numbers of sperm and eggs produced by meiosis are haploid. Such as Isatis indigotica Fort. There are 7 chromosomes in the somatic cells of haploid plants, namely n=X=7.

细胞内含有两组染色体的个体称为二倍体（用 2n 表示）。减数分裂前的细胞或由两性生殖细胞结合后发育产生的营养体细胞，染色体数目为二倍的，即含有两组染色体即二倍体。如水稻植株体细胞有 24 个染色体为二倍体，即 2n=2X=24，而单倍体细胞内仅有 12 个染色体，即 n=X=12。菘蓝二倍体植株的体细胞有 14 个染色体，即 2n=2X=14。

Diploid: An individuals with two sets of chromosomes in a cell is called diploids (represented by 2n). The cells before meiosis or the vegetative cells produced by the combination of bisexual germ cells have twice the number of chromosomes, that is, they contain two sets of chromosomes, namely diploids. For example, 24 chromosomes in somatic cells of rice are diploid, that is, 2n=2X=24, while there are only 12 chromosomes in haploid cells, namely n=X=12. The somatic cells of diploid plants of Isatis indigotica have 14 chromosomes, namely 2n=2X=14.

细胞内含有三组以上染色体的个体称为多倍体。多倍体广泛存在于植物界中，被子植物中大约有一半是多倍体。当植物细胞进行分裂时，受到温度、湿度、紫外线等的频繁刺激，细胞核中的染色体数目可能会加倍，这样的细胞继续繁殖分化，就能形成多倍体植物。这种受自然条件刺激所形成的多倍体植物称自然多倍体植物。自然多倍体植物长期以来被栽培利用，如三倍体的香蕉；四倍体的马铃薯；六倍体的小麦。延胡索属植物也存在多倍体化系列，例如全叶延胡索 *Corydalis repens* Mandl. et Mühld、齿瓣延胡索 *C. remota* Fisch. ex Maxim. 为二倍体，即 2n=2X=16；延胡索 *C. yanghusuo* W. T. Wang、夏天无 *C. decumbens*（Thunb.）Pers. 为四倍体，即 2n=4X=32；圆齿延胡索 *C. remota* Fisch. ex Maxim. var. *rotundiloba* Maxim. 为六倍体，即 2n=6X=48。

Polyploid: An individuals with more than three sets of chromosomes in a cell is called polyploids. Polyploidy exists widely in the plant kingdom, and about half of angiosperms are polyploid. When plant cells divide, the number of chromosomes in the nucleus may double when they are frequently stimulated by temperature, humidity and ultraviolet rays. Such cells continue to multiply and differentiate, and polyploid plants can be formed. This kind of polyploid plant which is stimulated by natural conditions is called natural polyploid plant. Natural polyploid plants have been cultivated for a long time, such as triploid bananas, tetraploid potatoes and hexaploid wheat. Polyploidy series also exist in the genus Corydalis, such as C. repens Mandl. et M ü hld, C. remota Fisch. Ex Maxim. There are also polyploid series in the genus Corydalis. It is diploid, that is, 2n=2X=16; C. yanghusuo W. T. Wang, C.decumbens (Thunb.) Pers. It is tetraploid, that is, 2n=4X=32, C. remota Fisch. ex Maxim. var. rotundiloba Maxima is hexaploid, namely 2n=6X=48.

人们为了获得优良性状的植物，在细胞分裂时，利用物理刺激（紫外线、X 射线照射、高温、低温处理等）或化学药物（秋水仙素等）处理的方法，诱导植物产生多倍体，称人工多倍体植物。药用植物多倍体育种已经取得了不少成绩，如菘蓝的四倍体 2n=4X=28 新品系与二倍体相比，叶中靛蓝的含量在收获期可成倍增加，靛玉红含量也有显著提高。牛膝 *Achyranthes bidentata* Bl. 的多倍体和二倍体相比，根肥大，木质化程度轻，产量高。此外，胡椒薄荷 *Mentha piperita* L. 的多倍体（2n=144）品系，不但挥发油含量高，而且抗旱、耐寒、抗病力强。毛曼陀罗 *Datura innoxia* Mill. 的三倍体平均生物碱的得率含量为二倍体的 4 倍，为四倍体的 3 倍。多倍体品系产量一般较高，品质较好。

In order to obtain plants with good characters, during cell division, plants are induced to produce polyploids by means of physical stimulation (ultraviolet, X-ray irradiation, high temperature, low temperature treatment, etc.) or chemical treatment (colchicine, etc.). It's called artificial polyploid plant. Many achievements have been made in polyploid breeding of medicinal plants. For example, compared with diploid, the content of indigo in leaves of the new tetraploid 2n=4X=28 of Isatis indigotica can be doubled at harvest time, and the content of indirubin is also significantly increased. Compared with diploid, the root in polyploid of Achyranthes bidentata Bl. is hypertrophy, the degree of lignification is light, and the yield is high. In addition, the polyploid (2n=144) strain of Mentha piperita L. not only has high volatile oil content, but also has strong drought resistance, cold resistance and disease resistance. The average alkaloid yield of triploid of Datura innoxia Mill. was 4 times higher than that of diploid and 3 times that of tetraploid. Polyploid lines generally have higher yield and better quality.

重 点 小 结
Summaries

植物细胞特有的结构有细胞壁、液泡和质体。植物液泡中储存的细胞后含物与植物次生代谢产物往往与药用植物、药材的鉴定和活性成分有关；植物细胞壁的特化是细胞长期进化过程中与功能相适应的结果，有助于了解植物不同部位细胞的发育。细胞周期常分为 4 个时期：DNA 合成前期（G₁ 期）、DNA 合成期（S 期）与 DNA 合成后期（G₂ 期）、有丝分裂期（M 期），其中前三个时期又合称分裂间期。根据细胞内含有的染色体组数分为单倍体、二倍体和多倍体等。

The unique structures of plant cells include cell walls, vacuoles, and plastids. The ergastic substances and secondary plant metabolites stored in plant vacuoles are often related to the identification and active substances of medicinal plants or medicinal materials; the cell wall specialization is a result of adaptation to function during the long-term evolution of cells which is helpful to understand the development of cells in different parts of the plant. The cycle is usually divided into interphase and mitosis, mitosis itself being subdivided into four phases. These intervals are designated as G_1, S and G_2, respectively. It can be divided into haploid, diploid and polyploid according to the number of chromosome groups contained in the cell.

目 标 检 测
Questions

1. 细胞单层膜、双层膜结构分别有哪些？

What are the structures with monolayer or bilayer cell membrane?

2. 细胞后含物主要存在于细胞的哪个部位？有哪些类型？与生理活性物质的区别是什么？

In which part of the cell the ergastic substances can be found? What types are they? What are the differences between physiologically active substances and them?

3. 细胞纹孔的类型有哪些？

What are the types of cell pits?

题库

4. 细胞壁特化的定义与类型分别是什么？

What are the definition and type of cell wall specialization?

5. 细胞周期的定义是什么？常分为哪 4 个时期？

What is the definition of cell cycle? What are the four periods usually be divided into?

第十四章 植物的组织
Chapter 14 Plant Tissue

许多来源相同、形态结构相似、机能相同而又紧密联系的细胞所组成的细胞群，称为组织。仅有一种细胞类型的组织称简单组织，由多种类型细胞构成的组织称复合组织。低等植物无组织分化，高等植物开始出现组织分化，植物进化程度越高，组织分化越明显。不同组织有机结合、相互协调有紧密联系，构成植物的器官。

A group of closely related cells of the same origin, similar morphology, structure and function, is called tissue. Tissue with only one cell type is called simple tissue, and tissue with multiple cell types is called composite tissue. The lower the level of plant tissue differentiation, the more vascular plant the plant tissue differentiation, the more obvious. Different tissues are closely related to each other and form the organs of plants.

第一节 组织的类型
1 Type of plant tissue

PPT

根据发育程度、形态结构和生理功能不同，将植物组织分为分生组织和成熟组织。成熟组织由分生组织的细胞分裂和分化所形成，根据形态结构和功能又可分为薄壁组织、保护组织、机械组织、输导组织和分泌组织。同一类型的组织在不同植物中常具不同的构造特征，可用于药材、饮片和某些中成药的鉴定。

Plant tissues are divided into meristematic tissue and mature tissue according to their development, morphological structure and physiological functions. Mature tissue is formed by cell division and differentiation of meristematic tissue, and can be divided into parenchyma, protective tissue, mechanical tissue, conducting tissue and secretory tissue according to morphology and function of the cell. The same type of tissue often has different structural characteristics in different plants, which can be used for the identification of Chinese herbal medicine, Processed materia medica and some Chinese patent medicines.

一、分生组织
1.1 Meristem

　　植物体中具有分裂能力的细胞群称分生组织。分生组织细胞一般体积较小，排列紧密，无胞间隙，仅具初生壁，细胞质浓，细胞核大；原生质分化程度低，尚无明显液泡和质体分化。分生组织的细胞具有旺盛的分生能力，这些细胞不断分生新细胞，其中一部分连续保持高度的分裂能力，另一部分经过分化，形成不同的成熟组织，使植物体不断生长。

The group of cells in a plant body that is capable of division is called meristematic tissue. In general, meristematic cells are small in size, closely arranged, without intercellular space, with only primary wall, dense cytoplasm and large nucleus, with low degree of protoplasm differentiation and no obvious vacuole and plastid differentiation. The cells of meristematic tissue have a vigorous ability to divide. These cells are constantly producing new cells. Some of these cells maintain a high degree of continuous division, while others are differentiated to form different mature tissues that allow the plant to grow.

视频

　　1. 按其来源和性质分类　原分生组织源于种子的胚，具旺盛的分裂能力；位于根、茎生长点最先端，是其他组织的最初来源。

　　(1) Classification by origin and nature　**Promeristem:** The promeristem tissue is derived from the embryo of the seed with vigorous division and is located at the apex of the root and stem and the primary origin of other tissues.

　　初生分生组织由原分生组织分裂衍生而来的细胞组成，位于原分生组织之后，这部分细胞一方面仍保持分裂能力，另一方面已开始分化。如茎的初生分生组织可分化为原表皮层、基本分生组织和原形成层。由这 3 种初生分生组织再进一步分化发育形成其他各种组织构造。

Primary meristem: Any of a group of cells derived from the division of promeristem, after which the cells retain the ability to divide and begin to differentiate. For example, the primary meristems of stem can be differentiated into protoderm, ground meristems and procambium, from which the three primary meristems can further differentiate and develop into other structures.

　　次生分生组织是由已经分化成熟的薄壁组织细胞（如皮层、中柱鞘、髓射线等）重新恢复分生能力而形成的组织。常与轴向平行排列成环状，但不是所有植物都有次生分生组织。其与裸子植物和双子叶植物根、茎的增粗和次生保护组织的形成有关，其组织活动的结果是形成植物的次生构造，即次生保护组织和次生维管组织，使根、茎不断增粗。

Secondary meristem: Secondary meristem is a meristem formed by the regeneration of mature parenchyma cells (such as cortex, pericycle, pith rays, etc.). Often arranged in rings parallel to the axis, but not all plants have secondary meristem. They are related to the thickening of the roots and stems of gymnosperms and Dicotyledones plants and the formation of secondary protective tissues, the result of which is the formation of the secondary structures of the plants, that is, the secondary protective tissues and the secondary vascular tissues, make root, stem continuously coarser.

2. 按分生组织所处的位置分类　顶端分生组织又称生长锥，是位于根尖、茎端的分生组织，包括原分生组织和初生分生组织，二者之间无明显分界线。这部分细胞能较长期保持旺盛的分生能力，使根或茎不断伸长生长。

(2) By the position of the meristematic tissues　**Apical meristem:** Apical meristems, also known as growth cones, are meristems located at the root tip and end of the stem, including promeristem and primary meristems, with no obvious dividing line between the two. The cells in this part can maintain the exuberant meristematic ability for a longer period of time, which make the root or the stem grow continuously elongate.

侧生分生组织来源于成熟组织脱分化恢复分裂能力产生的次生分生组织，主要存在于裸子植物和双子叶植物的根或茎中，包括维管形成层和木栓形成层。细胞多呈长纺锤形，液泡较发达，细胞与器官长轴平行，细胞分裂方向与器官长轴方向垂直，其活动使根或茎不断增粗。

Lateral meristem: Lateral meristem is a secondary meristem derived from mature tissues and found mainly in the roots or stems of gymnosperms and dicotyledonous plants, including the vascular cambium and cork cambium. The cells are long spindle shaped, vacuole is more developed, the cells are parallel to the organ long axis; The direction of cell division is perpendicular to the organ's long axis, which causes the root or stem to grow continuously coarser.

居间分生组织是从顶端分生组织保留的一部分分生组织，属初生分生组织，仅保持一定时间的分生能力，以后则完全转变为成熟组织。常位于茎、叶、子房柄、花柄等的成熟组织之间。如薏苡、水稻、小麦等禾本科植物的拔节或抽穗，韭菜叶子上部被割后还能长出新的叶片，都是居间分生组织作用的结果。

Intercalary meristem: An intermediate meristem is a part of a meristem that is retained from the apical meristem. It remains its ability to divide for a certain period of time before it is fully transformed into a mature tissue. It usually lies between the mature tissues of a stem, leaf, ovary stalk, flower stalk, etc. For example, *Coix lacryma-jobi*, *Oryza sativa*, *Triticum aestivum* and other gramineous plants have internode meristem at the lower part of the internode. When the top of the meristem is differentiated into young ear, it can still use the mitotic activity of intercalary meristem to jointing or heading, and can also make the lodging stalk gradually regain its upright position. With the upper part of the leaf of *Allium fistulosum* cut, new leaves can grow out, which is the result of the action of the intermediate meristems.

二、薄壁组织
1.2　Parenchyma

薄壁组织又称基本组织，在植物体中分布最广，是植物体的重要的组成部分，担负着吸收、同化、贮藏、通气、传递等功能。

Parenchyma, also called ground tissue, is widely distributed in plants and is an important part of plants. It is responsible for absorbing, assimilating, storing, ventilating and transmitting.

薄壁组织细胞的细胞壁薄，只有初生壁而无次生壁；排列疏松，细胞间隙发达，具有分生潜力。一定条件下经脱分化可转变成分生组织或分化成其它组织。按其生理功能和细胞结构不同可分为基本薄壁组织、吸收薄壁组织、同化薄壁组织、贮藏薄壁组织、通气薄壁组织等（图 14-1）。

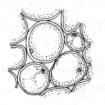  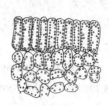 

基本薄壁组织	通气薄壁组织	同化薄壁组织	贮藏薄壁组织
Ordinary parenchyma	Aerenchyma	Assimilation parenchyma	Storage parenchyma

图 14-1　薄壁组织的类型
Fig.14-1　Types of the parenchyma

The parenchyma cells have thin cell wall, only primary wall but no secondary wall, loose arrangement, well-developed intercellular space, and have potential for regeneration. Under certain conditions, it can be transformed into meristematic tissue or other tissue by dedifferentiation(Fig.14-1).

1. **基本薄壁组织**　指植物体内主要起填充和联系其它组织作用的薄壁组织，也称填充薄壁组织。横切面观细胞呈类球形或多角形。一定条件下可转化为次生分生组织，如根、茎的皮层和髓部的薄壁组织。

(1) Ordinary parenchyma　The ordinary parenchyma refers to the tissue in a plant that primarily fills and connects with other tissues. Also called filling parenchyma. The cells were spherical or polygonal in cross section. Under certain conditions, it can be transformed into secondary meristem, such as cortex of root, stem and parenchyma of pith.

2. **吸收薄壁组织**　指位于根尖根毛区的表皮细胞和由其外壁向外延伸形成的管状结构——根毛，主要起吸收水分和无机养分的作用；根毛的存在增加了与土壤接触面积从而增大吸收面积。

(2) Absorbtive parenchyma　Refers to the epidermal cells located in the root tip and the tubular structure formed by the outer wall extending outward-root hair, which is mainly responsible for absorbing water and substances dissolved in water. The presence of root hairs increased the contact area with the soil and thus increased the absorption area.

3. **同化薄壁组织**　指能进行光合作用的薄壁组织，常见于植株绿色部分，又称绿色薄壁组织。细胞含大量的叶绿体，液泡化程度较高。如茎的幼嫩部分、绿色萼片及果皮，尤以叶肉的薄壁组织最典型。

(3) Assimilation parenchyma　Refers to the photosynthetic parenchyma, commonly found in the green part of the plant, also known as the green parenchyma. The cells contain a large number of chloroplasts and have a high degree of vacuolation. Such as the young part of the stem, green sepals and pericarp, especially the leaf parenchyma is most typical.

4. **贮藏薄壁组织**　指贮藏积聚有营养物质的薄壁组织，常见于根和茎的皮层、髓部，以及块根、地下茎、果实和种子的胚乳或子叶等器官。细胞体积较大，贮藏质主要有淀粉、蛋白质、脂肪、油和糖类等。芦荟、龙舌兰、仙人掌、景天等植物有大量的贮水薄壁组织，也属于贮藏薄壁组织。

(4) Storage parenchyma　Refers to the parenchyma that stores nutrients. It is found in the cortex and pith of roots and stems, as well as in the endosperm or cotyledons of roots, rhizomes, fruits and seeds. The cell volume is bigger, the storage quality mainly has the starch, the protein, the fat, the oil and the sugar and so on. *Aloe vera*, *Agave Americana*, *Opuntia stricta var. dillenii* and *Sedum erythrostictum* have a large number of water storage parenchyma, termed aqueous parenchyma.

　　5. **通气薄壁组织**　指水生和沼生植物体内的薄壁组织。细胞间隙发达，彼此联接形成网节状的气腔或气道，在体内形成一个发达的通气系统。该构造利于细胞呼吸和气体交换，也有利于植物体的漂浮。如莲的叶柄和根状茎、灯芯草茎髓的薄壁组织。

　　(5) Aerenchyma　Refers to the well-developed intercellular spaces of parenchyma in aquatic and marsh plants, which are connected with each other to form a netlike air cavity or airway, forming a well-developed ventilatory system in the body. This structure is conducive to cellular respiration and gas exchange in plants, and it is also effective in resisting the mechanical stress faced by plants in the water environment, which is conducive to the floating of plants. Such as the parenchyma of the petiole and rhizome of the *Nelumbo Adans*, and stem pith of *Juncus effuses*.

三、保护组织
1.3　Protective tissue

　　保护组织指包被在植物各器官表面，由一层或数层细胞组成，担负保护作用的组织。其能防止水分过度蒸腾，控制气体交换，抵御病虫侵害和机械损伤。按其来源和结构分为表皮和周皮。

　　Protective tissue is a kind of tissue which is covered by one or several layers of cells on the surface of every plant organ. It can prevent excessive water transpiration, control gas exchange, resist pests and diseases and mechanical damage. According to its origin and structure, it is divided into epidermis and periderm.

　　（一）表皮
1.3.1　Epidermis
　　表皮由原表皮分化而来，位于幼茎、叶、花和果实表面，由表皮细胞、气孔器、毛茸等组成，其主体是表皮细胞。表皮细胞常为一层生活细胞，也有少数植物的表皮由 2~3 层生活细胞组成，称复表皮，如夹竹桃、印度橡胶树的叶。

　　The epidermis is differentiated from the protoepidermis and located on the surface of young stems, leaves, flowers and fruits. The epidermis is often a layer of living cells. There are also a few plants in which the epidermis of certain organs consists of two or three layers of living cells, called polyepidermis, such as the leaves of *Nerium indicum* and *Ficus elasica* Roxb. *var. elastica*.

　　1. **表皮细胞**　是生活细胞，常呈扁平而不规则，镶嵌排列紧密，无胞间隙，液泡发达，不含叶绿体。横切面观，多呈长方形或方形，内壁较薄，外壁较厚，角质化并在表面形成角质层。有些植物（如冬瓜、葡萄等）在角质层外还有一层蜡被，使表面不易浸湿，利于防止真菌孢子萌发（图 14-2）。有些表皮细胞壁矿质化，增强机械支持作用，如木贼茎和禾本科植物叶。

　　(1) Epidermal cells　They are living cells, often flat and irregular, inlaid closely, with no intercellular space, developed vacuoles, or chloroplasts. The epidermal cells grow rectangular or square, the inner wall is thinner, the outer wall is thicker, keratinized and cuticle formed on the surface. Some plants (such as winte melon, grape, etc.) have a layer of wax outside of the cuticle, which makes the surface not easily wet and helps prevent fungal spore germination (Fig.14-2). Some epidermal cell walls are mineralized to enhance mechanical support, such as the stems of *Equisetum hyemale* and leaves of gramineae plants.

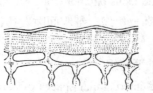
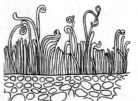

表皮及角质层　　　　　　　　表皮上的杆状蜡被
Epidermis and cuticle　　Rod-shaped wax quilt on the epidermis

图 14-2　角质层与蜡
Fig.14-2　Cuticle and wax

2. 气孔　双子叶植物叶表皮上常有两个特化成半月形的细胞凹入面相对形成的孔隙，这两个特化的表皮细胞称为保卫细胞，保卫细胞形成的孔隙称气孔（图 14-3）。气孔是植物体表面进行气体交换的通道，能控制气体交换和调节水分蒸散。气孔的张合受温度、湿度、光照和二氧化碳浓度等的影响。有些植物的保卫细胞周围还有 1 个或多个与表皮细胞形状不同的细胞，称副卫细胞。副卫细胞的形状、数目及排列方式与植物种类有关。保卫细胞与副卫细胞的排列关系称气孔轴式或气孔类型。双子叶植物常见的气孔类型有以下几种（图 14-4）。

(2) stomata　In the epidermis of the Dicotyledones leaf, there are one pore formed by the concave surface of the cells, which are called guard cells and the pores formed by guard cells are called stomata(Fig.14-3). Stomatas are gas exchange channels on plant surface, which can control gas exchange and regulate water evapotranspiration. Stomatal opening and closing are affected by temperature, humidity, light and carbon dioxide concentration. Some plants also have one or more cells around the guard cells that are different in shape from the epidermal cells, called subsidiary cell. The shape, number, and arrangement of subsidiary cell are related to plant species. The arrangement relationship between guard cells and subsidiary cells is called stomata shaft type or stomata type. The common types of stomata in dicotyledons are as follows(Fig.14-4).

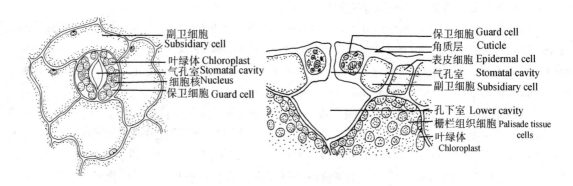

图 14-3　叶的表皮与气孔器
Fig.14-3　The epidermis and stomatal apparatus of a leaf

平轴式　　　　　直轴式　　　　　不等式　　　　　不定式　　　　　环式
Paracytic type　Diacytic type　Anisocytic type　Anomocytic type　Actinocytic type

图 14-4　气孔的类型
Fig. 14-4　Types of the stomata

平轴式：副卫细胞 2 个，其长轴与保卫细胞长轴平行。如茜草、菜豆、落花生、番泻和常山等植物的叶。

Paracytic type: There were 2 subsidiary cells around the guard cells, whose long axis is parallel to that of the guard cells. This type can be found on the leaves of plants such as *Rubia cordifolia*, *Phaseolus vulgaris*, *Arachis hypogaea*, *Cassia acufifolia* and *Dichroa febrifuga*.

直轴式：副卫细胞 2 个，其长轴与保卫细胞长轴垂直。如石竹科、爵床科和唇形科等植物的叶。

Diacytic type: 2 subsidiary cells around the guard cells, whose long axis is perpendicular to that of the guard cell. It can be found on the leaves of Caryophyllaceae, Acanthaceae, and Labiatae plants.

不等式：副卫细胞 3~4 个，大小不等，其中 1 个明显较小。如十字花科、茄科烟草属和茄属等植物的叶。

Anisocytic type: There are 3~4 subsidiary cells around the guard cells, one of which is obviously smaller. It is often distributed on the *leaf* surface of *Cruciferae*, *Solanum* and *Tobacco* plants.

不定式：副卫细胞数目不定，其大小基本相同，形状也与其它表皮细胞相似。如艾叶、桑叶、枇杷叶、洋地黄叶等。

Anomocytic type: The number of subsidiary cells varies, but they are basically the same size and shape as other epidermal cells. Such as the leaves of *Artemisia argyi*, *Morus alba*, *Eriobotrya japonica* and *Digitalis perpurea*.

环式：副卫细胞数目不定，其形状比其它表皮细胞狭窄，环状围绕保卫细胞排列。如茶叶、桉叶。

Actinocytic type: The number of subsidiary cells is variable, their shape is narrower than other epidermal cells, and they are arranged in a ring around the guard cells. Such as the leaves of *Camellia sinensis* and *Eucalyptus robusta*.

单子叶植物的气孔类型也很多，如禾本科和莎草科植物的保卫细胞呈哑铃型，其细胞壁在两端球状部分较薄，而中间窄的部分较厚，该特点使保卫细胞易因膨压改变而引起气孔开闭。保卫细胞两侧还有两个平行排列、略呈三角形的副卫细胞，对气孔的开启有辅助作用，如淡竹叶、玉蜀黍叶等。

The stomatal types of monocotyledonous plants are also various. For example, the guard cells of gramineae and Seperaceae plants are dumbbell type. The cell walls of the guard cells are thinner in the spherical parts at both ends and thicker in the narrow part in the middle. On both sides of guard cells and two parallel arrangements are triangular, slightly accessory cells of stomatal opening which have a supplementary role, such as *Phyllostachys glauca*, *Zea mays* L., etc.

各种植物具有不同类的气孔轴式，而在同一植物的同一器官上也常有两种或两种以上类型，且分布情况也不同，对植物分类鉴定和药材鉴定有一定价值。

All kinds of plants have different stomatal types, and there are often two or more types in the same organ of the same plant, and the distribution is also different, which is of certain value for plant classification and herbal medicine identification.

3. **毛茸**　植物体表面还存在有多种类型的毛茸，它们是由表皮细胞特化而形成的突起物，具有保护、减少水分蒸发、分泌物质等作用，是植物抗旱的形态结构。此外，还有保护植物免受动物啃食和帮助种子撒播的作用。根据其结构和功能常分为腺毛和非腺毛两种类型。

(3) Trichomes　There are also many types trichomes on the surface of plants. They are the outcrops formed by the specialization of epidermal cells, which have the functions of protecting, reducing water evaporation and secreting substances, and are the morphological structures of plants resistant to drought. In addition, they also protect plants from being eaten by animals and help spread seeds. According to the structure and function of trichome, they can be divided into glandular hairs and non-glandular hairs.

腺毛是具有分泌功能的毛茸，由腺头和腺柄组成，分泌物包括挥发油、树脂、粘液等。腺头常呈圆形，由 1 至多个分泌细胞组成，分泌物向外渗出，对植物具有保护作用；腺柄常有 1 至多个细胞，如薄荷、车前、洋地黄、曼陀罗等叶上的腺毛。薄荷、筋骨草等唇形科植物叶表面还有一种无柄或柄极短的腺毛，称腺鳞；腺头 8 个或 4~7 个分泌细胞组成。少数植物薄壁组织内部的细胞间隙存在腺毛，称间隙腺毛，如广藿香茎、叶和绵马贯众叶柄及根茎（图 14-5）。

Glandular hair: is a kind of pilose hair with secretory cells and functions. It is composed of the glandular head and the glandular stalk. The glandular head is usually round and consists of 1 or more secretory cells. The secretion initially accumulates between the cell wall and the stratum corneum, and then the stratum corneum ruptures and exudes, which has protective effects on plants. The glandular stalk usually has 1 or more cells, such as the glandular hairs on the leaves of mint, plantar, digitalis, mandala, etc. There are also sessile or very short sessile glandular hairs on the surface of the leaves of the Labiaceae plants such as *Mentha haplocalyx* and *Ajuga decumbens*. The glandular scale is formed by eight or four to seven secretory cells in the glandular head. A few plant parenchyma inside the intercellular existence of the glandular hairs, is called intercellular glandular hairs, such as *Pogostemon cablin* and D*ryopteris crassirhizoma* (Fig.14-5).

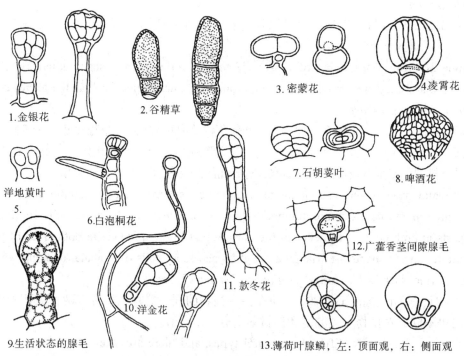

1. 金银花　2. 谷精草　3. 密蒙花　4.凌霄花　洋地黄叶 5.　6.白泡桐花　7.石胡荽叶　8.啤酒花　12.广藿香茎间隙腺毛　11. 款冬花　10.洋金花　9.生活状态的腺毛　13.薄荷叶腺鳞，左：顶面观，右：侧面观

1. *Lonicera japonica*; 2. *Eriocaulon buergerianum*; 3. *Buddleja officinalis*; 4. *Campsis grandiflora*; 5.The leaf of *Digitalis purpurea*; 6. The flower of *Paulownia fortunei*; 7.The leaf of *Centipeda minima*; 8.*Humulus lupulus*; 9.Living glandular hair; 10.*Datura metel*;11.*Tussilago farfara*; 12. Interstitial glandular hair in the stem of *Pogostemon cablin*;13.Glandular scale on the leaf of *Mentha haplocalyx*: left. top view; right:side view

图 14-5　腺毛与腺鳞
Fig.14-5　Gladular hair and glandular scale

非腺毛是无头、柄之分，且无分泌功能的毛茸，由 1 至多个细胞组成，末端常尖狭，起保护作用。非腺毛形态多样，其特征可用于药材鉴定。常见的形状有线状、分枝状、丁字形、星状、鳞片状等，有的非腺毛的细胞内有晶体沉积（图 14-6）。

Non-glandular hair: is a non-secretory pilose, consisting of 1 or more cells, with a protective, often narrow, terminal end. Non-glandular hair are varied, the common shapes are linear, branched, T-shaped, star-shaped, scale-shaped, and some non-glandular hair cells with crystal deposition (Fig.14-6).

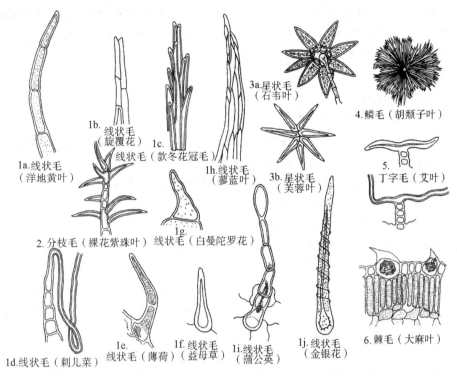

Linear hair: (1a.*Digitalis purpurea* ,1b.*Inula japonica*,1c.*Tussilago farfara*,1d.*Cirsium setosum*,1e.*Mentha haplocalyx*, 1f.*Leonurus artemisia*,1g.*Datura metel*,1h.*Polygonum tinctorium*,1i.*Taraxacum mongolicum*,1j.*Lonicera japonica*); 2.Branched hair; Stellate hair: 3a.*Pyrrosia lingua*,3b.*Hibiscus mutabilis*; 4.Squamous hair: *Elaeagnus pungens*; 5.T-shaped hair: *Artemisia argyi*; 6. Spiny hairs: *Cannabis sativa*

图 14-6 各种非腺毛
Fig.14-6 Types of non-gladular hair

（二）周皮
1.3.2 Periderm

当植的根和茎在加粗生长开始后，表皮逐渐被破坏，植物体相应地形成次生保护组织周皮来代替表皮行使保护功能。周皮是由木栓层、木栓形成层和栓内层组成的复合组织，由木栓形成层的分裂活动形成（详见第十五章）。

When plants undergo secondary growth, the epidermis is gradually destroyed after the thickening of roots and stems, and the plant body correspondingly forms secondary protective tissue periderm to replace the epidermis to perform protective functions. Periderm is composed of cork layer, cork cambium and cork inner layer (see Chapter 15 for details).

四、机械组织
1.4 Mechanical tissue

机械组织是细胞壁发生不同程度的加厚，起支持和巩固作用的一类成熟组织。植物幼嫩器官没有机械组织或机械组织不发达，随器官逐渐成熟后，其内部逐渐分化出机械组织。按细胞形态和壁增厚的方式不同，分为厚角组织和厚壁组织。

The mechanical tissue is a kind of mature tissue which the cell wall thickens in different degree and plays the role of support and consolidation. The young organs of plants have no mechanical tissue or the mechanical tissue is not developed. The inner part of the young organs gradually differentiates into the mechanical tissue as the organs mature. It can be divided into collenchyma and sclerenchyma according to the cell morphology and the way the cell walls thickening.

（一）厚角组织

1.4.1　Collenchyma

厚角细胞是生活细胞，内含原生质体，常含叶绿体；具有分生潜力，能参与木栓形成层的形成。细胞壁不均匀增厚，主要成分是纤维素和果胶质，不木质化。机械组织具一定韧性，可塑性和延伸性；既支持植物直立，又能随器官伸长而延伸。

Collenchyma cells are living cells that contain protoplasts and chloroplasts, and have the potential for involving in the formation of the cork cambium. The cell walls are inhomogeneously thickened and composed mainly of cellulose and pectin, not lignified. The cells are flexible, malleable and extensible, supporting the plant to stand upright while extending with the organs.

厚角组织多位于表皮下方，呈层状、环状或成束分布。如益母草、薄荷、芹菜、南瓜等植物茎的棱角处就是厚角组织集中分布的区域。厚角组织常存在于双子叶草本植物茎和尚未进行次生生长的木质茎中，以及叶片主脉上下两侧、叶柄、花柄的外侧部分，根内很少形成厚角组织，但如果暴露在空气中，则可发生。

Most of collenchyma tissue is located under the epidermis and distributed in layers, rings or bundles. As for *Leonurus heterophyllus*, *Mentha haplocalyx*, *Apium graveolens* and *Cucurbita moschata* and other plants, the collenchyma are distributed concentrately at the corner of the stems. Collenchyma is usually found in dicotyledonous herbaceous and woody stems that have not yet undergone secondary growth, as well as in the upper and lower sides of the main veins, petioles, and lateral parts of the flower petioles.

根据厚角组织细胞壁加厚方式的不同，常分为三种类型（图14-7）：①真厚角组织，又称角隅厚角组织，是最常见的类型；壁的增厚部位在几个相邻细胞的角隅处，如薄荷属、桑属和蓼属植物等。②板状厚角组织，又称片状厚角组织，细胞壁增厚部分主要在内、外切向壁上，如大黄属、细辛属、地榆属、泽兰属和接骨木属植物等。③腔隙厚角组织，细胞增厚发生在发达的胞间隙处，面对胞间隙部分细胞壁增厚，如夏枯草属、锦葵属、鼠尾草属等。

According to the thickening ways of the cell wall of the collenchyma, it can be divided into three types(Fig.14-7): ① Angular collenchyma, also known as corner collenchymas, is the most common type. Wall thickening occurs at the corners of several adjacent cells, such as *Mentha*, *Morus* and *Polygonum* plants. ② Plate-like collenchyma, also termed as flaky collenchymas. The thickened part of the cell wall is mainly on the inner and outer tangential walls, such as *Rheum*, *Asarum*, *Sanguisorba*, *Eupatorium* and *Sambucus* plants. ③ Lacunate collenchyma, cell thickening occurs in the well-developed intercellular space, and some cell walls are thickened in the intercellular space, such as *Prunella*, *Malva* and *Salvia* plants.

（二）厚壁组织

1.4.2　Sclerenchyma

厚壁组织是植物主要的支持组织，细胞具有全面增厚、木质化的细胞壁，壁上具层纹和纹孔，胞腔小，成熟后成为死细胞。厚壁组织细胞可单个或成群分散在其它组织之间，按细胞形态

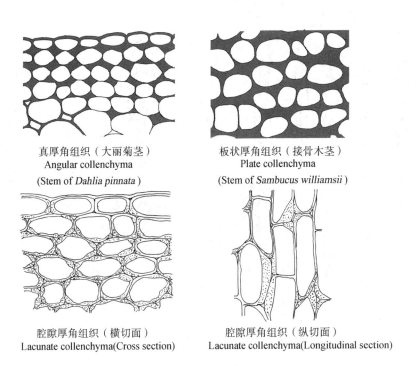

真厚角组织（大丽菊茎）
Angular collenchyma
(Stem of *Dahlia pinnata*)

板状厚角组织（接骨木茎）
Plate collenchyma
(Stem of *Sambucus williamsii*)

腔隙厚角组织（横切面）
Lacunate collenchyma(Cross section)

腔隙厚角组织（纵切面）
Lacunate collenchyma(Longitudinal section)

图 14-7 厚角组织的类型
Fig. 14-7 Types of collenchyma

不同，可分为纤维和石细胞。

Sclerenchyma is the main supporting tissue of plants. The cells have thickened and lignified cell walls with lamellae and pits on the walls. The cell cavity is small, after the cell matures it becomes the dead cell. Sclerenchymas may be separated into individual cells or groups of other tissues, and may be divided into fiber and stone cells according to cell morphology.

1. **纤维** 纤维是两端尖斜的长梭形细胞，次生壁明显，加厚的成分主要为纤维素或木质素，壁上有少数纹孔，细胞腔小。纤维单个或彼此嵌插成束分布于植物体中（图 14-8）。按其在植物体中分布和壁特化程度不同，纤维可分为木纤维和木质部外纤维，木质部外纤维又常称韧皮纤维。

(1) Fiber Fiber is a long spindle cell with sharp and oblique ends. The secondary wall is obvious. The thickened component is mainly cellulose or lignin. Fibers are distributed in plants individually or interspersed into bundles(Fig.14-8). According to their distribution in plants and the degree of wall specialization, the fiber can be divided into xylem fiber and extraxylary fiber, which is often called phloem fiber.

木纤维分布于被子植物木质部，较韧皮纤维短，长约 1mm。其细胞壁木化程度高，胞腔小，壁上具各式具缘纹孔或裂隙状单纹孔，坚硬而无弹性，脆而易断，机械巩固较强；壁增厚程度随植物种类和生长部位与生长时期不同而异。如黄连、大戟、川乌、牛膝等有一些薄壁木纤维，而栎树、栗树的木纤维则强烈增厚。春季生长的木纤维壁较薄，而秋季的则较厚。一些植物的次生木质部具有一种纤维，细胞细长，像韧皮纤维，壁厚并具裂缝状单纹孔，纹孔较少，称韧型纤维，如沉香、檀香等的木纤维。

Xylem fiber: Xylem fiber is distributed in the xylem of angiosperms, shorter and about 1mm longer than the bast fiber. The cell wall has high degree of xylization, small cell cavity, various marginal pits or fissure-shaped single pits. The cell wall is hard and inelastic, brittle and easy to break, with strong

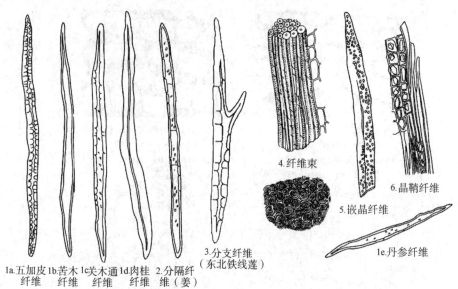

4.纤维束
6.晶鞘纤维
5.嵌晶纤维
3.分支纤维（东北铁线莲）
1e.丹参纤维
1a.五加皮纤维 1b.苦木纤维 1c.关木通纤维 1d.肉桂纤维 2.分隔纤维（姜）

Fiber in (1a. Cortex of *Acanthopanar gracilistμlus*, 1b.*Picrasma quassioides*, 1c.*Aristolochia manshuriensis*, 1d.*Cinnamomum cassia*, 1e.Salvia miltiorrhiza); 2. Septate fiber in *Zingiber officinale*; 3.Branched fiber in *Clematrs mandshusica*; 4.Fiber bundle; 5. Intercalary crystal fiber; 6. Crystal fiber

图 14-8　纤维束及纤维类型
Fig. 14-8　Fiber bundles and types of the fiber

mechanical consolidation. For some xylem fibes, the cell walls are thin as in *Coptis chinensis*, *Euphorbia pekinensis*, *Aconitum carmichaelii* and *Achyranthes bidentata*. but for some others, they are strongly thickened as in *Quercus* plants and *Castanea mollissima*. The xylem fiber walls are thinner in spring and thicker in autumn. The secondary xylem of some plants contains a type of fiber with long, thin cells, like phloem fibers, thick walls, and less pits, called libriform fiber, such as xylem fibers of *Aquilaria sinensis* and *Santalum album*.

木纤维仅存在于被子植物的木质部中，为木质部的主要组成部分；裸子植物的木质部没有纤维，主要由管胞组成，管胞同时具有输导和机械作用，也是裸子植物原始于被子植物的特征之一。

Xylem fiber exists only in the xylem of angiosperms and is the main component of the xylem. However, there is no fiber in the xylem of gymnosperms and it is mainly composed of tracheids, which have both conducting and mechanical functions and are also one of the characteristics of the origin of gymnosperms in angiosperms.

木质部外纤维分布在木质部以外，包括皮层、髓部、韧皮部和维管束周围。由于常分布在韧皮部，也称韧皮纤维。其细胞多呈长纺锤形，细胞壁富含纤维素，木化程度低，坚韧而有弹性，纹孔较少常成缝隙状。各种植物的韧皮纤维长度不一，木化程度各异。一些藤本双子叶植物茎的皮层、髓部，常具环状排列的皮层纤维、环髓纤维以及靠近维管束的环管纤维（又称周维纤维）等。部分单子叶植物，如禾本科植物茎中的维管束周围有纤维呈环状分布形成维管束鞘。药材鉴定中，常见以下几种特殊类型纤维：①晶鞘纤维（晶纤维），是纤维束及其周围含有晶体的薄壁细胞所组成的复合体；薄壁细胞含有方晶，如黄柏、甘草等；或簇晶，如石竹、瞿麦等；或石膏结晶，如柽柳等。②嵌晶纤维，纤维的次生壁外层镶嵌有细小的草酸钙方晶和砂晶；如南五味子根皮的纤维嵌有方晶，草麻黄茎的纤维嵌有的砂晶。③分隔纤维，是一种胞腔中具有菲薄横膈膜的纤维，可长期保留原生质体，并贮藏有淀粉、油类和树脂等。如姜、葡萄属、金丝桃属等植物等。④分枝纤维，是一种细胞呈长梭形顶端端具有明显分枝的纤维，如东北铁线莲根中的纤维。

Extraxylary fiber: Extraxylary fiber is found outside the xylem, including cortex, pith, phloem and around vascular bundle. Because it's often distribute in phloem, it's also called phloem fiber. Most of the cells are long spindle-shaped, the cell wall is rich in cellulose, low degree of lignification, tough and elastic, the pits are less often into gaps. The length of phloem fiber and the degree of lignification are different in different plants. In the cortex and pith of some vine dicotyledones, there are often cortical fibers arranged in a ring, annulus-medullary fibers and annulus fibers (also called circumferential fibers) near vascular bundles. Some monocots, such as the vascular bundles in the stems of grasses, are ringed with fibers that form the vascular bundle sheath. In the identification of herbal medcines, the following special types of fibers are common.

Crystal fiber is a complex of fiber bundles and their surrounding parenchyma cells containing crystals. In parenchyma cells, containing square crystals, such as *Phellodendron chinense*, *Glycyrrhiza uralensis*, etc. or clustered crystalls, such as *Dianthus chinensis*, *Dianthus superbus*, etc.; sometimes gypsum crystals, as in *Tamarix chinensis*.

Intercalary crystal fiber is a fiber whose secondary wall is covered with tiny calcium oxalate square crystals and sand crystals. For example, the root bark of *Kadsura longipedunculata* is embedded with square crystal, while the root fiber of *Ephedra sinica* Stapf. is embedded with sand crystal.

Septate fiber is a fiber with a thin diaphragm in the cell cavity, which can retain protoplast for a long time and store starch, oil and resin. Such as *Zingiber officinale*, *Vitis* sp, *Hypericum* and other plants.

Branch fiber (branched fiber) is a kind of cells with obvious branches at the top of the spindle fibers, as seen in the root of *Clematrs mandshusica* root.

2. 石细胞　一般由薄壁细胞的细胞壁强烈增厚形成，或由分生组织衍生细胞产生，是植物体内特别硬化的厚壁细胞。多为椭圆形、类圆形、类方形、不规则形等近等径的类型，也有分枝状、星状、柱状、骨状、毛状等类型；细胞壁强烈次生增厚并木化，呈现同心环状层纹，壁上有许多单纹孔，细胞腔极小，成熟后为死细胞。石细胞广泛分布于植物体内，单生或聚生于根、茎、叶、果皮和种皮内，其存在部位和形状是药材鉴定的重要特征之一（图 14-9）。如三角叶黄连、白薇等髓部具石细胞，黄柏、黄藤、肉桂的树皮具石细胞；黄芩、川乌根中的石细胞呈长方形、类方形、多角形，厚朴、黄柏中的石细胞为不规则状。一些植物的果皮和种皮中石细胞常构成坚硬的保护组织，如椰子、核桃、杏等坚硬的内果皮，以及菜豆、栀子的种皮等。

(2) Sclereid or stone cell　Stone cell is generally formed by the strong thickening of the cell wall of parenchyma, it can also be produced by meristematic derived cells. Most of stone cells are oval, round, square or irregular, sometimes branching, stellate, columnar, bone or hairy.

The cell wall was thickened and lignified with concentric ring lamellae and many single pits on the wall, and the cell cavity was very small. Stone cells are widely distributed in plants. They are solitary or aggregated in roots, stems, leaves, pericarp and seed coat. The location and shape of stone cell is one of the important characteristics for herbal medcine identification(Fig.14-9). For instance, the stone cells could be observed in the pith of *Coptis deltoidea* and *Cynanchum atratum* as well as the bark of *Phellodendron amurense*, *Daemonorops margaritae* and *Cinnamomum cassia*. The stone cells in the roots of *Scutellaria baicalensis* and *Aconitum carmichaelii* are rectangular, square-like and polygonal, while those in the barks of *Magnolia officinalis* and *Phellodendron chinense* are irregular. Stone cells in some plants often form hard protective tissues, such as the hard endocarp of *Cocos nucifera*, *Juglans regia*, *Armeniaca vulgaris* and the seed coats of *Phaseolus vulgaris* and *Gardenia jasminoides*.

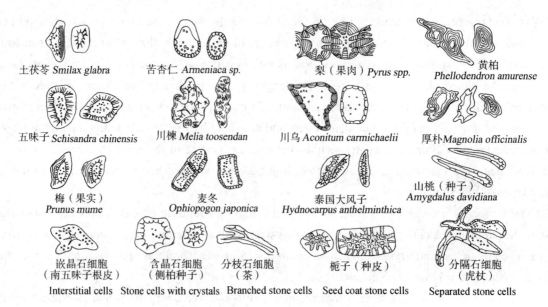

土茯苓 *Smilax glabra*　　苦杏仁 *Armeniaca sp.*　　梨（果肉）*Pyrus spp.*　　黄柏 *Phellodendron amurense*

五味子 *Schisandra chinensis*　　川楝 *Melia toosendan*　　川乌 *Aconitum carmichaelii*　　厚朴 *Magnolia officinalis*

梅（果实）*Prunus mume*　　麦冬 *Ophiopogon japonica*　　泰国大风子 *Hydnocarpus anthelminthica*　　山桃（种子）*Amygdalus davidiana*

嵌晶石细胞（南五味子根皮）Interstitial cells　　含晶石细胞（侧柏种子）Stone cells with crystals　　分枝石细胞（茶）Branched stone cells　　栀子（种皮）Seed coat stone cells　　分隔石细胞（虎杖）Separated stone cells

图 14-9　石细胞的类型
Fig.14-9　Types of different stone cells

五、分泌组织
1.5　Secretory tissue

植物在新陈代谢过程中，一些细胞能分泌某些特殊物质，如挥发油、乳汁、黏液、树脂、蜜液、盐类等，这种细胞称为分泌细胞，由分泌细胞所构成的组织称为分泌组织。分泌组织所产生的分泌物作用多样，如蜜汁、芳香油能引诱昆虫，以利传粉和果实、种子传播；分泌的盐分使植物免受高盐危害；有些分泌物能抑制病原菌和其他生物，防止组织腐烂、帮助创伤愈合、免受动物啃食，以保护自身；许多分泌物可作药用（如乳香、没药、樟脑、薄荷油等）、香料或其他工业原料。

In the process metabolism, some cells can secrete certain special substances, such as volatile oil, milk, mucus, resin, honey, salt, etc. These cells are called secretory cells, the tissue composed of secretory cells is termed secretory tissue. The secretions act in a variety of ways, e.g. honeydew and aromatic oil can attract insects, in order to pollinate and spread fruit or seed. Secretion of salt can protect plants from high salt harm; some secretions can inhibit pathogens and other organisms, protect itself by preventing tissue from decay, healing wounds, and being eaten by animals; many secretions can be used medicinally (e.g. frankincense, myrrh, camphor, peppermint oil, etc.), perfumery, or other industrial raw materials.

根据分泌细胞排出的分泌物是积累在植物体内还是排出体外，可将分泌组织分为外分泌结构和内分泌结构。

Secretory tissue can be divided into exocrine structure and endocrine structure according to whether the secretions are accumulated in the plant body or discharged from the body.

1. 外部分泌结构　分布在植物体表，其分泌物排出植物体外，如腺毛、蜜腺等。

(1) Exocrine tissue　The exocrine structure is distributed on the plant surface, and its secretion is excreted out of the plant body, such as glandular hairs, nectary and so on.

腺毛见保护组织 - 表皮。

Glandular hairs See 1.3 Protective tissue.

蜜腺是能分泌蜜液的腺体，由一群表皮细胞或其下面数层细胞特化组成。与相邻细胞相比，腺体细胞的细胞壁较薄，无角质层或角质层很薄，细胞质较浓。细胞质产生蜜液并通过角质层扩散或经腺体表皮上的气孔排出。一般位于花萼、花冠、子房或花柱的基部，为花蜜腺。若存在于茎、叶、托叶、花柄处，为花外蜜腺，如蚕豆的托叶，桃和樱桃叶片基部均具蜜腺；枣、白花菜和大戟属花序中也有不同形态的蜜腺。

Nectaries: Nectaries are nectariferous glands that are specialized in a group of epidermal cells or several layers below them. Compared with the adjacent cells, the gland cells have thinner cell wall, thinner and thicker cytoplasm without cuticle or cuticle. The cytoplasm produces nectaries and diffuses through the stratum corneum or through the stomata on the epidermis of the gland. Usually located at the base of calyx, corolla, ovary, or style. Nectaries existing in the stem, leaf, stipule and petiole are called extrafloral nectaries. For example, the stipules of *Vicia faba*, the base of the leaves of *Amygdalus persica* and *Cerasus pseudocerasus* all have nectaries; the inflorescences of *Brassica napus*, *Fagopyrum esculentum*, *Ziziphus jujuba. var. spinosa*, *Sophora japonica* have nectaries of different forms.

2. **内部分泌结构**　位于植物体内，其分泌物也积存在体内。根据形态结构和分泌物的不同，分为分泌细胞、分泌腔、分泌道和乳汁管（图 14-10）。

(2) Internal secretory structure　The internal secretion structure is located in the plant body, and its secretion also accumulates in the plant. According to the differences of morphology and its secretion, the internal secretory structure is divided into secretory cell, secretory cavity, secretory canal and laticifer. (Fig.14-10)

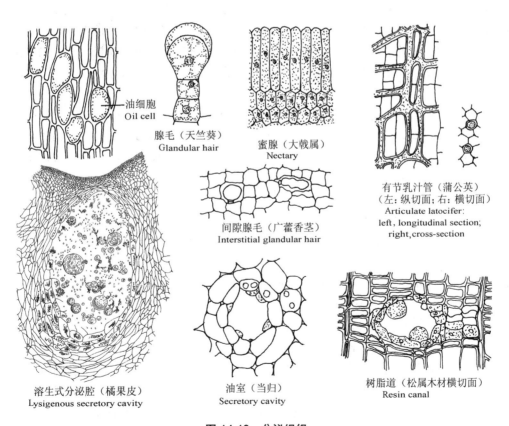

油细胞
Oil cell

腺毛（天竺葵）
Glandular hair

蜜腺（大戟属）
Nectary

有节乳汁管（蒲公英）
（左：纵切面；右：横切面）
Articulate latocifer:
left, longitudinal section;
right, cross-section

间隙腺毛（广藿香茎）
Interstitial glandular hair

溶生式分泌腔（橘果皮）
Lysigenous secretory cavity

油室（当归）
Secretory cavity

树脂道（松属木材横切面）
Resin canal

图 14-10　分泌组织
Fig.14-10　Secretory structure

分泌细胞是分布在植物体内具有分泌能力的细胞，通常比周围细胞大，它们并不形成组织，以单个细胞或细胞团（列）分散在其他组织中。按贮藏的分泌物不同分为油细胞（木兰科、樟科），黏液细胞（如半夏、玉竹、山药、白及），单宁细胞（豆科、蔷薇科、壳斗科、冬青科、漆树科），芥子酶细胞（十字花科、白花菜科），含晶细胞（蔷薇科、桑科、景天科）等。

Secretory cell are the specialized secretory cells in plants, usually larger than the surrounding cells. They do not form tissue, but are dispersed in other tissues as single cells or clusters of cells (columns). According to the secretions stored in cells, they are divided into oil cells (Magnoliaceae, Lauraceae), mucous cells (*Pinellia ternata*, *polygonatum odoratum*, *Dioscorea oppositae.*, *Bletilla striata*), tannic cells (Leguminosae, Rosaceae, Fagaceae, Aquifoliaceae and Aanacardiaceae), myrosinase cells (Cruciferae, Capparidaceae), crystalline cells (Rosaceae, Moraceae, Crassulaceae) and so on.

分泌腔又称分泌囊或油室。在形成之初，原来有一群分泌细胞，由于这些细胞中分泌物积累增多，最后使细胞壁破裂溶解，在体内形成一个含有分泌物的腔穴，这种分泌腔称溶生式分泌腔，如陈皮、橘叶等。有些植物的分泌腔是由于分泌细胞彼此分离，胞间隙扩大而形成的，称裂生式分泌腔，如金丝桃的叶以及当归的根等。

Secretory cavity also known as secretory sac or oil chamber. At the beginning of the formation, there was a group of secretory cells. Due to the increased accumulation of secretions in these cells, the cell wall was finally ruptured and dissolved, forming a cavity containing secretions in the body. This kind of secretion cavity is called a lysogenic secretion cavity, as seen in the leaves of *Citrus* plantorange. The secretory cavity of some plants is formed by the separation of secretory cells from each other and the expansion of the intercellular space. It is called a schizogenous secretory cavity, such as the leaves of *Hypericum* plants and the roots of *Angelica* plants.

分泌道是由一群分泌细胞彼此分离形成的一个长管状胞间隙腔道，腔道围绕的分泌细胞称上皮细胞，上皮细胞产生的分泌物贮存在腔道中。贮藏树脂的为树脂道，如松树茎中的分泌道；贮藏挥发油的称油管，如小茴香果实的分泌道；黏贮藏黏液的称黏液道或黏液管，如美人蕉和椴树的分泌道。

Secretory canal is a kind of long tubular intercellular cavity formed by a group of secretory cells separated from each other. The secretory cells surrounding the cavity are called epithelial cells, and secretions produced by epithelial cells are stored in the cavity. Mostly are resin canal as seen in Pinus stem, vitta as seen in the fruit of *Foeniculum vulgare*, and slime canal as seen in the stems of *Canna indica* and *Tilia* plants.

乳汁管是分泌乳汁的长管状单细胞，常具分枝，或由一系列细胞合并，横壁消失连接而成，常在植物体内形成系统。乳汁常呈白色或乳白色，少数为黄色、橙色或红色；有橡胶、糖类、蛋白质、生物碱、苷类、酶、单宁等物质。

Laticifer Long: Tubular, single cells that secrete milk, often with branches, or by merging a series of cells, resulting in the disappearance of the transverse wall, often in the formation of systems in plants. The secretions are usually white or milky white, a few yellow, orange or red, mostly are rubber, sugar, protein, alkaloids, glycosides, enzymes, tannin and other substances.

按乳汁管的发育和结构，可分为两种类型：①无节乳汁管，由一个细胞发育而成，随着植物的生长不断延长和分枝，贯穿植物体内，长者可达数米以上，如夹竹桃科、萝藦科、桑科以及大戟科大戟属等植物的乳汁管。②有节乳汁管，是由许多长管状细胞连接而成，连接处的细胞壁溶解贯通，成为多核巨大的管道系统，乳汁管分枝或不分枝，如菊科、桔梗科、罂粟科、旋花科等植物的乳汁管。

According to the development and structure of the laticifer, it can be divided into two types. ①Nonarticulate laticifer: Plants of Asclepiadaceae, Moraceae and Euphorbiaceae have the laticiferous tubules that develop from a single cell and extend and branch through plants for several meters or more. ②Articulate laticifer, which are made up of many long tubular cells connected together, and are formed by the dissolution of the cell wall at the junction, forming a large polynuclear duct system with or without branching laticifers, as seen in Compositae, Campanulaceae, Papaveraceae, Convolvulaceae and other plants.

六、输导组织
1.6 Conducting tissue

输导组织是维管植物体内运输水分和养料的组织。其细胞一般呈管状，上下相接，贯穿于整个植物体内。蕨类植物、裸子植物和被子植物具有输导组织。根据输导组织的构造和运输物质的不同，可分为两类，一类是木质部中的导管和管胞，主要运输水和无机盐；另一类是韧皮部中的筛管和筛胞，主要运输有机营养物质。

Conducting tissue is a tubular structure that transports water, inorganic salts and organic matters over long distances in vascular plants, which forms a continuous conducting system between different organs. Its cells are generally tubular, connected from top to bottom, throughout the whole plant body. It can be divided into two categories according to the different transport substances in the conducting tissues, vessels and tracheids in xylem, which mainly transport water and inorganic salts, and the sieve vessels and sieve cells in phloem, which mainly transport postal nutrients.

1. 导管和管胞
(1) Vessels and tracheids

导管是被子植物主要的输水组织，少数原始的和一些寄生被子植物则无导管，如金粟兰科草珊瑚属植物；而少数进化的裸子植物如麻黄科植物，以及蕨类中较进化的少数真蕨类植物有导管。导管是由一系列长管状或筒状的死细胞连接而成，横壁溶解形成穿孔，具有穿孔的端壁称穿孔板（图 14-11）。穿孔的形成使导管连成为管道系统。

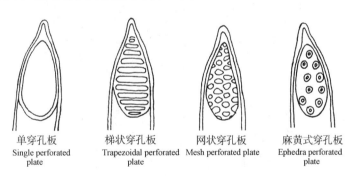

单穿孔板 Single perforated plate　梯状穿孔板 Trapezoidal perforated plate　网状穿孔板 Mesh perforated plate　麻黄式穿孔板 Ephedra perforated plate

图 14-11 导管分子的穿孔板类型
Fig.14-11 Different perforation plate of the vessels

导管侧壁的次生增厚是不均匀增厚，木质化，形成了不同的纹理和纹孔。根据导管增厚所形成的纹理不同，可将导管分为5种类型（图 14-12、图 14-13）。

Vessels are long tubes composed of individual cells called vessel elements that are open at each end. As each vessel element develops, the perforation plate, in some instances, can become barlike strips of wall material that extend across the openings. However, the flow of fluid through the vessels is not blocked by the strips. The vessels can be observed in most developed angiospermae plants and some evolved gymnosperms such as Ephedreacea plants (Fig.14-11), it is absent in a few primitive and some parasitic angiosperms, such as the plants from *Sarcandra* of Chloranthaceae. There are five basic patterns of vessel element side wall thickenings(Fig.14-12、Fig.14-13).

环纹导管：管壁呈环状的木质化增厚，这种增厚的环纹之间仍有薄壁，有利于生长而伸长。环纹导管常出现在器官的幼嫩部分，如南瓜茎、凤仙花的幼茎、半夏的块茎中。

Annular: A vessel element with annular secondary thickenings has individual rings of lignied secondary cell wall. is is the simplest and most economical form of vessel element wall patterning. It occurs in the young parts of the organ, such as stems of *Cucurbita moschata* and *Impatiens*

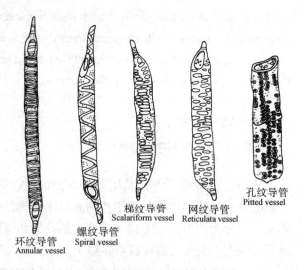

图 14-12　导管细胞类型
Fig. 14-12　Types of the vessel

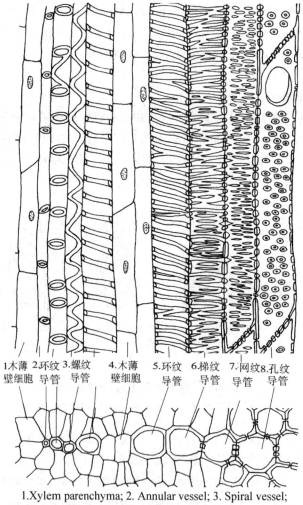

1.木薄壁细胞 2.环纹导管 3.螺纹导管 4.木薄壁细胞 5.环纹导管 6.梯纹导管 7.网纹导管 8.孔纹导管

1.Xylem parenchyma; 2. Annular vessel; 3. Spiral vessel;
4.Xylem parenchyma; 5. Annular vessel; 6. Scalariform vessel;
7. Reticulate vessel; 8. Pitted vessel

图 14-13　半边莲属植物的初生木质部（示导管）
Fig.14-13　The primary xylem of the Lobelia plants, which shows the vessels

balsamina, and the tubers of *Pinellia ternata*.

螺纹导管：侧壁上有一或多条成螺旋带状木质化增厚的次生壁，容易与初生壁分离，不碍导管的生长。也存在于植物器官的幼嫩部分，如南瓜茎、天南星块茎。"藕断丝连"的丝就是常见的螺纹导管。

Spiral: A spiral vessel element wall pattern is one in which secondary wall thickenings wrap around the cell. The vast majority of the cell wall is primary, with only the thickened rings made of secondary wall. A vessel element with a helical pattern has the same strength, water flow, and extensibility characteristics as found in annular patterns and would be found in young, expanding tissues, as seen in stems of *Cucurbita moschata* and rhizome *Arisaema heterophyllum* Blume. The filament of the lotus root is the phenomenon of secondary wall separation.

梯纹导管：侧壁上呈几乎平行的横条状木化增厚，并与未增厚的初生壁相间排列成梯形，如葡萄茎、常山根中的导管。

Scalariform: thickenings run in a transverse direction, like the rungs of a ladder, and cover large portions of the vessel element side. Because the ends of the "rungs" are attached to vertical portions of the wall, a scalariform vessel element cannot stretch in an axial direction; therefore, scalariform vessel elements are found in tissues that have ceased elongation, as seen in the stem of *Vitis vinifera* L. and the root of *Dichroa febrifuga* Lour.

网纹导管：侧壁呈网状木化增厚，网孔是未增厚的初生壁。如大黄、苍术根状茎中的导管。

Rreticulate: vessel element wall pattern is an irregular, netlike combination of different secondary cell layers of slightly different orientations. Vessel elements with a reticulate wall pattern are not capable of axial extension, as the vessel in the rhizome of *Rheum palmatum* and *Atractylodes lancea*.

孔纹导管：侧壁大部分木化增厚，未增厚部分形成单纹孔或具缘纹孔。如甘草、赤芍根中的导管。

Pitted: the secondary lateral walls of tracheary elements may be more or less continuous, interrupted only by pits, Because pits allow for cell-to-cell water movement, and each cell has a cell wall, pits almost always exist as a pit pair. Such as the vessel in the roots of *Glycyrrhiza uralensis* and *Paeonia veitchii*.

上述五种导管类型中，环纹导管和螺纹导管是原始的初生类型，在器官的形成过程中出现较早，常存在于植物幼嫩器官的部分，能随器官生长而延长；因管径小，输导能力差。网纹和孔纹导管是进化的次生类型，在器官中出现较晚，多存在于器官的成熟部分；管径大，导管分子较短，管壁较硬，抗压能力强，输导效率高。

Of the five types of vessels mentioned above, the first two vessels appeared earlier and often occurred in the organs at the early stage of growth. The last three are mainly differentiated and formed at the later stage of organ growth, with large tube diameter, short vessel molecules, hard tube wall, strong compression resistance and high translocation efficiency.

当新导管形成后，早期形成的导管邻接薄壁细胞膨胀，通过导管壁上未增厚部分或纹孔，连同其内含物侵入导管腔内形成大小不同的囊状突出物，称侵填体。侵填体含有单宁、树脂、晶体和色素等物质，能起抵御病菌侵害的作用，其中也有植物药的活性物质，使导管相继失去输导能力。

The carrying capacity of a vessel is not permanent, and its validity varies with plant species, with perennials ranging from years to decades. After the formation of the new vessel, the vessel formed in the early stage was adjacent to parenchyma cells, which expanded, penetrated into the vessel cavity with its contents and formed sac-like protrusions of different sizes through the unthickened part or the perforation on the vessel wall, and was called tylosis. The tylosis contains tannins, resins, crystals, pigments and other substances, which can resist the invasion of bacteria, including active substances of plant drugs, so that the vessel has lost the ability to conducting.

管胞是绝大部分蕨类植物和裸子植物的输水组织，被子植物的叶柄、叶脉处也可见到。管胞为长管状细胞，两端斜尖，两管胞间不形成穿孔，相邻管胞通过侧壁上的纹孔对输导水分，所以输导能力较导管低，是较原始的输导组织。在其发育过程中细胞壁次生加厚并木化，最后使原生质体消失而成死细胞。其次生壁也形成类似导管的环纹、螺纹、梯纹、孔纹等类型（图14-14）。

Tracheids, which, like vessel elements, are dead at maturity and have relatively thick secondary cell walls, are tapered at each end, the ends overlapping with those of other tracheids. Tracheids have no openings similar to those of vessels, but there are usually pairs of pits present wherever two tracheids are in contact with one another. Pits are areas in which no secondary wall material has been deposited. They allow water to pass from cell to cell. Tracheids can be observed in most ferns and gymnosperms. They are also found in the petioles and veins of angiosperms (Fig.14-14).

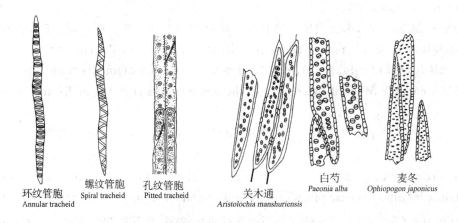

环纹管胞　螺纹管胞　孔纹管胞　　关木通　　白芍　　麦冬
Annular tracheid　Spiral tracheid　Pitted tracheid　Aristolochia manshuriensis　Paeonia alba　Ophiopogon japonicus

图 14-14　管胞类型和药材粉末中的管胞碎片
Fig.14-14　Tracheid type and tracheid fragments in herbal medicine powder

2. 筛管、伴胞和筛胞　筛管存在被子植物的韧皮部中，是运输有机物质的管状结构。筛管由多个细胞（筛管分子）纵向连接而成（图 14-15）。

(2) Sieve tube, companion cell and sieve cell　The sieve tube mainly exists in the phloem of angiosperms and is a tubular structure for transporting organic materials produced by photosynthesis. It is a structure formed by the longitudinal connection of several long tubular living cells. Each cell is called a sieve tube element(Fig.14-15).

筛管仅具由纤维素和果胶组成的初生壁，端壁上形成的许多小孔称筛孔，具有筛孔的区域称筛域。分布有一个或多个筛域的端壁称筛板，仅有一个筛域的筛板称单筛板，如南瓜的筛管；具多个筛域的筛板称复筛板，如葡萄的筛管，通过筛孔的原生质丝较胞间连丝粗大称联络索。联络索使筛管分子间彼此相连贯通，有些植物筛管侧壁上还可见筛孔，侧壁上的筛孔使相邻筛管彼此联系，从而实现植物体内有机物的有效输导。

The sieve tube only has primary wall composed of cellulose and pectin. Many small holes formed on the end wall are called sieve pores, and the area with sieve pores is called sieve area. The end wall of

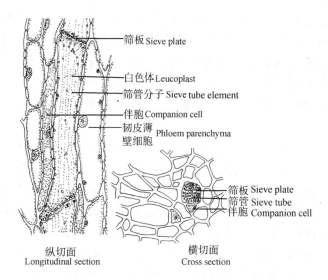

图 14-15　烟草韧皮部（示筛管及伴胞）
Fig.14-15　Tobacco phloem（showing sieve tube and companion cell）

the sieve plate with one or more sieve areas is distributed, and the sieve plate with only one sieve area is called the single sieve plate, such as the sieve tube of the pumpkin. A sieve plate with multiple sieve areas, such as a *Vitis vinifera*. sieve, is called a connecting strand. The connecting strand connects the sieve tube molecules to each other. In some plants, the sieve tube can also be seen on the side wall of the sieve tube.

筛管分子发育过程，早期有细胞核，随后细胞核逐渐溶解而消失，发育成熟后成无核的生活细胞（图 14-16）。

In the development process of sieve tube molecular, early nuclear, cytoplasm dense, then the nucleus gradually dissolved and disappeared, cytoplasm reduced, mature into non-nuclear living cells. It is also believed that the sieve tube molecules mature into a multicore structure, which is not easy to observe because the nuclei are small and dispersed (Fig.14-16).

筛孔的四周围绕联络索可逐渐积累一些特殊的碳水化合物，称为胼胝质；随着筛管成熟老

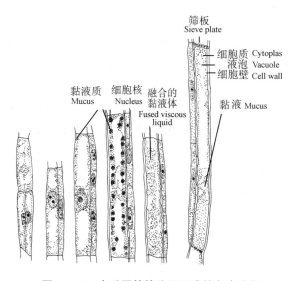

图 14-16　南瓜属筛管分子形成的各个阶段
Fig.14-16　Formation stages of the sieve tube in pumpkin

化，胼胝质不断增多，最后在整个筛板上形成垫状物，称为胼胝体。胼胝体一旦形成，筛孔被堵塞，联络索中断，筛管也就失去运输功能。单子叶植物筛管的输导功能可保持到整个生活期，而一些多年生双子叶植物在冬季来临前形成胼胝体，筛管暂时丧失输导作用，来年春季，胼胝体溶解，筛管又逐渐恢复其功能。但部分较老的筛管形成胼胝体后，将永远丧失输导能力而被新筛管取代。此外，当植物受到机械或病虫害等外界刺激时，胼胝体能迅速形成，封闭筛孔，以阻止营养物流失。

Special carbohydrates, called callose, can be accumulated around the sieve pore with the connecting strand. As the sieve tube matures and ages, the callose increases and eventually forms a pad on the sieve plate, called the callus. Once the callus is formed, the sieve pore is blocked, the connecting strand is broken, and the sieve tube loses its conducting function. In monocotyledons, the function of the sieve tube can be maintained throughout the whole life period, while in some perennial dicotyledons, the corpus callosum is formed before winter, and the sieve tube temporarily loses its function. In the following spring, the corpus callosum dissolves, and the sieve tube gradually recovers its function. However, after the formation of the corpus callosum, part of the older sieve tube will lose the ability to carry blood forever and will be replaced by the new sieve tube. In addition, when plants are stimulated by external stimuli such as machinery or pests and diseases, the corpus callosum can rapidly form and close the sieve to prevent nutrient loss.

伴胞和筛管是由同一母细胞分裂发育而成，二者之间存在发达的胞间连丝，共同完成有机物的输导。伴胞小型、细长，细胞壁薄，细胞核大，细胞质浓，液泡小，线粒体丰富。伴胞含有多种酶类物质，生理活动旺盛，筛管运输能力和伴胞代谢密切相关，伴胞会随着筛管的死亡而失去生理活性。

Companion cell is a small, slender living cell with a thin cell wall, large nucleus, dense cytoplasm, small vacuoles, and abundant mitochondria. Sieve tube molecules in angiosperms often have one or more companion cells. The companion cell and the sieve tube are formed by the division and development of the same mother cell. The companion cell contains a variety of enzymes, with vigorous physiological activities. The conducting capacity of the sieve tube is closely related to the metabolism of the companion cell, and the companion cell will lose its physiological activity with the death of the sieve tube.

（3）筛胞　筛胞是蕨类植物和裸子植物运输有机养料的组织。筛胞是单个狭长的细胞，无伴胞，直径较小，端壁倾斜，没有筛板，侧壁上具不明显的筛域。筛胞不像筛管那样首尾相连，而是彼此相重叠而存在，靠侧壁上筛域的筛孔运输，所以输导功能较差，属较原始的运输结构。

Sieve cell is the conducting molecule of ferns and gymnosperms. Sieve cell is a single long and narrow cell, without companion cell, small diameter, both ends acuminate and inclined, no sieve plate, the side wall without obvious sieve area. The sieve cells are not connected end to end like the sieve tube, but overlap with each other and exist, relying on the sieve pores on the side wall for conducting, so the conducting function is poor, belonging to a more primitive conducting structure.

PPT

第二节　维管束及其类型
2 Vascular bundle

维管束是蕨类植物、裸子植物、被子植物等维管植物的输导系统。维管束是一种束状结构，贯穿于整个植物体的内部，除了具有输导功能外，同时对植物体还起到支持作用。

Vascular bundles are the delivery systems of vascular plants such as ferns, gymnosperms, and angiosperms. Vascular bundle is a bundle-like structure that runs through the entire plant body. It has a transduction function and also supports the plant body.

一、维管束的组成
2.1　Vascular bundle composition

维管束主要由韧皮部与木质部组成。在被子植物中，韧皮部是由筛管、伴胞、韧皮薄壁细胞和韧皮纤维组成，木质部由导管、管胞、木薄壁细胞和木纤维组成；裸子植物和蕨类植物的韧皮部主要是由筛胞和韧皮薄壁细胞组成，木质部由管胞和木薄壁细胞组成。

Vascular bundles are mainly composed of phloem and xylem. The phloem of angiosperms is composed of sieve tubes, companion cells, phloem parenchyma cells, and phloem fibers. Xylem is composed of vessels, tracheids, wood parenchyma cells and wood fibers. The phloem of gymnosperms and ferns is mainly composed of sieve cells and phloem parenchyma cells, and the xylem is composed of tracheids and wood parenchyma cells.

裸子植物和双子叶植物的维管束在木质部和韧皮部之间常有形成层存在，能持续不断的分生生长，所以这种维管束称为无限型维管束或开放型维管束；蕨类植物和单子叶植物的维管束中没有形成层，不能进行不断的分生生长，所以这种维管束称为有限型维管束或闭锁型维管束。

The vascular bundles of gymnosperms and dicotyledones often have cambium between xylem and phloem, which can grow continuously, so they are called infinite vascular bundles or open vascular bundles. The vascular bundles of ferns and monocots have no cambium and can not grow continuously, so they are called finite vascular bundles or closed vascular bundles.

视频

二、维管束的类型
2.2　Type vascular bundle

根据维管束中韧皮部与木质部排列方式的不同，以及形成层的有无，将维管束分为下列几种类型（图 14-17、图 14-18）。

Vascular bundles are classified into the following types according to the arrangement of phloem and xylem(Fig.14-17, Fig.14-18).

1. 有限外韧型维管束　韧皮部位木质部外侧，中间没有形成层。如单子叶植物茎的维管束。

(1) Closed collateral vascular bundle　The phloem is lateral to thexylem and has no cambium in

医药大学堂
WWW.YIYAODXT.COM

305

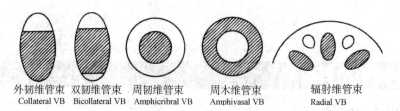

外韧维管束　双韧维管束　周韧维管束　周木维管束　辐射维管束
Collateral VB　Bicollateral VB　Amphicribral VB　Amphivasal VB　Radial VB

图 14-17　维管束的类型模式图
Figure 14-17　Type diagram of vascular bundle

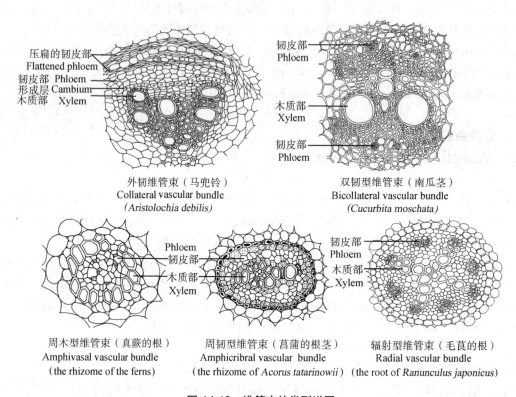

压扁的韧皮部
Flattened phloem
韧皮部　Phloem
形成层 Cambium
木质部　Xylem

韧皮部
Phloem

木质部
Xylem

韧皮部
Phloem

外韧维管束（马兜铃）
Collateral vascular bundle
(Aristolochia debilis)

双韧型维管束（南瓜茎）
Bicollateral vascular bundle
(Cucurbita moschata)

Phloem
韧皮部
木质部
Xylem

韧皮部
Phloem
木质部
Xylem

周木型维管束（真蕨的根）
Amphivasal vascular bundle
(the rhizome of the ferns)

周韧型维管束（菖蒲的根茎）
Amphicribral vascular bundle
(the rhizome of Acorus tatarinowii)

辐射型维管束（毛茛的根）
Radial vascular bundle
(the root of Ranunculus japonicus)

图 14-18　维管束的类型详图
Figure 14-18　Detailed view of the type of vascular bundle

the middle. Like the vascular bundles of a monocots stem.

2. **无限外韧型维管束**　韧皮部位于木质部外侧，两者之间有形成层，可使植物逐渐增粗生长。如裸子植物和双子叶植物茎中的维管束。

(2) **Open collateral vascular bundle**　The phloem is located on the outside of the xylem and there is a cambium between them, which can make the plant grow coarsely. Such as the vascular bundles in the stems of gymnosperms and dicotyledones.

3. **双韧型维管束**　木质部内外两侧都有韧皮部，而无形成层。常见于茄科、葫芦科、夹竹桃科、萝藦科、旋花科等植物的茎中。

(3) **Bicollateral vascular bundle**　There are phloems on the inner and outer sides of the xylem. It is commonly found in the stems of Solanaceae, Cucurbitaceae, Apocynaceae, Asteraceae and Convolvulaceae plants.

4. **周韧型维管束**　木质部位于中间，韧皮部围绕在木质部的四周，无形成层。如百合科、禾本科、棕榈科、蓼科及蕨类的某些植物。

(4) Amphicribral vascular bundle　The xylem is located in the middle, and the phloem is around the xylem, as seen in certain plants of Liliaceae, Poaceae, Palmaceae, Polygonaceae and some ferns.

5. **周木型维管束**　韧皮部位于中间，木质部围绕在韧皮部的四周。常见于少数单子叶植物的根状茎，如菖蒲、石菖蒲、铃兰等。

(5) Amphivasal vascular bundle　The phloem is located in the middle, and the xylem surrounds the phloem. It is common in rhizomes of a few monocotyledons, such as *Acorus calamus, A. tatarinowii* and *Convallaria majalis.*

6. **辐射型维管束**　韧皮部和木质部相互间隔交互呈辐射状排列。根初生构造的维管束属于该类型。

(6) Radial vascular bundle　The phloem and xylem alternate with each other in a radial arrangement. It is the primary structure of the root.

重 点 小 结
Summary

植物组织
- 分生组织：按位置（顶端、侧生、居间），按来源（原生、初生、次生）
- 成熟组织
  - 薄壁组织：基本、同化、贮藏、吸收、通气
  - 保护组织：表皮、周皮
  - 机械组织：厚角、厚壁（纤维、石细胞）
  - 输导组织：管胞、导管；筛管、伴胞、筛胞
  - 分泌组织：外部（腺毛、蜜腺）、内部（分泌细胞、分泌腔、分泌道、乳汁管）

维管束：木质部（导管、管胞、木薄壁细胞、木纤维），韧皮部（筛管、伴胞，筛胞；韧皮薄壁细胞，韧皮纤维）。

Plant tissue
- Meristem tissue
  - Apical meristem, lateral meristem, intercalary meristem
  - Promeristem, primary meristem, secondary meristem
- Mature tissue
  - Parenchyma: basic, assimilation, storage, absorption, ventilation
  - Protective tissue: epidermis, periderm
  - Mechanical tissue: collenchyma, sclerenchyma
  - Conducting tissue: vessel and tracheid, sieve tube, companion cell, sieve cell
  - Secretory tissue: external (glandular hair, nectary), internal (secretory cell, secretory cavity, secretory canal, laticifer)

Vascular bundle refers to the structure of vascular tissue of vascular plants, which consists of xylem and phloem arranged in bundles. Vascular tissues are connected to each other to form a vascular system. Its main role is to conduct water, inorganic salts and organic nutrients for plants, and also to support plants. In addition, the same type of tissue often has different structural characteristics in different plants, which can be used for microscopic identification of traditional Chinese medicine.

题库

目 标 检 测
Questions

1. 简述植物分生组织的分类方法。

Please describe the types of the plant meristem.

2. 植物中常见的薄壁组织类型有哪些？

What are the common types of parenchyma in plants?

3. 绘图说明常见的气孔轴式类型有哪些。

Please draw and explain the common types of stoma type.

4. 简述厚角组织和厚壁组织的主要区别。

Please describe the main differences between collenchyma and sclerenchyma.

5. 导管按其加厚形式不同，可分为哪些类型？

According to the different thickening forms, what types can vessel be divided into?

6. 植物中常见的内部分泌组织有哪些类型？

What are the common types of internal secretory tissues in plants?

第十五章　根、茎、叶的组织结构

Chapter 15　Structure of Root, Stem and Leaf

学习目标 | Learning goals

1. **掌握**　双子叶植物和单子叶植物根的初生结构特征；双子叶植物根的次生结构特征；双子叶植物茎的初生构造和次生构造特征及变化过程；叶的组织结构特征。

2. **熟悉**　根尖的结构及根的异常构造；根的初生结构向次生结构变化的过程。

● Know the primary structural characteristics of the roots of dicotyledonous and monocotyledonous plants; the secondary structural characteristics of the roots of dicotyledonous plants; the primary and secondary structural characteristics of the stems of dicotyledonous; the anatomical characteristics of leaves.

● Be familiar with the structure of the root tip and the abnormal structure of the root; the process of the change from the primary structure of the root to the secondary; the differences in the structure of stems of monocotyledon and dicotyledon.

第一节　根的组织结构

1 Anatomy of the root

一、根尖的构造
1.1　Root tips

根尖指从根的顶端到着生根毛的部分，长 4~6mm，是根的生命活动最活跃的部位。损伤根尖会直接影响根的生长、发育和吸收。根尖常分为根冠、分生区、伸长区和成熟区（图 15-1）。

The root tip refers to the part from the top of the root to the root hair, which is 4 to 6 mm long and is the most active part of the root. Damaged root tips will directly affect the growth, development and absorption of roots. The growing root tip may be further developing into four zones: the root cap, the zone of division, the zone of elongation, and the zone of maturation(Fig.15-1).

1. **根冠**　位于根尖最先端，呈帽状，由多层不规则排列的薄壁细胞组成，有保护作用。根

PPT

视频

冠外层细胞损坏后形成黏液，有助于根向前延伸。绝大多数植物的根尖都有根冠，但寄生和有菌根的植物通常无根冠。

(1) The root cap The root cap is composed of a thimble-shaped mass of parenchyma cells covering the tip of each root. One of its functions is to protect from damaging the delicate tissues behind it as the young root tip pushes through soil particles. The dictyosomes of the root cap's outer cells secrete and release a slimy substance that lodges in the walls and eventually passes to the outside. The cells, which are replaced from the inside, constantly slough off, forming a slimy lubricant that facilitates the root tip's movement through the soil. This mucilaginous lubricant also provides a medium favorable to the growth of beneficial bacteria that add to the nitrogen supplies available to the plant.

2. **分生区** 位于根冠之内，呈圆锥状，长1~3mm，为顶端分生组织所在部位，是细胞分裂最旺盛的区域，故又称为生长锥。分生区最先端的一群细胞来源于种子的胚，属于原分生组织。分生区细胞不断分裂，细胞数目增加，将来进一步分化形成根的表皮、皮层和中柱。

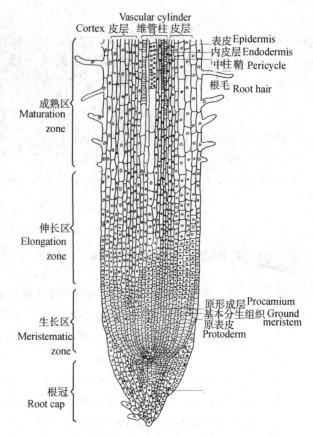

图 15-1　根尖的构造
Fig.15-1　Root tip structure

(2) The division zone The devision zone Which arises directly from the root apical meristem (RAM), is just behind the root cap and generates cells of both the root cap and the root proper. While it may be largely a semantic distinction, the RAM may be thought of as the quiescent center and its immediate, undifferentiated daughter cells. Those products subsequently differentiate into the protoderm, procambium, and the ground meristem, and it is those three meristems that make up the zone of division.

3. **伸长区** 位于分生区后方到出现根毛处，一般长2~5mm，其细胞分裂活动逐步减弱，分化程度逐渐增高，细胞纵向伸长。从生长锥分裂出来的细胞，在此迅速伸长，沿根的长轴方向显著延伸，使根尖不断伸入土壤中。同时，细胞开始分化，相继出现环纹导管和筛管。

(3) The elongation zone At the upper end of the zone of cell division (i.e., toward the stem) is a gradual change into a zone of elongation. True to its name, cells in this zone undergo elongation prior to full differentiation, and this elongation is responsible for an increase in length of the root. Little to no secondary cell wall is laid down yet, and the cells do not function in water or ion uptake.

4. **成熟区** 位于伸长区后方，由伸长区细胞进一步分化而来。细胞分化成熟，并形成各种初生组织。最显著的特点是，表皮细胞的外壁向外突出形成根毛，故又称根毛区；根毛增加了根的吸收面积，使该区成为根部行使吸收作用的主要部位。水生植物一般无根毛。

(4) The maturation zone The zone of maturation is where differentiation of tissues takes place, involving vascular tissues, cortical parenchyma, and the rhizodermis. Stele formation begins with the

医药大学堂
WWW.YIYAODXT.COM

maturation of primary xylem, primary phloem, and the endodermis allowing for the uptake, and transport, of water and ions. Root hairs, which play a critical role in water uptake, begin to appear. Root hairs can only develop after root elongation has ceased, or else they would be torn off as the root advanced through the soil. The growth of root hairs greatly increases the area of the root allowing for effcient uptake of water and mineral substances, which can be delivered to the stele.

二、根的初生生长和初生结构
1.2　Primary growth and structure of roots

（一）根的初生生长
1.2.1　Primary growth of the roots

根的发育起源于生长锥。由其原分生组织分裂出来的细胞，逐渐分化出原表皮层、基本分生组织和原形成层。最外层的原表皮层细胞垂周分裂，进一步分化成表皮；中间的基本分生组织细胞垂周分裂和平周分裂，增大体积，分化为根的皮层；最内的原形成层分化成由中柱鞘、初生木质部和初生韧皮部组成的维管柱。这种直接来源于顶端分生组织细胞的增生和成熟，使根延长的生长称为初生生长。由初生生长所产生的各种成熟组织，称为初生组织。由初生组织所形成的表皮、皮层和维管柱组成的结构称为根的初生结构。

The root apical meristem is the source of new cells for the root. The cells split from the original meristem gradually differentiate into the protoderm, the ground meristem and the protocambium. The cells of the outer protodermal layer divided vertically and further differentiate into the rhizodermis. The ground meristem cells in the middle divided vertically and horizontally and differentiate into the cortex, and the cells of the innermost protocambium differentiate into a vascular column, which is composed of the pericycle, the primary xylem and the primary phloem. After that, the primary structure of the root then formed. Therefore, the primary structure of the root includes the rhizodermis, the cortex and the vascular column.

（二）双子叶植物根的初生结构
1.2.2　Primary structure of the eudicot roots

从双子叶植物根尖的成熟区作一横切面，从外到内依次为表皮、皮层和维管柱三部分。

1. 表皮　位于根的最外方，由单层细胞组成，细胞排列整齐、紧密，无细胞间隙，细胞壁薄，不角质化，富有通透性，无气孔。一部分细胞的外壁向外突出，形成根毛（图15-2）。

(1) Rhizodermis　It is more appropriate to utilize the term "rhizodermis" rather than "epidermis" for the outermost layer of root tissues in the primary state of growth (Fig.15-2). Since this tissue in root systems functions very differently from that of stems and leaves. In its underground state, the rhizodermis has no stomata, it is specialized for the absorbance of water and mineral substances, it produces mucigel as a lubricant, and it is never covered by a cuticle but develops short-living root hairs. Upon further development, multiple rhizodermal layers may arise, as is common in the irises, such as epiphytic orchids. The multiple layers are not a true periderm because they are the products of cell divisions in the rhizodermis, not in the phellogen.

2. 皮层　位于表皮内方，由多层薄壁细胞组成，在成熟区中占有最大的部分。细胞排列疏松，细胞间隙明显，由外向内依次分为外皮层、皮层薄壁组织（中皮层）和内皮层。

(2) Cortex　The root cortex, limited by the endodermis, is a site of storage and oxygen transport. The root cortex is parenchymatous tissue derived from the ground meristem, which in keeping with its

311

role in storage, is often much larger than the stem cortex. The cortex may become heavily scleried in older monocot roots.

外皮层为邻接表皮的1层或几层细胞，排列整齐、紧密，无细胞间隙。在表皮被破坏后，此层细胞壁常增厚并栓质化，起保护作用。

The exodermis lies between the rhizodermis and cortex. Not all roots have an exodermis.The exodermis is a functional equivalent of the endodermis. It may be suberized, cutinized, and contain casparian strips on the anticlinal walls. Like with the endodermis, water and ions must pass into the symplast of the exodermis cells to enter the root cortex.

皮层薄壁组织（中皮层）细胞壁薄，排列疏松，细胞间隙明显，既能将根毛吸收的溶液转送到根的维管柱，又可将维管柱内的养料转送出来，还有贮藏作用。

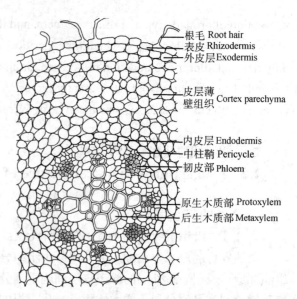

图 15-2　双子叶植物根的初生构造（毛茛）
Fig.15-2　The primary structure of the dicotyledon roots (*Ranunculuss sp.*)

Cortex parenchyma, also known as the mesothelium, is the layers of cells between the rhizodermis and the endodermis. The cell wall is thin and the space among cells is obvious. It can not only transfer the nutrient solution absorbed by the root hair to the vascular column, but also transport the nutrient in the vascular column to other cells or secret to the soil, and has a storage effect.

内皮层为皮层最内方的一层细胞，排列紧密整齐，无细胞间隙。内皮层的细胞壁有时会沿径向壁（侧壁）和上下壁（横壁）局部木质化或木栓化增厚，增厚部分呈带状，称凯氏带（图15-3）；从横切面观，增厚部分成点状，故又叫凯氏点。部分单子叶植物内皮层细胞出现全面木栓化增厚，在正对原生木质部角处留存细胞壁不增厚的内皮层细胞，称通道细胞，起控制皮层与维管束间物质转运的作用。

The endodermis is a single layer of cells that does not permit the free flow of water and ions between the individual cells but rather forces such materials to cross a cell membrane. This is due to presence of a Casparian strip which is a band of suberized lignin and proteins that covers the radial and transverse (anticlinal) walls of the endodermis (Fig.15-3). It is a component of the primary cell wall. Its presence forces the water and solutes to pass from the apoplast through the plasma membrane via a symplastic route to cross the endodermis. In monocots, the entire cell may develop thick deposits of

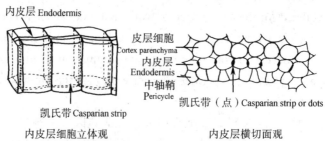

图 15-3　内皮层及凯氏带
Fig.15-3　The endodermis and casparian strip

suberin as the root matures. After entering the endodermal cells, the water can then move freely to the xylem where it is swept upward to the leaves in a stream of transpirational flow. Passage cells (sometimes called transfusion cells) lack the heavy suberization and provide for a diffusional pathway with less resistance. They tend to be located at the ends of the xylem poles.

3. **维管柱**　是内皮层以内的所有部分，位于根中央，所占比例最小，也称中柱。包括中柱鞘、初生木质部和初生韧皮部三部分，有的植物还具有髓部。

(3) The vascular cylinder　The central vascular cylinder of the root is also called the stele. The root stele is bound by the endodermis to the exterior and contains xylem, phloem, parenchyma, and a meristematic layer, the pericycle, which is the source of lateral roots.

中柱鞘是内皮层以内维管柱最外方的 1 层细胞，常为薄壁细胞；少数植物有 2 层至多层细胞。中柱鞘细胞排列整齐、无细胞间隙，其分化程度较低，具有潜在的分生能力。

The pericycle: The pericycle is a single layer of parenchyma cells lying just inside the endodermis in all roots. Pericycle cells are meristematic and the source of all lateral roots. They also contribute to the vascular cambium and cork cambium in those roots that exhibit secondary growth.

初生维管束位于根的最内方，是根的输导系统，由原形成层直接分化而成，包括初生木质部和初生韧皮部。一般初生木质部分为几束，呈星芒状，初生韧皮部排列在初生木质部星芒状之间，所以又称为"辐射维管束"。初生木质部和初生韧皮部辐射状相间排列是根的初生构造的特点。

初生木质部分化成熟的顺序是自外向内发育，称外始式。最先分化成熟的木质部称原生木质部，位于初生木质部的外方，导管直径小，多环纹或螺纹，位于木质部的角隅处；后分化成熟的初生木质部称后生木质部，其导管直径较大，多梯纹、网纹或孔纹。这种分化成熟的顺序体现了植物形态构造和生理机能的统一性。

根的初生木质部束横切面观呈星芒状，其束的数目随植物种类而异。一般来说，同种植物其初生木质部束是相对稳定的，如十字花科、伞形科和多数裸子植物的根中，只有两束木质部，称为二原型；毛茛科的唐松草属有三束，称三原型；葫芦科、杨柳科及毛茛科毛茛属的一些植物有四束，称四原型；如果束数更多，多于六束则称多原型。被子植物的初生木质部由导管、木薄壁细胞和木纤维组成；裸子植物的主要是管胞。

初生韧皮部发育成熟的方式也是外始式，外方先分化成熟的韧皮部为原生韧皮部，内方后分化成熟的为后生韧皮部。在同一植物的根内，初生韧皮部束的数目和初生木质部束的数目相同；被子植物的初生韧皮部一般由筛管、伴胞、韧皮薄壁细胞和韧皮纤维构成；裸子植物的初生韧皮部主要是筛胞。

一般双子叶植物初生木质部一直分化到根的中心，一般不具髓部。部分单子叶植物根的木质部一般不分化到中心，有发达的髓部，如百部块根。

The primary vascular bundles: All of the root vasculature is a more or less solid cylinder of xylem in the middle with phloem to the exterior. thus, most eudicot and monocot steles are protosteles. Pith is rare in eudicots as it gets crushed early on in root development but somewhat common in monocots.

Root xylem is arranged in a spoke-like pattern with the number of spokes, or "xylem poles," being characteristic of the taxon. Eudicots typically have discrete xylem poles. A two-arm pattern is called diarch, followed by triarch, tetrarch, pentarch, and up to as many as eight poles. Monocots are typically polyarch with numerous xylem poles arranged in a more or less ring shape at the periphery of the stele and a central pith. Larger monocot roots have a polyarch with so many xylem poles that they may be called an atactostele.

In terms of phloem distribution, primary phloem in eudicot roots is positioned between the xylem arms, as seen in Ranunculus. Monocot distribution is similar but the polystelic nature of monocot roots results in individual groupings of xylem with phloem arranged in patches in between the groupings or scattered to the outside of the xylem.

Monocot and eudicot roots both have an exarch pattern of vascular development with protoxylem to the exterior and metaxylem to the interior. Note that stem vascular development is endarch.

（三）单子叶植物根的结构

1.2.3 Root structure of monocotyledons

单子叶植物和双子叶植物根的初生构造相似，也分为表皮、皮层和维管柱三部分，但有所不同。

Monocotyledonous plants have the same primary structure as dicotyledonous plants and can also be divided into three parts: epidermis, cortex and vascular column. But they are different in the following aspects.

1. 内皮层 大部分细胞在发育后期常呈五面增厚，其两侧径向壁、上下壁及内切向壁（内壁）均显著增厚并木栓化，只有外切向壁（外壁）未加厚，横切面观，内皮层细胞壁增厚部分呈马蹄形。正对初生木质部束的内皮层细胞常停留在凯氏带阶段，称为通道细胞（图15-4），一般认为通道细胞是维管柱内外物质运输的主要途径，有些植物根内皮层无通道细胞。

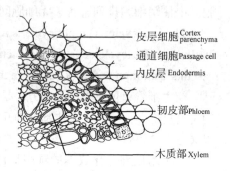

图 15-4 通道细胞
Fig.15-4 Passage cell

(1) Endodermis The walls of most of the cells in the endodermis often show five-sided thickening in the late stage of development. The radial wall on both sides, the upper and lower walls and the inner tangential wall (inner wall) are significantly thickened and corked, while only the outer tangential wall (outer wall) is not thickened. The thickened part of the wall in endodermis cell is horseshoe shaped when viewed in transverse section. The endodermal cells at the radiation angle of the primary xylem are often not thickened and are called channel cells(Fig.15-4). It is generally believed that channel cells are the main way of material transport inside and outside the vascular column, but there are no channel cells in the endodermis of some plants.

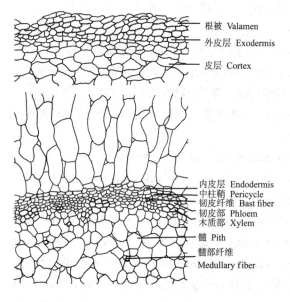

图 15-5 直立百部块根横切面
Fig.15-5 Cross section of the tuber of *Stemona sessilifolia*

2. 维管柱 中柱鞘细胞在发育后期部分或全部成木化厚壁组织，如竹类、菝葜等。维管柱多为多原型，至少是六原型（图15-5）。髓部发达，由薄壁细胞组成，也有髓部细胞增厚木化成厚壁组织的，如鸢尾。

(2) Vascular cdumn In the late stage of root development, the scabbard is usually partly or completely woody. In monocotyledonous plants, the scabbard is thick-walled tissue, such as bamboo, smilax and sarsaparilla.The vascular column of the root of monocotyledonous plant is at least 6 bundles(Fig.15-5), namely 6 prototypes, generally 8 to 30 bundles, some plants in the palm family can reach hundreds of bundles. In the center is a well developed pith, composed of parenchyma cells. There are also pith cells thickened and woody to form thick-walled tissue, such as iris.

（四）侧根的发生

1.2.4　Origin of the lateral roots

侧根起源于中柱鞘。当侧根发生时，部分中柱鞘细胞脱分化恢复分裂能力（图 15-6）。首先进行平周分裂，使细胞层数增加，并向外突起；然后进行平周分裂和垂周分裂，产生一团新细胞，形成侧根原基，其顶端分化为生长锥和根冠，生长锥细胞继续进行分裂、生长和分化，逐渐伸入皮层，并突破皮层和表皮，形成侧根。

Lateral roots originate in the pericycle and push through the cortex. Lateral (or branch) roots develop off of an existing root, starting just behind the zone where root hairs senesce(Fig.15-6). Primary laterals arise from a taproot, secondary laterals arise from primary, tertiary from a secondary, and so on. In contrast to the shoot where branches originate exogenously from the apical meristem, lateral (or branch) roots are initiated endogenously in the pericycle without any relation to the apical meristem.

Typically, the point of origin of a lateral root is opposite to the xylem in eudicots and opposite to the phloem in monocots. The lateral root primordial growth must penetrate the endodermis, cortex, exodermis, and the rhizodermis in order to appear on the surface of parental root.

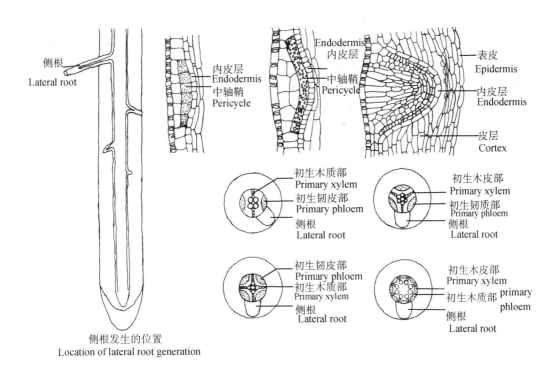

图 15-6　侧根的发生和形成

Fig.15-6　Occurrence and formation of lateral roots

三、根的次生生长和次生结构
1.3　Transition from primary to secondary growth in roots

大多数蕨类植物和单子叶植物的根一直保持着初生构造。双子叶植物和裸子植物的根在初生生长基础上，经次生生长，形成次生构造。

The roots of gymnosperms and woody eudicots exhibit secondary growth. Doing so requires the development of two meristems—a vascular cambium (to produce secondary xylem and phloem) and a phellogen (to replace the rhizodermis with a corky periderm) —both of which must form a continuous cylinder or meristematic tissue. The phloem and the pericycle play roles in the origin of these secondary meristems.

1. **维管形成层的产生及其活动**　当根进行次生生长时，在初生木质部和初生韧皮部之间的一些薄壁细胞恢复分裂能力，进行平周分裂，转变成形成层，并逐渐向中柱鞘部位扩展，使相连接的中柱鞘细胞也开始分化成为形成层的一部分，这样形成层就由片段状连成一个凹凸相间的形成层环（图 15-7）。

(1) Formation and activity of vascular cambium　The first step is the differentiation of phloem cells at the inner edge of the protophloem strands into cambial initials capable of generating xylem to the inside and phloem to the outside. Subsequently, pericycle cells to the exterior of the protoxylem poles also divide to produce more new cells of the vascular cambium. Eventually, the two groups of dividing cells will merge to form a circular meristem and complete an encompassment of the primary xylem(Fig.15-7).

形成层细胞不断进行平周分裂，向内产生新的木质部，加于初生木质部的外方，称次生木质部，包括导管、管胞、木薄壁细胞和木纤维；向外产生新的韧皮部，加于初生韧皮部的内方，称次生韧皮部，包括筛管、伴胞、韧皮薄壁细胞和韧皮纤维。此时，维管束中的木质部和韧皮部已由初生构造的间隔排列转变为木质部在内方、韧皮部在外方的内外排列。次生木质部和次生韧皮部合称为次生维管组织，是次生构造的主要部分。

形成层细胞活动时，在一定部位也分化出一些沿径向延长的薄壁细胞，呈辐射状排列，贯穿于次生维管组织中，称维管射线，亦称次生射线。位于木质部的称木射线，位于韧皮部的称韧皮射线。射线具有横向运输水分和养料的功能。维管射线组成根维管组织的径向系统，而导管、管胞、筛管、伴胞、纤维等组成维管组织的轴向系统。

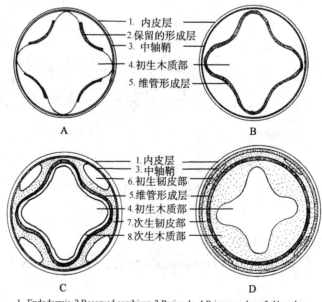

1. Endodermis；2.Reserved cambium；3.Pericycle；4.Primary xylem；5. Vascular cambium；6.Primary phloem；7.Secondary phloem；8.Secondary xylem

图 15-7　维管形成层的发生过程
Fig.15-7　Generation process of vascular cambium

次生生长过程，新生次生维管组织不断添加在初生韧皮部的内方，初生韧皮部遭挤压而被破坏，成为没有细胞形态的颓废组织。由于形成层产生的次生木质部的数量较多，并添加在初生木质部之外，因此，粗大的树根主要是木质部，非常坚固。

在根的次生韧皮部中，常有各种分泌组织分布，如马兜铃根有油细胞，人参根有树脂道，当归根有油室，蒲公英根有乳汁管。有的薄壁细胞（包括射线细胞）中常含有晶体及贮藏物质，如生物碱、糖类等，多为药用成分。

All of the tissues outside of the pericycle, which include the rhizodermis, exodermis, cortex, and endodermis, eventually die and are shed. As the vascular cambium becomes active at the start of each growing season, the production of new xylem and phloem increases the girth of the root. With multiple layers of xylem, the root may also become woody and reveal growth rings as in stem wood.

The rhizodermis in the primary state of growth is quite different anatomically from that in the secondary state of growth. Primary growth represents an opportunity for roots to increase in length in search of exploitable resources of water and minerals. Therefore, they are specialized for uptake with root hairs being a major route for the entry of water and minerals. Secondary growth involves strengthening the root to provide maximum anchorage.

2. **木栓形成层的产生及其活动**　由于形成层的活动，根不断加粗，表皮及部分皮层不能相应加粗而破坏。这时中柱鞘细胞恢复分裂能力形成木栓形成层，它向外分生产生木栓层，覆盖在根外层起保护作用。向内分生产生栓内层，栓内层为数层薄壁细胞，排列较疏松；有的栓内层比较发达，称为"次生皮层"。栓内层、木栓形成层和木栓层三者合称为周皮。植物学上的根皮指周皮，而药材学所称的根皮类药材，如香加皮、地骨皮、牡丹皮等，则是指形成层以外的部分，主要包括韧皮部和周皮（图 15-8）。

(2) Generation and activity of cork cambium　Root thickening accompanied by vascular cambium activity, the outer epidermis and part of the cortex were destroyed because they could not be correspondingly thickened. At this point, the cells in the pericycle resume their ability to divide and form the cork cambium. It produces the cork layer outwards and covers the root. There are 1~2 layers of cells in the inner layer of the cork, with thin cell walls and loose arrangement. Some plug more developed inner layer, become secondary cortex. Cork inner layer, cork cambium and cork layer are collectively called pericarp. All the tissues (epidermis and cortex) outside the periderm have withered away due to the loss of moisture and nutrients within. Therefore, there is no epidermis and cortex in the secondary structure of the root, but the periderm instead. The root bark in botany refers to the part of pericarp, while the root bark in the medicinal materials, such as rhizome bark, rhizome bark, peony bark, etc., refers to the part outside the cambium, mainly including phloem and pericarp(Fig.15-8).

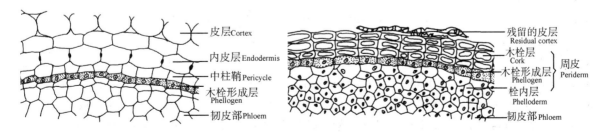

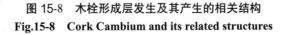

图 15-8　木栓形成层发生及其产生的相关结构
Fig.15-8　Cork Cambium and its related structures

单子叶植物的根没有形成层，不能加粗，也没有木栓形成层，不能形成周皮，而由表皮或外皮层行使保护机能。也有一些单子叶植物，如百部、麦冬等，表皮分裂成多层细胞，细胞壁木栓化，形成"根被"的保护组织。

The roots of monocotyledonous plants have no cambium, cannot be thickened, and cannot have a cork cambium, and cannot form a pericarp, which is protected by the epidermis or outer cortex. There are also some monocotyledonous plants, such as hundred parts, ophiopogonis, etc., the epidermis divided into multiple cells, cell wall cork, forming a protective tissue called "velalame".

3. 根的次生结构　根的维管形成层与木栓形成层的活动产生了根的次生结构，主要包括周皮、次生韧皮、维管形成层、次生木质部和维管射线（图15-9）。在次生结构中，最外侧是周皮，起保护作用。次生韧皮部呈间断连续的筒状，由筛管、伴胞、韧皮纤维和韧皮薄壁细胞组成。次生木质包括导管、管胞、木薄壁细胞和木纤维，导管多为梯纹、网纹和孔纹导管。维管射线径向排列，横贯次生韧皮部和次生木质部。

(3) The secondary structure of the root　The activity of the vascular cambium and the cork cambium of the root forms the secondary structure of the root, which mainly includes the periderm, secondary phloem, secondary xylem, vascular cambium and vascular ray(Fig.15-9). In the secondary structure of the root, the outermost part is the cork. The cells in the cork layer of the peririderm are arranged very neatly in the radial direction, and the inner layer is the cork cambium, and the phelloderm is located at the innermost side. The secondary phloem is in a continuous tubular shape, which contains sieve, companion cells, phloem fibers and phloem parenchyma cells. The secondary xylem has vessel element with different diameters, most of which are annular vessel, spiral vessel, scalariform vessel, reticulate vessel,pitted vessel. In addition to the vessels, xylem fibers and parenchyma cells are also visible. Vascular rays are arranged in a radial direction across the secondary phloem and secondary xylem. And the vascular rays have the function of transversely transporting water and nutrients.

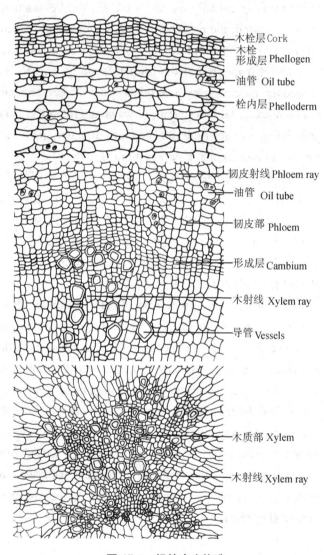

图 15-9　根的次生构造
Fig.15-9　The secondary structure of the root of *Saposhnikovia divaricata*

4. 根的异常结构　一些双子叶植物的根，除正常的次生构造外，还产生一些特有的维管束，称异形维管束，形成根的异常构造，也称为三生构造（图15-10）。常见有以下几种类型。

(4) Anomalous structure of root　In addition to the normal secondary structure, the roots of some

dicotyledonous plants also produce some unique vascular bundles, called anomalous vascular bundles, forming the anomalous structure of the roots, also known as the tertiary structure(Fig.15-10). The main difference between the tertiary structure and the secondary structure lies in the continuous generation of new cambium rings in the cortex and the formation of new anomalous vascular bundles. There are several common types as follow.

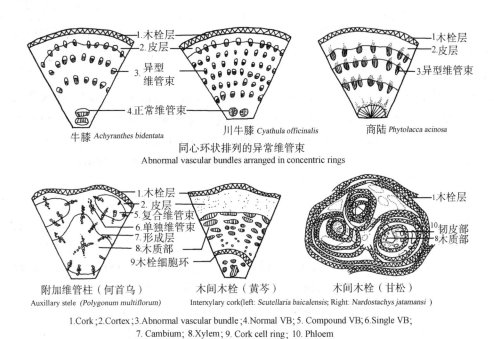

同心环状排列的异常维管束
Abnormal vascular bundles arranged in concentric rings

1.Cork；2.Cortex；3.Abnormal vascular bundle；4.Normal VB；5. Compound VB；6.Single VB；
7. Cambium；8.Xylem；9. Cork cell ring；10. Phloem

图 15-10　根的异常构造示例

Fig.15 -10　The abnormal structure of the root

　　同心环状排列的异型维管束：一些双子叶植物的根，初生生长和早期次生生长都发育正常。当次生生长发育到一定阶段，形成层常失去分生能力，在相当于中柱鞘部位的薄壁细胞恢复分生能力，形成新的形成层，向外分裂产生大量薄壁细胞和一圈异型维管束，如此反复多次，形成多圈异型维管束，其间有薄壁细胞相隔，呈同心环状排列。这种异常构造常见于苋科、商陆科、紫茉莉科等植物中。如牛膝根中央正常维管束的外方形成数轮多数小型的异型维管束，排列成 3~4 个同心环；川牛膝的异型维管束排列成 5~8 个同心环。商陆的横切面上的异形维管束形成多轮凹凸不平的同心环状层纹，习称"罗盘纹"。

　　Anomalous vascular bundles: Anomalous vascular bundles arranged in concentric rings are normal for both primary and early secondary growth in the roots of some dicotyledonous plants. When completed secondary growth, the vascular cambium often lose meristematic ability. Parenchyma cells in the pericycle recover meristematic ability and form new cambium. It produces a large number of parenchyma cells and a circle of closed collateral vascular bundle outward. And there are many different vascular bundle formed repeatedly which separated by parenchyma cells and arranged in concentric rings. This anomalous structure is common in such plants as Amaranthaceae, Phytolaccaceae, Nyctaginaceae. A typical example is the vascular bundles in the root of the niuxi (*Achyranthes bidentata*). On the outside of the central normal vascular bundle, several rounds of most small anomalous vascular bundles are formed.The vascular bundles are point-shaped and arranged in consecutive circles in the cross section. The central normal vascular bundle is larger. The anomalous vascular bundles in the root of *Cyathula*

officinalis are arranged in 5~8 concentric rings. And the anomalous vascular bundles in the root of the *Achyranthes bidentata* are arranged into 3~4 concentric rings. The number of anomalous vascular bundles and the number of concentric rings are different, which can be distinguished from the two herbs. Another example is *Phytolacca acinosa*. There are a number of uneven concentric rings on the cross section of its root, which can be used as an important characteristics in the identification of traditional Chinese medicine.

附加维管柱：一些双子叶植物的根，皮层部分薄壁细胞转化为多个新的形成层环，对于原有形成层环而言是异心的，由此分生出一些大小不等的异型维管束，形成异常构造。如何首乌的"云锦花纹"。

Auxillary stele: Roots of some dicotyledonous plants form normal vascular bundles. And then the parenchyma cells in cortex form several new cambium rings. They are different centers from the original cambium ring and produce anomalous vascular bundles, called auxillary stele. And it can be also used as an important characteristics in the identification of traditional Chinese medicine, such as *Polygonum multiflorum*.

木间木栓：一些双子叶植物的根，次生木质部薄壁组织细胞脱分化，恢复分生能力，在次生木质部内再形成木栓带，称木间木栓。如黄芩的老根中央可见木栓环，新疆紫草根中央也有木栓环带。甘松根中的木间木栓环包围一部分韧皮部和木质部而把维管柱分隔成2~5束。

Interxylary cork or included periderm: The secondary xylem of roots in some dicotyledonous plants also forms a cork belt, called interxylary cork or included periderm. And it is usually formed by parenchyma cell of secondary xylem, such as the old root of *Scutellaria baicalensis* and the central part of the root of *Arnebia euchroma*. The interxylary cork encloses part of the phloem and xylem and divides the vascular cylinder into 2–5 bundles. And they break off into separate bundles in the older parts of the root.

第二节　茎的组织结构
2 Anatomy of the stem

PPT

一、茎尖
2.1　Stem tip

茎尖是指茎或枝的顶端，为顶端分生组织所在部位，由生长锥（分生区）、伸长区、成熟区三部分组成。茎尖结构与根尖相似，但无类似根冠的构造，而由幼叶或芽鳞紧紧包裹。顶端分生组织活动形成叶原基和芽原基，后发育成叶或腋芽，腋芽发育成枝；成熟区的表皮常有气孔和毛茸。过茎尖的纵切面，自上而下可分为分生区、伸长区和成熟区三部分。

There is an apical meristem in which cells actively divide at the tip of each stem, and it is this meristem that contributes to an increase in the length of the stem. The cells of the apical meristem undergo mitosis, and soon differentiate into protoderm, procambium, and ground meristem. The growing stem tip may be further developing into three zones: meristematic zone (the growth cone), elongation zone, and mature zone. Stems represent the main axes of plants, being distinguished into nodes and

医药大学堂
WWW.YIYAODXT.COM

internodes, and bearing leaves and axillary buds at the nodes. The buds grow out into lateral shoots, inflorescences or flowers.

二、双子叶植物茎的初生生长及其结构
2.2 Primary structure of dicotyledons

由生长锥分裂的细胞逐渐分化形成原表皮层、基本分生组织、原形成层等初生分生组织，进而分化形成茎的初生构造。将双子叶植物茎的成熟区作一横切面，从外到内分别为表皮、皮层、维管柱三部分（图 15-11）。表皮具气孔和毛茸。皮层占较小比例，由薄壁组织和厚角组织构成，无内皮层，少数植物茎皮层具有淀粉鞘。维管柱包括维管束、髓部、髓射线等。初生维管束由初生韧皮部、初生木质部、束中形成层构成。

The protoderm gives rise to the stem epidermis. The procambium produces water conducting primary xylem cells and primary phloem cells that have several functions, including the conduction of food. The ground meristem produces two tissues (pith and cortex) composed of parenchyma cells(Fig.15-11). All five of the tissues produced by the apical meristem complex (epidermis, primary xylem, primary phloem, pith, and cortex) arise while the stem is increasing in length and are called primary tissues. The primary structure of the stem is also the region where the mature zone of the stem is located, which may be divided into three zones: the epidermis, the cortex and the vascular cylinder. Dicotyledonous stems do not remain in the primary state for long, and some secondary growth in thickness is usually present in most species.

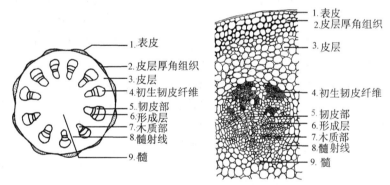

视频

1. Epidermis; 2. Parenchyma in cortex; 3.Cortex; 4.Primary phloem fiber; 5. Phloem; 6.Cambium; 7. Xylem; 8. Medullary ray; 9. Pith

图 15-11 双子叶植物茎的初生构造
Fig.15-11 The primary structure of dicotyledon stem

1. 表皮 位于茎的最外方，由一层长方形、扁平、排列整齐的细胞组成，无细胞间隙；一般不含叶绿体，有些含花青素；有各式气孔，毛茸；细胞外壁厚，长具角质层，少数具蜡被。

(1) Epidermis The outermost layer of stem tissues in the primary state of growth is called epidermis. The cells may be similar in form to those of the leaf of the same species. It consists of a layer of rectangular, flat, neatly arranged cells, no cell gaps, thick cytoderm, keratinized, being kinds of stomata, sometimes also having hairs, and a few plant epidermis are waxed.

2. 皮层 是表皮和维管柱之间的部分，由多层生活细胞构成，占较小比例，不如根的皮层发达。皮层主要由薄壁组织构成，细胞大、壁薄，排列疏松，具细胞间隙；但在近表皮部分常有厚

角组织，有的还有分泌组织、纤维、石细胞等；无内皮层，故皮层与维管区域间无明显分界，少数植物茎皮层最内一层薄壁细胞含丰富的淀粉粒，称淀粉鞘，如蚕豆、蓖麻等。

(2) Cortex　The cortex can be very narrow, and composed of few cell layers, or wide and multilayered, not as well developed as root cortex. Cells are large, thin walls, loosely arranged, with intercellular spaces. The cortical zone is traditionally regarded as extending from epidermis or hypodermis to an inner boundary inside which vascular bundles are present. The inner boundary is often very indistinct. Chloroplasts may be present in the collenchyma cells of the outer cortex, or in more or less well-defined layers of parenchymatous cells. Fibres and sclereids are a prominent feature in the cortex of many species. Often the grouping of fibres into strands with well-defined cross-sectional outlines and in characteristic positions in the cortex will help in the identification of a plant. There is no endodermis. The innermost parenchymal cells in the stem cortex of a few plants contain a large number of starch granules, called starch sheaths.

3. 维管柱　是皮层以内的所有区域，包括环状排列的维管束、髓部和髓射线等，占有较大比例；和根的构造不同，无明显的内皮层和中柱鞘，因此维管柱和皮层无明显界线。

(3) Vascular cylinder　The stem vascular cylinder, which occupies a large proportion in the cross section, is bound by the cortex to the exterior and contains vascular bundles, pith, and medullary rays, with no obvious endodermis and pericycle, so there is no obvious boundary between the vascular column and cortex (in some but not all cases). if present, Primary xylem, primary phloem, and the pith make up a central cylinder called the stele in most younger and a few older stems.

初生维管束包括初生韧皮部、初生木质部和束中形成层。木本植物维管束排列紧密，束间区域较窄，似乎连成一圆环状；大多数藤本和草本植物维管束之间的束间区域较宽。

Primary vascular bundles include primary phloem, primary xylem, and fascicular cambium. The vascular bundles of woody plants are closely arranged, and the area between the bundles is narrow, shaped in a ring. The area between the vascular bundles of vines and herbs is wide. The plant vascular system plays a pivotal role in the delivery of nutrients to distantly located organs.

初生木质部分化成熟的顺序是自内向外的内始式；被子植物的初生木质部由导管、管胞、木薄壁细胞和木纤维组成。初生韧皮部分化成熟的顺序是自外向内的外始式；在外方先分化成熟的原生韧皮薄壁细胞会发育形成纤维束位于初生韧皮部外侧，称为初生韧皮纤维，可加强茎的韧性；被子植物的初生韧皮部一般由筛管、伴胞、韧皮薄壁细胞和韧皮纤维构成。束中形成层位于初生韧皮部和初生木质部之间，由1~2层具有分生能力的细胞组成，有潜在的分生能力。

Primary xylem is composed of protoxylem, in which the tracheary elements usually have helical (spiral) or annular wall thickenings. In metaxylem the wall thickenings can be more extensive, and breached by pits (with membranes) arranged in scalariform, alternate, reticulate or less regular ways. The protoxylem has to be capable of considerable extension, without breaking, during the first phases of primary growth in length of the stem. The order of primary xylem differentiation and maturity is from the inside to the outside (endarch). The primary xylem of the angiosperm consists of vessels, tracheids, wood parenchyma cells and wood fibers. The xylem of gymnosperms stem is composed mainly of tracheids with conspicuous bordered pits.

The primary phloem is partially mature and the sequence is from outside to inside (exarch). The protophloem parenchymal cells that have differentiated and matured outside will develop into clusters of fibers located outside the primary phloem, called the primary phloem fiber bundles, which can strengthen the toughness of the stem. The primary phloem of angiosperms is generally composed of sieve tubes, companion cells, phloem parenchyma cells, and phloem fibers.

The fascicular cambium in the vascular bundle is located between the primary phloem and the primary xylem. It consists of one or two layers of cells with meristematic capacity, which can make the stem thicker continuously.

髓由基本分生组织薄壁细胞构成，位于茎的中心。木本植物茎的髓部一般较小，草本植物茎的髓部较大；有些植物的髓部在发育过程中消失形成中空的茎，如芹菜；有些植物茎髓部局部破坏形成横髓隔，如胡桃；有些木质茎髓部较大，如接骨木；有些植物的髓部围绕一层紧密排列且壁较厚的小型细胞，称为环髓区，如椴树。

The pith is composed of ground meristem parenchymal cells, and is located in the center of the stem. The pith of the woody plant stems is generally small, the herbaceous plants' is relatively large. Some plant stems disappear during the development process to form hollow stems, such as *Apium graveolens* L.; in some plants, the pith of the stems is partially damaged to form a transverse pith, Such as *Juglans regia* L.; some woody stems' pith has a large proportion, such as *Sambucus williamsii* Hance; some plant stems have a pith around a layer of densely arranged small cells with thick walls, called perimedullary region, such as *Tilia tuan* Szyszyl.

髓射线是位于初生维管束之间的薄壁组织，内通髓部，外达皮层，具分生能力。一定条件下，髓射线细胞会分裂产生不定芽、不定根；在次生生长开始时，邻接束中形成层的髓射线细胞可以转化为束间形成层。在茎的横切面上，髓射线呈放射状，具运输和贮藏作用。

Medullary rays are parenchymatous tissues with meristem ability located among the primary vascular bundles, inside of the cortex, toward the center pith. Under certain conditions, medullary ray cells can divide to produce adventitious buds and adventitious roots; at the beginning of secondary growth, the medullary ray cells that adjoin with fascicular cambium can be transformed into cambium among bundles, called interfascicular cambium. On the cross-section of the stem, the medullary rays are radial, having transport and storage effects.

三、双子叶植物茎的次生生长及其结构
2.3　Secondary structure of dicotyledons

双子叶植物茎在初生构造形成后，由于维管形成层和木栓形成层细胞的分裂、分化，不断产生新的组织，使茎不断加粗，这种使茎增粗的生长称为次生生长，由次生生长所产生的组织称次生组织，由此形成的结构称次生结构。木本植物的次生生长可持续多年，因此次生构造发达。茎次生生长时，邻接束中形成层的髓射线恢复分生能力转变为束间形成层，并和束中形成层连接，形成完整的形成层环，产生次生木质部、次生韧皮部、维管射线；多数植物茎由表皮内侧皮层薄壁组织细胞恢复分裂机能而产生周皮。

The stems of dico exhibit secondary growth. A narrow band of cells between the primary xylem and the primary phloem may retain its meristematic nature and become the vascular cambium. The cells of the cambium continue to divide indefinitely, with the divisions taking place mostly in a plane parallel to the surface of the plant. The secondary tissues produced by the vascular cambium add to the girth of the stem instead of to its length. The transition from primary to secondary growth in stems involves the development of two new meristema vascular cambium (to produce secondary xylem and phloem) and a phellogen (to replace the outer bark with a corky periderm) —both of which must form a continuous cylinder or meristematic tissue. The phloem and the pericycle play roles in the origin of these secondary meristems. Cells produced by the vascular cambium and phellogen continuously generate new tissues, becoming secondary structures, such as vessels, tracheids, sieve tube members, etc., and the stem is continuously thickened. This growth that makes the stem thicker is called secondary growth. The secondary growth of woody plants can last for many years, so their secondary structures are developed.

（一）木质茎

2.3.1　Wooden stem

1. 维管形成层及其活动　茎在次生生长时，束间形成层和束中形成层连接形成一个完整的形成层环。形成层细胞（纺锤原始细胞，呈纺锤形，液泡明显）切向分裂，向内产生次生木质部，增添于初生木质部外方；向外产生次生韧皮部，增添于初生韧皮部内侧，并将初生韧皮部挤到外侧。通常次生木质部数量比次生韧皮部多。同时，少数近等径的形成层细胞（射线原始细胞）也进行分裂产生次生射线细胞，存在于次生木质部和次生韧皮部，形成横向的联系组织，称为维管射线。茎加粗生长的同时，形成层细胞本身也进行径向或横向分裂，增加数量，扩大圆周，以适应内方木质部的增大，形成层的位置也逐渐外移。

　木本植物次生生长时，束间形成层部分分裂分化形成维管组织，部分则形成维管射线，故木本植物维管束之间距离变窄；藤本植物束间形成层只分化成薄壁组织，故维管束间距离较宽。

(1) Vascular cambium and its activities　In woody plants, the obvious differences begin to appear as soon as the vascular cambium and the cork cambium develop. During secondary growth, the fascicular and interfascicular cambium are connected to form a complete cambium ring (viewed on the cross-section). Then, cambial cells (spindle shaped, with obvious vacuole) began to undergo tangential division, producing secondary xylem inward and adding to the outside of the primary xylem; producing secondary phloem outward and adding to the inside of the primary phloem, squeezing the primary phloem to the outside.Usually, the number of secondary xylem is much larger than the secondary phloem. At the same time, a few nearly equal-diameter cambial cells (ray primordial cells) also divide to produce secondary ray cells existing in the secondary xylem and secondary phloem, forming a horizontally connected tissue, called vascular rays. Simultaneously, the cambial cells themselves also undergo a radial or lateral division, increasing cell numbers, expanding circumference, to accommodate the increase of the inner xylem, and the position of cambium gradually moving outward. During secondary growth of woody plants, part of the interfascicular cambium divides and differentiates to form vascular tissues, and part of them forms vascular rays, so the distance between the vascular bundles of woody plants becomes narrower; the interfascicular cambium of vine plants only differentiates into parenchymal cells, and do not differentiates into vascular tissues, so the distance between vascular bundles is wide.

2. 次生木质部　是木质茎次生构造的主要部分，由导管、管胞、木薄壁细胞、木纤维和木射线组成。形成层的纺锤原始细胞发展形成次生木质部中的纵向疏导系统如导管、管胞、木薄壁细胞和木纤维等；射线原始细胞径向延长形成维管射线，位于次生木质部内的称为木射线。

(2) Secondary xylem　The secondary xylem, one of the products of the vascular cambium, is the main part of the secondary structure of the woody stem. It consists of vessels, tracheids, wood parenchymal cells, wood fibers and wood rays. The cambial spindle protocells develop to form the longitudinal support and conduction systems, including vessels, tracheids, parenchymal cells, fibers and so on in the secondary xylem; the ray primitive cells extend radially to form vascular rays, and that located inside the secondary xylem is called xylem ray.

　次生木质部在木质茎次生生长中变化明显。形成层的活动受季节气候或其它环境影响很大，温带春季或热带雨季，气候温和，雨量充沛，形成层活动旺盛，所形成的次生木质部其细胞径大壁薄、质地疏松、色泽较淡，称早材或春材；温带的夏末秋初或热带的旱季，形成层活动减弱，所形成的次生木质部细胞径小壁厚、质地紧密、色泽较深，称晚材或秋材。年轮即是当年的秋材与次年的春材间形成的界限分明的同心环层（图 15-12）。

The most conspicuous differences involve the secondary xylem. The activities of the vascular cambium is greatly affected by seasonal climate or other environmental stresses. When the vascular cambium of a tree becomes active in the temperate spring or tropical rainy season, it usually produces relatively large vessel elements of secondary xylem, having a large cell diameter, loosen texture, lighter color; such xylem is referred to as early wood or spring wood. In temperate summer and early autumn or tropical dry season, the activity of the vascular cambium weakened, and the secondary xylem formed with small cell diameter, dense texture, darker color, called late wood or autumn wood. Over a period of years, the result of this type of switch between the early spring and the summer growth one year's growth of xylem is called an annual ring(Fig.15-12). The annual ring is a series of alternating concentric rings of dark and light cells between the autumn wood of the current year and the spring wood of the next year.

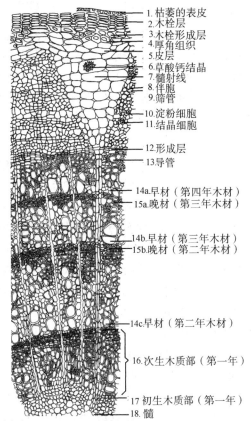

1. 枯萎的表皮
2. 木栓层
3. 木栓形成层
4. 厚角组织
5. 皮层
6. 草酸钙结晶
7. 髓射线
8. 伴胞
9. 筛管
10. 淀粉细胞
11. 结晶细胞
12. 形成层
13. 导管
14a. 早材（第四年木材）
15a. 晚材（第三年木材）
14b. 早材（第三年木材）
15b. 晚材（第二年木材）
14c. 早材（第二年木材）
16. 次生木质部（第一年）
17 初生木质部（第一年）
18. 髓

1.Withered epidermis; 2.Cork; 3. Phellogen; 4. Collenchyma; 5. Cortex; 6.Cacium oxalate crystals; 7. Medullary ray; 8. Companion cell; 9. Sieve tube; 10.Starch grains; 11. Crystal cell; 12. Cambium;13. Vessels; 14. Spring wood (14a. The forth year xylem, 14b.The third year , 14c.The second year xylem); 15.Summer wood(15a. The third year xylem, 15b. The second year xylem); 16. The first year secondary xylem; 17. The first year primary xylem; 18.Pith

图 15-12　四年生木质茎的构造
Fig.15-12　Structure of four-year-old woody stems

次生木质部是木材的主要来源。在木质茎（木材）横切面上，靠近形成层的部分颜色较浅，质地松软，为边材；中心部分颜色较深，质地坚固，为心材，心材常含挥发油、单宁、树胶、色素等代谢产物，是茎木类药材的主要药用部位，如降香、沉香、檀香、苏木等。

Secondary xylem, or wood, is put to an extremely wide variety of uses. On the cross-section of the wood, The lighter, still-functioning xylem closest to the cambium is called sapwood, while the older, darker wood at the center is called heartwood. The heartwood often contains volatile oils, tannins, gums, pigments, etc. metabolites, and is the main medicinal parts of stem and wood medicinal materials, such as Lignum Dalbergiae Odoriferae, Lignum Aquilariae Resinatum, Lignum Santali Albi, Lignum Sappaan.

鉴定木类药材时，常采用横切面、径向切面、切向切面进行比较观察。横切面是与纵轴垂直的切面，可见射线的长度和宽度，射线呈辐射状排列，年轮呈同心环状，导管、管胞、木纤维、木薄壁细胞呈大小不一、细胞壁厚薄不同的类圆形或多角形；径向切面是通过茎的直径的纵切面，可见射线的高度和长度，射线横向分布，年轮呈平行垂直的带状，导管、管胞、木纤维等均为纵长细胞，呈纵长筒状，次生壁的增厚纹理清晰可见；切向切面是不通过茎的中心而垂直于茎的半径的纵切面，可见射线的宽度和高度以及细胞列数和两端细胞的形状，射线细胞群呈纺锤状不连续纵行排列，年轮呈 U 型波纹，导管、管胞、木纤维等的形态与径向切面相似。在木材的三切面中，射线的形状最为突出，可作为判断切面类型的重要依据。

When identifying wood medicinal materials, three specific planes of section from a block of wood are often used for comparison and observation. These are the transverse section (TS), the radial longitudinal section (RLS) and the tangential longitudinal section (TLS). TS is a section perpendicular to the longitudinal axis. The annual rings are concentric rings, the rays are arranged in a radial pattern. The length and width of the rays, and the vessels, tracheids, wood fibers, and wood parenchymal cells between the two rays are shown, ceslls are round or polygonal shapes; RLS is a longitudinal section made by the diameter of the stem, annual rings are parallel and vertical strips, the rays are distributed laterally. The height and length of rays, vessels, tracheids and wood fibers are shown, cells are elongated cannular and the thickened texture of the secondary wall is clearly visible; TLS is a vertical section that does not pass through the center of the stem and is perpendicular to the radius of the stem. Annual rings are U-shaped corrugated. The ray cell group is arranged in a discontinuous row in a spindle shape. The width and height of rays, the number of cell columns, the shape of the cells at both ends are shown, and the shapes of vessels, tracheids, and wood fibers are similar to RLS. Among the three sections of wood, the shape of the rays is the most prominent, which can be used as an important basis for determining the type of section. Rays are variable, from one to many cells high and of variable width. Wider rays may be associated with resin canals.

3. 次生韧皮部　维管形成层向外分裂形成次生韧皮部，同时初生韧皮部被挤压到外方，使筛管、伴胞及其它薄壁细胞破坏，细胞界线不清，形成颓废组织。由于形成层向外分裂的次数远不如向内分裂的次数，故次生韧皮细胞数量较次生木质部少。次生韧皮部常由筛管、伴胞、韧皮纤维、韧皮薄壁细胞组成，有的还具石细胞，如肉桂、杜仲，有的具乳汁管，如夹竹桃。次生韧皮部薄壁细胞常含有多种生理活性物质和营养物质。维管射线位于次生韧皮部的部分，细胞壁不木质化，形状也没有木射线那样规则，这些薄壁组织称为韧皮射线。

(3) Secondary phloem　The vascular cambium is producing secondary phloem to the outside, while the primary phloem is squeezed to the away side, causing the sieve tube, companion cells, and other parenchymal cells to be destroyed. So that cell boundaries are unclear, forming decadent tissue. Because the number of outward divisions is far less than the number of inward divisions, the number of secondary phloem cells is less than that of secondary xylem. Secondary phloem is often composed of sieve tubes, companion cells, phloem fibers, and phloem parenchyma cells. It also has sclereid, such as *Cinnamomum cassia*, *Eucommia ulmoide*; and some have laticifer, such as *Nerium oleander*. The parenchyma cells of the secondary phloem often contain a variety of physiologically active substances and nutrients. Vascular rays located in secondary phloem are called phloem rays, which cell walls are not lignified and the shape is not as regular as xylem rays.

4. 木栓形成层的产生及其活动　次生生长使茎不断加粗，外方的表皮一般不能相应增粗而死亡，此时表皮内侧皮层薄壁组织细胞恢复分生机能形成木栓形成层，它向外分生产生木栓层，向内产生栓内层，三者合称周皮，代替表皮行使保护作用。一般木栓形成层的活动仅数月，多数树木又可在其内方依次产生新的木栓形成层，发生的位置内移，形成新的周皮，老周皮内方的组织被新周皮隔离后由于失去水分和营养的联系而逐渐枯死，这些周皮及被隔离的死亡组织的综合体常常会剥落，称落皮层，如白皮松、白桦等（图 15-13）。

(4) The generation and activities of the cork cambium Secondary growth make the stem increase in girth continuously, while the epidermis cannot generally thicken and die accordingly. At this moment, a second cambium arises within the cortex or, in some instances, develops from the epidermis or phloem. This is called the cork cambium, or phellogen. The cork cambium may also produce parenchyma-like phelloderm cells to the inside and phellem layer to the outside, collectively named periderm, that makes up the outer bark of woody plants. Generally, The cork cells, which are produced annually in cylindrical layers, die shortly after they are formed. The activity of the cork cambium is only a few months, and most trees can generate the new in order, so that the location of the occurrence will move inward to form a new periderm. After being isolated by neo-periderm, the older periderm gradually die due to the loss of water and nutrition. These died cork tissues would be soon sloughed off, so it is called rhytidome, such as *Pinus bungeana, Betula platyphylla*(Fig.15-13).

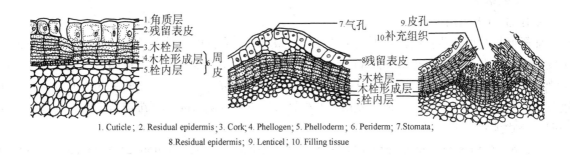

1. Cuticle；2. Residual epidermis；3. Cork；4. Phellogen；5. Phelloderm；6. Periderm；7.Stomata；
8.Residual epidermis；9. Lenticel；10. Filling tissue

图 15-13　双子叶植物茎木栓形成层分裂产生周皮及皮孔示意图
Fig.15-13　Diagram of periderm and lenticels from stem cork cambium in Dicotyledones

植物学上的树皮有两种概念，狭义的树皮即落皮层（外树皮）；广义的树皮指维管形成层以外的所有组织，包括落皮层、木栓形成层以内的次生韧皮部（内树皮），皮类药材如厚朴、杜仲、肉桂、黄柏等均指广义树皮。

The term bark has two concepts. The narrow bark is the rhytidome (outer bark); the generalized bark is usually applied to all the tissues outside the cambium, including the phloem. Some scientists distinguish between the inner bark, consisting of primary and secondary phloem, and the outer bark (periderm), consisting of cork tissue and cork cambium. The medicinal materials such as Cortex Magnolia Officinalis, Cortex Eucommia, Cortex Cinnamomi, Cortex Pellodendri Chinensis, etc. all refer to generalized bark.

（二）草质茎

2.3.2　Herbaceous stem

草质茎生长期短，次生生长有限，次生构造不发达，木质部量少，质地较柔软。最外层为表皮，常附有各式毛茸、气孔、角质层、蜡被等；少数在表皮下方产生木栓形成层，向外产生 1~2 层木栓细胞，表皮仍未破坏；有些种类仅有束中形成层而无束间形成层，有些甚至束中形成层也不明显；髓部发达，髓射线较宽，有些种类的髓部中央破裂成空洞状（图 15-14）。

Herbaceous stems have a short growth period, limited secondary growth, underdeveloped secondary structures, a small amount of xylem, and a soft texture. The structural characteristics of herbaceous stems are: the outermost is the epidermis, often with a variety of fur, stomata, cuticles, wax, etc.; some species only have fascicular cambium but no interfascicular cambium, and some have not fascicular cambium, the interfascicular cambium is not obvious; the pith is developed, the pith rays are wide, and the center of

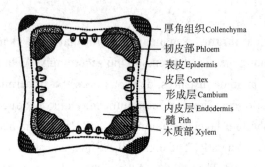

图 15-14　薄荷茎的横切面简图
Fig.15-14　A brief drawing of the cross section
of a mint stem

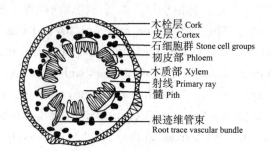

图 15-15　黄连根状茎横切面简图
Fig.15-15　Cross section diagram of rhizome
of *Coptis chinensis*

some plants' pith ruptures into a hollow shape(Fig.15-14).

（三）根状茎

2.3.3　Rhizome

双子叶植物根状茎一般指草本双子叶植物的根状茎，它的贮藏薄壁组织发达，机械组织大多不发达，其构造与地上茎类似（图 15-15）。表面具木栓组织，少数具表皮或鳞叶；皮层中常有根迹维管束和叶迹维管束斜向通过，根迹维管束是茎中维管束与不定根中维管束相连的维管束，叶迹维管束是茎中维管束与叶柄维管束相连的维管束；皮层内侧有时具纤维或石细胞；维管束为外韧型，环状排列；髓部明显。

Dicotyledonous rhizomes generally refer to the herbaceous dicotyledonous rhizomes. Its storage parenchymal cells are more developed, and the mechanical tissues are mostly underdeveloped. Its structure is similar to that of aerial stems(Fig.15-15), and its structural characteristics are: cork tissues on the surface, a few with epidermis or scale leaf; cortex often has root trace and leaf trace vascular bundles. Root trace vascular bundles are those connect stem and adventitious root vascular bundles. Leaf trace vascular bundles are those connect stem and petiole vascular bundles; sometimes there are fibers or sclereid on the inside of the cortex; vascular bundles are collateral and arranged in a ring; the pith is obvious.

四、单子叶植物茎和根状茎的构造
2.4　Structure of stems and rhizomes of monocotyledons

单子叶植物茎一般没有形成层和木栓形成层，终身只具初生构造，不能无限加粗。少数单子叶植物如龙血树、朱蕉、丝兰等的茎有形成层，具次生生长，但这种形成层的起源和活动与双子叶植物不同。

Monocotyledonous stems generally do not form vascular and cork-like cambium, only having a primary structure throughout their lifetime. A few monocots, such as plants from the genus *Dracaena Yucca smalliana* and *Cordyline fruticosa*, have cambium with secondary growth,but the origin and activity of the cambium is different from that of dicotyledones.

1. 单子叶植物茎的构造　大多数单子叶植物茎最外层是由一列细胞构成的表皮，通常不产生周皮；表皮以内为基本薄壁组织和散布其中的维管束，维管束为有限外韧型，没有皮层和髓及髓射线之分。禾本科植物茎秆表皮下方常有数层厚壁细胞，以增强支持作用。

(1) The stem structure of monocotyledons　Most monocotyledonous stems generally have neither a vascular cambium or a cork cambium and thus produce no secondary vascular tissues or cork, and have only a primary structure throughout their lifetime. The surfaces of the stems are covered by an epidermis. The parenchyma tissue between the vascular bundles is not separated into cortex and pith. The xylem and phloem tissues produced by the procambium appear in cross section as discrete vascular bundles scattered throughout the stem instead of being arranged in a ring. The vascular bundles are closed collateral. In gramineae plants' stem, there are more bundles just beneath the surface than there are toward the center. Also, a band of sclerenchyma cells, usually two or three cells thick, develops immediately beneath the epidermis. These all contribute to giving the stem the capacity to withstand stresses.

2. **单子叶植物根状茎的构造**　表面为表皮或木栓化皮层细胞，少数有周皮，如射干。皮层占较大比例，常分布有叶迹维管束，维管束大多为有限外韧型，也有周木型，如香附子；有的兼有周木型和有限外韧型两种，如石菖蒲。内皮层大多明显并具凯氏带，如姜。有些植物根状茎在皮层靠近表皮部位的细胞形成木栓组织，如姜，有的皮层细胞转变为木栓细胞形成"后生皮层"，以代替表皮行驶保护作用，如藜芦。

(2) The structure of rhizomes of monocotyledons　The surface of monocotyledonous rhizomes is epiderm or suberification cortex cells, and a few have periderm, such as *Belamcanda chinensis*. The cortex occupies a large proportion, and leaf trace vascular bundles are often distributed, that most of the vascular bundles are closed collateral, and some are amphivasal like *Cyperus rotundus*, and some have both amphivasal and closed collateral vascular bundles, like *Acorus tatarinowii*. The endodermis is mostly obvious and has casparian strip, like *Zingiber officinale*. The cortex cells next to the epidermis of some rhizomes could form cork tissues, like *Zingiber officinale*, and some are transformed into cork cells to form an "metaderm", to replace epidermal driving protection, like *Veratrum nigrum*.

五、茎的异常构造
2.5　Anomalous structure of stem

某些双子叶植物的茎和根状茎除形成正常构造外，部分薄壁细胞会恢复分生能力转化成形成层，通过这些形成层活动产生多数异型维管束，形成异常构造。常见的有以下几种类型。

Some dicotyledonous stems and rhizomes have some thin-walled cells that restore meristematic capacity and transform into cambium. Most of these heterotypic vascular bundles are produced by these cambiums. There are several common types below.

1. **髓维管束**　在茎或根状茎的横切面上可见除正常排列成环状的维管束外，髓中还有异形维管束数个，这些髓中的维管束即髓维管束，如掌叶大黄根状茎的横切面上可见髓部有许多星点状的异型维管束（图 15-16）。

(1) Vascular bundles in pith　In the transverse sections of the dicotyledonous stem or rhizome, there are several irregularly shaped vascular bundles distributed in the pith in addition to the normal

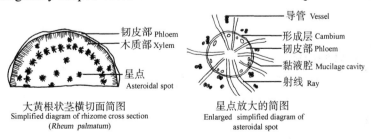

韧皮部 Phloem
木质部 Xylem
星点 Asteroidal spot

导管 Vessel
形成层 Cambium
韧皮部 Phloem
黏液腔 Mucilage cavity
射线 Ray

大黄根状茎横切面简图
Simplified diagram of rhizome cross section
(*Rheum palmatum*)

星点放大的简图
Enlarged simplified diagram of
asteroidal spot

图 15-16　大黄根状茎横切面简图

Fig.15-16　Transverse section diagram of rhizome of rheum

vascular bundles arranged in a circular shape. For example, there are many star-shaped vascular bundles in the pith of the rhizome of *Rheum palmatum*(Fig.15-16).

2. **同心环状排列的异常维管组织** 在一些双子叶植物茎内，正常的次生生长发育到一定阶段，次生维管柱的外围又形成多轮同心环状排列的异常维管组织。如密花豆的老茎（鸡血藤）横断面，韧皮部可见 2~8 个红棕色至暗棕色环带，与木质部相间排列，最内一圈为圆环，其余为同心半圆环。

(2) The anomalous vascular tissues with concentric annular arrangement In the stems of some dicotyledons, the primary and early secondary growth are normal. When the normal secondary growth develops to a certain stage, there are several anomalous vascular tissues formed again in the periphery of the secondary vascular cylinder. For example, in the cross-section of *Spatholobus suberectus* climing stem, the phloem shows 2 to 8 red-brown to dark brown annulus bands, which are arranged alternately with the xylem. The innermost is a ring, and the rest are concentric semi-rings.

3. **木间木栓** 有些双子叶植物的根状茎，在次生木质部内也形成木栓带称木间木栓或内涵周皮，通常由次生木质部薄壁组织细胞分化形成。如甘松根状茎的横切面，可见木间木栓呈环状，包围一部分韧皮部和木质部而把维管柱分隔为数束。

(3) Interxylary cork In the rhizomes of some dicotyledonous plants, there are cork bands formed within the secondary xylem, called interxylary cork or included periderm, which are usually formed from the differentiation of the secondary xylem parenchyma cells. For example, in the cross-section of the rhizome of *Nardostachys chinensis* Bat. We can see that the interxylary cork is in a ring shape, surrounding part of the phloem and xylem and dividing the vascular cylinder into several bundles.

PPT

第三节　叶的组织结构

3 Leaf anatomy

叶由茎尖生长锥后方的叶原基发育而成，叶的初生组织分为原表皮层、基本分生组织和原形成层，幼叶在发育过程中已完全成熟，不再保留原分生组织，因此叶的解剖结构中，叶柄构造和茎相似，但叶片和托叶构造与茎有明显差异。

The leaf is developed from the leaf primordium behind the growth cone at the tip of the stem. The primary tissue of the leaf is divided into the original epidermal layer, the basic meristem and the original formation layer. The young leaves have fully matured during the development process and no longer retain the original meristem. Therefore, in the anatomical structure of the leaf, the petiole structure is similar to the stem. However, the structure of laminas and stipules is obviously different from that of stems.

一、双子叶植物叶的构造
3.1　Structure of dicotyledonous leaves

（一）叶柄的结构
3.1.1　The petiole
叶柄横切面呈半圆形、圆形（图15-17）。最外层为表皮；表皮以内为皮层，皮层中有多层厚

医药大学堂
WWW.YIYAODXT.COM

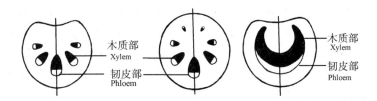

图 15-17　叶柄结构

Fig.15-17　the structure of petiole

角组织，或有厚壁组织；维管束呈半圆形或圆形排列在薄壁组织中，和幼茎中的维管束相似。木质部在近轴面，韧皮部在远轴面，两者间往往具一层形成层，形成层活动期短。维管束因合并或分裂，使维管束的数目和排列变化很大。

Cross-section is semicircular and circular(Fig.15-17). The outermost layer is the epidermis; the inner layer of the epidermis is the cortex. There are multiple layers of thick-angled tissues or thick-walled tissues in the cortex. The xylem is on the paraxial surface, and the phloem is on the far-axis surface. There is often a layer of cambium between them, and the cambium has a short active period. The number and arrangement of vascular bundles vary greatly due to merging or splitting of vascular bundles.

（二）叶片的结构

3.1.2　The structure of lamina

双子叶植物叶片常由表皮、叶肉和叶脉三部分组成（图 15-18）。

A typical mature leaf has an adaxial (upper) epidermis, an abaxial epidermis, a zone of photosynthetic tissue in between called the mesophyll, and vascular system of leave consists of veins(Fig.15-18).

1. 表皮　表皮覆盖整个叶片的表面，分上表皮和下表皮，上表皮在叶片的上表面（近轴面），下表皮在叶片的下表面（远轴面）。表皮常由一层扁平的生活细胞组成，常不含叶绿体，无胞间隙，排列紧密。顶面观，表皮细胞呈不规则形，彼此嵌合；横切面观呈方形或长方形，外壁较厚，角质化并具角质层，有的具蜡质。少数植物表皮由多层细胞组成，称复表皮，如夹竹桃叶、海桐叶等。

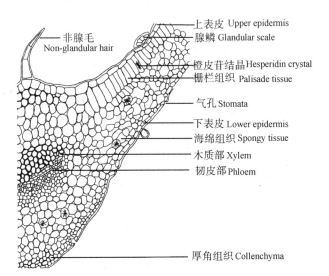

图 15-18　双子叶植物叶（薄荷）横切面

Fig.15-18　Cross section of dicotyledonous leaves (*Mentha haplocalyx*)

视频

331

(1) The epidermis The epidermis forms the boundary between the atmosphere and the underlying mesophyll and vascular and non-vascular tissues. The majority of dicotyledons tend to have epidermal cells of irregular shape and size. They have straight, curved or sinuous anticlinal walls. The epi dermis may be one or more layers thick, there may be either a thick or a thin cuticular covering. Sometimes the cuticle are masked by a covering of wax.

叶片表皮上常具有气孔；一般下表皮的气孔较上表皮多；气孔的数目、形态结构和分布因植物种类和环境条件而异，为叶类生药的重要鉴别特征。有些植物的叶尖和叶缘尚有一种排水的孔状结构，称水孔，如番茄、禾本科植物幼苗。除了气孔，表皮上还有毛状体等附属物；毛的有无和毛的结构因植物种类而异。有些植物叶表皮细胞中可见碳酸钙结晶。

Stomatas are often distributed on the epidermis of the leaves. The stomata on the lower epidermis are usually more than those on the upper epidermis. The number, shape, structure and distribution of stomata vary with plant species and environmental conditions, which are important basis for identification of leafy herbs. Some plants have a pore-like structure at the tip and margin of the leaf, called a drainage pore, such as the seedings of tomatoes and grasses. In addition to stomata, there are also trichomes and other appendages on the epidermis of the leaves. The presence or absence of hair varies with plant species. Calcium carbonate crystals are sometimes found in the epidermal cells of the leaves of the same plant.

2. **叶肉** 是叶上、下表皮之间的部分，薄壁细胞含有叶绿体，是植物进行光合作用的主要场所。按叶肉薄壁组织的细胞形态和排列方式不同，分为栅栏组织和海绵组织，栅栏组织位于上表皮之下，细胞呈长圆柱形，其长径与表皮垂直，排列紧密而整齐，呈栅栏状，胞间隙小，呈纵向排列，利于气体的交换；细胞所含叶绿体较多，所以叶片上表面绿色较深；各种植物栅栏组织细胞的层数不一样，有时可作为叶类药材鉴别的特征之一，如枇杷叶、冬青叶的栅栏组织有两层。海绵组织位于栅栏组织与下表皮之间或栅栏组织之间，细胞圆形或不规则形，排列疏松，胞间隙发达，细胞所含叶绿体少，所以叶下面的颜色较浅。叶肉组织在上、下表皮的气孔处有较大空隙，称孔下室。有些植物叶肉中含有分泌腔，如桉叶；有的含石细胞，如茶的叶；有的含结晶体，如曼陀罗叶肉中含有砂品、方晶和簇晶。叶肉组织明显分化为栅栏组织和海绵组织的叶，称两面叶或异叶面；上、下表皮内方都有栅栏组织或栅栏组织和叶肉组织分化不明显者，称等面叶，如番泻、桉的叶（图 15-19）。

(2) The mesophyll The mesophyll usually consists of the thin-walled parenchymatous cells containing chloroplasts, called chlorenchyma, and other thin-walled cells concerned with water, food or ergastic or so-called 'waste product' (e.g. crystals, tannins) storage. Dicotyledons generally have a mesophyll which is composed of two differing photosynthetic cell types – palisade and spongy mesophyll

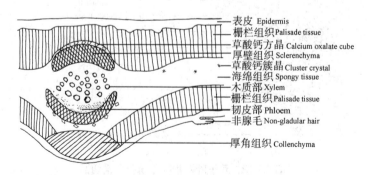

表皮 Epidermis
栅栏组织 Palisade tissue
草酸钙方晶 Calcium oxalate cube
厚壁组织 Sclerenchyma
草酸钙簇晶 Cluster crystal
海绵组织 Spongy tissue
木质部 Xylem
栅栏组织 Palisade tissue
韧皮部 Phloem
非腺毛 Non-gladular hair
厚角组织 Collenchyma

图 15-19 番泻叶横切面简图（等面叶）
Fig.15-19 Cross section diagram of Senna Leaf (isofacial leaf)

cells; parenchyma cells may be present between these.

Because some leaves lack a distinction of layers and others have very well marked layers, the mesophyll can be used as an aid to identification. For example, palisade cells can be present next to the upper or to the lower surface, or to both. There are, however, striking changes that can occur to the layers themselves. In some cases, the numbers of layers of palisade cells have been counted and this figure used as a diagnostic character(Fig.15-19).

In some plants, the mesophyll consists of radiating, elongated mesophyll cells surrounding a (usually) parenchymatous but often lignified bundle sheath, which, in turn, surrounds the vascular bundles. The radiating mesophyll is chloroplast-rich. The parenchymatous bundle sheath cells on the other hand usually contain large, prominent, generally agranal chloroplasts.

In many gymnosperms and some angiosperms the mesophyll cells are plicate, with inwardly directed wall foldings. The infoldings increase cell wall surface area and probably therefore make up, to some extent, for the smaller number of chlorenchyma cells that are often found in such leaves.

3. 叶脉 是叶片中的维管束，具有输导和支持作用。主脉和各级侧脉结构有所不同。主脉和较大侧脉由维管束和机械组织组成，维管束结构和茎相同，木质部在近轴面，由导管和管胞组成；韧皮部在远轴面，由筛管和伴胞组成；形成层活动有限。主脉维管束的上、下表皮内方常有厚壁组织或厚角组织，在叶背面较发达。主脉处上、下表皮内方常不分化出叶肉组织，少数有1至几层栅栏组织，如番泻叶、石楠叶。较细的叶脉位于叶肉组织中，维管束外面常包围着1层或多层排列紧密的大型细胞，称维管束鞘。叶脉越细结构越简单，先是形成层简化，机械组织减少，细脉末端仅1~2个短的螺纹管胞，韧皮部分仅存在短而狭的筛管分子和增大的伴胞。小叶脉处常有特化的传递细胞，能有效地从叶肉组织输送光合作用产物到达筛管。

(3) Vein The specialized cells which conduct water and salts upwards from the roots, and the cells involved in the transport or translocation of the substances synthesized in the leaf mesophyll and other tissues, are grouped together in well-defined strands called vascular bundles. In the leaf, these are seen as the midrib and vein system. The major veins in dicotyledonous foliage leaves occupy much of the cross-sectional area of the leaf. Vascular bundles are continuous, either directly or if developed, through the petiole, with the primary system of vascular tissue in the stem. Alternatively, if secondary growth has occurred, leaf bundles may be continuous with the secondary xylem and phloem.

In most dicotyledonous leaves, the phloem pole in a vascular bundle faces the lower (abaxial) leaf surface and the xylem pole the adaxial surface. Broadly, the phloem consists of a series of conducting elements associated with vascular parenchyma elements. In angiosperms, the conducting elements are referred to as sieve tube elements or sieve tube members, and these are almost always associated with specialized parenchyma cells called companion cells. Among the gymnosperms and ferns, the phloem is com-posed of less specialized conducting cells, called sieve cell.

Besides, mature leaves may contain additional marginal strands of sclerenchyma, and some fibre strands or girders may be associated with the vascular bundles.Collenchyma is frequently present in the raised ribs above and below the midrib bundle, and is also occasionally found in similar positions in relation to the large and intermediate vascular bundles in monocotyledons.

二、单子叶植物叶的构造
3.2 Structure of monocotyledonous leaves

单子叶植物叶的结构与双子叶植物相似，也由表皮、叶肉组织和叶脉三部分组成，但各部分的结构又有其特点。现以禾本科植物叶片的结构为例进行介绍。

The structure of the monocots is similar to that of the dicotyledones, which consists of epidermis, mesophyll and venation, but each part has its own characteristic structure. Taking the leaves of gramineae plants as an example, the structure of the leaves of monocots is described as follows.

1. 表皮 表皮由长细胞与短细胞组成。长细胞呈长方形，其长径与叶的纵轴平行，横切面近方形，外壁角质化，并含有硅质；短细胞分为硅质细胞和栓质细胞，位于叶脉外侧呈纵向相隔排列，硅质细胞腔内充满硅质体，栓质细胞的壁木栓化。表皮还存在一些大形细胞，细胞壁薄，液泡发达，横切面上略呈扇形排列，称泡状细胞（图 15-20）；干旱失水时，这类细胞能使叶子卷曲，减少水分蒸发，又称运动细胞。气孔呈纵向排列，保卫细胞呈哑铃型。

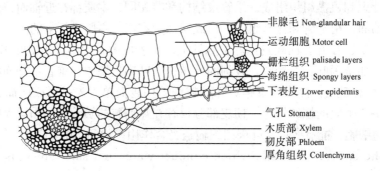

图 15-20 单子叶植物叶（淡竹叶）横切面
Fig.15-20 **The cross section of monocotyledonous leaves (*Lophatherum gracile*)**

(1) The epidermis Anticlinal walls of the epidermal cells of both monocotyledons and dicotyledons can be very thin and hardly visible from the surface, or they may range through degrees of thickness to very thick, so that the lumen of the cells appears from the surface to be very reduced. In monocotyledons, especially the grasses, the inter-cellular spaces are greatly reduced, particularly in more xerophytic species. Many leaves that are capable of rolling up in dry, unfavorable conditions, and reopening again under conditions when there is no water stress, have special, thin-walled water-containing cells that enable them to make these movements. These are the bulliform or motor cells(Fig.15-20). The shape, size and disposition of such cells can be used as an aid to classification and identification.

2. 叶肉 叶肉无明显栅栏组织和海绵组织分化，叶肉细胞间隙小，孔下室较大。一些植物的叶肉组织明显分化成栅栏组织和海绵组织的两面叶，如淡竹叶。

(2) The mesophyll In many monocotyledonous plants, the mesophyll is not differentiated into spongy and palisade layers.

3. 叶脉 有限外韧型维管束近平行排列。主脉维管束的上下表皮内方常有厚壁组织分布，增强了机械支持作用。维管束外围常有 1、2 层或多层细胞的维管束鞘，如玉米、甘蔗为 1 层较大薄壁细胞，水稻、小麦为 1 层薄壁细胞和 1 层厚壁细胞组成。维管束鞘常作为禾本科植物的鉴别特征。

(3) The vascular bundles are surrounded by an initially parenchymatous bundle sheath, which may undergo lignification as the cells mature.

Most monocotyledonous foliage leaves are basically parallel-veined, but large numbers of cross-veins serve to interconnect the parallel vein system. Generally, in monocotyledons, the phloem within mature bundles is composed of functional metaphloem sieve elements, associ ated with vascular parenchyma cells, including companion cells. The phloem may contain specialized, late-formed metaphloem sieve tubes, which appear to lack the companion cell associations that exist with the early, thin-walled sieve tubes. The late-formed metaphloem sieve tubes have thickened, usual ly cellulosic walls, which, in some cases (e.g. barley and wheat) may undergo lignification. The thick-walled sieve tubes usually border on, or occur in close proximity to, the metaxylem vessels within leaf blade bundles.

The phloem and xylem are not the only tissues present in the veins. They form the central core, around which sheaths of specialized cells are formed, which separate the vascular tissues from the mesophyll. Two principal types of sheath exist, namely sclerenchymatous and parenchymatous sheaths. There may also be parenchyma associated with the phloem or xylem, and the phloem may contain fibres.

三、叶的显微常数
3.3　Microscopic index of leaf

叶片微形态特征可用于探讨植物种间、属间甚至科的分类和系统演化关系，具有重要的分类价值。叶的显微常数是较稳定的显微特征，是分类鉴别的重要参数，主要有气孔指数、栅表比、脉岛数等。

The microscopic index of leaves are relatively stable microscopic characteristics and are important parameters for classification and identification. The main methods for determining the microscopic constants of leaf of the traditional Chinese medicine include stomatal index, palisade ratio and vein islet number.

1. 气孔指数　指一种植物叶的单位面积（mm^2）上气孔数与表皮细胞数的比例，即气孔指数＝（单位面积上的气孔数）/（单位面积上的气孔数 + 单位面积上的表皮细胞数）×100%。在全草类、叶类中药鉴定中，一特定植物的气孔指数在一定范围且相当恒定，如蓼蓝 *Polygonum tinctorium* Ait. 叶片上、下表皮的气孔指数分别是 8.4%~11.4%、22.4%~28.0%。

此外，还有气孔数（stomatal number）和气孔比率（stomatal ratio）。气孔数，指单位面积表皮面积上的气孔平均数，常用于两种亲缘关系较远的植物或药材鉴别。气孔比率，指叶的上下表皮气孔数之比，常用于两种亲缘关系较近的植物或药材鉴别。

(1) Stomatal index　Refers to the ratio of stomatal number to epidermal cell number per unit area (mm^2) of a plant leaf. Stomatal index=(number of stomata per unit area)/(number of stomata per unit area+ number of epidermal cells per unit area)×100%. In the identification of medicinal herbs, the stomatal index of a certain plant is constant, such as in *Polygonum tinctorium* Ait. The stomatal index of of upper and lower epidermis are 8.4%~11.4% and 22.4%~28.0% respectively. In addition, there are also other indices such as Stomatal number and Stomatal ratio. The former refers to the average number of stomata per unit area on epidermis, it is often used to identify two different plants or medicinal herbs with distant relationship. The latter, the ratio of stomatal number of upper and lower epidermis of leaves, is often applied to identify two closely related plants or medicinal herbs.

2. **栅表比** 指叶片上栅栏细胞与表皮细胞之间的比例关系，即一个表皮细胞下栅栏细胞数目的均值。因不同植物的栅表比常较恒定，所以它是叶类或全草类中药的鉴定特征之一，常用以区别一些同属不同种的物种，如尖叶番泻 *Cassia acutifolia* Delile 叶片的栅表比为 1∶（4.5~18）。

(2) Palisade ratio the proportional relationship between palisade cells and epidermal cells on a leaf, that is, the mean number of palisade cells under one epidermal cell. This is particularly useful in defining small leaf fragments in powdered leaf products. This measure indicates the number of palisade cells that can be seen beneath an epidermal cell in surface view. An average figure is produced after many cells are counted. A statistically sound count will produce a fairly reliable typification and hence identification of the material.

3. **脉岛数** 叶脉中最微细的叶脉所包围的叶肉组织称为脉岛，脉岛数指单位面积（mm^2）叶片上脉岛的数目。同种植物的叶，脉岛数较恒定，且不受植物生长年龄和叶片大小而变化。因此，脉岛数可作为全草类、叶类中药的鉴定特征之一。如杜虹花 *Callicarpa formosana* Rolfe. 叶（紫珠叶）的脉岛数为（11.31+1.82）个 /mm^2，大叶紫珠 C. *macrophylla* Vahl. 叶的脉岛数为（3.82+1.44）个 /mm^2，华紫珠 C. *cathayana* H.T.Chang 为（4.66+1.73）个 /mm^2。

(3) Vein islet number The mesophyll tissue enclosed by the most delicate veins of the leaf blade is called a vein island. Vein-islet number is the number of vein island per unit area (mm^2) on a leaf. The vein-islet number of the same plant species is constant, and it does not change with the age of plant and the size of the leaf. Therefore, the number of islets can be used as one of the identification characteristic for herbs. Such as *Callicarpa formosana*, the vein-islet number is (11.31+ 1.82)/mm^2, while the number of the vein island of *C. macrophylla* and *C.cathayana* is (3.82+1.44)/ mm^2 and (4.66+1.73)/ mm^2, respectively.

重 点 小 结
Summaries

　　根尖分为根冠、分生区、伸长区和成熟区四部分。分生区是细胞分裂最活跃的部分，伸长区细胞的生长使根向前推进，成熟区形成了根的初生结构。根的初生构造包括表皮、皮层、维管柱三部分。表皮具根毛，是重要的吸收结构。皮层细胞多为排列疏松的薄壁组织，内皮层明显，具有凯氏带或凯氏点。维管柱的初生木质部束与初生韧皮部束相间排列，呈辐射状结构，外面包被明显的中柱鞘。根次生生长时，初生木质部与初生韧皮部间保留的薄壁细胞及靠近木质部脊的中柱鞘细胞恢复分生能力后，形成维管形成层，并产生次生木质部、次生韧皮部、维管射线；中柱鞘细胞分裂衍生形成木栓形成层，产生周皮。维管形成层与木栓形成层的异常发生与活动产生了异常维管组织，包括同心环状排列、异心环状排列的维管组织及木间木栓等。

　　The growing root tip may be further developing into four zones: the root cap, the zone of division, the zone of elongation, and the zone of maturation. The primary structure of the root is also the region where the mature zone of the root is located, which may be divided into three zones: the root rhizodermis, the root cortex and the vascular cylinder. The activity of the vascular cambium and the cork cambium of the root forms the secondary structure of the root. Monocotyledonous plants generally have no vascular cambium and cork cambium, so they generally have only primary structures. Some dicotyledonous plants also produce some unique vascular bundles, called anomalous vascular bundles, forming the anomalous structure of the roots.

双子叶植物茎的初生构造，从外到内分为表皮、皮层、维管柱三部分，维管柱包括维管束、髓部、髓射线等，初生维管束由初生韧皮部、初生木质部、束中形成层构成；双子叶植物木质茎的次生构造，从外到内分为表皮、周皮、皮层、维管柱，周皮由木栓层、木栓形成层、栓内层组成，次生维管束由次生韧皮部、次生木质部、束中形成层构成，次生木质部是木质茎次生构造的主要部分，由导管、管胞、木薄壁细胞、木纤维和木射线组成，次生韧皮部常由筛管、伴胞、韧皮纤维、韧皮薄壁细胞组成；某些双子叶植物的茎和根状茎会形成异常构造，常见的有髓维管束、同心环状排列的异常维管组织、木间木栓。单子叶植物茎一般没有形成层和木栓形成层，终身只具有初生构造，其最外层是由一列细胞构成的表皮，表皮以内为基本薄壁组织和散布其中的维管束，维管束为有限外韧型。

The primary structure of a dicotyledonous plant stem is divided into three parts: epidermis, cortex, and the vascular cylinder which includes vascular bundles, pith, and pith rays. The primary vascular bundle is composed of the primary phloem, primary xylem, and fascicular cambium; the secondary structure of the dicotyledonous woody stem is divided into epidermis, periderm, cortex, and vascular cylinder from the outside to the inside. The periderm is composed of a phellem layer, cork cambium and phelloderm. Vascular bundles are composed of secondary phloem, secondary xylem, and the fascicular cambium. Secondary xylem is the main part of secondary structure of woody stem. It is composed of vessels, tracheids, wood parenchymal cells, wood fibers and wood rays. The phloem is often composed of sieve tubes, companion cells, phloem fibers, and phloem parenchyma cells; the stems and rhizomes of some dicotyledonous plants can form abnormal structures, like "Vascular bundles in pith, The anomalous vascular tissues with concentric annular arrangement, Interxylary cork". Monocotyledonous plant stems generally do not form cambium and a cork cambium, and have only a primary structure throughout life. The outermost layer is the epidermis composed of a row of cells. The inner part of the epidermis is the parenchymal tissue and the closed collateral vascular bundles distributed therein.

单子叶植物和双子叶植物叶片的都可分为表皮、叶肉和叶脉，两者又有区别。叶的表皮细胞、气孔、表皮附属物、栅栏组织、海绵组织、叶脉维管束等等的结构特点，都可作为植物种类鉴别的依据。

The leaf structure of both monocotyledonous and dicotyledonous plants can be composed of epidermis, mesophyll and vein. The structural characteristics of epidermis, stomata, epidermal appendages, palisade tissue, spongy tissue, vascular bundle and so on can be used as the basis for the identification of plant species.

目 标 检 测
Questions

1. 双子叶植物和单子叶植物根的初生构造有什么异同点？

How does the primary structure of the dicotyledonous stem and root differ?

2. 双子叶植物根的初生构造和次生构造有何区别？次生构造是如何形成的？

What are the differences between the primary and the secondary structure of dicotyledon root? How is the secondary structure formed?

3. 双子叶植物根的异常构造是如何形成的？

题库

How is the abnormal structure formed?

4. 双子叶植物茎与根的初生构造有何不同？

What are the difference of the primary structure between dicotyledon root and monocotyledon root?

5. 比较单子叶植物茎与双子叶植物茎的构造的区别点。

What is the difference between monocotyledonous and dicotyledonous stem structures?

6. 在木材三切面上见到的导管、管胞、木纤维、木射线和年轮有何特征？

What are the characteristics of the vessels, tracheids, wood fibers, wood rays and annual rings seen on the three sections of the wood?

7. 一般植物叶下表皮气孔多于上表皮，这有何优点？

What are the advantages of more stomata in the lower epidermis than in the upper epidermis?

8. 叶的表皮细胞一般呈透明状，细胞液无色，这对叶的生理功能有何意义？

The epidermal cells of the leaves are generally transparent and the cell fluid is colorless. What is the significance of this to the physiological function of the leaves?

9. 总结双子叶植物叶的解剖构造。

Summarize the structure character of dicotyledonous leaves.

10. 试述禾本科植物叶的解剖构造。

Discuss the structure of plant leaves of Gramineae.

专业词汇英汉对照
The English to Chinese Terminology

A

achene 瘦果

achlamydeous flower 无被花

acicular crystal 针晶

actinomorphic flower 辐射对称花

adventitious roots 不定根

aerial roots 气生根

aeurone grain 糊粉粒

aggregate fruit 聚合果

akinete 厚壁孢子

algal layer 藻胞层

algin fucoidin 褐藻糖胶

allophycocyanin 别藻蓝素

alternation of generation 世代交替

amitosis 无丝分裂

amphicribral vascular bundle 周韧型维管束

amphivasal vascular bundle 周木型维管束

androecium 雄蕊群

angiosperms 被子植物

annual ring 年轮

annulus 菌环

anomalous structure 异常结构

antheridiophore 雄器托

antheridium 精子器

apical meristerm 顶端分生组织

aplanospore 不动孢子

apothecium 子囊盘

archegomiatae 颈卵器植物

archegoniophore 雌器托

archegonium 颈卵器

arthrospore 节孢子

ascocarp 子囊果

Ascolichens 子囊衣纲

ascospore 子囊孢子

ascus 子囊

asymmetric flower 不对称花

atactostele 散生中柱

autospore 拟亲孢子

autotrophic plant 自养植物

autumn wood 秋材

axillary bud 腋芽

B

bark 树皮

basidiocarp 担子果

basidiospore 担孢子

basidium 担子

berry 浆果

bicollateral vascular bundle 双韧型维管束

Binomial nomenclature 双命名法

bisexual flower 单性花

blade 带片

blastospore 芽生孢子

blue green algae 蓝绿藻

bordered pit 具缘纹孔

brown algae 褐藻

bud scale scar 芽鳞痕

bulb 鳞茎

bulbil 小鳞茎

bulliform cell 泡状细胞

C

calcium carbonate crystal 碳酸钙结晶

calcium oxalate crystal 草酸钙结晶

calyx 花萼

capitulum 头状花序

capsule 蒴果

339

carporgonium　果胞

carpospores　果孢子

carposporophyto　果孢子体

caryopsis　颖果

Casparian strip　凯氏带

catkin　柔荑花序

caudicle　花粉块柄

cell cycle　细胞周期

cellulose　纤维素

centrol body　中心体

centroplasm　中心质

chitin　几丁质

chromatoplasm　色素质

chromosome　染色质

circinate　拳卷

cladophylls　叶状茎

clamp connection　锁状联合

cleavage polyembryony　裂生多胚现象

cleistothecium　闭囊壳

clignification　木质化

climbing stem　攀援茎

closed collateral vascular bundle　有限外韧维管束

cluster crystal　簇晶

collective fruit/multiple fruit　聚花果

collenchyma　厚角组织

columnar crystal　柱晶

companion cell　伴胞

complete flower　完全花

complex tissue　复合组织

compound umbel　复伞形花序

conducting tissue　输导组织

conjugation　接合生殖

conjugatophyceae　结合藻纲

cork cambium　木栓形成层

corm　球茎

cormophytes　茎叶体植物

corolla　花冠

cortex　皮层

corymb　伞房花序

cremocarp　双悬果

crista　嵴

crustose lichens　壳状地衣

cryptogamia　隐花植物

crystal sand　砂晶

cutinization　角质化

cyanophycean starch　蓝藻淀粉

cyanophycin granules　蓝藻颗粒

cypsela　连萼瘦果

cystocarp　囊果

cytoplasm　细胞质

D

dehiscent fruit　闭果

dicotyledons　双子叶植物

dictyostele　网状中柱

diploid　二倍体

double flower　两性花

double perianth flower　重被花

drupe　核果

E

early wood　早材

endarch　内始式

endodermis　内皮层

endspore　内生孢子

epidermis　表皮

equisetum　木贼属

erect stem　直立茎

ergastic substance　后含物

eustele　真中柱

exarch　外始式

exospore　外生孢子

F

fascicular cambium　束中形成层

fat and fatty oil　脂肪和脂肪油

female cone　雌球花

fern　羊齿植物；蕨类植物

fertile frond　能育叶

fiber　纤维

fibrous root system　须根系

filament　丝状体

fine root　纤维根

fixed roots　定根

fleshy fruit　干果

fleshy root　贮藏根

floridean starch　红藻淀粉

floridose　红藻糖

flower diagram　花图式

flower formula　花程式

foliage leaf　营养叶

foliose lichens　叶状地衣

follicle　蓇葖果

forma　变型

fruticose lichens　枝状地衣

fucoxanthin　墨角藻黄素

G

gametangium　配子囊

gamete　配子

gametophyte generation　配子体世代

gametophyte　配子体

gelatinous corona　胶质冠

gelatinous sheath　胶质鞘

gills　菌褶

gymnosperms　裸子植物

gynoecium　雌蕊群

gynostegium　合蕊冠

gynostemium　合蕊柱

H

half bordered pit　半缘纹孔

haploid　单倍体

heart wood　心材

herbaceous stem　草质茎

hesperidium　柑果

heterocyst　异形胞

heterogay　异配

heteromerous lichens　异层地衣

heteromorphic alternation of generations　异形世代交替

heteromorphic leaf　异型叶，两型叶

heterotrophy　异养型

higher plant　高等植物

holdfast　固着器

homoenmerous lichens　同层地衣

homogonium　藻殖段

homomorphic leaf　同型叶，一型叶

hymenium　子实层

hypanthodium　隐头花序

hyphae　菌丝

I

incomplete flower　不完全花

indefinite inflorescence　无限花序

indehiscent fruit　不裂果

intercalary meristem　居间分生组织

intercellular layer　胞间层

interfascicular cambium　束间形成层

internode　节间

inulin　菊糖

isogamy　同配

isomorphic alternation of generation　同形世代交替

isospore　孢子同型

L

laminarin　褐藻淀粉

late wood　晚材

lateral meristem　侧生分生组织

lateral roots　侧根

laticifer　乳汁管

leaf gap　叶隙

leaf scar　叶痕

legume　荚果

liver starch　肝糖

liverworts　苔类

lower plant　低等植物

M

macrophyll　大型叶

macrospore　大孢子

male cone　小孢子叶球

mechanical tissue　机械组织

medullary ray　髓射线

megasporangium　大孢子囊

meiosis　减数分裂

meristem　分生组织

mesophyll　叶肉

microphyll　小型叶

microsporangium　小孢子囊

microsporophyll　小孢子叶

microsporum　小孢子

mineralization　矿质化

mitosis　有丝分裂

multiple epidermis　复表皮

monochasium　单歧聚伞花序

mosses　藓类

mucilagization　黏液质化

mucopolysaccharide　黏多糖

mycelium　菌丝体

myxomycophyta　黏菌门

N

nectary　蜜腺

node　节

nonseptate hypha　无隔菌丝

nuclevid　拟核

nut　坚果

O

oogamy　卵配

oospore　卵孢子

open collateral vascular bundle　无限外韧维管束

organelle　细胞器

ovulate strobilus　大孢子叶球

ovules　胚珠

P

palisade ratio　栅表比

palisade tissue　栅栏组织

paraphysis　侧丝

parasitic roots　寄生根

parasitism　寄生

parenchyma　薄壁组织

pectic acid　果胶酸

pedicel　花梗

pepo　瓠果

peptidoglycan　肽聚糖

perianth　花被

perichaetium　雌器苞

periderm　周皮

perigonium　雄器苞

perimedullary region　环髓区

perine　孢子周壁，周壁

periplasm　周质

perithecium　子囊壳

phloem　韧皮部

photosynthetic lamellae　光合片层

phycobilin　藻胆素

phycobiliprotein　藻胆蛋白

phycobilisome　藻胆体

phycocyanobilin　藻蓝素

phycoerythrobilin　藻红素

pileus　菌盖

pinna rachis　羽轴

pinna　羽片

pistil　雌蕊

pit　纹孔

pith　髓

plasmodesmata　胞间连丝

plastid　质体

plurilocular sporangium　多室孢子囊

pollen chamber　贮粉室

pollen grain　花粉

pollen sac　花粉囊

pollen tetrads　四合花粉

pollinarium　花粉器

pollinium　花粉块

polyembryony　多胚现象

polyploid　多倍体

pome　梨果

primary meristem　初生分生组织

primary mycelium　初生菌丝

primary phloem　初生韧皮部

primary root system　直根系

primary roots　主根

primary wall　初生壁

primary xylem　初生木质部

prokaryon　原核

promeristem　原分生组织

prop roots　攀援根

propagule　繁殖枝

prosenchyma　疏丝组织

protective tissue　保护组织

prothallus　原叶体

protonema　原丝体

protoplast　原生质

protostele　原生中柱

pseudobulb　假鳞茎

pseudoparenchyma　拟薄壁组织

pteridophyte　蕨类植物

pyrenoid　蛋白核

R

raceme　总状花序

radial section　径向切面

radial vascular bundle　辐射型维管束

radicle　胚根

receptacle　花托

repent stem　平卧茎

resin duct　树脂道

respiratory roots　呼吸根

retinaculum　着粉腺

rhizoid　假根

rhizome　根状茎

rhizomorph　根状菌索

rhytidome　落皮层

root system　根系

S

samara　翅果

sap wood　边材

saprophytism　腐生

sclereid　石细胞

sclerenchyma　厚壁组织

sclerotium　菌核

secondary meristem　次生分生组织

secondary metabolites　次生代谢产物

secondary mycelium　次生菌丝

secondary phloem　次生韧皮部

secondary wall　次生壁

secondary xylem　次生木质部

secretory canal　分泌道

secretory cavity　分泌腔

secretory cell　分泌细胞

secretory tissue　分泌组织

seed plant　种子植物

septate hypha　有隔菌丝

shoot thorn　刺状茎

shoot　枝条

sieve cell　筛胞

sieve tube　筛管

silicuela　短角果

siliqua　长角果

simple perianth flower　单被花

simple polyembryony　简单多胚现象

simple tissue　简单组织

siphonostele　管状中柱

solitary crystal　方晶

sorus　孢子囊堆

spadix　肉穗花序

spermatophyta　种子植物

spermattangium　精子囊

spike　穗状花序

spongy tissue　海绵组织

sporangiorus　孢子囊群

sporangium　孢子囊

spore plant　孢子植物

sporophore　子实体

sporophyll　孢子叶

sporophyll spike　孢子叶穗

sporophyte　孢子体

sporophyte generation　孢子体世代

spring wood　春材

spurious fruit/ false fruit　假果

sporocarp　孢子果，孢子荚

stamen　雄蕊

staminate strobilus　雄球花

starch sheath　淀粉鞘

starch　淀粉

stele　中柱

sterigma　小梗

sterile frond　不育叶

sticky disk　粘盘

stilt roots　支持根

343

...pe　叶柄

stipule scar　托叶痕

stolon　匍匐茎

stomatal index　气孔指数

stomatal band　气孔带

strobilus　孢子叶球

stroma　子座

suberization　木栓化

subspecies　亚种

substomatic chamber　孔下室

succulent　肉质茎

symbiosis　共生

T

tangential section　切向切面

tendril and hook　茎卷须和钩状茎

teriary mycelium　三生菌丝

terminal bud　顶芽

tetraspore　四分孢子

thallophyte　原植体植物

thylakoid　类囊体

tracheid　管胞

translater　载粉器

translater handle　载粉器柄

transverse section　横切面

trichogyne　受精丝

true fruit　真果

tuber　块茎

twining stem　缠绕茎

U

umbel　伞形花序

unilocular sporangium　单室孢子囊

universal veil　外菌幕

utricle　胞果

V

varietas　变种

vascular bundle　维管束

vascular cylinder　维管柱

vascular plant　维管植物

vascular ray　维管射线

vaseular bundle sheath　维管束鞘

vegetative organs　营养器官

vein islet number　脉岛数

verticillaster　轮伞花序

vesicle　泡囊

vessel　导管

volva　菌托

W

woody stem　木质茎

water bloom　水华

X

xylem　木质部

Z

zoospore　游动孢子

zygomorphic flower　两侧对称花

zygospore　接合孢子

Zygote　合子

参 考 文 献
References

1. David F. Cutler, Ted Botha, et al. Plant Anatomy-An Applied Approach［M］. Hoboken (New Jersey): Wiley Blackwell, 2008.

2. 詹姆斯·吉·哈里斯，米琳达·沃尔芙·哈里斯. 图解植物学词典［M］. 北京：科学出版社，2001.

3. 刘春生. 药用植物学［M］.（4 版）. 北京：中国中医药出版社，2016.

4. 严铸云，郭庆梅. 药用植物学［M］.（2 版）. 北京：中国医药科技出版社，2018.

5. 林莺. 中药植物学［M］. 北京：中国医药科技出版社，2014.

6. Chang Zhangfu. Chinese materia medica［M］. 北京：人民卫生出版社，2014.

7. Walter S. Judd. Plant Systematics［M］. Sunderland (Massachusetts): Sinauer Associates, Inc. 2016.

8. 陆树刚. 蕨类植物学［M］. 北京：高等教育出版社，2007.

9. Richard Crang, Sheila Lyons-Sobaski, Robert Wise. Plant Anatomy［M］. Heidelberg: Springer, 2018.